T0310133

**Spectrum Sharing**

# Spectrum Sharing

The Next Frontier in Wireless Networks

*Edited by*

*Constantinos B. Papadias*
The American College of Greece
Athens
Greece

*Tharmalingam Ratnarajah*
University of Edinburgh
Edinburgh
UK

*Dirk T.M. Slock*
EURECOM
Sophia Antipolis
France

This edition first published 2020
© 2020 John Wiley & Sons Ltd

All rights reserved. No part of this publication may be reproduced, stored in a retrieval system, or transmitted, in any form or by any means, electronic, mechanical, photocopying, recording or otherwise, except as permitted by law. Advice on how to obtain permission to reuse material from this title is available at http://www.wiley.com/go/permissions.

The right of Constantinos B. Papadias, Tharmalingam Ratnarajah and Dirk T.M. Slock to be identified as the authors of the editorial material in this work has been asserted in accordance with law.

*Registered Offices*
John Wiley & Sons, Inc., 111 River Street, Hoboken, NJ 07030, USA
John Wiley & Sons Ltd, The Atrium, Southern Gate, Chichester, West Sussex, PO19 8SQ, UK

*Editorial Office*
The Atrium, Southern Gate, Chichester, West Sussex, PO19 8SQ, UK

For details of our global editorial offices, customer services, and more information about Wiley products visit us at www.wiley.com.

Wiley also publishes its books in a variety of electronic formats and by print-on-demand. Some content that appears in standard print versions of this book may not be available in other formats.

*Limit of Liability/Disclaimer of Warranty*
While the publisher and authors have used their best efforts in preparing this work, they make no representations or warranties with respect to the accuracy or completeness of the contents of this work and specifically disclaim all warranties, including without limitation any implied warranties of merchantability or fitness for a particular purpose. No warranty may be created or extended by sales representatives, written sales materials or promotional statements for this work. The fact that an organization, website, or product is referred to in this work as a citation and/or potential source of further information does not mean that the publisher and authors endorse the information or services the organization, website, or product may provide or recommendations it may make. This work is sold with the understanding that the publisher is not engaged in rendering professional services. The advice and strategies contained herein may not be suitable for your situation. You should consult with a specialist where appropriate. Further, readers should be aware that websites listed in this work may have changed or disappeared between when this work was written and when it is read. Neither the publisher nor authors shall be liable for any loss of profit or any other commercial damages, including but not limited to special, incidental, consequential, or other damages.

*Library of Congress Cataloging-in-Publication Data applied for*

Hardback ISBN: 9781119551492

Cover Design: Wiley
Cover Image: © Ivision 2u/Shutterstock

Set in 9.5/12.5pt STIXTwoText by SPi Global, Chennai, India

Printed and bound by CPI Group (UK) Ltd, Croydon, CR0 4YY

10  9  8  7  6  5  4  3  2  1

*For Maria-Anna, Anna, Billy and Dimitri, C.B.P.*
*For the memory of my father, Dr. D. Tharmalingam and brother, D. Varatharajah, T.R.*
*For Aida, my parents, our families, and my students, D.T.M.S.*

# Contents

**15       Performance Analysis of Spatial Spectrum Reuse in Ultradense Networks**   *305*
*Youjia Chen, Ming Ding, and David López-Pérez*

**16       Large-scale Wireless Spectrum Monitoring: Challenges and Solutions based on Machine Learning**   *321*
*Sreeraj Rajendran and Sofie Pollin*

# About the Editors

**Constantinos B. Papadias** is the Executive Director of the Research, Technology and Innovation Network (RTIN) of The American College of Greece, where he is also a faculty member, since Feb. 1, 2020. Prior to that, he was the Scientific Director / Dean of Athens Information Technology (AIT), in Athens, Greece, where he was also Head of the Broadband Wireless and Sensor Networks (B-WiSE) Research Group. He is currently an Adjunct Professor at Aalborg University and at the University of Cyprus. He received the Diploma of Electrical Engineering from the National Technical University of Athens (NTUA) in 1991 and the Doctorate degree in Signal Processing (highest honors) from the Ecole Nationale Supérieure des Télécommunications (ENST), Paris, France, in 1995. He was a researcher at Institut Eurécom (1992–1995), Stanford University (1995–1997) and Lucent Bell Labs (as Member of Technical Staff from 1997–2001 and as Technical Manager from 2001–2006). He was Adjunct Professor at Columbia University (2004–2005) and Carnegie Mellon University (2006–2011). He has published over 200 papers and 4 books and has received over 9000 citations for his work, with an h-index of 43. He has also made standards contributions and holds 12 patents. He was a member of the Steering Board of the Wireless World Research Forum (WWRF) from 2002–2006, a member and industrial liaison of the IEEE's Signal Processing for Communications Technical Committee from 2003–2008 and a National Representative of Greece to the European Research Council's IDEAS program from 2007–2008. He has served as member of the IEEE Communications Society's Fellow Evaluation and Awards Committees, as well as an Associate Editor for various journals. He has contributed to the organization of several conferences, including, as General Chair, the IEEE CTW 2016 and the IEEE SPAWC 2018 workshops. He has acted as Technical Coordinator in several EU projects such as: CROWN in the area of cognitive radio; HIATUS in the area of interference alignment; HARP in the area of remote radio heads and ADEL in the area of licensed shared access. He is currently the Research Coordinator of the European Training Network project PAINLESS on the topic of energy autonomous infrastructure-less wireless networks as well as the Technical Coordinator of the EU CHIST-ERA project FIRE-MAN on the topic of predictive maintenance via machine learning empowered wireless communication networks. His distinctions include the Bell Labs President's Award (2002), the IEEE Signal Processing Society's Young Author Best Paper Award (2003), a Bell Labs

Teamwork Award (2004), his recognition as a "Highly Cited Greek Scientist" (2011), two IEEE conference paper awards (2013, 2014) and a "Best Booth" Award at EUCNC (2016). He was a Distinguished Lecturer of the IEEE Communications Society for 2012–2013. He was appointed Fellow of IEEE in 2013 and Fellow of the European Alliance of Innovation (EAI) in 2019.

**Tharmalingam Ratnarajah** is currently with the Institute for Digital Communications, the University of Edinburgh, Edinburgh, UK, as a Professor in Digital Communications and Signal Processing. He was a Head of the Institute for Digital Communications during 2016–2018. Prior to this, he was with McMaster University, Hamilton, Canada, (1997–1998), Nortel Networks (1998–2002), Ottawa, Canada, University of Ottawa, Canada, (2002–2004), Queen's University of Belfast, UK, (2004–2012). His research interests include signal processing and information theoretic aspects of 5G and beyond wireless networks, full-duplex radio, mmWave communications, random matrices theory, interference alignment, statistical and array signal processing and quantum information theory. He has published over 400 publications in these areas and holds four U.S. patents. He has supervised 15 PhD students and 20 post-doctoral research fellows, and raised $11 million+ USD of research funding. He was the coordinator of the EU projects ADEL in the area of licensed shared access for 5G wireless networks, HARP in the area of highly distributed MIMO, as well as EU Future and Emerging Technologies projects HIATUS in the area of interference alignment and CROWN in the area of cognitive radio networks. Dr Ratnarajah was an associate editor IEEE Transactions on Signal Processing, 2015–2017 and Technical co-chair, The 17th IEEE International workshop on Signal Processing advances in Wireless Communications, Edinburgh, UK, 3–6, July, 2016. Dr Ratnarajah is a member of the American Mathematical Society and Information Theory Society and Fellow of Higher Education Academy (FHEA).

**Dirk T.M. Slock** received an electronics engineering degree from Ghent University, Belgium in 1982. In 1984 he was awarded a Fulbright scholarship for Stanford University, USA, where he received the MSEE, MS in Statistics, and PhD in EE in 1986, 1989 and 1989 resp. While at Stanford, he developed new fast recursive least-squares algorithms for adaptive filtering. In 1989–91, he was a member of the research staff at the Philips Research Laboratory Belgium. In 1991, he joined EURECOM where he is now professor. At EURECOM, he teaches statistical signal processing (SSP) and signal processing techniques for wireless communications. His research interests include SSP for wireless communications (antenna arrays for (semi-blind) equalization/interference cancellation and spatial division multiple access (SDMA), space-time processing and coding, channel estimation, diversity analysis, information-theoretic capacity analysis, relaying, cognitive radio, geolocation), and SSP techniques for audio processing. He invented semi-blind channel estimation, the chip equalizer-correlator receiver used by 3G HSDPA mobile terminals, spatial multiplexing cyclic delay diversity (MIMO-CDD) now part of LTE, and his work led to the Single Antenna Interference Cancellation (SAIC) integrated in the GSM standard in 2006. Recent research keywords are MIMO interference channel, multi-cell, distributed resource allocation, variational and empirical Bayesian techniques, large random matrices, stochastic geometry, audio source separation, location estimation and exploitation.

In 25 years, he has graduated over 35 PhD students, 9 of which are in academia (6 professors), and about 10 others are in research in industry. His research led to: h-index: 41, total citations: 8800, 10 book chapters, 50 journal papers, 500 conference papers. In 1992 he received one best journal paper award from IEEE-SPS and one from EURASIP. He is the coauthor of two IEEE Globecom'98, one IEEE SIU'04, one IEEE SPAWC'05, one IEEE WPNC'16 and one IEEE SPAWC'18 best student paper award, and an honorary mention (finalist in best student paper contest) at IEEE SSP'05, IWAENC'06, IEEE Asilomar'06 and IEEE ICASSP'17. He has been an associate editor for various journals, and conference organizer of SPAWC'06, IWAENC'14, EUSIPCO'15. He was a member of the IEEE-SPS Awards Board 2011–13 and of the EURASIP JWCN Awards Committee. Over the past 10 years he has participated in the French projects ERMITAGES, ANTIPODE, PLATON, SEMAFOR, APOGEE, SESAME, DIONISOS, and DUPLEX (which he coordinated), MASS-START and GEOLOC, summing to over 2M€ in funding, and in the European projects K-SPACE, Newcom/++/#, WHERE(2), CROWN, SACRA, ADEL and HIGHTS summing up to over 2.5M€ in funding. He has also had a number of direct research contracts with Orange (6), Philips, NXP, STEricsson, Infineon, and Intel, and scholarships for 10 PhD students. He cofounded in 2000 SigTone, a start-up developing music signal processing products, and in 2014 Nestwave, a start-up developing Ultra Low-Power Indoor and Outdoor Mobile Positioning. He has also been active as a consultant on xDSL, DVB-T and 3G systems. He is a Fellow of IEEE and EURASIP. In 2018 he received the URSI France medal.

# List of Contributors

**Dani Anderson**
Department of Electronic and Electrical
Engineering
University of Strathclyde
Glasgow
United Kingdom

**Adrish Banerjee**
Department of Electrical Engineering
Indian Institute of Technology Kanpur
Kanpur
India

**Sudip Biswas**
Indian Institute of Information Technology
Guwahati
India

**M. Majid Butt**
Nokia Bell Labs
Paris-Saclay
France

**Ali Cagatay Cirik**
Ofinno Technologies
USA

**Youjia Chen**
Fuzhou University
Fuzhou
P.R. China

**David Crawford**
Department of Electronic and Electrical
Engineering
University of Strathclyde
Glasgow
United Kingdom

**Ming Ding**
Commonwealth Scientific and Industrial
Research Organisation (CSIRO)
Eveleigh
Australia

**María Dolores (Lola) Pérez Guirao**
Sennheiser Electronic GmbH & Co. KG
Wedemark
Germany

**Miltiades C. Filippou**
Intel Deutschland GmbH
Neubiberg
Germany

**Kalyana Gopala**
Institut Eurecom
Communication Systems Department
Biot Sophia Antipolis
France

**Abhishek K. Gupta**
Department of Electrical Engineering
Indian Institute of Technology Kanpur
Kanpur
India

**Tero Henttonen**
Nokia Bell Labs CTO
Espoo
Finland

**Eduard A. Jorswieck**
TU Braunschweig
Braunschweig
Germany

**Faheem Khan**
School of Computing and Engineering
University of Huddersfield
Queensgate
Huddersfield
United Kingdom

**Vireshwar Kumar**
Virginia Tech
Arlington
USA

**Markku Kuusela**
Nokia CSD Digital Automation
Lahti
Finland

**Daniela Laselva**
Nokia Bell Labs
Aalborg
Denmark

**William Lehr**
Massachussetts Institute of Technology
Cambridge
USA

**Fan Liu**
Department of Electronic & Electrical
Engineering
University College London
London
United Kingdom

**David Lópéz-Pérez**
Nokia Bell Labs
Dublin
Ireland

**Christos Masouros**
Department of Electronic & Electrical
Engineering
University College London
London
United Kingdom

**António J. Morgado**
Instituto de Telecomunicações
Aveiro
Portugal

**Markus Mueck**
Intel Deutschland GmbH
Neubiberg
Germany

**Konstantinos Ntougias**
University of Cyprus
Nicosia
Cyprus

**Taiwo Oyedare**
Virginia Tech
Arlington
USA

**Constantinos B. Papadias**
Research, Technology and Innovation
Network
The American College of Greece
Athens
Greece

**Georgios K. Papageorgiou**
Heriot-Watt University
Edinburgh
United Kingdom

**Jung-Min (Jerry) Park**
Virginia Tech
Arlington
USA

**David Lópéz-Pérez**
Nokia Bell Labs
Dublin
Ireland

**Marius Pesavento**
Darmstadt University of Technology
Darmstadt
Germany

**Sofie Pollin**
KU Leuven
Heverlee
Belgium

**Sreeraj Rajendran**
KU Leuven
Heverlee
Belgium

**Rao Yallapragada**
Intel Corp.
San Diego
USA

**Tharmalingam Ratnarajah**
University of Edinburgh
Edinburgh
United Kingdom

**Mika Rinne**
Nokia Technologies
Espoo
Finland

**Claudio Rosa**
Nokia Bell Labs
Randers
Denmark

**Mathini Sellathurai**
School of Engineering & Physical Sciences
Heriot-Watt University
Edinburgh
United Kingdom

**K.A. Shruthi**
Department of Electronic and Electrical
Engineering
University of Strathclyde
Glasgow
United Kingdom

**Dirk T.M. Slock**
EURECOM
Communication Systems Department
Biot Sophia Antipolis
France

**Srikathyayani Srikanteswara**
Intel Corp.
OR
USA

**Christian Steffens**
Hyundai Mobis
Frankfurt
Germany

**Robert W. Stewart**
Department of Electronic and Electrical
Engineering
University of Strathclyde
Glasgow
United Kingdom

**Andrew Stirling**
Larkhill Consultancy
Surrey
United Kingdom

**Richard Womersley**
LS Telcom
Germany

# Preface

Our efforts over the years to tame the air as a communication medium have been hampered by the electromagnetic spectrum's limiting nature since the early days of radio. Unlike wired communication over, for example, copper wires or fiber, where new channels can be added simply by using more cables, wireless communication systems and networks have always had to struggle to fit as many communication links as possible into a given geographic area through the same medium. Given the finite available spectrum (due to nature, regulation and to the transmitter and receivers' capabilities) and Shannon's fundamental law of channel capacity, electromagnetic spectrum management has become a crucial ongoing need that accompanys all types and generations of wireless systems and networks.

The canonical paradigm in spectrum allocation has been to provide orthogonal channels to the different users in a given geographic area – and then of course to reuse the same spectrum in other geographic areas. This simple principle, including a careful frequency planning and dimensioning of the resulting interference, has allowed cellular networks to develop rapidly since the late 1980s all the way to today's phenomenal success of 4G and emerging 5G networks, which have impacted all types of human activity and have changed the way we interact, do business, and provide various services to citizens. In order to meet the cellular networks' growing demands in data rates, capacity, and quality-of-service (QoS) requirements, more and more spectrum keeps being allocated, typically through government-based licensing that provides exclusive (often national level) rights of use to a number of operators, usually for a high fee, following the orthogonal allocation paradigm mentioned earlier. The orthogonal model has permitted operators to provide QoS guarantees to their users.

However, in parallel with the strict paid licensing model mentioned above, unlicensed use of the spectrum has been also allowed for a number of applications that do not need to provide QoS guarantees to their users and whose range and user density are smaller than that of cellular networks. Such applications included, in the early years, amateur radio, cordless phones, and even non-communication uses such as microwave ovens and other appliances. A big boost to the unlicensed use of spectrum was undoubtedly given by the proliferation of wireless local area networks (LANs) that rely on Wi-Fi-type systems. In spite of the lack of QoS guarantees (and benefiting from continuously improved protocols), Wi-Fi has become a huge success, largely due to its fee-free use and little interference in several, typically static, environments (such as the home or the office). As a result, these networks carry an amount of wireless data that is comparable to that of their cellular counterparts.

In parallel with the above core models of spectrum usage (licensed and unlicensed), a third paradigm has emerged over the last two decades, wherein unlicensed operators would make use of licensed spectrum. This concept originated with the advent of cognitive radio and has gone through various phases since. It relies on the key requirement that the operator who does not hold a license should not interfere with the ones who do. This may be easier in cases of sparse usage as well as when the licensed spectrum is largely unused, but is much more challenging in dense usage and crowded spectrum situations; hence, in order to succeed, this model requires a very good awareness of the spectrum activity in a given area (attained via either spectrum sensing or geolocation databases, or both), as well as of course a careful design of the wireless communication protocol used.

Collectively called "spectrum sharing," these techniques are gaining increased traction and have evolved significantly over the last decade. This is largely due to the continued (exponential-like) growth of wireless service demands, the "addiction" of users to unlicensed broadband access, the saturation of existing licensed spectrum usage in many areas, the emergence of new types of operators and service models, the proliferation of research activity in spectrally efficient technologies, and the rather slow and bureaucratic nature of spectrum auctioning.

The purpose of this book has been to collect, in a single volume, the key technologies and approaches related to spectrum sharing, dating back to the inception of the cognitive radio concept and going all the way to today's novel approaches and emerging research concepts. Our goal has been to capture all the related dimensions, including the technical, key regulatory, standardization, and financial aspects.

We have been privileged to collaborate in the context of two important collaborative research projects that have received funding from the European Commission (under its 7th Framework Program), whose generous support is herein gratefully acknowledged. These projects are FET Open project CROWN (Cognitive Radio Oriented Wireless Networks) which ran from 2009 to 2012, and Future Networks project ADEL (Advanced Dynamic spectrum 5G mobile networks Employing Licensed shared access), which ran from 2013 to 2016. Key spectrum sharing concepts were introduced in these projects ahead of their time (such as that of horizontal sharing even within the same operator suggested in CROWN, now used in LTE Licensed Assisted Access (LAA), and sensing-assisted Licensed Shared Access proposed in ADEL, now used in the Spectrum Access System (SAS) in the USA). These projects allowed us not only to participate in the fascinating research on spectrum sharing, introducing to it several PhD students and young researchers, but also to stay in touch with the most current trends, interact with all types of stakeholders (from industrial to regulatory to end users), and contribute to exciting proof-of-concept demos of emerging solutions. They also helped us to establish numerous research collaborations with a growing number of research teams that have continued and expanded beyond these projects and due to which this endeavor is largely owed.

Given the spurt of activity in spectrum sharing and our personal involvement and interactions, we felt that the time was right for a comprehensive edited volume on the topic, written by some of the top experts in all related areas. We were highly encouraged by the many positive responses for chapter contributions and are grateful to all the authors for their inputs and for allowing us to cover all the topics that we deemed important, including very recent ones such as full duplex-based spectrum sharing, communication-radar coexistence, mmWave, massive MIMO, and machine learning-based spectrum monitoring, among others.

Our addressable audience includes readers from the academic (students, professors), industrial (engineers, practitioners), as well as regulatory/standardization sectors, who share an interest on how spectrum has been used to date and how it can be best used and shared in the coming years.

To the extent that the interested reader will find the answers they are looking for and acquire a well-rounded knowledge of spectrum sharing technology and its surrounding ecosystem, our goal will have been met. We hope that all readers will do so and that this book becomes a useful item of their library and a reference for years to come!

*Constantinos B. Papadias*
Athens, Greece
*Tharmalingam Ratnarajah*
Edinburgh, United Kingdom
*Dirk T.M. Slock*
Sophia Antipolis, France

---

Dedicated to the many researchers and engineers whose contributions over the years have made this book possible.

---

# Abbreviations

| | |
|---|---|
| 3D | three-dimensional |
| 3G | third generation |
| 3GPP | 3rd Generation Partnership Project |
| 4G | fourth generation |
| 5G | fifth generation |
| 5GS | 5G system |
| AAE | adversarial autoencoder |
| ADC | analog-to-digital converter |
| ADEL | advanced dynamic spectrum 5G mobile networks employing licensed shared access |
| AI | artificial intelligence |
| AMC | automatic modulation classification |
| AMPS | advanced mobile phone system |
| AP | access point |
| API | application programming interface |
| APT | Asia Pacific Telecommunity |
| ASA | authorized shared access |
| ASE | area spectral efficiency |
| ATC | air traffic control |
| AUL | autonomous uplink transmission |
| AWGN | additive white Gaussian noise |
| BC | broadcast channel |
| BF | beacon falsification |
| BF | beamformer/beamforming |
| B-IFDMA | block-interleaved frequency division multiple access |
| BnB | branch-and-bound |
| BNetzA | German Regulation Administration |
| BPDN | basis pursuit denoising |
| BPSK | binary phase shift keying |
| BS | base station |
| BSS | basic service set |
| BWA | broadband wireless access |
| CAPEX | capital expenditure |

| | |
|---|---|
| Cat2 | Category 2 LBT |
| Cat4 | Category 4 LBT |
| CBF | coordinated beamforming |
| CBRS | Citizens Broadband Radio Service |
| CBSD | Citizens Broadband Service device |
| CCA | clear channel assessment |
| CCC | control channel corruption |
| CCD | complementary cumulative distribution |
| CCI | co-channel interference |
| CD | code-division multiple access |
| CEPT | European Conference of Postal and Telecommunication Administration |
| CFAR | constant false-alarm rate |
| CI | constructive interference |
| CITEL | Inter-American Telecommunication Commission |
| CMC | constant-modulus constraint |
| CNN | convolutional neural network |
| CoBF | coordinated beamformer/beamforming |
| CoMP | coordinated multi-point |
| COT | channel occupancy time |
| CPE | customer premise equipment |
| CR | cognitive radio |
| C-RAN | Cloud RAN |
| CRS | common reference signals |
| CRSS | communication and radar spectrum sharing |
| CSI | channel state information |
| CSIR | channel state information at the receiver |
| CSI-RS | Channel State Information-Reference Signals |
| CSIT | channel state information at the transmitter |
| CSMA/CA | carrier sense multiple access with collision avoidance |
| CU | central unit |
| CWSC | University of Strathclyde's Centre for White Space Communications |
| D2D | device-to-device |
| DAC | digital-to-analog converter |
| DAPA | database access protocol attack |
| dB | decibel |
| DFH | dynamic frequency hopping |
| DFRC | dual-functional radar communication |
| DIA | database inference attack |
| DL | downlink |
| DMTC | discovery reference signal measurement timing configuration |
| DoA | direction of arrival |
| DoD | Department of Defense |
| DoD | direction of departure |
| DoF | degree of freedom |
| DoS | denial of service |

| | |
|---|---|
| DR | dynamic range |
| DRS | discovery reference signal |
| DSA | dynamic spectrum access |
| DSP | digital signal processing |
| DSS | dynamic spectrum sharing |
| DSSS | direct-sequence spread spectrum |
| DTV | digital television |
| DVB-T | digital video broadcasting — terrestrial |
| EC | European Commission |
| ECC | Electronic Communications Committee |
| EC/CEPT | European Conference of Postal and Telecommunication Administration |
| ED | energy detection |
| EIRP | equivalent isotropically radiated power |
| eLAA | enhanced licensed assisted access |
| eLSA | evolved licensed shared spectrum |
| eMBB | enhanced mobile broadband |
| eNB | evolved node B |
| EPC | evolved packet core |
| ESC | environmental sensing capability |
| ESIP-WSR | expected signal and interference power |
| ETEB | estimated time to empty buffers |
| ETSI | European Telecommunications Standards Institute |
| EU | European Union |
| EWSMSE | expected weighted sum mean squared error |
| EWSR | expected (or ergodic) weighted sum rate |
| FCC | Federal Communications Commission |
| FCC | first coefficient constraint |
| FD | full duplex |
| FDD | frequency-division duplex |
| FDMA | frequency-division multiple access |
| FFT | fast Fourier transform |
| FHSS | frequency hopping spread spectrum |
| FIS | forward inter-system |
| FrFT | fractional Fourier transform |
| FS | frame structure |
| FSS | fixed satellite services |
| FSS | fixed satellite system |
| FTP | file transfer protocol |
| GAA | general authorized access |
| GDD | geolocation database dependent |
| GHz | gigahertz |
| GNSS | global navigation satellite system |
| GPS | global positioning system |
| GRE | generic routing encapsulation |
| GSM | global system for mobile |

| | |
|---|---|
| HA | hex-antenna |
| HARQ | hybrid automatic-repeat-request |
| HD | half duplex |
| HMM | hidden Markov model |
| HT | hypothesis testing |
| IBC | interfering broadcast channel |
| IA | interference alignment |
| ICA | independent component analysis |
| ICD | initial commercial deployment |
| ICI | inter-cell interference |
| ICPA | interference-constrained PA |
| ICSI | interfering channel state information |
| ICT | information computing and telecommunications |
| IEEE | Institute of Electrical and Electronics Engineers |
| IETF | Internet Engineering Task Force |
| IFC | interference channel |
| IFFT | inverse fast Fourier transform |
| i.i.d. | independent and identically distributed |
| IMT | international mobile telecommunications |
| InfoGAN | Information Maximizing Generative Adversarial Networks |
| InH | indoor hotspot |
| IoT | Internet of Things |
| IP | Internet protocol |
| IPC | interference-power constraint |
| IPSec | IP security |
| IPT | interference power threshold |
| IQ | in-phase and quadrature phase |
| ISM | industrial, scientific, and medical |
| ISP | Internet service provider |
| ISS | inter-satellite service |
| ITM | international mobile telecommunications |
| ITRSSL | interference threshold restricted sharing of spectrum licenses |
| ITU | International Telecommunications Union |
| IU | incumbent user |
| JRC | Joint Research Center of the European Commission |
| KKT | Karush–Kuhn–Tucker |
| KPI | key performance indicator |
| LAA | licensed assisted access |
| LAN | local access network |
| LBT | listen-before-talk |
| LMDS | local multipoint distribution service |
| LMI | linear matrix inequality |
| LMMSE | linear minimum mean squared error |
| LoS | line-of-sight |
| LPI | low-probability-of-intercept |

| | |
|---|---|
| LS | least squares |
| LSA | licensed shared access |
| LSTM | long short-term memory |
| LTE | long-term evolution |
| LTE-A | long-term evolution advanced |
| LTE-LAA | long-term evolution – licensed assisted access |
| LTE-U | LTE in unlicensed spectrum |
| LU | licensee user |
| LWA | LTE-WLAN (radio) aggregation |
| LWAAP | LWA adaptation protocol |
| LWIP | LTE WLAN radio level integration with Internet protocol security tunnel |
| LWIPEP | LWIP encapsulation protocol |
| MAC | media access control |
| MaMIMO | massive multiple input multiple output |
| MED | maximum-eigenvalue-based detection |
| MF | matched filter |
| MFCN | mobile/fixed communications network |
| MIMO | multiple input multiple output |
| MISO | multiple input single output |
| ML | machine learning |
| MMSE | minimum mean squared error |
| mmWave | millimeter-wave |
| MNO | mobile network operator |
| MOP | multi-objective programming |
| MRC | maximal ratio combining |
| MRT | maximum ratio transmission |
| MS | mobile stations |
| MSE | mean squared error |
| MU | multi-user |
| MUI | multi-user interference |
| MU-MIMO | multi-user MIMO |
| MVNO | mobile virtual network operator |
| NaaS | network as a service |
| NBS | Nash bargaining solution |
| NE | Nash equilibrium |
| NEWSR | naive expected (or ergodic) weighted sum rate |
| NG-RAN | next generation (5G) radio access network |
| NHN | neutral host networks |
| NI | National Instrument |
| NLoS | non-line-of-sight |
| NOI | notice of inquiry |
| NPRM | notice of proposed rulemaking |
| NR | new radio |
| NRA | national regulatory agency |
| NRA | national regulation administration |

| | |
|---|---|
| NRA | national regulatory authority |
| NR-U | new radio in unlicensed spectrum |
| NSF | National Science Foundation |
| NSP | null-space projection |
| OA&M | operations, administration, and management |
| OAM&P | operations, administration, management, and provisioning |
| Ofcom | Office of Communications |
| OFDM | orthogonal frequency division multiplexing |
| OFDMA | orthogonal frequency division multiple access |
| OOB | out of band |
| OPEX | operating expenditure |
| ORAN | Open Radio Access Network |
| OSA | opportunistic spectrum access |
| OSDaaS | Open Spectrum Data as a Service |
| OTA | over-the-air |
| P2MP | point to multi-point |
| P2P | point-to-point |
| PA | power amplifier |
| PA | priority access |
| PAL | priority access license |
| PAWS | protocol to access white space |
| PCA | partly calibrated array |
| PCS | personal communication service |
| PCI | physical cell-identity |
| PDCP | packet data convergence protocol |
| PDF | probability density function |
| PDU | protocol data unit |
| PHY | physical |
| PIM | pulse interval modulation |
| PMSE | program making and special events |
| POE | power over Ethernet |
| PoP | point of presence |
| PPA | PAL protection area |
| PPDR | public protection and disaster relief |
| PPP | Poisson point process |
| PRB | Physical Resource Block |
| PRF | pulse repetition frequency |
| PRI | pulse repetition interval |
| PS | primary system |
| PSD | power spectral density |
| PSS | primary synchronization signal |
| PSK | phase shift keying |
| PU | primary user |
| PUSCH | physical uplink shared channel |
| QAM | quadrature amplitude modulation |

| | |
|---|---|
| QCQP | quadratically constrained quadratical programming |
| QoS | quality of service |
| QPSK | quadrature phase shift keying |
| R&O | report and order |
| RA | resource allocation |
| RadioML | radio machine learning |
| RAN | radio access network |
| RAT | radio access technology |
| RB | resource blocks |
| RCC | Regional Commonwealth in the Field of Communications |
| RCS | radar cross-section |
| RF | radio frequency |
| RFID | radio frequency identification |
| RIS | reverse inter-system |
| RLC | radio link control |
| RLS | radio location services |
| RMSE | root mean square error |
| RRM | radio resource management |
| RRS | reconfigurable radio systems |
| RSI | residual self-interference |
| RSPG | Radio Spectrum Policy Group |
| RSRP | reference signal received power |
| RSRQ | reference signal received quality |
| RSS | received signal strength |
| RSSI | received signal strength indicator |
| RSSL | restricted sharing of spectrum licenses |
| RV | random variable |
| Rx/RX | receive/receiver/reception |
| RZF | regularized zero forcing |
| SAIFE | spectrum anomaly detector with interpretable features |
| SAS | spectrum access system |
| SBW | small back-off window |
| SC | similarity constraint |
| SCH | superframe control header |
| SCN | small cell network |
| SD | sensing device |
| SDMA | space-division multiple access |
| SDP | semi-definite programming |
| SDR | software-defined radio |
| SDR | semi-definite relaxation |
| SE | spectral efficiency |
| SeGW | security gateway |
| SI | self-interference |
| SIC | successive interference cancellation |
| SINR | signal-to-interference-plus-noise ratio |

| | |
|---|---|
| SIR | signal-to-interference ratio |
| SIMO | single input multiple output |
| SINR | signal-to-interference-plus-noise ratio |
| SISO | single input single output |
| SLNR | signal-to-leakage-plus-noise ratio |
| SND | simultaneous non-unique decoding |
| SNR | signal-to-noise ratio |
| SON | self-organizing network |
| SP | spectrum provider |
| SPC | sum-power constraint |
| SR | sum-rate |
| SRM | secure radio middleware |
| SS | spectrum sharing |
| SS | secondary system |
| SSC | WInnForum's Spectrum Sharing Committee |
| SSDF | spectrum sensing data falsification |
| SSR | spatial spectrum reuse |
| SSS | secondary synchronization signal |
| STA | station |
| SU | secondary user |
| SU | spectrum user |
| SULI | spectrum utilization-based location inference |
| SVD | singular-value decomposition |
| SVM | support vector machine |
| sZF | statistical zero-forcing |
| TC | Technical Committee |
| TCP | transmission control protocol |
| TD | time division |
| TDD | time-division duplex |
| TDMA | time-division multiple access |
| TDOA | time difference of arrival |
| TIN | treating interference as noise |
| TIP | Telecom Infra Project |
| TPC | transmission power constraint |
| TR | Technical Report |
| TRAI | Telecom Regulatory Authority of India |
| TVHT | television very high throughput |
| TVWS | TV white space |
| Tx/TX | transmit/transmitter/transmission |
| TxOP | transmission opportunity period |
| UAS | user associated strategy |
| UAV | unmanned aerial vehicle |
| UCI | uplink control information |
| UDN | ultradense network |
| UDP | user datagram protocol |

| | |
|---|---|
| UE | user equipment |
| UHF | ultrahigh frequency |
| UKPM | UK Prediction Model |
| UL | uplink |
| ULA | uniform linear array |
| UMI | Urban Micro |
| UPT | user perceived throughput |
| USRP | universal software radio peripheral |
| USSL | uncoordinated sharing of spectrum licenses |
| UWB | ultrawide band |
| V2X | vehicle-to-everything |
| VAE | variational autoencoder |
| VHF | very high frequency |
| VoIP | voice over IP |
| WF | water-filling |
| Wi-Fi | wireless fidelity |
| WiGig | Wireless Gigabits Alliance |
| WiMAX | Worldwide Interoperability for Microwave Access |
| WInnForum | Wireless Innovation Forum |
| WLAN | wireless local area network |
| WRAN | wireless regional area network |
| WRC | World Radiocommunication Conferences |
| WSD | white space device |
| WSDB | white space database |
| WSMSE | weighted sum mean squared error |
| WSN | wireless sensor networks |
| WSR | weighted sum rate |
| WT | WLAN termination |
| ZF | zero forcing/forced |

# 1

# Introduction: From Cognitive Radio to Modern Spectrum Sharing

*Constantinos B. Papadias[1*], Tharmalingam Ratnarajah[2], and Dirk T.M. Slock[3]*

[1] *The American College of Greece, Greece*
[2] *University of Edinburgh, UK*
[3] *Institut Eurecom, France*

## 1.1 A Brief History of Spectrum Sharing

Limited spectrum availability is a real constraint for existing and future wireless systems. Spectrum scarcity is one key factor that prevents operators from meeting the increasing user demands in capacity and quality of service (QoS) and induces additional expenditures (capital expenditure and operating expenditure) that network operators reflect in the service prices to their customers. The introduction of novel spectrum management paradigms can address the spectrum crunch issue. Furthermore, it allows new types of players (operators, also called "users") who might not otherwise be able to afford or wish to have an exclusive/national-level license to provide service with QoS guarantees to their clients through a substantially smaller investment.

The use of the spectrum in commercial applications is typically either licensed or license-exempt. Spectrum sharing, wherein both licensed and license-exempt (or other types of non-exclusively licensed) users co-exist within the same frequency bands in a given geographic location, first explored via the concept of cognitive radio (CR), is an alternative approach in spectrum usage. CR is traditionally thought of as a technology that enables non-licensed secondary users (SUs) to make use of idle spectrum without causing harmful interference to licensed primary users (PUs). As such, it was regarded with suspicion by mobile broadband operators, who were reluctant to allow the use of their expensively acquired spectrum by any SU that claimed they would respect the regulatory CR policies. This reluctance on the side of legacy operators was accentuated by the fact that, in its original form, CR, which was first considered for the so-called TV white space (TVWS) spectrum freed by former analog TV providers, relied heavily on spectrum sensing in order to avoid causing interference to PUs. This was clearly insufficient due to the low levels of sensing sensitivity, the well-known hidden node problem, etc. The architectural (supported by regulation) addition of using a spectrum registry (database) in order to better/further prevent harmful interference to the PUs improved the situation,

---

*This work was performed when Dr. Papadias was with Athens Information Technology.

*Spectrum Sharing: The Next Frontier in Wireless Networks,* First Edition.
Edited by Constantinos B. Papadias, Tharmalingam Ratnarajah, and Dirk T.M. Slock.
© 2020 John Wiley & Sons Ltd. Published 2020 by John Wiley & Sons Ltd.

but was still insufficient to make CR take off as a service paradigm. Traditional CR was also problematic from the SUs viewpoint, as it could only guarantee a QoS level similar (at best) to unlicensed access, i.e., with no guarantees.

The next important milestone emerged in early 2011, when Nokia and Qualcomm formally introduced the concept of authorised shared access (ASA), also known as licensed shared access (LSA), which is described by the EU Radio Spectrum Policy Group (RSPG) as, *"An individual licensed regime of a limited number of licensees in a frequency band, already allocated to one or more incumbent users, for which the additional users are allowed to use the spectrum (or part of the spectrum) in accordance with sharing rules included in the rights of use of spectrum granted to the licensees, thereby allowing all the licensees to provide a certain level of QoS."* By establishing formal contractual agreements between license holders and "licensees" (amounting to some type of spectrum leasing), the first step of bringing incumbent operators and new entrants closer together was achieved, with the latter no longer considered as unreliable or "rogue." On the technical front, the LSA system architecture relies on both a spectrum registry (LSA repository) where incumbents declare their spectrum occupancy, and a control unit (LSA controller) that handles the spectrum management and compliance. On the legal front, a legal framework was postulated in order to handle any kind of misbehavior of the licensees. Furthermore, it was the first time that QoS guarantees were given to the licensee. The introduction of ASA/LSA can therefore be viewed as an important breakthrough to make spectrum sharing a commercial reality.

As could be expected of course, the initial adoption of LSA was rather limited. For example, the initial version of LSA adopted by the European Conference of Postal and Telecommunication Administration (EC/CEPT) excluded concepts such as opportunistic spectrum access (OSA), typically secondary use or secondary service where the applicant has no protection from the PU. Moreover, according to this version, LSA applies only when the incumbent user(s) and the LSA licensees are of different natures (e.g., governmental versus commercial), operate different types of applications, and are subject to different regulatory constraints. Furthermore, the original version of LSA was geared mostly towards traditional mobile network operators (MNOs) as typical licensees, neglecting the various emerging vertical applications and new types of networks prescribed in fifth-generation (5G) technology. This was later improved by the introduction of evolved LSA (eLSA), which prescribes local area networks for use in cases such as industrial automation, e-health, and emergency services, among others (see Chapter 2).

The next important step came with the opening of the Citizens Broadband Radio Service (CBRS) in the frequency band 3.55–3.7 GHz by the Federal Communications Commission (FCC) in the USA, intended for spectrum sharing via a combination of licensed and unlicensed spectrum use. The corresponding system, pushed by both the Wireless Innovation Forum (WInnForum) and 3rd Generation Partnership Project (3GPP), is called the spectrum access system (SAS) and prescribes three tiers of users (operators): incumbents (such as radar systems), who enjoy exclusive spectrum usage, priority access license (PAL) users, who have exclusive access in the absence of the incumbent, and general authorized access (GAA) users, who have sensing-assisted unlicensed access in the absence of the incumbent (similar to traditional CR users). The availability of the released spectrum, backing from FCC, and inclusion of all three tiers of users make the use of SAS in the CBRS spectrum a strong contender for spectrum sharing-based access, in spite of the various remaining challenges and specifications that need to be met.

A brief comparison of the two dominant emerging types of spectrum sharing described above can be found below:

- LSA (EU version)
    - Pushed by CEPT, ETSI, 3GPP
    - Two-tier model: incumbents, licensees
    - Spectrum sensing is country-wide
    - Incumbent protection through database
- SAS (USA)
    - Pushed by FCC, 3GPP, WInnForum
    - Three-tier model: incumbents, PAL, GAA
    - Spectrum sensing in reduced areas (e.g., census tracks of 4000 people)
    - Interference mitigation across census tracts
    - Sensing-based protection of incumbents

More recently, another important trend arose: the coexistence of long-term evolution (LTE) and Wi-Fi. Trying to solve this and other important challenges, there has been a recent explosion of spectrum-sharing concepts: LTE in unlicensed bands (LTE-U), license-assisted access (LAA) in LTE advanced (LTE-A), LTE wireless local area network (WLAN) aggregation (LWA), LTE-WLAN radio level integration with Internet protocol security tunnel (LWIP), MulteFire, Wi-Fi in licensed band (Wi-Fi-Lic), Wi-Fi Boost, etc. (see Chapters 4 and 14). Given the availability of the corresponding LTE and Wi-Fi technologies, this approach is also well poised to affect spectrum access in the immediate future.

The culmination of these trends over the last decade constitutes a significant technology evolution (or possibly revolution) which we believe will affect the way the spectrum is accessed and used for a variety of applications and players in the forthcoming years, affecting both the economy and society. This edited volume is our attempt to collect the key concepts and emerging approaches, as well as to hint at the future impact of the important emerging field of spectrum sharing.

## 1.2  Background

The editors' joint involvement with spectrum sharing started with our collaboration in the context of the European Commission's (EC) Future and Emerging Technologies collaborative project, entitled Cognitive Radio Oriented Wireless Networks (CROWN, see https://cordis.europa.eu/project/rcn/90432/factsheet/en), which ran from 2009 to 2012 under the EC's 7th Framework Program (FP7). In CROWN we explored heavily, among others, the use of the spatial dimension (enabled by antenna arrays) in various cognitive radio setups, with contributions at both the physical (PHY) layer and the media access control (MAC) layer, introducing probably one of the first directional-based MAC protocols. The project was one of the first to propose horizontal spectrum sharing (i.e., between the same type of users) with joint spectrum access (i.e., without any operator vacating the band for another) and encompassed the emerging (at the time) concept of database-assisted sharing. It also introduced collaborative/distributed sensing and provided proof-of-concept experimentation in an over-the-air LTE demo.

Our next collaboration was in the context of another FP7 collaborative project called Advanced Dynamic Spectrum 5G Mobile Networks Employing Licensed Shared access (ADEL, see https://cordis.europa.eu/project/rcn/189128/factsheet/en), which ran from 2013 to 2016. In ADEL, we studied, along with other partners, an enhanced LSA system that incorporates an opportunistic spectrum access in order to radically improve the capacity of the system by exploiting as much unused bandwidth as possible, and we investigated additional business cases, such as the scenario where both the incumbent and the LSA licensee are MNOs. ADEL aimed to overcome one of the main challenges of such an enhanced LSA system, that is, to make the sharing conditions sufficiently attractive and predictable (spectrum without unacceptable interference, enough spectrum availability, etc.) to enable the LSA licensee to invest in network equipment and licensing fees. As stated earlier, although the concept of flexible spectrum access has been researched and developed for quite some time, its adoption by the wireless industry and regulators is still timid (and was even more so when the ADEL project started). This is due to several reasons, including a number of technical barriers but most importantly the lack of an attractive business case. As we believed that the business need for additional wireless network capacity was increasingly making LSA-type spectrum sharing a necessity, our work in ADEL aimed to demonstrate the feasibility of QoS provisioning in dynamic spectrum access under an enhanced LSA regime, for a number of practical scenarios, thus contributing to setting the path for standardization and regulatory adoption of enhanced LSA paradigms (a trend that was soon after adopted in practice with the emergence of eLSA).

As part of ADEL's dissemination activities, we also organized a project booth at the EuCNC2016 conference in Athens, Greece, with an intermediate version of the LSA Proof of Concept demo. The ADEL booth won the Best Booth Award, as voted for by the conference participants. This was another confirmation of the mounting awareness of the importance of spectrum sharing by the telecom community. We also organized the ADEL Indian Summer School on Spectrum Aggregation and Sharing for 5G Networks (SS-SAS5G) at EURECOM, France, in October 2016 (see http://www.euracon.org/index.php/2013-02-12-09-41-49/sssas5g). This successful event also contributed to the widening perception that the CR concept has given way to more advanced spectrum sharing paradigms that are likely to affect the spectrum landscape in the near future.

Our joint involvement in the above research and dissemination activities, which allowed us to become exposed to the key facets of spectrum sharing technology (ranging from hands-on research to prototyping/demonstration to regulation), as well as the realization that a collection of all these aspects and recent technology components lack a single volume in the literature, is what prompted us to put together this book. Our desire to do so was further enhanced by the frenzy of activity in the area of spectrum sharing described in the previous section. Our intention was to collect the key attributes of spectrum sharing, including both its early beginnings and historical evolution, the state-of-the-art, and the key emerging trends. As stated above, we aim to capture both the key technological components, and also the evolving regulatory and business environment. On the technology front, we include theoretical techniques, as well as trials, demos, prototypes, and performance analysis studies, whereas on the regulatory and business front we address both the evolution of standards and inputs on the current market and its future outlook. We have tried our best to provide the interested reader with a collection of contributions and trends from numerous sources. We believe that this will be a valuable reference for state-of-the-art and emerging technologies in the area of spectrum sharing for graduate students and researchers working in the areas of wireless communications and signal processing engineering. It will also be an important reference for radio communications engineers and practitioners, especially all those who deal with spectrum management aspects, including not only designers of radio communication systems, but also spectrum owners and policy regulators. In the following, we provide a brief description of the contents of the remainder of the book.

## 1.3 Book overview

The continuing story of the evolution of spectrum sharing is developed in the chapters that follow this introductory chapter.

### 2 Regulation and Standardization Activities Related to Spectrum Sharing

*Markus Mueck[1], María Dolores (Lola) Pérez Guirao[2], Rao Yallapragada[3], and Srikathyayani Srikanteswara[3]*

[1] *Intel Deutschland GmbH, Germany*
[2] *Sennheiser, Germany*
[3] *Intel Corporation, USA*

This chapter focuses on the standardization and regulatory landscape of spectrum sharing. A historical perspective is provided, starting with the early days of CR and TVWS and moving onto the more recent trends of LSA and CBRS, which are both described in detail. The emphasis then shifts to the current status, including 5G and Wi-Fi evolution standards. The discussion throughout the chapter is cast in a regulatory framework, referring to the latest trends in the US (FCC), Europe (Ofcom and national telecom authorities), and other regions.

### 3 White Spaces and Database-assisted Spectrum Sharing

*Andrew Stirling*

*Larkhill Consultancy, UK*

This chapter provides an overview of the evolution of electromagnetic spectrum sharing, building on the emergence of the cognitive radio concept, to enable more efficient access to the spectrum without the need for conventional licensing. The role of regulation is emphasized and the key concept of white space spectrum is discussed. The chapter continues with a presentation of the three-tier access model and introduces the opportunistic spectrum access and margins of protected service. It then discusses the basics of license-exempt access and dynamic spectrum access (DSA)/CR, including the key aspects of receiver sensing sensitivity, the hidden node problem, and the geolocation/spectrum database, for which a detailed architecture and spectrum-sharing options are provided. The main techniques for spectrum sensing are also described, followed by the presentation of cooperative sensing as a way to avoid the hidden node problem, as well as the use of dedicated beacons for spectrum awareness. The chapter also discusses software-defined radio (SDR) as an implementation approach and concludes with some suggestions on how to use the spectrum more flexibly in the future.

## 4 Evolving Spectrum-sharing Methods, Standards, and Trials: TVWS, CBRS, MulteFire, and more

*David Crawford, Dani Anderson, K.A. Shruthi, and Robert W. Stewart*
*University of Strathclyde, UK*

This chapter stems from the University of Strathclyde's Centre for White Space Communications (CWSC) involvement in numerous spectrum-sharing trials and testbeds over the last decade. It presents a selection of developing spectrum-sharing technologies and provides key outcomes from shared spectrum trials that have been carried out across the UK. The evolution of shared spectrum is presented, including developing and emerging technologies, across multiple operating frequency bands. Beginning with TVWS, as discussed in Chapter 3, it provides details of several UK-based TVWS projects and trials, completed by the CWSC and various partners, as tangible examples of shared spectrum implementation. The chapter goes on to provide a survey of emerging technologies in the 3.5-GHz (with emphasis on CBRS) and 5-GHz bands, including LTE in unlicensed spectrum (LTE-U), LWA, LWIP, LAA, and MulteFire; a cellular technology developed by an international consortium in early 2017.

## 5 The Spectrum Landscape above Radio and up to mmWave Bands

*Abhishek K. Gupta and Adrish Banerjee*
*Indian Institute of Technology (IIT), Kanpur, India*

This chapter is devoted to the spectrum landscape above radio bands, up to millimeter-wave (mmWave), where several target bands are currently being studied for service. These bands are expected to become increasingly important in the future given the upcoming spectrum crunch, which will eventually limit the continued growth of wireless services and applications. The chapter starts with the key differentiators of these bands with respect to sub-6 GHz bands, i.e., their sensitivity to blockages, which affect the interference environment, and their use of highly directional antennas, which are needed to compensate for the path loss at these frequencies. The benefits of the spectrum-sharing approach over the conventional exclusive licensing model are then presented in this context, with emphasis on the key observation that spectrum sharing may provide gains over exclusive spectrum usage even without interference coordination (a requirement that cannot be relaxed as easily in the sub-6 GHz bands). The authors go on to provide details about several bands above 6 GHz [namely, 24 GHz, local multipoint distribution service (LMDS), and 37/39-, 42-, 57/64-, and 70/80-GHz bands], preceded by a discussion of the radio spectrum regions as defined by the International Telecommunications Union (ITU) and the mmWave allocations given by the various relevant regulating authorities across the globe [e.g., the RSPG, CEPT, the World Radiocommunication Conference (WRC), Asia Pacific Telecommunity (APT), Inter-American Telecommunication Commission (CITEL) and Regional Commonwealth in the Field of Communications (RCC)]. A discussion on whether to share or not to share in mmWave bands is presented, taking into account several factors. The conventional approaches of exclusive licensing and unlicensed spectrum access are contrasted to three key models for spectrum sharing in mmWave (uncoordinated, static, and dynamic), for which some performance results are also provided. The chapter continues with a description of various shared licensing approaches and ends with a discussion of secondary licenses and markets.

## 6 The Licensed Shared Access Approach

*António J. Morgado*

*Instituto de Telecomunicações, Aveiro, Portugal*

Licensed (or authorized) shared access is one of the most recent trends of spectrum sharing and constitutes an important milestone in that it provides certain guarantees of QoS to the users of both incumbent and licensee operators. After a brief historical review of spectrum sharing, this chapter presents the basics of the licensed/authorized (shared) access approach, including key definitions, early regulatory actions and system architectures, with emphasis on the key network architectural elements required to support these types of spectrum sharing. It also presents in some detail the corresponding standardization actions in Europe, as promoted by the Electronic Communications Committee (ECC) of CEPT, including the definition of the designated bands in Europe for LSA, with emphasis on mobile broadband licensee operators (first targeting the 2.3–2.4-GHz band and more recently the 3.6–3.8-GHz band). The chapter then goes on to describe a new proposed system architecture for LSA that stemmed from the EU project ADEL, whose purpose has been to offer more dynamic spectrum sharing to potentially several licensee users that may belong either to the same or to different service classes. The suggested architecture includes, among others, an LSA band manager for handling the resource allocation actions with QoS guarantees to all users, a number of sensing networks for better spectrum awareness, and a radio environment map to assist the band manager in implementing the dynamic spectrum sharing. The chapter then provides a detailed discussion on how the ADEL system can provide dynamic LSA to three key spectrum sharing scenarios and concludes with a summary of the key benefits of the approach.

## 7 Collaborative Sensing Techniques

*Christian Steffens[1] and Marius Pesavento[2]*

*[1] Hyundai Mobis, Frankfurt, Germany*
*[2] Darmstadt University of Technology, Darmstadt, Germany*

Spectrum sensing is a key technology in order to allow spectrum sharing and was in fact a key ingredient of cognitive radio in its early days. While the evolved spectrum-sharing technologies no longer rely solely on spectrum sensing, the latter remains an important ingredient that can further improve performance (e.g., on top of database-assisted or regulation-based sharing). In certain systems it is even prescribed by the current regulation. After presenting an overview of the key conventional spectrum-sensing signal processing techniques (i.e., energy detection, matched filtering, and cyclostationarity-based, see also Chapter 3), the authors bring our attention to the key ingredients of advanced spectrum sensing: (i) multi-antenna receivers, (ii) collaborative sensing, and (iii) techniques based on sparse signal reconstruction. Then they go on to present their novel approach that combines all these three elements, wherein the collaborative sparse sensing takes place at a fusion center after being fed measurements collected from the sensing nodes. Their approach is evaluated via numerical simulations and compared against both centralized and distributed sensing. Finally, the technique's ability to estimate node location is demonstrated.

## 8 Cooperative Communication Techniques

*Faheem Khan[1], Miltiades C. Filippou[2], and Mathini Sellathurai[3]*

[1] *University of Huddersfield, UK*
[2] *Intel Deutschland GmbH, Germany*
[3] *Heriot-Watt University, UK*

This chapter focuses on cooperative communication techniques for spectrum sharing that aim to fully exploit the sensing information available from the sensing stage, such as described in Chapter 7. The chapter starts with an overview of cooperative techniques that are relevant for spectrum sharing and a discussion on how they relate to the sensing process. It presents the three key spectrum-sharing categories (interleave, underlay, and overlay communication) and then describes in detail two techniques for underlay downlink communication. In the first one, a multiple input single output (MISO) primary transmitter/receiver pair shares its spectrum with a MISO secondary pair with the goal of maximizing the secondary user's average data right, subject to a corresponding data rate constraint for the primary user. The considered scenario would be applicable to an enhanced mobile broadband (eMBB) class of service for both primary and secondary transmission. Each transmitter knows perfectly its own channel and has only statistical knowledge of the other channels involved (in terms of channel covariance). A distributed cooperative technique is derived under the above assumptions and its performance is demonstrated to be far superior to the standard interference temperature-based approach. Then, a single input multiple output (SIMO) uplink scenario is considered with similar channel knowledge assumptions, targeting a primary link with a requirement for high reliability, assuming that the secondary system is provided with the primary system's traffic statistics. A hybrid interweave/underlay scheme is adopted and a joint optimized sensing and reception scheme is developed. Its performance is evaluated, showing its superiority over either standalone underlay or interweave communication in terms of the average rate achieved by the secondary system, both as a function of the primary user's occupancy and its outage performance. The chapter concludes with a summary of the open research challenges in this area.

## 9 Reciprocity-based Beamforming Techniques for Spectrum Sharing in MIMO Networks

*Kalyana Gopala and Dirk T.M. Slock*

Institut Eurecom, France

The effectiveness of several spectrum sharing techniques, especially those used for horizontal sharing (i.e., between users of the same type, in the same band, at the same time), relies heavily on the accurate and timely knowledge of the involved user and interference channels. These techniques are often based on beamforming by multiple antennas to separate users by exploiting the spatial dimension (see also Chapter 12). The requirements for channel state knowledge at the transmitter (CSIT) become crucial in massive multiple input multiple output (MaMIMO), which is a key ingredient of 5G and is also well suited for LSA as it offers higher spatial resolution. However, MaMIMO is hard to implement in frequency division duplex (FDD) systems due to the complications imposed by the feedback channel. An alternative approach is to consider time division duplex (TDD) systems, where, in theory at least, the forward and reverse channels are equal, hence not requiring a feedback channel. This chapter deals with the actual case in which reciprocity is only obtained after calibration of the radio frequency (RF) parts in the transmitter and receiver chains. An overview of the state of the art in reciprocity calibration techniques is first provided, with an emphasis on internal calibration usable in MaMIMO. A number of promising reciprocity-based techniques are then presented for the design of transmit precoders for spectrum sharing between incumbents and licensees. In particular, the concept of naive uplink/downlink duality is presented, which allows the additional information exchange required for utility optimization or to deal with non-cooperative nodes to be reduced further.

## 10 Spectrum Sharing with Full Duplex

*Sudip Biswas[1], Ali Cagatay Cirik[2], Miltiades C. Filippou[3], and Tharmalingam Ratnarajah[4]*

[1] Indian Institute of Information Technology Guwahati, India
[2] Ofinno Technologies, USA
[3] Intel Deutschland GmbH, Germany
[4] University of Edinburgh, UK

This chapter focuses on the advantages of using full duplex (FD) in spectrum sharing technologies such as CR. By transmitting in FD mode, a CR can simultaneously transmit and sense the transmission status of other nodes (refer to Chapter 8 for joint sensing and reception in non-FD mode), which makes it suitable to combat numerous issues at the medium access control layer, such as hidden terminals, large delays, and congestion. Starting by introducing the motivation for using FD in CRs from the perspective of both cellular systems (CS) and the Internet of Things (IoT), this chapter overviews the design of FD transceivers and accordingly analyses the fundamental requirements for the co-implementation of the two technologies and the corresponding benefits obtained over transmission through traditional half duplex. It then postulates the necessary mathematical framework, including the optimization problems associated with both CS and IoT.

Next, detailed steps for the conversion of the problems into tractable form are illustrated along with efficient transceiver design algorithms. Finally, comprehensive numerical results are provided to justify the use of FD in CRs and open problems in the field of CRs transmitting in FD mode are presented to summarize the chapter.

## 11 Communication and Radar Systems: Spectral Coexistence and Beyond

*Fan Liu and Christos Masouros*

University College London, UK

This chapter focuses on recent progress in the area of communication and radar spectrum sharing (CRSS), which not only presents advantages in enabling the efficient usage of the spectrum, but also provides a new way to design novel systems that can benefit from the cooperation of radar and communications, thus introducing a new spectrum sharing paradigm. Starting by introducing the motivation for CRSS from both civilian and military perspectives, this chapter overviews the applicable scenarios and analyses the fundamental requirements for sharing the spectrum between the two systems. It then provides general definitions and mathematical models, and further introduces the associated key performance metrics for radar and communication systems. As a step further, the chapter provides an overview of the state of the art for CRSS, from the coexistence of individual radar and communication devices, to the design of the dual-functional system that enables simultaneous communication and remote sensing (a topic also dealt with, in different contexts, in Chapters 8 and 10). Finally, the discussion is summarized by reviewing the open problems in the research field of CRSS.

## 12 The Role of Antenna Arrays in Spectrum Sharing

*Constantinos B. Papadias[1], Konstantinos Ntougias[2], and Georgios K. Papageorgiou[3]*

[1] The American College of Greece, Greece
[2] University of Cyprus, Cyprus
[3] Heriot-Watt University, UK

By offering the so-called spatial dimension, antenna arrays are an important enabler of spectrum sharing for all types of wireless systems. This chapter reviews the basic attributes of antenna arrays that allow them to reuse the spectrum efficiently, handle the interference environment, and even aid spectrum policy enforcement. Emphasis is placed on antenna array-based spectrum sharing that is applicable to the technologies in current wireless standards [such as the use of multiple input multiple output (MIMO) and coordinated multi-point (CoMP) in fourth-generation (4G) devices] or has the potential of impacting spectrum sharing networks in the near future. The chapter starts with a review of the key attributes of spectrum sharing, both from a physical and a regulatory viewpoint, and continues with an overview of the key attributes of antenna arrays and a discussion on the beneficial synergy of the two technologies. It then goes on to present a novel technique for antenna-array-aided spectrum sharing, based on coordinated linear precoding, as well as a new approach to spectrum sensing that relies on low complexity (parasitic) antenna arrays, for which over-the-air results are presented. The chapter concludes with a summary of its key findings, pointing to the continued beneficial use of antenna arrays in future spectrum sharing systems.

## 13 Resource Allocation for Shared Spectrum Networks

*Eduard A. Jorswieck[1] and M. Majid Butt[2]*

[1] TU Braunschweig, Germany
[2] Nokia Bell Labs, France

The coexistence of devices in dense wireless networks requires careful design resource allocation algorithms for spectrum sharing. In particular, the conflicting interests of heterogeneous devices and their service level requirements in terms of data rate, reliability, latency, security, and energy efficiency lead to complicated resource assignment and allocation problems. In this chapter, some recent resource assignment and trading algorithms for spectrum sharing are reviewed and their properties in terms of computational and implementation complexity are analyzed. The chapter begins with the observation that the key limiting factor in spectrum sharing is the interference observed at the physical layer and proposes corresponding resource allocation remedies. It then goes on to provide an information-theoretic background that allows the achievable rate regions of the underlying interference channel model to be quantified. After presenting a brief classification of the main considered types of spectrum sharing in terms of operators and radio access technology (RAT), the resource allocation problem is introduced and the key targeted challenges are defined. These three challenges (regarding multi-objectives, conflicting utilities, and distributed implementation) are explained and approached by a multi-objective programming (MOP) problem framework, game theoretic approaches, and stable matching-based resource allocation. The resource trading approach to spectrum sharing is then reviewed (see also Chapter 18) and the chapter concludes with a summary of its findings and the future outlook.

## 14 Unlicensed Spectrum Access in 3GPP

*Daniela Laselva[1], David López-Pérez[2], Mika Rinne[3], Tero Henttonen[4], Claudio Rosa[1], and Markku Kuusela[5]*

[1] Nokia Bell Labs Aalborg, Denmark
[2] Nokia Bell Labs Dublin, Ireland
[3] Nokia Technologies, Finland
[4] Nokia Bell Labs CTO, Finland
[5] Nokia CSD Digital Automation, Finland

With the emergence of spectrum sharing techniques that may operate in an unlicensed spectrum, there has been an increased interest in providing access in such a spectrum via existing wireless protocols (see also the discussion in section 1.1). This chapter reviews some important recent advances of this type, focusing on unlicensed spectrum access from the 3GPP standard viewpoint, namely LWA, LWIP, and LAA. In a comparative manner, the chapter reviews how each of these technologies, with the design choices of protocol architectures, procedures, mobility, and security provisioning, enables flexible usage of both licensed and unlicensed spectra as well as fair coexistence in the unlicensed spectrum with other wireless systems. It first describes LTE-WLAN aggregation (LWA), focusing on how WLAN (with emphasis on the current 802.11ac version) can be aggregated to operate under the control of LTE and how the aggregation of traffic flows works for the radio bearers in the LTE convergence protocol. The necessary network interfaces are described in detail. The alternative of LWIP (LTE-WLAN over an IP secured tunnel) is then presented, followed by a description of the LTE LAA (License-Assisted Access) mode, including a discussion on the changes required in order to enable its operation in an unlicensed band. These three technologies are then evaluated via a performance analysis, which also addresses their coexistence. The chapter concludes with an outlook of the anticipated research and standardization directions in the context of new radio (NR) operations in the unlicensed spectrum before summarizing its key findings.

## 15 Performance Analysis of Spatial Spectrum Reuse in Ultradense Networks

*Youjia Chen[1], Ming Ding[2], and David  Lópéz-Pérez[3]*

[1]*Fuzhou University, P.R. China*
[2]*Commonwealth Scientific and Industrial Research Organisation (CSIRO), Eveleigh, Australia*
[3]*Nokia Bell Labs, Dublin, Ireland*

Aggressive spatial spectrum reuse (SSR) by network densification using smaller cells has successfully driven the wireless communication industry onward in the past decades. In our future journey toward ultra-dense networks (UDNs), a fundamental question needs to be answered. Is there a limit to SSR? In other words, is activating all base stations (BSs) on the same frequency spectrum always the best strategy? Chapter 15 presents a theoretical analysis to answer this question. It starts with a definition of the network scenario and system model, including line-of-sight (LoS) and non-LoS transmission, antenna heights, user equipment (UE) densities, and active/sleep BSs. It continues with an analytical derivation of the coverage probability and the area spectral efficiency (ASE) expressions for the considered UDNs, based on stochastic geometry theory. While studying how aggressive the SSR approach can be (i.e., activating as many BSs as possible in the same time/frequency resource), a milder approach (called multi-channel spectrum sharing) is also considered, wherein BSs are uniformly allocated to a given number of channels. The two approaches are then compared numerically, leading to the conclusion that multi-channel spectrum sharing strategy can greatly boost coverage probability and the ASE due to the enhanced signal-to-interference-plus-noise ratio (SINR) when the network is ultradense, and there is a channelization that optimizes the ASE. Since the study points to the existence of an optimal SSR density that can maximize the network capacity, it is hence suggested that this limitation should be considered in the operation of future UDNs.

## 16 Large-scale Wireless Spectrum Monitoring: Challenges and Solutions Based on Machine Learning

*Sreeraj Rajendran and Sofie Pollin*
*KU Leuven, Belgium*

This chapter focuses on the need for large-scale spectrum monitoring to enable spectrum sharing. Manual wireless spectrum management and analysis will be inefficient and suboptimal in the dense and heterogeneous wireless environments that are encountered in today's (e.g., 5G) and future generations of wireless networks (see also Chapter 15). Furthermore, unauthorized wireless spectrum usages and anomalies are increasing every year in the form of uncertified wireless devices, fake BSs, unintentional transmitter leakages, and easily available spectrum jammers. It is hence becoming clear that the monitoring of this complex and dense electromagnetic environment will have to be automated, reliable, and cost-efficient. Automated monitoring of wireless spectrum over frequency, time, and space is, however, still challenging. Some approaches to address the key challenges of large-scale wireless spectrum monitoring are proposed, starting with an architecture for large-scale crowd-sourced data collection. New machine learning models are then presented which can be used to interpret sensed spectrum data effectively in terms of anomaly detection and signal classification. A semi-supervised deep learning setup is then presented that is based on the latest deep learning research and achieves performance close to fully supervised models with only 20% of the labeled samples. The performance of quantized models is analyzed and it is shown that, in principle, considerable computational performance reduction can be achieved at a cost of 10% classification accuracy loss. These findings underline the emerging importance of deep learning methods for wireless applications, which are expected to become much more prominent in the forthcoming years and for which some future research directions are suggested.

## 17 Policy Enforcement in Dynamic Spectrum Sharing

*Jung-Min (Jerry) Park, Vireshwar Kumar, and Taiwo Oyedare*
Virginia Tech, USA

To ensure the viability of the spectrum sharing paradigm, it is critical to identify potential threats and security vulnerabilities that may undermine the harmonious operation of a spectrum sharing ecosystem and adopt effective measures to counter them. A framework and mechanisms for enforcing spectrum sharing policies and/or rules which have been prescribed by the relevant regulator is hence required (see, for example, Chapters 2 and 6). Policy enforcement is needed to thwart noncompliant (or rogue) transmitters and to minimize the probability of potential harm to compliant users of a spectrum sharing ecosystem. For example, policy enforcement is particularly important when sharing government (including military) spectrum with non-government (commercial) systems (as considered in Chapter 11). This chapter discusses, with representative examples, security and policy violation threats that impact the stakeholders of a spectrum sharing ecosystem, with a focus on the spectrum sharing environment of the USA. A taxonomy for classifying the threats is first presented, considering fundamental mechanisms for enabling coexistence, as well as the points of attack with respect to the five-layer protocol stack. Policy enforcement mechanisms are discussed in the context of two categories: ex ante (preventive) and ex post (punitive) enforcement. The chapter concludes by discussing the research and regulatory challenges that need to be addressed to ensure policy enforcement in dynamic spectrum sharing.

## 18 Economics of Spectrum Sharing, Valuation, and Secondary Markets

*William Lehr*
Massachusetts Institute of Technology, USA

Advances in wireless technology that increase spectrum agility provide much of the technical foundation needed for spectrum sharing among heterogeneous networks, as described in earlier chapters. New technology gives rise to new market opportunities and business models, which in the case of wireless requires new spectrum management frameworks. Ensuring that the valuable resource of spectrum is used efficiently and directed toward its highest value uses for society will require the co-evolution of technology, wireless markets, and regulatory policies. The focus of this chapter is understanding what advances in information, computing and telecommunications (ICT) technologies, leading us toward 5G, imply for the economics and future of spectrum management. The chapter reviews the basic economics of spectrum as a resource and explains how regulatory and technical trends have increased both the need and opportunities for sharing spectrum more intensively. It provides a review of the different ways in which spectrum may be valued in dollar terms, the challenges to using such estimates, and the factors that contribute to making some spectrum usage rights more valuable than others. When viewed as an economic asset, spectrum is best understood as a bundle of property or usage rights that establish the terms under which potential users of the spectrum may use it. These usage rights are given form by the technologies, regulatory policies, and markets that comprise the wireless ecosystem. They may be altered and transferred by changes in technology, regulatory policies, or markets. Altering these rights and managing how they are used is central to understanding how spectrum may be shared. The chapter discusses the need for and some of the challenges associated with the rise of more robust markets for secondary spectrum trading.

### 19 The Future Outlook of Spectrum Sharing

*Richard Womersley*

LS Telcom, Germany

This final chapter of the book provides a number of conclusions and forecasts regarding the evolution of today's spectrum sharing landscape. In this sense it can be seen as complementary to Chapter 2. The provided outlook focuses mostly on commercially available systems and their anticipated evolution, taking into account the current regulatory and standardization trends. The chapter starts with a brief overview of spectrum sharing in practice from an operator's viewpoint and continues with a description of how regulatory agencies have been warming up to spectrum sharing recently. It continues with a factual discussion on spectrum demand, as well as a discussion on the impact of sharing on spectrum demand. It then brings up the important issue of the need for authorization for spectrum sharing and concludes with an outlook and some predictions for the future of spectrum sharing.

## 1.4 Summary

As indicated by the summary of chapter contributions described above, this book constitutes an ambitious endeavor to showcase, within a single volume, all the critical aspects of the spectrum sharing paradigm. Starting with a description of the spectrum sharing landscape as it looks today and a historical yet detailed review of the key technologies and their evolution in time, reaching up to mmWave frequencies, it then moves towards the most recent spectrum sharing paradigms, such as LSA, and promising technology enablers, such as collaborative sensing and cooperative communication. A number of advanced upcoming technologies that can further assist spectrum sharing are then described, including reciprocity-based transmission, full duplex communication, coexistence of communication and radar systems, antenna arrays, and advanced resource allocation. The paradigm of unlicensed spectrum access is then discussed in the context of 3GPP, capturing a key emerging standardization trend. The role of spatial spectrum reuse in ultra dense networks is then analyzed. The importance of machine learning is highlighted in terms of its ability to improve large-scale spectrum monitoring. The key regulatory aspect of policy enforcement is then discussed along with an analysis of the economics of spectrum sharing, with emphasis on secondary markets. The book concludes with a description of the future outlook of spectrum sharing. We feel strongly that by bringing all these aspects together, legacy and emerging approaches, promising technologies, standardization, regulation and economics, as well as by capturing some key trends in both wireless technology (e.g., mmWave, full duplex, ultradense networks) and ICT overall (e.g., sparse modelling, machine learning, security), this volume will provide to the interested reader a picture of this fascinating emerging technology that is as comprehensive as possible, explaining where it has come from, where it stands today, and where it is likely to be going. We hope that our effort in putting this book together, with the invaluable help of the individual chapters' highly qualified authors, who come from all corners of the globe, will serve its purpose of providing a well-rounded view of spectrum sharing to the interested reader. In this way, it could become a useful reference for this emerging technology that is poised to affect the way spectrum is accessed in the future.

# 2

# Regulation and Standardization Activities Related to Spectrum Sharing

*Markus Mueck[1], María Dolores (Lola) Pérez Guirao[2], Rao Yallapragada[3], and Srikathyayani Srikanteswara[3]*

[1] *Intel Deutschland GmbH, Germany*
[2] *Sennheiser, Germany*
[3] *Intel Corporation, USA*

## 2.1 Introduction

Fifth-generation (5G) communication systems are designed to provide a 1000× to 10 000× capacity increase compared to legacy fourth-generation (4G) technology. Such objectives imply the need for substantial additional spectral resources which are made available through multiple strategies. First, the usage of spectrum is investigated in upper frequency bands, typically centimeter-wave and millimeter-wave bands (typically, 10 GHz and above). In particular in the high gigahertz (or even terahertz) region, bands are available or can be repurposed at reasonable expenditure. While this approach may be useful for some applications, not all applications are compatible with technology characteristics (such as wireless propagation properties) in high gigahertz bands. Because of this, a second direction is investigated in parallel: enabling more efficient usage of spectrum below 6 GHz. Indeed, traditional cellular spectrum below 6 GHz is likely to still play a key role in the future 5G ecosystem and beyond. In recent years, a new technology has been introduced targeting shared exploitation of television broadcast bands, also known as TV white spaces (TVWSs). In particular the US [1] and UK [2] administrations have driven the introduction of a suitable regulation framework. The basic principle relates to the idea of allowing secondary devices to access spectrum at specific geographic locations and during specific time intervals when the spectrum is not occupied by the incumbent (primary user), i.e., TV broadcast services in the TVWS case. While TVWS systems sufficiently protect incumbents, a suitable level of quality of service (QoS) typically cannot be guaranteed, mainly because management mechanisms between secondary systems themselves are lacking. Furthermore, the actual availability of TVWS spectrum is uncertain, particularly in densely populated areas. These issues have hindered the commercial success of TVWS technology until recently. With the lessons learned from the definition, deployment, and operation of TVWS systems, a second-generation spectrum sharing technology is being developed in Europe and the USA with the objective of eventually providing global coverage in applicable bands. The European Telecommunications Standards Institute (ETSI) and the

*Spectrum Sharing: The Next Frontier in Wireless Networks,* First Edition.
Edited by Constantinos B. Papadias, Tharmalingam Ratnarajah, and Dirk T.M. Slock.
© 2020 John Wiley & Sons Ltd. Published 2020 by John Wiley & Sons Ltd.

European Conference of Postal and Telecommunication Administration (CEPT) have developed a number of deliverables enabling the usage of the so-called licensed shared access (LSA) scheme in Europe in the 2.3–2.4 GHz long-term evolution (LTE) time division duplex (TDD) band 40 [3]. The actual deployment of LSA, however, is still at an early stage in Europe. One reason for this may be the fact that the solution has been designed specifically for usage by mobile network operators (MNOs) and their investment strategy is typically focused on dedicated licensed spectrum for cellular deployment and, to some extent, to fully unlicensed bands for systems such as Wi-Fi. The Federal Communications Commission (FCC) issued a report and order enabling the operation of the so-called Citizen Broadband Radio System (CBRS) in the USA in frequency band 3.55–3.7 GHz (LTE TDD bands 42 and 43) [4] on 17 April 2015. Rules governing the CBRS are found in Part 96 of the Commission's rules. These systems are expected to provide a key component for future generation spectrum management. Interestingly, the CBRS solution is not limited to a single stakeholder group, but targets established players (classical MNOs) as well as new entrants (such as small business owners, factory sites, etc.). The roll-out of CBRS is ongoing in the USA and is currently attracting more substantial investment compared to LSA. Following the growing market acceptance of CBRS in the USA and taking the learnings from the earlier LSA activity, ETSI has started to develop the so-called "evolved licensed shared access (eLSA)" concept [5] with the objective of providing solutions that are designed for user groups beyond the classical stakeholders. eLSA is indeed suited to meet the requirements of an announced new spectrum regulation regime in 3.7–3.8 GHz as proposed by the German Regulation Administration (BNetzA, [6]) and which targets the needs of professional stakeholders [including the industrial automation industry, program making and special events (PMSE) industry, etc.] in particular. Other administrations are expected to follow the approach taken by BNetzA. A brief overview on LSA, eLSA, and CBRS is given in Figure 2.1.

The following sections give an overview of the above-mentioned standards and regulation activities. In particular, it is explained how ETSI and CEPT collaborate in order to ensure that the LSA standardization output and the regulation framework are well aligned and suitably complement each other. Today, corresponding standards are indeed readily available for applying spectrum sharing in the 2.3–2.4 GHz band which can be rapidly applied by European Regulation Administrations in their respective countries at any time. The feasibility has repeatedly been demonstrated through trials, as reported, for example, by the

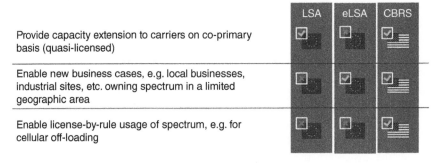

**Figure 2.1** A high-level comparison of LSA, eLSA, and CBRS.

Joint Research Center of the European Commission in 2016 [7]. The US flavor of spectrum sharing, CBRS, is defined by the Wireless Innovation Forum (WInnForum) and the CBRS Alliance. The split of activities between the two organizations is further outlined in the sections below. Finally, an overview of the regulation framework in Europe and the USA is given, provided by CEPT and the FCC, respectively.

## 2.2  Standardization

### 2.2.1  Licensed Shared Access

The LSA system is designed to meet the needs of the following stakeholders:

- **incumbent user(s):** primary users with the ability to sub-license spectrum to LSA licensees under certain conditions
- **LSA licensee(s):** operating a wireless system under a sharing agreement, typically a MNO providing 3GPP LTE services
- **national regulation administrations (NRAs):** control and monitor spectrum sharing activities.

In comparison to legacy TVWS communication systems, the upper guidelines imply a substantial change. The LSA approach establishes a clear business case: a long-term rental relationship is established between incumbents and LSA licensees which leads to a defined money flow and LSA licensees obtaining guaranteed QoS conditions in a given geographic area, frequency band, and time period. At the same time, the operation of incumbent systems is equally protected such that their respective QoS requirements are fulfilled. TVWS solutions offer neither a comparable clear business model for all stakeholders nor a guaranteed level of QoS, which may at least partly give some indications for the technology's limited (commercial) success. In Europe, the 2.3–2.4 GHz band has been selected for an initial deployment of LSA [3], as illustrated by Figure 2.2. This band corresponds to LTE TDD band 40 and is used in other regions as dedicated licensed LTE spectrum. ETSI's Reconfigurable Radio Systems (RRS) Technical Committee has developed corresponding system requirements [8] and system architecture [9] documents, defining the key building blocks and interfaces related to the upper framework. Although the original specifications are defined for a given band, the inherent solutions are rather frequency agnostic and generally applicable to any target bands. Corresponding revisions of the standards may be implemented at any suitable time.

The LSA repository is an entity comprising database and other functionalities; further details are given below. In the European LSA context, the LSA repository takes an important role because essential information related to expected future spectrum occupancy is provided by the incumbent(s) to the database. The US model follows a different strategy for users entering coastal protection zones: all such information needs to be derived by an environmental sensing capability (ESC) and needs to meet strict confidentiality requirements, as detailed later in this in chapter. In this context, the European Horizon 2020 project ADEL (Advanced Dynamic Spectrum 5G Mobile Networks Employing Licensed Shared Access) [10] made a substantial contribution. Indeed, an explicit sensing reasoning module was proposed which relies on (possibly dedicated) sensing networks and cooperated with a database

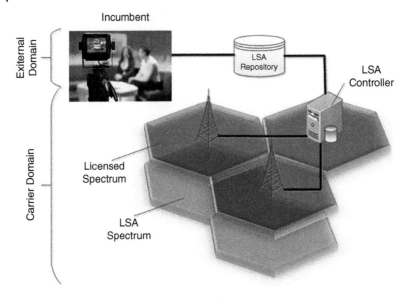

**Figure 2.2** LSA architecture reference model.

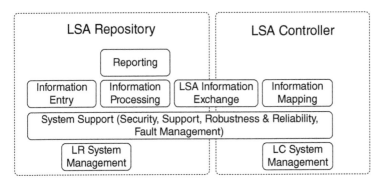

**Figure 2.3** Mapping of high-level functions and function groups to logical elements.

(repository). In this sense, it encompasses both the (e)LSA and the CBRS approaches. The LSA controller provides processing and decision-making capabilities building on the data elements made available by the LSA repository. The LSA controller will exchange information with an MNO's operations, administration, and management (OA&M) framework in order to indicate spectrum availability, and request short-term vacating of the spectrum and other functions as illustrated in Figure 2.3. In accordance to the definitions in [9], the LSA repository and LSA controller components are also detailed in Figure 2.3.

Figure 2.3 introduces high-level functions that are derived from the ETSI requirements specification [8]:

- The entry and storage of information is managed by the information entry function; this is essential for the operation of the LSA system.
- The derivation of LSA spectrum resource availability information for each licensee is managed by the information processing function and is provided to the information

exchange function to be further forwarded to the respective information mapping function of the LSA licensee. The information entry function provides corresponding input data for this function. It furthermore provides support for multiple incumbents and multiple LSA licensees, scheduled and on-demand modes of operation, and logging of processing information.

- LSA spectrum resource availability information is received by the information mapping function, its reception is confirmed, and respective operations are initiated in the mobile/fixed communications network (MFCN). Furthermore, an acknowledgment is provided to the information exchange function (for forwarding to the information processing function) when changes in the MFCN are processed.
- Creating and providing reports regarding the LSA system operation are managed by the reporting function and forwarded to administration/NRA, incumbent(s), LSA licensee(s), and possibly other relevant stakeholders.
- Communication mechanisms, internal to the LSA system, are provided by the LSA information with the capability to exchange LSA spectrum resource availability and related acknowledgement information.
- The system support functions group provides functions such as authentication and authorization, failure detection, etc.
- The system management functions group includes OA&M features in the LSA system.

It is expected that this system approach is able to meet the needs of concerned stakeholders, such as incumbents, LSA licensees, NRAs, and others such that:

- the incumbent(s) will have the potential to monetize spectrum which is available for secondary usage in a specific geographic area, a specific frequency band, and during a specific time period
- the LSA licensee will have the potential to access additional spectrum enjoying guaranteed QoS conditions
- the NRAs can monitor and possibly interfere in order to ensure the best possible usage of previously allocated spectrum.

Due to the extended feature set over legacy TVWS systems, a clearer business case and thus broad commercial success is expected for the LSA approach.

## 2.2.2 Evolved Licensed Shared Access

The ETSI Technical Committee (TC) RRS is currently working on an evolved version of the ETSI LSA framework (eLSA). The objective of this work is to facilitate spectrum access to local high-quality wireless networks operated by verticals. The concept of local high-quality wireless networks was introduced in ETSI Technical Report (TR) 103 588 [11] and describes local area networks serving applications that require predictable levels of QoS, such as those typical in the industrial automation, audio-visual content production, public protection and disaster relief (PPDR), and e-health vertical sectors. A similar term to refer to local high-quality wireless networks has been coined in the research community: the micro-operator concept [11]. While the ETSI TC RRS definition of local high-quality wireless networks is technology and frequency agnostic, the micro-operator concept has a focus on 5G technology and its frequency ranges. Although LSA did not explicitly exclude verticals in its role model, use cases were not foreseen. In that case, LSA overlooked a large stakeholder group that will be willing to rely on licensed shared spectrum to secure their businesses, since they typically are not able to afford exclusive spectrum under the current individual authorization regime. TR 103 588 [11] identified several high-level use cases in the scope of local high-quality wireless networks and three feasible spectrum access schemes. Common to all use cases is their demand for predictable QoS levels at all operation times in local environments within short-term to long-term deployments (i.e., from days to years).

The three spectrum access schemes identified in TR 103 588 can be summarized:

- **local licensing:** i.e., spectrum can be locally licensed to local high-quality wireless networks
- **leasing/subleasing:** i.e., incumbents and/or eLSA licensees (e.g., MNOs) can lease/ sublease out part of their spectrum to local high-quality wireless networks
- **spectrum as a service:** an MNO is willing to provide both spectrum access and automatic interference management services to local high-quality wireless networks deployed by vertical sector operators.

Table 2.1 provides a comparison of the three spectrum sharing schemes identified in [11]. Depending on the use case, local high-quality wireless networks might require the parallel use of these different spectrum access schemes. For eLSA to facilitate those spectrum access schemes, the following functional enhancements with regards to LSA need to be introduced, at a minimum:

- eLSA explicity extends the scope of the LSA licensee role to include vertical sector operators
- eLSA expands the temporal and spatial granularity of LSA enabling licensed spectrum sharing relationships covering short-term to long-term periods (e.g., from days to years) in locally confined areas such as factories, campuses, theaters, sport arenas, etc.
- eLSA allows for additional licensed sharing methods like local licensing and leasing/subleasing of spectrum resources. For instance, in the leasing/subleasing case an incumbent and/or an eLSA licensee (e.g., an MNO) may be allowed to lease/sublease spectrum resources to a vertical sector operator running a local high-quality wireless network or to a third party acting on behalf of a vertical sector operator.
- eLSA provides a general automated technical approach for licensed spectrum sharing applicable under any feasible national spectrum regulatory framework.

Note that the upper functionality extensions imply a substantial change over the existing LSA business case with a focus on MNOs as LSA licensees and long-term nationwide rental relationships. The eLSA scheme aims to open a path for verticals to access affordable spectrum with QoS guarantees while keeping their ability to deploy their own network infrastructure. To this end, eLSA represents a technical development to facilitate automated licensed spectrum sharing, including local site licensing and leasing agreements, between vertical sector operators and incumbents, in both international mobile telecommunications (IMT) and non-IMT bands – and/or between vertical sector operators and the NRA, for instance in those bands where no incumbents exist. It should be noted that the interest on IMT bands from the vertical sector operators side is motivated by the wide availability of wireless equipment in those bands and their economies of scale. In particular for 5G, an important obstacle to the successful integration of vertical industries use cases into the ecosystem is currently the lack of viable and sustainable business models. In this sense, eLSA could help to lower the entry level in the business model development process and speed up the deployment of 5G technology for vertical use cases. Here, it is important to stress that verticals, with their professional applications, would expand the customer basis of 5G significantly. However, this will only be the case if the necessary adaptations of 5G technologies to meet the requirements of vertical use cases are implemented, and specific interfaces and chipsets are available for the vertical industries. ETSI TC RRS is working on eLSA as a frequency agnostic technological approach, meaning that ETSI TC RRS is not targeting a concrete frequency band, as it did with LSA in the 2.3–2.4 GHz band. While at this point Germany and Sweden have recently announced their plans to make available frequencies in the range 3.7–3.8 GHz for local and regional allocations, other countries could follow with different frequency ranges [6]. eLSA should be applicable independently to the targeted frequency range. An comparative overview is given in Table 2.1.

| | Local licensing | Leasing/subleasing | Spectrum as a service |
|---|---|---|---|
| Spectrum | Granted by LSA repository to the vertical operator | Leased from incumbent (e.g., MNO) to the vertical operator | Granted by MNO to the vertical operator |
| Network infrastructure | Private to vertical operator | Private to vertical operator | From public MNO (e.g., network slicing) Private to vertical operator (MNO network densification) |
| Radio access technology | Any, following harmonized standards (e.g., HEN) | Any, following harmonized standards (e.g., HEN) | Used by MNO (e.g., NG-5G) |
| Network management | Local stand-alone | Local stand-alone | MNO supported |

Today, eLSA is still at an early standardization stage. Currently, ETSI TC RRS is finalizing stage 1 of a three-phase normative standardization process by defining the eLSA system requirements [12]. During stage 1, ETSI TC RRS seeks to answer following questions:

- What are the general working assumptions for eLSA?
- What are the new requirements and amendments to the LSA spectrum sharing concept to enable eLSA?

New requirements and amendments to LSA include:

- facilitation of further appropriate licensed frequency bands to be shared via eLSA methods
- extension of the LSA role concept to include vertical sector operators

- adaptation of the LSA role concept to allow a more flexible mapping of roles to the respective operation options of the evolved LSA system (i.e., evolved LSA controller and evolved LSA repository) according to the needs of vertical sector operators, for example an MNO may act as a third party to provide a repository service of its spectrum resources to a local high-quality wireless network operated by a vertical sector operator
- the spectrum sharing arrangement may contain respective practical details for a given eLSA spectrum resource when used by a local high-quality wireless network in a detached mode, i.e., without having a permanent connection to the eLSA system
- facilitation of the handling of a high number of neighboring local high-quality wireless networks, i.e., the eLSA system has to provide means to secure coexistence between incumbents and a potentially high number of eLSA licensees according to the agreed sharing conditions.

Based on the system requirements identified in Stage 1, system architecture and high-level procedures (in stage 2) as well as protocols and information elements (in stage 3) will follow.

### 2.2.3 Citizen Broadband Radio System

Initially, the innovative shared spectrum model adopted by the US FCC for the CBRS made a bold and historic shift in spectrum allocation. The FCC originally finalized the rules for spectrum sharing in the CBRS band in April 2016, making 150 MHz available for mobile broadband and other commercial applications. The FCC created a unique sharing paradigm for the CBRS 3.5 GHz band in the USA that builds on a combination of licensed and unlicensed spectrum called general authorized access (GAA)-designated spectrum. A seemingly complex framework needs to be put into place that includes three tiers: a tier for incumbents, a priority access license (PAL) tier, and a GAA tier. The incumbents get the top priority on the spectrum and the lower tiers are required to vacate the spectrum as soon as the incumbents start accessing them. The second tier or the PAL can use the spectrum exclusively, similar to licensed access. However, this can happen only when the spectrum is temporarily unused by incumbents. The third tier or GAA operates in a similar way to unlicensed spectrum users with the caveat that operations need to cease when an incumbent starts accessing the spectrum. The three tiers are coordinated through dynamic spectrum access system (SAS) administrators that allocate spectrum to the PALs and GAAs. ESC operators provide essential information on the actual availability of the band to the SAS, who in turn will inform the PAL and GAA base stations or access points to cease operation. The PAL part of the band has yet to be fully addressed by the FCC as it considers new rules; industry stakeholders are currently in the process of launching commercial services in the GAA portion. There is indeed an increasing need for innovative wireless spectrum access solutions to make new bands available which would otherwise be reserved for other applications. However, there numerous challenges, technical and political, that still need to be overcome. In its Report and Order [4], the FCC provides high-level functionalities of the CBRS, including the SAS entity (the SAS coordinates and authorizes access across users) and the ESC that is needed for transmitting inside an exclusion/protection zone (where incumbents needs to be protected). The WInnForum has published corresponding specifications with the support of its members from industry (equipment and device manufacturers, and service providers) and the Department of Defense (DoD). The latter is required since the target frequency band, 3.55–3.7 GHz, is used in particular by naval radar systems. A possible approach for the architecture and interface definition is shown in Figure 2.4.

The SAS interfaces with other SASs, with the ESC, and with the FCC's databases. Additionally, incumbents are able to report their band usage either directly to the SAS or via a database. This requires interfaces between SAS entities and registered citizens broadband radio service devices (CBSDs). Additionally, special types of SASs may be introduced that serve the needs of the service provider and additionally have to interface with other SASs for interference and incumbent protection. The industry consortium WInnForum is currently defining and developing the requirements and technical specifications to enable the foundational CBRS framework for SAS-to-CBSD communications, ESC sensing, etc. The Forum completed the CBRS baseline industry standards in 2018 [13, 14]. WInnForum's Spectrum Sharing Committee (SSC) handles FCC Part 96 rules related to CBRS and has close interactions with the US Government. The first set of FCC-certified devices were made available by Ericsson, Nokia, Sercomm, and Ruckus networks. WInnForum's work products are intended to be agnostic of the specific radio access technology (WIMAX, LTE, CDMA, etc.) that might be implemented; however, challenges related to controversial topics include, for example, specific scenarios such as coexistence of LTE deployments and users. The 3GPP standards body and the processes are typically restricted and broader in scope. The FCC issued a public notice in July 2018 seeking proposals for initial commercial deployment (ICD) by conditionally approved SAS administrator(s) [15]. ICDs are short-term, limited geographic commercial deployments prior to full-scale deployment of these systems. The purpose of an ICD is to prove the ability of SASs and CBSDs to perform properly in the field. The FCC is also considering appeals that request licenses to be renewed every 3 years instead of being re-auctioned.

### 2.2.4 CBRS Alliance

The CBRS Alliance is devoted to introducing specific LTE-based solutions in USA which will be optimized for operation in the 3.5 GHz band, utilizing shared spectrum to enable capacity expansion and coverage both indoors and outdoors on a massive scale. In order to maximize the full potential of spectrum sharing, the CBRS Alliance enables a robust ecosystem through development of standards specifications, testing, and certification programs. The essential purpose of the CBRS Alliance is to drive momentum in creating standards, overcoming regulatory hurdles, and enabling deployments in the 3.5 GHz CBRS band. The work of the CBRS Alliance was defined to be complementary to and an extension of WInnForum's activities. The CBRS Alliance was aimed at enabling and optimizing LTE operation in the CBRS band while leveraging the work of WInnForum and other organizations such as 3GPP. In parallel, WInnforum continues its work on developing standards primarily in the area of specification of SAS. The LTE standard in 3GPP is designed for mobile network operators deploying in exclusively licensed spectrum whereas CBRS is lightly licensed shared spectrum that provides opportunities for new use cases not addressed by 3GPP (e.g., neutral host networks). The CBRS Alliance was formally unveiled in August 2016 by six companies:

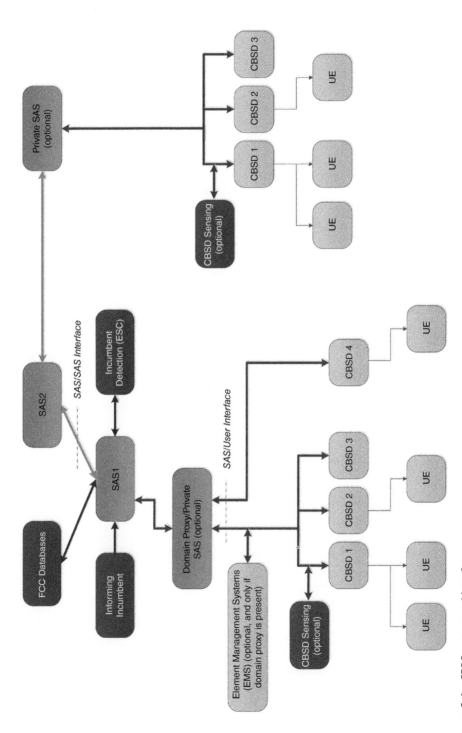

**Figure 2.4** CBRS system and interfaces.

Google, Federated Wireless, Nokia, Qualcomm, Intel, and Ruckus Wireless. As of September 2018, the organization has grown to include more than 100 companies, including all four nationwide US wireless operators. The CBRS Alliance came together with the basic objective of providing a forum to support the common interests of members, developers, and users in the application of innovative technology solutions and to address LTE-specific deployment issues in the US 3.5 GHz shared spectrum band. The CBRS Alliance has become an industry forum to champion LTE in the CBRS band (3.55–3.7 GHz) in accordance with FCC Part 96 Rules. The issue at hand was to find a way out of the shortcomings in existing standards which did not necessarily identify the intricacies in handling shared spectrum. With various factions and different interests, the CBRS Alliance has tasked itself to build consensus on technical solutions and eventually to develop a standardization strategy with an underlying plan to collaborate with other standards bodies (3GPP, WInnForum, ATIS, MuLTEfire etc.). The CBRS Alliance identified the need to drive the technology developments necessary to fulfill the mission, including multi-operator LTE capabilities. It should be noted that CBRS networks must comply with WInnForum requirements (which include FCC rules) and 3GPP standards as the CBRS Alliance defines new specifications. One of the key tasks of the CBRS Alliance is to evangelize LTE-based CBRS technology, use cases, and business opportunities, and to identify required advocacy steps related to marketing, promotion, certification, branding, regulation, etc., catalyzing action in these areas. On the other hand, there were multiple companies already trialing their products using LTE technology in 3.5 GHz as early as 2016 in multiple venues and waiting for industry and regulator guidance. The CBRS Alliance decided to establish an effective product certification program for LTE devices in the US 3.5 GHz band ensuring multi-vendor interoperability, an enterprise similar to that of Wi-Fi certification in the WiFi Alliance. The underlying theme is that such an experiment is first carried out in the USA. Primarily the vision of the CBRS Alliance was put into action realizing the full market potential of 3.5 GHz CBRS for the benefit of:

- augmenting traditional MNO networks
- enabling enterprises and other new entrants to build their own neutral host or private LTE networks
- "enabling last mile" connectivity via fixed wireless applications.

If the USA is successful in deploying spectrum-sharing technologies in the 3.5 GHz band, this effort could set a global precedent and have a huge impact in other parts of the world.

**CBRS-specific Spectrum Issues**

It is important to note that the US 3.5 GHz band falls right in between 3GPP bands 42 and 43. There was an immediate need to redefine the US 3.5 GHz band such that it would be consistent with FCC requirements, which are much stricter than existing 3GPP band 42/43 definitions. One of the first initiatives of the CBRS Alliance member companies was to drive a consensus across other players to define a new band in 3GPP keeping in mind channel capability, possible channel assignment modes, and carrier aggregation modes. There are also operational aspects that need to be considered in defining the new band, for example avoiding unnecessary random access channeling by band 42/43 user equipment (UE) that does not support US 3.5 GHz. CBRS Alliance members lead the initiative in 3GPP standards for a new band 48; further work was completed in December 2016 as part of Release 14 for PAL and GAA operation that covers the full range of spectrum from 3.55 to 3.7 GHz, compliant with FCC emissions requirements. 3GPP subsequently defined supporting channel bandwidth of 5, 10, 15, and 20 MHz and necessary signaling to indicate UE should meet additional FCC requirements, particularly with respect to UE power reduction for 15 and 20 MHz channel bandwidth. Also, all 3GPP base station classes classified by the output power supported by band 48 were defined and the UE Conformance Test Specs (RAN5) approved in June 2017.

**Technical Objectives for CBRS Alliance**
The main objective of the CBRS Alliance was not to reinvent the wheel in any manner but simply to address technical issues that are not or cannot be addressed effectively or in a timely manner by existing bodies like WInnForum, 3GPP, and other forums, and to build technical consensus among CBRS proponents quickly. The objective was to develop contributions in technical reports, recommendations, guidelines, and specifications pertaining to CBRS agreements and cosigned by many advocates as inputs to various relevant standard bodies. The technical objectives of the CBRS Alliance were mainly concerned with the following:

- to develop LTE-specific solutions to meet the challenges involved in CBRS spectrum sharing challenges, particularly with respect to defining the CBSD LTE-specific measurements and mechanisms in reporting to SAS
- to develop technical enhancements to mitigate incumbent interference
- to address LTE configuration issues primarily in terms of determining a common timing source for synchronization between CBSDs in the same cluster and across different clusters (with and without GPS availability)
- to address LTE TDD configuration parameter alignment across different deployments to minimize guard band requirement and synchronize silence intervals to aid in received signal strength indicator (RSSI) measurements.

The CBRS Alliance marked a major milestone in December 2017 by publishing the Network and Coexistence Baseline Specifications that will enable deployment and coexistence of LTE in the 3.5 GHz CBRS band. The networking and coexistence specifications are a critical foundation to ensure seamless interoperability between CBRS Alliance-certified CBSDs when operating in CBRS.

## 2.3 Regulation

### 2.3.1 European Conference of Postal and Telecommunications Administrations

From a regulatory perspective for LSA, the European CEPT organization has acted following investigations and a corresponding mandate by the European Commission [10–12]. In this process, CEPT and ETSI have closely collaborated; in particular, ETSI has established a so-called System Reference Document [3] which is the official vehicle for providing industry agreed requirements to CEPT for further consideration. CEPT has received the ETSI inputs and subsequently produced a number of Reports, Recommendations and Decisions [13–17], and has finally closed the corresponding working groups. From a CEPT perspective, the work is complete and the actual usage of the LSA band in Europe now depends on NRAs to enable the usage of spectrum sharing in national territories. Concerning the evolution of LSA, i.e., eLSA, the regulation framework is currently under development in Europe. The German Administration, BNetzA, has recently taken the initiative to provide a regulation proposal for the 3.7–3.8 GHz band [6] that specifically addresses the needs of vertical stakeholders, such as industrial automation, the PMSE industry, etc. It is expected that the novel framework will serve as an example that is taken up by further administrations and will finally be available across Europe. Its exact regulation framework is currently still under debate; it is closely monitored by all involved stakeholders since the allocation of specific bands to vertical stakeholders will certainly impact the business models of MNOs who intend to compete in this field with specific tailored services for verticals through network slices.

### 2.3.2 Federal Communications Commission

Following an Notice of Inquiry (NOI) and a Notice of Proposed Rulemaking (NPRM), the FCC formally released the Report and Order (R&O) to the CBRS 3.5 GHz band in April 2015 [4]. The FCC outlined a three-priority access system for sharing the band with incumbents. It requires protection of incumbent military radar and fixed satellite services.

- **Incumbents:** These are the current users of the spectrum. They can use the spectrum that they have been hereto using without any limitations. The main incumbent is the DoD with naval shipborne radars. Other incumbents include fixed satellite systems, radio location services, and terrestrial wireless systems. The incumbents get interference protection from the lower two tiers.
- **Priority access (PA):** This is similar to a licensed spectrum that can be won in an auction. However, PA users have to vacate the spectrum for an incumbent should they need to use it. PA operations receive protection from GAA operations. PALs, defined as an authorization to use a 10 MHz channel in a single census tract for 3 years, will be assigned in up to 70 MHz of the 3.55–3.65 GHz portion of the band [4].
- **GAA:** GAA is allowed throughout the 3.55–3.7 GHz band but gets no interference protection from other CBRS users (PA and incumbent). It is guaranteed at least 80 MHz of spectrum.

The deployment of systems is divided into two phases with the first phase using the SAS to coordinate spectrum access outside the specific zones in which incumbents needs to be protected. In phase two, the ESC coordinates transmissions inside the exclusion/protection zones (where incumbents need to be protected). The ESC is expected to be a form of sensor network. The FCC outlined the high-level SAS architecture in the R&O, as shown in Figure 2.5.

The main functions of the SAS include incumbent protection and protection of PALs from GAA following a hierarchical approach. Comparing SAS and LSA, the SAS entity can be interpreted to be the LSA controller counterpart. However, ETSI defined in ETSI TS 103 235 [9] that this entity must be within the MNO network. In SAS, the interference coordination

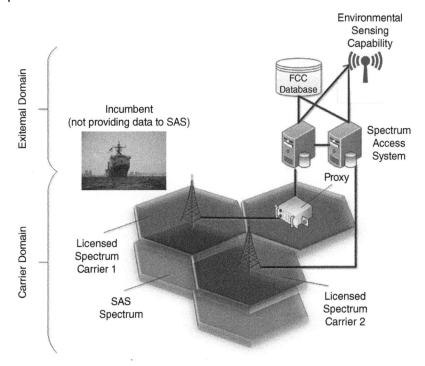

**Figure 2.5** The FCC's CBRS architecture.

across multiple networks is expected to require a SAS entity that is at least partly located outside a specific MNO network domain. The FCC defines three types of devices categories: Category A with a maximum EIRP of 30 dBm/10 MHz, a slightly higher power Category B (non-rural) with an EIRP of 40 dBm/10 MHz, and a Category B device with an EIRP of 47 dBm/10 MHz. The emission mask itself is specified as shown in Figure 2.6, where the out of band emissions are limited to −13 dBm at the adjacent channel and −25 dBm at the alternate adjacent channel. There is a special requirement of −40 dBm (20 MHz away) at the two edges of the 3.5 GHz band.

### 2.3.3   A Comparison: (e)LSA vs CBRS Regulation Framework

From a use and business case perspective, LSA targets a straightforward application of shared bands to classical MNO services based, for example, on 3GPP LTE technology. CBRS includes similar possibilities but also allows an extended application to new stakeholder groups; the underlying system concept is thus more complex but also allows more flexibility. Due to the recent success of CBRS in the USA, ETSI has been evolving the LSA framework towards eLSA, which addresses the specific needs of vertical industries such as industrial automation, PMSE, etc. Technically speaking, the LSA system is based on two tiers—, incumbents and LSA licensees, —which each get exclusive use of the spectrum while they are using it. Furthermore, the incumbent populates a database indicating when the LSA licensee can access the spectrum in a given geographic area, a given frequency band, and a given period of time. The 3.5 GHz spectrum has three tiers with a (modified)

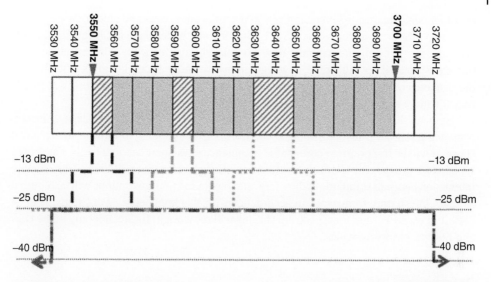

**Figure 2.6** 3.5 GHz emission mask.

unlicensed component to it (requiring communication capabilities with the SAS entities for interference mitigation, etc.) which does not exist in the LSA system. However, the most notable difference is probably the fact that the DoD will not populate any databases giving usage information and it has to be entirely determined by sensing. It puts accurate and reliable sensing technologies at the forefront, unlike LSA, where sensing could be used to improve network performance but is not essential for accessing the band. The interference mitigation problem is also enhanced in the 3.5 GHz system for two reasons. First, the size of a census tract (i.e., the minimum geographic area which can be auctioned/used independently for each 10 MHz band) is based on population and not area. As a result, in densely populated urban areas these census tracts could be as small as a few blocks and greatly increase the coordination needed for interference mitigation along each of these boundaries. Second, GAA users need to be actively managed to prevent interference to PAL users; something that is not needed in LSA.

### 2.3.4 Conclusion

In this chapter we have outlined the second generation of spectrum sharing technology after a previous but commercially unsuccessful initiative in the context of TVWS technology. The industry and regulation administrations have indeed learnt from that experience and developed new regulation rules as well as a technological framework that overcomes the insufficiencies of TVWS technology, in particular in terms of a clear business model and service provision with guaranteed QoS. The technological solutions have taken different paths in the USA and Europe. Europe initiated the rally with the development of LSA technology; unlike the US solution, CBRS, LSA is not yet fully deployed and operational despite the availability of ETSI standards and clear regulation rules. One reason for this is that the target stakeholder of LSA is still focused on dedicated licensed spectrum. CBRS, on the other hand, addresses a broader group of potential addressees and enjoys a wider acceptance and proliferation across the USA. ETSI has recently developed LSA further towards eLSA in order to extend its customer base to vertical industries, such as industrial automation, PMSE, etc. At the same time, new regulations are under discussion to specifically meet the demands of such user groups [6]. It is expected that the new framework will finally trigger a more active usage of spectrum sharing in Europe.

Despite all this progress, spectrum sharing technology is still in its early stages. The corresponding roll-out is just beginning in the USA and still pending in Europe, as explained above. It will be important to observe the acceptance of the new services by the concerned stakeholder groups and to openly address the insufficiencies that will likely arise. A close interaction between standards bodies, fora, and regulation administrations will be required to adapt the sharing framework in accordance with this knowledge. Work remains to be done, but we are certain that spectrum sharing will see a bright future, not only in the USA and Europe, but world-wide.

# References

**1** Allen Y. Overview of FCC's new rules for TV white space devices and database updates, ITU-R SG 1/WP 1B Workshop: Spectrum Management Issues on the use of White Spaces by Cognitive Radio Systems, 2014.

**2** ETSI EN 301 598 v1.0.0: White space devices (WSD); wireless access systems operating in the 470 MHz to 790 MHz frequency band; harmonized EN covering the essential requirements of article 3.2 of the R&TTE directive, 2014.

**3** ETSI TR 103 113 v1.1.1: Electromagnetic compatibility and radio spectrum matters (ERM); system reference document (SRdoc); mobile broadband services in the 2 300 MHz to 2 400 MHz frequency band under licensed shared access regime, 2013.

**4** Report and order and second further notice of proposed rulemaking, FCC 15-47, adopted April 17, 2015; released April 21, 2015.

**5** ETSI TR 103 558: Reconfigurable radio systems (RRS); feasibility study on temporary spectrum access for local high-quality wireless networks, 2018.

**6** Anhörung zur lokalen und regionalen Bereitstellung des Frequenzbereichs 3.700 MHz bis 3.800 MHz für den drahtlosen Netzzugang [Call for comments on local and regional allocation of the spectrum band 3.700 MHz to 3.800 MHz for wireless access]. Bundesnetzagentur (German regulation administration), 2018.

**7** Joint Research Center of the European Commission (JRC), Modern spectrum management experience with license shared access process, 2016.

**8** ETSI TS 103 154 v1.1.1: Reconfigurable radio systems (RRS); system requirements for operation of mobile broadband systems in the 2 300 MHz to 2 400 MHz band under licensed shared access (LSA), 2014.

**9** ETSI TS 103 235 v1.1.1: Reconfigurable radio systems (RRS); system architecture and high level procedures for operation of licensed shared access (LSA) in the 2300 MHz to 2400 MHz band, 2015.

**10** Advanced dynamic spectrum 5G mobile networks employing licensed shared access. doi: https://cordis.europa.eu/project/rcn/189128/factsheet/en.

**11** ETSI TR 103 588, v1.1.1: Reconfigurable radio systems (RRS); feasibility study on temporary spectrum access for local high-quality wireless networks, 2018.

**12** ETSI TS 103 652-1: Reconfigurable radio systems (RRS); evolved licensed shared access (ELSA); part 1: System requirements, 2019.

**13** https://www.wirelessinnovation.org/assets/work products/specifications/winnf-15-s-0112-v1.0.0.

**14** https://www.wirelessinnovation.org/assets/work products/specifications/winnf-16-s-0016-v1.0.0.

**15** https://docs.fcc.gov/public/attachments/DA-18-783A1.pdf.

**16** Report on cus and other spectrum sharing approaches, RSPG (radio spectrum policy group), 2011. doi: RSPG11-392.

**17** RSPG opinion on licensed shared access, RSPG (radio spectrum policy group), 2013; doi: RSPG13-538.

**18** European Commission (EC) mandate to CEPT for the 2300–2400 MHz frequency band in the EU, 2014.

**19** ECC report 205: Licensed shared access (LSA), 2014.

**20** ECC recommendation ECC/REC/(14)04 on cross-border coordination for MFCN and between MFCN and other systems in the frequency band 2300-2400 MHz, 2014.

**21** ECC decision ECC/DEC/(14)02 on harmonized conditions for MFCN in the 2300-2400 MHz band, 2014.

**22** CEPT ECC report 55: Report A from CEPT to the European Commission in response to the mandate on harmonised technical conditions for the 2300–2400 MHz ('2.3 GHz') frequency band in the EU for the provision of wireless broadband electronic communications services; technical conditions for wireless broadband usage of the 2300–2400 MHz frequency band, 2014.

**23** CEPT ECC report 56: Report B1 from CEPT to the European Commission in response to the mandate on harmonised technical conditions for the 2300–2400 MHz ('2.3 GHz') frequency band in the EU for the provision of wireless broadband electronic communications services; technological and regulatory options facilitating sharing between wireless broadband applications (WBB) and the relevant incumbent service/application in the 2.3 GHz band, 2015.

# 3

# White Spaces and Database-assisted Spectrum Sharing

*Andrew Stirling*

*Larkhill Consultancy, UK*

This chapter provides an overview of the evolution of electromagnetic spectrum sharing, building on the emergence of the cognitive radio (CR) concept, to enable more efficient access to spectrum without the need for conventional licensing. The key concepts of white space spectrum and database-assisted spectrum sharing systems will be presented and discussed.

## 3.1 Introduction

Across industry sectors, companies are looking to use data to reach new levels of productivity and maintain their competitive edge. As a key enabler of the data economy, the communications industry is focused on provisioning ever-greater network capacity to meet the growing demand. Spectrum has a vital role in enabling this performance to be delivered cost effectively and flexibly.

As the worldwide race to deploy 5G continues, there is a desire to deliver more seamless and higher performance connectivity to user devices which are increasingly tetherless and mobile. This inevitably leads to increasing integration of wireless with other networking technologies in the communication infrastructure, with wireless links bridging (and sometimes even replacing) the gap between end-user devices and fiber access points, as well as directly between end-user devices.

However, the electromagnetic spectrum, which is a fundamental resource for communications, is still managed in a bureaucratic way, limiting how quickly innovative services and technologies can be brought to market.

The system is based on issuing licenses (permissions) to users who have been approved or selected (in the case of competition). It can require considerable investment to acquire spectrum rights, well before revenues from operating wireless services can start to flow.

Spectrum sharing has traditionally been heavily constrained, with major established wireless applications (such as TV broadcasting and mobile communications) enjoying their own dedicated frequencies within internationally agreed frameworks. With an

*Spectrum Sharing: The Next Frontier in Wireless Networks,* First Edition.
Edited by Constantinos B. Papadias, Tharmalingam Ratnarajah, and Dirk T.M. Slock.
© 2020 John Wiley & Sons Ltd. Published 2020 by John Wiley & Sons Ltd.

increasing number of applications benefiting from the flexibility that wireless connectivity delivers, dedicated spectrum is gradually giving way to a more open, application-agnostic approach to spectrum allocation.

The scope for spectrum sharing, and speed of innovation, has been determined by regulation as much as by technology capability, linked to the need to reach sharing arrangements which are acceptable to established spectrum users such as broadcasters and mobile operators.

## 3.2 Demand for Spectrum Outstrips Supply

Many applications across industry sectors are embracing the freedom that wireless flexibility provides. Some already have spectrum dedicated to them (including broadcasting and mobile communications) whilst others are looking for spectrum. In general, even applications that already have spectrum allocations are seeking more capacity, for example to support a move to using broadband data streams (including images) from narrow band data and voice channels that were previously the norm.

Since there is little unassigned spectrum to share between all the potential applications, it is necessary to make more efficient use of spectrum which already has been allocated and assigned. This is where dynamic spectrum sharing comes in, building on CR technology capabilities.

### 3.2.1 Making Room for New Wireless Technology

Governments have managed national spectrum use and coordinated with each other for many years. Historically and for simplicity, particular bands were allocated to particular applications. Services and applications could be grouped in such a way as to make it easier to determine the technical parameters required for coexistence.

Rapid advances in wireless networking technology and developments in applications using wireless connectivity, however, have challenged the rigid dedicated allocations of the past. Now, regulators have come to see that there is greater economic and social value to be gained by allowing more flexible use of spectrum [1], such that users/licensees can determine how best to use available capacity. The more successful applications and services are, the more capacity they should be able to access to meet their growing needs.

To encourage spectrum users to innovate and drive more economic value from their spectrum holdings, European regulators introduced the Wireless Access Policy for Electronic Communications Services (WAPECS) in the 2000s, under which EU national regulators were expected to award spectrum on a technology and service neutral basis [2].

Under this policy, promoting application and technology neutrality, only the minimum technical coexistence requirements to prevent harmful interference were to be laid down in new spectrum regulation.

A key requirement for spectrum efficiency is the emission mask of transmitting devices, as this determines how much of the transmitted energy leaks into adjacent channels and therefore how much impact there might be on reception of signals in those adjacent channels, in frequency and geographic terms.

Receiver sensitivities are also now of interest because the more effectively receivers can filter out adjacent received signals, the more efficiently the band can be shared. This is the reasoning behind the recently introduced European Radio Equipment Directive [3], which requires receiver devices (such as TVs) to meet more stringent selectivity requirements. It means that receivers need to be better at rejecting signals in channels neighboring the ones of interest to them. The better the receiver selectivity, the more tightly services can be packed together.

Studies on the coexistence of long-term evolution (LTE) with broadcast [and program making and special events (PMSE[1])] in the ultra high frequency (UHF) bands and for TV white space (TVWS) in the same bands have helped regulators evolve their methods for determining the coexistence rules that will facilitate more intensive spectrum sharing. Differences in network density add to the complexity of sharing, alongside differences in signal modulation, encoding etc. For example, the LTE networks in cleared bands tend to have more dense deployments of base stations than are used (for transmitters) in the more sparsely served TV networks. This is because TV network planners can generally use much higher transmission power and can assume that receivers are more sensitive than, for example, the mobile device clients that would typically be connected to an LTE network.

At the edge of coverage for a TV transmitter, reception is at risk from nearby LTE base stations, as, for example, in a crowded restaurant where someone talking on the same table can disrupt other table occupants' ability to listen to a more distant speaker.

### 3.2.2 Unused Spectrum

Although empty bands ready for new applications are scarce, there is a considerable amount of spectrum that lies unused. This spare capacity is the basis for new forms of shared access and is vital for enabling the future innovation and growth of wireless connectivity in general.

Spare capacity arises because licensees generally deploy and operate infrastructure only in locations where it is commercially viable, or where regulation insists that they do so.

White spaces are geographic regions where a particular band (range of frequencies) is not used by a recognized/licensed service or application. In the case of bands used for terrestrial TV broadcasting, this might be a single 8-MHz-wide channel in the UHF bands (IV and V), which forms the basic unit of assignment for digital terrestrial television broadcasting multiplexes. The term "white spaces" may also be used to refer to the blocks of frequencies which are available at any given location. It can additionally have a temporal dimension, if frequencies are only left unused for a portion of the time. Figure 3.1 shows an imagined small country with three frequencies in use (A, B, and C) for TV transmission. Shaded circles indicate the areas of coverage and the remaining white represents white space across all three frequencies.

There may be a wide frequency range available for opportunistic use (multiple consecutive channels) or it might be that unused channels are scattered across the band in question.

In Europe, because television broadcasting tends to be organized into regional and national networks, there is a high degree of reuse of broadcast spectrum across the

---

1 PMSE covers communication links for equipment such as wireless microphones, in-ear monitors etc.

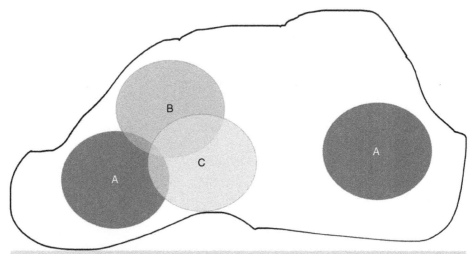

*TV coverage involves the reuse of multiple frequencies – especially for national/regional services. Transmitters using the same frequency (e.g. A) need to be spaced to avoid overlap. This leaves* **white spaces** *– geographic areas where frequencies are unused e.g. frequency A is unused outside the (blue) coverage areas.*

**Figure 3.1** An illustration of white space.

continent. This is needed to accommodate the multiplicity of terrestrial stations and results in fragmented white spaces. Exceptions can be found in rural fringes of the continent, such as the Western Isles of Scotland, where the prime limit on spectrum use is the commercial viability of TV broadcast infrastructure rather than the need to leave gaps between transmitters operating on the same frequency, as happens in more densely populated areas.

The fragmented nature of white spaces has been a deterrent to commercial applications. Major network operators have preferred the simplicity and lower infrastructure costs that come with deploying networks in contiguous (licensed) spectrum across large geographic regions. However, technology for aggregating bands has been developing as radio developers seek to support ever greater throughput, and now offers the possibility of commercial use of this "leftover" spectrum.

Spectral fragments are also more complex and costly for regulators to make available for other users, even if there is a clear demand for the capacity.

## 3.3 Three-tier Access Model

Access can be modeled as a pyramid, with the highest priority users at the top as the prime users. Lower priority users can make use of remaining spectrum capacity in the band provided that they do not cause harmful interference to higher priority users. Lower priority users cannot seek protection from harmful interference caused by higher priority users (assuming that higher priority user is operating within the terms of its license) (see Figure 3.2).

Prime users of the UHF bands include terrestrial broadcasters and, more recently, mobile operators (co-prime users), who moved into the top of the original broadcast band (700 MHz

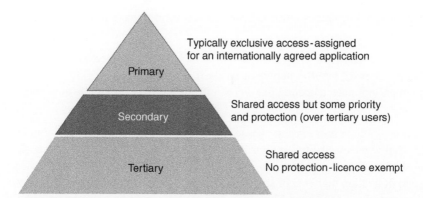

**Figure 3.2**  Spectrum usage pyramid: primary, secondary, and tertiary.

and above in Europe) from the mid-2000s onwards. The licenses they hold are exclusive and highly valuable, especially given the favorable coverage that can be achieved cost-effectively with this range of frequencies. As primary users, they enjoy protection against harmful interference from other primary users and lower tiers of users.

The rights awarded to secondary and tertiary users are illustrative. Since only the primary users are recognized in major international spectrum agreements, it is up to national administrations to manage the use of capacity left over by the prime users.

### 3.3.1  Secondary Users: Exploiting Gaps left by Primary Users

The spectral fragments left over from primary users (particularly terrestrial TV broadcasters) do not lie completely unused, at least in certain locations. Although TV broadcasting is internationally recognized as the primary use of the broadcast band (currently 470–790 MHz in the UK and soon to be 470–690 MHz), secondary uses are also permitted.

Secondary applications include wireless microphone links and other links supporting events, live performances, and studio productions offering performers untethered mobility. These applications are referred to collectively as PMSE. PMSE applications tend to be most intense around broadcast studios and major theatres during rehearsal and performances times.

In many countries, such secondary use for professional applications is license-exempt and it is uncertain where, how, and when the white spaces are being used. In the UK, this use is licensed through Ofcom, to whom users pay an administration fee.

Although secondary use is very intensive in major cities around TV studios and theatres, for example, it can also occur, occasionally, in more remote and rural locations such as golf courses and outdoor performance stages at certain times of the year.

The number of secondary users is large compared to the number of primary users (broadcasters), with cost and complexity implications for regulators such as Ofcom, particularly in providing protection to these users (mainly from each other).

### 3.3.2  Passive Users: Vulnerable to Transmissions in White Space Frequencies

As well as services and applications which actively use white space spectrum to transmit signals under a license or through license exemption, there are others whose operations assume that usage of white spaces remains below certain thresholds. These passive users fall into two groups:

- incumbent licensed wireless users who benefit from increased market access (e.g., areas outside planned/licensed wireless network coverage)
- others who might not need a license, but whose infrastructure is vulnerable to wireless transmissions.

Incumbent users include broadcasters whose market reach is increased by reception in areas beyond what was expected when their license was acquired. This might be through increased sensitivity of receiving installations. For example, an "out-of-area" consumer might choose to fit a taller, higher gain antenna to receive services from a TV transmitter that was not intended to serve them. The motivation for this could be to receive a wider selection of TV channels or different content (regional variations). Out-of-area usage is not protected by the regulator, so increased spectrum sharing might disrupt reception.

Other types of passive user include cable TV operators, whose infrastructure might have weaknesses, particularly in the region of customer premises, where cable screening weaknesses and poor termination issues might arise. New shared users of spectrum might interfere with services carried over such vulnerable infrastructure. Since the cable operator does not typically have a wireless license for the spectrum in question, it does not qualify for regulatory protection from licensed or license-exempt services and applications.

Regulators have looked at offering spectrum usage rights to passive users to enable their reliance to be recognized and protected. However, this concept has not advanced much beyond the discussion stage [4].

Meanwhile, passive users contribute to the friction against moving towards more intensive spectrum sharing because of the risk of losing benefits that they enjoy, which lack regulatory protection.

### 3.3.3 Opportunistic Spectrum Users

Aside from primary users, who are the concern of international agreements, and secondary users whose access is usually managed nationally, there are other, opportunistic users. They form the bottom of the access pyramid (shown in Figure 3.2), having the lowest priority and lacking any legal protection against other users.

The latest form of opportunistic access in the UHF is through license-exempt TV white space devices (WSDs). In general, these can make use of approved/validated geolocation databases to discover which channels are available in a particular location. However, there is no guarantee of continuity of access to spectrum for these devices. It depends on what licensed applications are present and the level of protection they are entitled to.

## 3.4 What is Efficient Use of Spectrum?

Given the growing importance of spectrum to national economies, regulators around the world have been pondering how to use it more efficiently. Related to this is the question of how efficiency can be defined and measured.

Efficiency of use can be defined in different ways, depending on whether you are looking at it as a user or as a regulator. The former seeks to maximize its own economic benefits, whilst the regulator needs to take a much broader view (the common good).

Signals are transmitted and received in support of an application which has economic or social value. Some bands, such as those supporting mobile broadband and mobile voice services, are intensively used in urban areas and less often in remote rural and marine areas.

Spectrum occupancy does not mean efficient use – the mere presence of a signal is of no value if there are no receivers capable of receiving it (e.g., if the signal is too weak for reliable reception) in a given area. For a broadcaster, it may be more cost-effective to have a transmitter reaching more homes, even if its coverage happens to include areas where no receivers are present/usable. However, that coverage overspill would not assist the wider economy if it displaced other applications (such as rural broadband) that could have exploited the spare spectrum in those areas.

If the transmission (and regulatory protection thereof) can be confined to a more focused coverage area (ideally a cable/waveguide) between the transmitter and the location of the intended receiver or receivers, then there is greater scope for other, non-conflicting, applications to share the spectrum in that location and thereby increase the efficiency of use.

### 3.4.1 Broadcasters prefer Large Coverage Areas with Lower Spectrum Reuse

Regulatory requirements ensuring the widespread availability of TV services and affordability of access (i.e., low-cost reception equipment) have led public broadcasters to evolve their distribution platforms around sparse networks of high-tower, high power transmitters. This network configuration meets coverage requirements in a cost-effective way. Coverage is easier to achieve when fixed receivers are the target, with simple roof-mounted antennae.

In broadcast services, the size of a coverage cell does not impact the bandwidth that can be delivered to each user. Since broadcast services are primarily receive-only (i.e., without an integrated return channel), the ideal distribution platform would use a single transmitter, visible from everywhere. Ultimately, this is what satellites provide,[2] offering far cheaper national distribution than typical multi-transmitter regional and national terrestrial networks [5].

However, if the reception requirements are more complex, then satellite may not be as good a fit as a terrestrial network. Examples include supporting reception on mobile devices indoors or dense urban areas, where satellite dishes might be shaded by neighboring buildings. Broadcast services may also need to partition their coverage geographically to enable local content variations, such as for advertising and news.

### 3.4.2 ISPs Respond to Growing Bandwidth Demand from Subscribers

User bandwidth expectations continue to rise as richer content and more capable personal devices offer greater value to consumers. Internet service providers (ISPs) need to keep evolving their infrastructure to accommodate the market expectations on broadband access

---

2 Some areas will be shaded from view of the satellite by mountains/high hills or buildings, especially at higher latitudes, assuming that the broadcast satellite is in the usual geostationary orbit.

speeds. From an ISP's perspective, smaller cell sizes/coverage areas can enable greater reuse of spectrum, offering more bandwidth to be shared with end-users, assuming that high-capacity backhaul, such as optical fiber, is available to feed differentiated content to each cell base station. For a given spectrum capacity, finer partitioning of the coverage area (smaller cell sizes) reduces the number of users sharing the capacity of each cell, enabling greater performance to be delivered to each. This higher density coverage, more suitable for mobile devices, costs considerably more than the broadcast distribution models, scaling with the number of users. However, the fixed/mobile operators' income also scales (from subscribers paying significant monthly fees).

### 3.4.3 Protection of Primary Users Defines the Scope for Sharing

Over many decades, across the world, television stations have made use of UHF bands (between 400 and 900 MHz) for terrestrial broadcasting to provide convenient cost-effective distribution to viewers' homes. Early transmissions used very high frequency (VHF) bands (between 100 and 300 MHz), but this lower frequency range did not have sufficient capacity to support the increases in picture quality and the introduction of color, alongside increasing station choice.

When TV broadcasting started, there was little other commercial demand for the bands they were assigned (which had been previously been reserved for military applications). Broadcast transmitter antennae are often mounted as high as possible, coupled to a high-power transmitter, e.g., 10–100 kW, to ensure coverage for as many homes as possible.

Service planners, together with regulators, determined what coverage could reasonably be assumed (and required to be maintained). Since coverage is a trade-off between transmission power output (including transmission antenna gain) and receiver system sensitivity, certain assumptions were made regarding the typical home receiver antenna height and gain.

Service planning assumptions on coverage in the UK were based on a receiver antenna with 10-dB gain mounted at a height of 10 m, corresponding to the roof level of a typical two-story house. In areas where the received TV signal strength is higher than the established minimum for reliable reception, generally closer to the transmitter, consumers enjoy extra flexibility in antenna arrangements, such as the ability to use a portable indoor antenna. This has been assisted by the use of orthogonal frequency division multiplexing (OFDM) and forward error correction techniques in the digital television transmission standards, which reduce the criticality of antenna gain and direction.

However, on the fringes of the covered area, where received signal level falls below the minimum, reception may still be possible, but is not guaranteed, for example:

- areas lying outside the area of intended coverage: this could work if viewers have increased the sensitivity of their receiver installation compared to the characteristics assumed by the planners,
- areas lying within the notional area of coverage, but where the received signal is diminished either by an external obstruction, e.g., a skyscraper, or by internal walls when using an internal/portable receiver antenna.

Figure 3.3 shows an imagined TV transmitter coverage (darker), with the fringes of coverage in a lighter shade, both external (A) and internal (B).

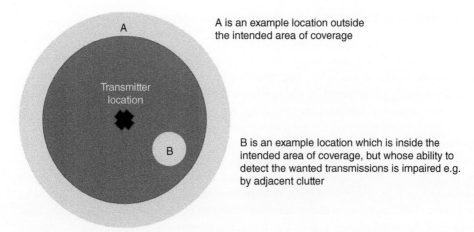

A is an example location outside the intended area of coverage

B is an example location which is inside the intended area of coverage, but whose ability to detect the wanted transmissions is impaired e.g. by adjacent clutter

**Figure 3.3**  On the margins of a protected service.

In protecting TV broadcast services, it is the receiver end of the transmission chain where the vulnerability is found. The lower the minimum threshold signal level assumed by the regulator, the greater the effective TV coverage will be and the lower the spectrum capacity available for sharing. Depending on assumptions about the likely separation of TV WSDs from TV receivers, the transmission power of the sharing device will be constrained to ensure that when a TV receiver is receiving at or above the minimum wanted TV signal level, its reception will be unimpaired by any sharing of the band by TV WSDs in the vicinity.

Due to their ubiquity, it is not easy to determine where every TV receiver is located, nor how close a given WSD might be to the TV's antenna. Although it can be reasonably assumed that TV receivers are not present in unpopulated areas, in other areas the regulator uses cautious assumptions on the distance between a WSD and a TV receiver antenna, as well as the effective gain of the TV antenna.

Fringe areas are places where reception is most vulnerable, if it is possible at all. They are therefore places that would cause the greatest constraints on spectrum sharing if regulators choose to protect reception there.

## 3.5  Tapping Unused Capacity: the Evolution of Spectrum Sharing

Spectrum sharing requires coordination between the users of a particular band to ensure that co-habiting applications do not disrupt each other, which we refer to as harmful interference.

Such coordination might draw on a combination of:

- quasi-static elements, which include assumed features of the coexisting services, such as the wireless communication technologies on which they are based (e.g., modulation characteristics, receiver sensitivities, etc.)
- dynamic elements, which could include receiver and transmitter locations, particularly in mobile/portable applications.

### 3.5.1 Traditional Coordination is a Slow and Expensive Process

For licensed applications, frequency coordination is typically a manual process, where different stakeholders have a chance to input their requirements to the regulators who lay down and enforce the rules. The rules would often be enshrined in the technical conditions which are required by the spectrum license. In the first place, it is necessary for governments to coordinate the use of spectrum with each other. This requirement is laid down in international treaties, which are administered through the International Telecommunications Union (ITU). Conferences are held periodically under ITU auspices to agree any proposed allocation changes such as increasing spectrum for mobile telecommunications. The most significant changes need to wait for the 4-yearly World Radio Conference and, in principle at least, need to be already on the agenda at the preceding conference. Thus, 8 years can elapse between raising an issue to be addressed and the issue being resolved. Further time is then needed to implement any agreed change at regional and national levels.

More detailed arrangements for implementing changes happen at regional conferences of the ITU. For example, the switchover from analogue terrestrial television to digital terrestrial television (in the UHF bands) was dealt with at a special ITU Region 1 conference for national administrations from Europe, the Middle East and Africa. Parallel regional conferences developed the switchover plans for Asia/Pacific and the Americas. In the Region 1 conference, a digital broadcast plan was developed that gave each country around seven to eight frequencies, but only covered transmitters of around 300 W and above. Finer grained coordination, involving lower power transmitters, was handled through bilateral discussions between adjacent countries.

It is ultimately the national governments which bear responsibility for making sure that any licenses they award do not lead to infringement of international agreements and will not cause any harmful interference to other licensees within their jurisdiction. This is managed through technical conditions imposed as part of the spectrum licenses.

In the case of license-exemption, the required technical conditions are imposed on equipment which is to be provided in their national market. Once the devices have entered the market, it is not normally possible to vary the technical conditions until the devices are no longer in use, which could be a decade or more after their introduction.

Whilst coordination on a national level can be handled more speedily than international coordination, it can still take days or weeks to respond to a new situation or requirement.

### 3.5.2 License-exempt Access as the Default Spectrum Sharing Mechanism

A long-established method of enabling open spectrum sharing has been through license exemption of certain spectrum bands. License-exempt use is permitted provided that the equipment meets technical conditions specified by the regulator. This approach is useful for spectrum sharing as it places the responsibility for meeting technical sharing requirements on the manufacturer/distributor of end devices (rather than their users). This enables new wireless technologies to reach a mass market and economies of scale to be realized quickly. Recent examples include IEEE 802.11ax (Wi-Fi 6) and Bluetooth Low Energy.

Such blanket power limitations have been characteristic of license-exempt applications, which have therefore often been referred to as short range.

However, the technical constraints typically imposed on license-exempt access have severely limited the applications that can be supported, e.g., in terms of the range that can be provided for wireless communication between devices. The reason for the severity of technical constraints (particularly on transmission power) is that under conventional license exemption, regulators have assumed that license-exempt devices might be used in the worst possible places (i.e., those where adjacent licensed applications are at their most vulnerable). For example, in city-centers, high housing density tends to reduce the distance between home Wi-Fi routers and therefore suggests a lower transmission power limit should apply. In rural settings, on the other hand, higher power could be safely used and would often be beneficial, but equipment has to conform with national requirements, which are determined to be safe in urban settings.

The power limitations can be an advantage for applications that only need short range (such as wireless car door locking or contactless card payments). However, this lack of flexibility has traditionally precluded wide area applications from benefiting from this means of spectrum sharing. An example would be delivering broadband access to homes and businesses in rural communities, where license-exempt access might be the only option for a small local ISP. The power restriction on license-exempt access results in denser infrastructure deployment being required to cover a given area and thus a higher cost per end user in areas where affordability is often a critical issue.

### 3.5.3   DSA offers Lower Friction and more Scalability

With CR, artificial intelligence (AI), and supporting technologies, it has become possible and necessary to make coordination much more responsive and efficient, serving the needs of a new generation of wireless applications.

If the spectrum coordination task is moved into the cloud or even down to the end user device, then access permissions can be granted far more quickly and even fleeting opportunities to apply the spectrum can be enjoyed.

- Spectrum access can become available opportunistically, on demand, e.g., facilitating temporary links between devices which might be mobile.
- Technical constraints on use of the spectrum can be varied within a timescale of hours or even minutes to manage the risk of harmful interference. Local spectrum congestion and changes in atmospheric conditions might feed into the process for determining the constraints.

Regulators still need to decide on the rules for coordination, implementing the policies which have been agreed between governments, e.g., giving one service priority over another service and the extent to which coverage is permitted to spill over an international border. The rules would be a framework for the operation of dynamic spectrum access (DSA), implemented through spectrum databases etc. By decoupling the slow, painstaking process of traditional rule development from the process of gaining access to spectrum, the new DSA technology enables finer grained and more timely adaptation than regulators/spectrum managers could currently handle.

### 3.5.3.1 Early days of DSA

In the late 1990s, an early form of DSA was used in the 5-GHz band, *U-NII*, to enable Wi-Fi to have access to frequencies which had been reserved for radar systems. The spectrum access was conditional on the license-exempt Wi-Fi devices vacating the specified band whenever a radar signal was detected. The method to achieve this, based on spectrum sensing, is called dynamic frequency selection (DFS). Although the intention behind the introduction of DFS was to improve capacity when the band was not required for radar, it has not been a great success.

Essentially DFS works as follows. Before using any of the affected 5-GHz channels, a Wi-Fi access point (AP) must check its availability (i.e., absence of radar signals) for a period of 60 s. If the channel is found to be available, then the AP may use it but has to check continuously for radar signals, and vacate the channel within 10 s if any are detected (UK/EU requirements).

This mechanism works, but because radar is much harder to detect than Wi-Fi emissions, it is common for APs with DFS to detect radar even when it is not present, defined as a false positive.

Broadband/Wi-Fi service providers report that there are too many false positives: Wi-Fi routers equipped with DFS often avoid use of the band even when radar is not present. This impacts Wi-Fi performance in a way which is hard to explain to end-users. Although in many cases end-users may not notice an impact, certain latency-sensitive applications, such as real-time voice and video links, are affected in a noticeable way.

### 3.5.3.2 CR: Towards Flexible, Adaptive, Ad Hoc Access

Over recent years, technology has evolved to provide license-exempt (or lightly licensed) access to unused spectrum in a more dynamic and flexible way, supporting regulatory goals of more efficient and safe sharing of available capacity.

The generic term for such technology is *cognitive radio (CR)*, which has been referred to under the more recent collective term, DSA. CR/DSA technology enables devices to determine which frequencies might be available in their location at a time when they need to create a wireless link.

Although the name suggests that devices could make their own decisions about which frequencies would be safe to use, this has generally not been allowed by regulators. A CR device, using in-built sensing capabilities, could have a much more detailed perception of local spectrum occupancy – in a geographic and temporal sense – than would be possible using monitoring tools presently available at national administration level. Results of trials by the Federal Communications Commission (FCC) as part of opening up TVWS also demonstrated that spectrum sensing had sufficient sensitivity to determine which frequencies would be safe to use [6].

To facilitate license-exempt access to TVWS spectrum, technology was developed with the capability of detecting TV signals over 1000 times (30 dB) weaker than the minimum level needed for reliable reception. The threshold for reliable reception is just below –80 dBm for digital terrestrial TV receivers, so this sensing equipment was able to detect broadcast signal presence at below –110 dBm.

However, concerns remained about the possibility of CR devices being shadowed by local clutter (see section 3.6.3.1), leading to possible requirements to sense even lower levels than were being demonstrated by technology developers. Increasing the spectrum sensitivity is costly, including in terms of energy usage, and also risks increasing the number of false positives, where the device wrongly concludes that a channel is occupied. In addition, a device cannot know how close it is to a potentially vulnerable receiver. Nor can it, by itself, know whether or not any signals it detects are those of services which are protected by the regulator in that location.

The FCC's regulations, issued in 2010, made provision for future introduction of sensing-only devices, but required their capability to be demonstrated first [7]. Sensing has been permitted within specified shared bands, such as in the 5-GHz band B (channels 100–140), where DFS is a condition of access to channels.

### 3.5.4 Spectrum Databases are Preferred by Regulators

Driven largely by broadcaster concerns, regulators have preferred to use a combination of device geolocation and spectrum databases to ensure that broadcast services are protected from the new WSDs. The regulator approved database does not depend on the spectrum sensing capabilities of the WSD or the nature of its location (e.g., hidden in clutter or in plain view) and provides certainty for both incumbents and new (WSD) spectrum users. In a similar way, the database protects licensed wireless microphone users in registered locations.

The regulator requirements vary from band to band and the type of service being permitted access. If a band is licensed, then a licensee may be allowed to deploy technology (including CR technology) to facilitate spectrum sharing between its users in the band. For example, a mobile broadband service may be in use by thousands of users at any given time, all of them sharing the spectrum that has been licensed by the mobile network operator. If, on the other hand, the band is to be shared with users of services other than those operated by the licensee(s), then the sharing mechanism needs to provide sufficient reassurance to incumbents.

Taking the specific case of TVWS, we will elaborate below on how what started as an autonomous mechanism, based on spectrum sensing, has evolved towards more centrally managed access using permitted spectrum databases to determine which frequencies are available in a given location, with defined constraints. This evolution towards centralized access control was required by regulators to address concerns of the licensed users of the TV band.

The US and UK approaches to developing the enabling regulation for effective sharing of the TV band differ in a way that reflects the way that broadcasting and its accompanying regulation and business models have developed in those countries (and the spectrum administrative regions they belong to) [8, 9]. In the USA, broadcasting tends to be city-centered, whereas in the UK (and other parts of Europe), broadcast services are licensed for regions or entire nations. This means that white spaces tend to be scarcer in US cities than they are in UK cities (except in the vicinity of studios and theatres where PMSE equipment is used intensively).

The models used for TV service planning also differ correspondingly, leading to different criteria for protecting the TV services from harmful interference. In Europe, the coexistence rules were developed by the Electronic Communications Committee, which is the association of European communication regulators tasked with preparing technical guidelines for implementing spectrum policy decisions across the region [10, 11].

## 3.6 Determining which Frequencies are Available to Share: Technology

### 3.6.1 CR: Its Original Sense

CR was a radical step forward, providing a vision for the future of devices which could safely determine which frequencies would be available in their location [12]. The concept originally applied to a much broader device awareness of resources, users' needs etc. However, the issue of spectrum resources has taken prime place in recent years as existing spectrum management mechanisms struggle to meet the growing demand for wireless communication bandwidth.

In the context of spectrum use, CR originally referred to devices making use of spectrum sensing to detect occupation, but it is now used to refer to a broad range of mechanisms (including spectrum databases) for DSA, enabling devices to find out which frequencies are available for their use and under what conditions (principally emission power limit).

### 3.6.2 DSA is more Pragmatic and Immediately Applicable

DSA is opportunistic access to available spectrum that can be handled directly and quickly by devices rather than waiting for human-mediated access (license applications etc.). As the name suggests, DSA is not fixed but can vary with time and place according to the regulatory priorities given to other entities which may be sharing the band(s) in question.

Three principal techniques have so far been considered by regulators for enabling access to unused or underused spectrum:

1. spectrum sensing, by which a device could determine whether a signal is present at a given frequency or not
2. beacons transmitting on agreed frequencies to provide a local information service on spectrum available to share
3. spectrum database, which is a key part of a spectrum authorization service (SAS).

### 3.6.3 Spectrum Sensing

Spectrum sensing is key to greater efficiency of spectrum use in the future. It enables much finer grained access and sharing to be facilitated, with fast adaptation to local conditions. Such micro-coordination is well beyond the scope of current regulator predictive models, which take a highly simplified and fairly static view of the world. It is at the core of coordination between license-exempt devices sharing a band, which often use a protocol referred to as "listen before talk" to avoid conflicting with one another.

Sensing technology has developed rapidly, with a range of techniques being applied to determine whether a signal is present and what type of signal it is. Since the signal of a service to be protected might be present only intermittently, spectrum sensing needs to be repeated at intervals to ensure that frequency availability is not assumed incorrectly or indefinitely.

Sensing techniques range from energy detection, where knowing the type of signal or its application is not important, to signal matching (correlation), where a particular type of signal can be positively identified (e.g., TV broadcast signals).

*Energy detection*, where the received energy in a given band (power spectral density) is measured and compared with a threshold, is the simplest approach. This method is agnostic to the application and does not protect any particular type of signal (e.g., TV transmissions). It means that signals which would not otherwise need protection from a regulatory point of view are included in the measurement and could therefore prevent other applications sharing the spectrum.

Energy detection is also vulnerable to uncertainty about noise levels, which can lead to false positives in which the device concludes incorrectly that the frequency is occupied. However, the simplicity of the energy detection method has favored its use in cooperative sensing in which multiple, distributed sensors pool the results of their individual measurements. *Eigenvalue detection* improves on the detection performance compared with energy detection, particularly with low signal to noise + interference ratio environments [13].

*Cyclostationary feature detection* looks at periodicity of signal statistics and is a compromise between the simpler energy detection techniques and full *signal matching*, where a particular signal type can be identified, such as digital video broadcasting — terrestrial (DVB-T), for TV broadcasting.

The more sophisticated techniques enable detection of the target signal type, even below the prevailing noise and interference level.

Further details on sensing methods can be found in [14].

### 3.6.3.1   Hidden Nodes: Limiting the Scope/Certainty of Sensing

A device may have very sensitive spectrum sensing capabilities, but these are not guaranteed to protect other users in the same band by themselves. The issue is that whilst it is possible to detect transmissions, the vulnerability of a service/application to spectrum sharing lies at the receiver end. In the case of broadcast services, the receiver is silent and therefore not intrinsically detectable. Therefore, it is also possible that a device that wants to start sharing could conclude from sensing that a particular frequency is available (e.g., if the device is shaded due to obstructions on the path between it and the transmitter of a protected service), but this may nonetheless cause problems for a protected service receiver which may be in its vicinity.

### 3.6.3.2   Overcoming the Hidden Node Problem: a Cooperative Approach

A single sensing device may have its ability to detect a protected service impaired due to obstructions between the device and the protected signal source, as explained above. However, if multiple sensing devices cooperate by sharing their data, then it is possible to mitigate the effect of obstacles. This can translate into sensing gain, in other words the sensitivity of each device may be lower because their results are combined. There has been considerable research into the "sensing gain" which can be achieved with multiple spectrum sensors instead of just a single one [14]. Figure 3.4 illustrates the advantages of a cooperative approach to spectrum sensing.

There are a few challenges with this approach. In the first instance, the cooperating devices need to be able to communicate with each other. This is straightforward when there is a wired/fiber connection between the devices, such as with remote radio heads. In the absence of a wire, the devices must communicate wirelessly in order to cooperate so must already know a frequency (or set of frequencies) that they may use legally. This requires spectrum capacity to be dedicated for administrative purposes, reducing the

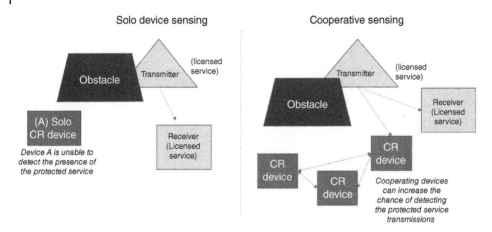

**Figure 3.4** Hidden node: solo versus cooperative sensing.

available capacity for application. A similar limitation is faced by operators of beacons (see section 3.6.4).

The other issue is that regulators find it difficult to decide what allowance could be made for the use of cooperative sensing as part of a regime to protect an incumbent service. For example, if it were decided that a single spectrum sensing device would need to be able to detect signal strengths of $-x$ dBm or lower then what threshold sensing level would apply in a cooperative sensing application?

It would be logical to accept lower sensitivity, say $-y$ dBm, where $y < x$, due to the cooperative gain. However, regulators have been reluctant to enumerate what "gain" might be allowed from this advancement of the sensing capability. They would first need to develop reference scenarios that could be agreed between incumbent user and (new) sharing user communities alike, with sufficient added capacity bonus for device manufacturers to be able to justify the extra implementation costs that cooperative mechanisms might bring.

### 3.6.4 Beacons

Beacons are signals that are broadcast, usually at pre-determined frequencies. Beacon broadcasts contain lists of frequencies (and possibly other data, such as the allowed power levels at the listed frequencies). This allows receiving devices to "know" which frequencies are allowed in that area.

Whilst beacons offer a potential solution to signposting unused spectrum, they have not found favor with implementors. The key problems include the following:

- The need to pay for a beacon network to be constructed and operated. Multiple beacon transmitting stations are likely be needed to achieve near-geographically comprehensive coverage.
- Beacon coverage would not be perfect: in some areas a beacon signal might not be available, in others a beacon might be detected and its data used outside of the area of its validity.
- Beacons would require a dedicated frequency or set of frequencies, which would then be denied to other wireless applications.

### 3.6.5   Spectrum Databases used with Device Geolocation

As an alternative to devices sensing spectrum directly to determine which frequencies are unused, the availability of spectrum in each location can be pre-calculated/determined by a spectrum administrator and held in a database. These databases can be approved by regulators and made accessible to end-user devices.

As Figure 3.5 illustrates, the basic sharing protection data (relating to TV broadcast services) could be provided or validated by regulators as an input to spectrum database services (via interface A). The database service may be permitted to "add value" through further processing (perhaps for PMSE registrations) before passing the data, on request, to end-user devices (via interface B).

In order to retrieve the correct frequency availability information, it is necessary for end-user devices to know where they are, i.e., they need to be able to detect their position using satellite positioning systems [such as a global positioning system (GPS)] or some other method. The database will also need to know the type of device and application that will be using the spectrum to make sure that the frequency and power limit data it supplies will be compliant with the spectrum coexistence requirements laid down by the regulator.

Use of a spectrum database requires the end-user device to connect to the database (normally via the Internet). Thus, it can work well in an Internet service provision role, for example, where Internet connectivity can be assumed. However, the requirement for a continuous connection is less helpful in remote areas, where there might be no wide-area network coverage. On the other hand, there is often abundant unused spectrum in such remote areas, which could still have a valuable role in facilitating local area connectivity. For example, such unused spectrum could be applied in precision agriculture and smart fishing to allow local communication networks to be formed. In such "off-grid" cases,

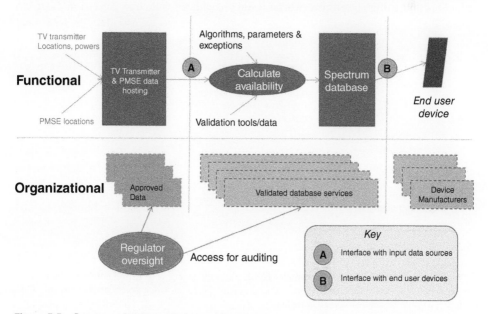

**Figure 3.5**   Spectrum database service architecture.

a combination of cached database and spectrum sensing could be sufficient to meet regulatory concerns if not compliant with the current regulations as framed.

The clear advantage of databases over autonomous spectrum sensing by single devices is that it provides certainty for incumbent spectrum users and control to regulators, who can more quickly diagnose causes of harmful interference and adjust the spectrum database contents appropriately. Databases also provide more certainty in terms of the capacity available for users of applications that are facilitated by dynamic sharing techniques. Users can check likely communications performance by entering the coordinates where they might want to use their devices and seeing what spectrum capacity is available at that location, at that time.

Advance retrieval of information from a spectrum database could streamline the process of ascertaining spectrum availability for particular applications. For example, on a journey by car or train, the available frequencies might change many times along the route. In this case, being able to prefetch frequency availability along a planned route might bring some connection continuity advantages within the time validity of the pre-fetched data.

Figure 3.6 illustrates the issue of spectrum data validity, which is a function of device position $(x, y)$ ($z$ is not shown) as well as time. The time validity for TVWS data is driven by PMSE requirements. Broadcast spectrum users are effectively static and therefore easy to plan around. However, licensed wireless microphone users (PMSE) are less predictable and stable in their spectrum needs, potentially requiring access at short notice[3] and receiving priority over license-exempt users.

In the case of TV spectrum, it is the secondary applications which are the limiting factor in time validity for any frequency availability information provided by a geolocation database.

Currently, spectrum databases are populated by calculations based on geographical factors, mature (simplistic) propagation models, and coexistence parameters specified by regulators and using algorithms, whose results have been validated. Spectrum databases

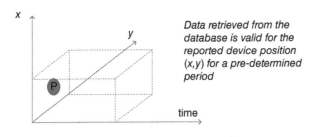

*Data retrieved from the database is valid for the reported device position (x,y) for a pre-determined period*

The position of the device at the time of requesting channel availability

**Figure 3.6** Spectrum data validity.

---

3 One of the main drivers for very short notice PMSE reservations (e.g. with 15 min notice to spectrum databases) is illegal use of the spectrum, resulting in potential conflict with an event reservation, might not be detected until just before the event goes live.

are thus currently stocked with data predicted from a regulatory model, rather than the real spectrum conditions pertaining at each location.

In the future, spectrum databases would be able to make use of widely distributed low-cost spectrum sensors, which report back to a cloud-based data hosting platform, as explored in the European Electrosense project [15]. Sensing could also be performed by devices which are clients of the database and the results could form the basis for coordination between secondary devices and between secondary and primary users to optimize the spectrum sharing potential [16]. The extent of cooperation between primary and secondary users will depend on how the secondary access is authorized. If the latter is on an opportunistic basis facilitated by regulators then any coordination with primary users will be driven entirely by what the regulators require, if anything. If the secondary access is arranged through commercial agreement (such as sub-licensing) then trust will be greater and so closer cooperation is more likely to be possible, accompanied by more detailed and efficient coordination.

Coupling the raw data from distributed sensing with AI tools would allow the calculations of spectrum database contents to be adapted dynamically to the radio environment and able to cope with variations in atmospheric conditions, local interference issues, etc.

## 3.7 Implementing Flexible Spectrum Access

### 3.7.1 Software-defined Radio Underpins Flexibility

In the early days of radio communications, the transmitter and receiver technology were limited in their capabilities. Radio equipment needed to be designed around specific bands so that filters and tuning components etc. could be optimized with reasonable cost and complexity. Early equipment tended to be bulky and expensive, being largely intended for professional/expert use and requiring a skilled operator.

Over the decades since its introduction, radio technology has advanced radically, enabling compact and low-cost implementations that can be integrated into other devices, such as mobile phones. Mobile phone technology has driven the development of highly sophisticated and compact two-way communications technology, with an increasing proportion of the functionality being implemented in software and reaching new heights of flexibility.

Instead of radios needing to be designed with fixed band and modulation standards, it is now possible to use software to provide adaptability over an increasing range of operating characteristics. This also provides scope for later upgrading, bug-fixing etc., extending device life in a rapidly changing service world.

The term software defined radio (SDR) has been used to define a new generation of wireless communication devices in which hardware is a minor part of the device implementation, with an ever more marginal role. The advantage of this approach is that new communication standards can be implemented flexibly and rapidly, with the ability to upgrade and patch devices to suit emerging application and location requirements.

Software can also enable more effective sharing of spectrum by allowing radio systems to coordinate with each other and adapt their behavior as market requirements evolve.

### 3.7.2 Regulation Needs to Adapt to the New Flexibility in Radio Devices

Device radio characteristics used to remain largely constant through their lifetime, resulting from their design and verified after production. For such "fixed" devices, the key characteristics can be verified through testing (e.g. during type approval) and can then be assumed for calculating coexistence criteria when sharing spectrum with other applications.

However, with software taking over an increasing share of device functionality, leading ultimately to SDRs, it has become possible to change the device behavior when the device is in the field and therefore subvert the regulator's requirements. The regulator can lay down ranges within which the key parameters should be set, but the device manufacturer would need to provide security measures to ensure that only regulator approved updates would be possible. Such measures might also raise competition issues by restricting the software that an end-user might wish to buy from other parties.

## 3.8 Foundations for More Flexible Access in the Future

### 3.8.1 Finer-grained Spectrum Access Management

Whereas regulators have traditionally followed a "manual" licensing process, whose complexity and duration varies with the value of the spectrum and degree of exclusivity required, cognitive access can be automated and very rapid. It can also be much finer-grained (in geographic terms) than is practical with the tools and techniques currently in use for licensing spectrum. This broadens the base of potential applications and ensures that spectrum can be used more fully. Applications for this flexibility include rural broadband and extension of mobile coverage [17].

The automation of access, using a combination of sensing and database technologies together with AI techniques, could be very helpful for supporting near-instantaneous spectrum allowance for ad hoc, mobile access requirements, such as might be needed for Internet of Things (IoT) applications.

Technical operating parameters, such as power limits, could be varied dynamically, for example to manage interference and offer emergency services priority.

Connectivity could be provided on demand, as and where needed. Imagine a fire crew fighting a major forest fire: there may be multiple vehicles at the scene and the crew members need to be able to stay in touch with each other. There might not be any networks (commercial or even public service) covering that area. Even if there is an emergency services network in place, it may be restricted to voice or narrowband data.

If a sufficiently wide band or collection of bands were available to access opportunistically, then the fire-fighting units could establish a network covering the area of operations (e.g., using mesh technology). An example of such a network was demonstrated by the Dutch Fire Service based on peer-to-peer and mesh networking techniques, using license-exempt spectrum [18].

### 3.8.2 More Flexible License Exemption

To improve the flexibility of license exemption it is necessary for regulators to know more about where and how license-exempt devices will be used and be able to track this over time for a potentially large, widely distributed, device population (such as in consumer applications). Conventional regulatory approaches, such as used for spectrum licensing, could not easily support this. The task would require significant data gathering and processing abilities, more like those of a large network operator than a rule-setting public body.

With the arrival of the Internet and particularly the IoT, such information as device location and other key properties can be readily harvested at consumer market scale and in much greater detail than was possible before. By enabling end-user devices to gather and communicate key parameters about spectrum occupancy, it becomes possible to manage interference better between adjacent users sharing the same frequency range and thus achieve greater efficiency.

The emerging ability to source such richly detailed data about spectrum usage fortunately coincides with the growing power of highly scalable cloud-based data hosting and processing platforms where AI tools could be deployed. AI could be harnessed to identify significant features from the potentially vast streams of data that comprise a live national or even international frequency occupation map. The results could be used to drive spectrum database updates, so that application demands can be met to the greatest extent possible at a given time in any location covered by the database.

The need for more rapid access to additional spectrum capacity, coupled with the slow pace of clearing older technologies out of the spectrum, has led even licensed operators to start factoring license-exempt bands into their network upgrade plans. Regulators have simply not been able to liberate spectrum quickly enough from older technologies/applications to meet market demands so the importance of license-exempt access has grown rapidly, particularly at the dense end of the network densification scale, represented by the millions of home network (Wi-Fi) APs using license exemption in the 2.4- and 5-GHz bands.

Chapter 19 looks in more detail at the potential for more sophisticated and efficient methods of sharing in the coming years.

### 3.8.2.1 Towards a UHF Spectrum Commons or Superhighway

Since the early 2000s there has been a progressive reduction in the span of UHF frequencies used for terrestrial television broadcasting. This has been in response to the growing demand for mobile data capacity. Since TVWS arises largely from the interleaved nature of TV coverage, rolling back spectrum used for broadcasting inevitably reduces the TVWS capacity too. However, the process of clearing out terrestrial broadcasting is politically sensitive and expensive, so we can expect TVWS capacity to remain significant for many years to come. When the point is reached where broadcasting no longer uses the UHF bands, it may be necessary to reserve part of these bands for opportunistic, flexible, shared use. If a sufficient capacity were reserved in this way, it could serve the needs of a wide range of applications, including rural broadband and IoT applications such as remote imaging and PMSE. Such a *spectrum common* (or *superhighway*, as some have termed it [19]) would also leave space for applications that are still to emerge, much as the Wi-Fi bands have been doing over the last couple of decades. The difference with UHF is that the applications using such spectrum commons could enjoy a much wider reach than is possible with the 2.4-GHz band for a given transmission power limit.

Providing the capacity and coverage continuity that wireless application users will expect in the future will require flexible access to a diverse range of frequencies, with lower frequencies providing reach and higher frequencies providing capacity uplift. With many different applications and their users competing for access there is scope for future advances in dynamic sharing to help make the best possible use of available spectrum.

## References

**1** Minaev, I. (2007) Flexible use of Spectrum. Presented at the ITU seminar on Economic Aspects of national radio frequency spectrum management. Kyiv, Ukraine, July 3–5, 2007. Available at: https://www.itu.int/ITU-D/finance/spectrum_management/Kyiv/Minaev1-EN.PDF.

**2** Commission of The European Communities (2007) Rapid Access to Spectrum for Wireless Electronic Communications Services through more Flexibility. Available: https://eur-lex.europa.eu/LexUriServ/LexUriServ.do?uri=COM:2007:0050:FIN:EN:PDF.

**3** European Parliament and Council (2014) Directive 2014/53/EU on the harmonisation of the laws of the Member States relating to the making available on the market of radio equipment and repealing Directive 1999/5/EC Text with EEA relevance.

**4** M. Cave and W. Webb (2012) The unfinished history of usage rights for spectrum. *Telecommunications Policy*, 36(4): 293–300.

**5** RTL Group Press Release (2013) DVB-T programme distribution expires at the end of 2014, 17th January 2013. Available: http://www.rtlgroup.com/en/news/2013/3/dvb-t_programme_distribution_e.cfm.

**6** Jones, S.K., Phillips, T.W., Van Tuyl, H.L., and Weller R.D. (2008) Evaluation of the Performance of Prototype TV-Band White Spaces Devices. Phase 2. OET Report FCC/OET 08-TR-1005, Federal Communications Committee, Office of Engineering Technology. Available: https://docs.fcc.gov/public/attachments/DA-08-2243A3.pdf.

**7** Federal Communications Commission (2010) Unlicensed Operation in the TV Broadcast Bands, Second Memorandum Opinion and Order. September, FCC. Available: https://docs.fcc.gov/public/attachments/FCC-10-174A1.pdf.

**8** Office of Communications (2015) Implementing TV White Spaces. Statement, 12th February 2015. Available: https://www.ofcom.org.uk/__data/assets/pdf_file/0034/68668/tvws-statement.pdf.

**9** Federal Communications Commission (2008) Report and order on TV white space. Available: http://hraunfoss.fcc.gov/edocs_public/attachmatch/FCC-08-260A1.pdf.

**10** Electronic Communications Committee (2011) Technical and operational requirements for the possible operation of cognitive radio systems in the 'white spaces' of the frequency band 470–790 MHz. ECC 159th Report. Available: https://www.ecodocdb.dk/download/be051b35-91e9/ECCREP159.DOC.

**11** Electronic Communications Committee (2013) Complementary Report to ECC Report 159, Further definition of technical and operational requirements for the operation of white space devices in the band 470–790 MHz, approved January 2013. ECC Report 185. Available: https://www.ecodocdb.dk/download/a1251422-659a/ECCREP185.DOCX.

**12** Mitola, J. (2009) Cognitive radio architecture evolution. Proceedings of the IEEE, 97(4): 626–641. Available: https://ieeexplore.ieee.org/abstract/document/4814771.

**13** Kortun, A., Ratnarajah, T., Sellathurai, M., Zhong, C.J., and Papadias, C.B. (2011) On the performance of eigenvalue-based cooperative spectrum sensing for cognitive radio. *IEEE Journal of Selected Topics in Signal Processing*, 5(1): 49–55. Available: https://ieeexplore.ieee.org/document/5549843.

**14** Tevfik Y. and Huseyin A. (2009) A survey of spectrum sensing algorithms for cognitive radio applications. *IEEE Communications Surveys & Tutorials*, 11(1): 116–130. Available: https://ieeexplore.ieee.org/document/4796930.

**15** Calvo-Palomino, R., Cordobès, H., Engel, M. *et al.* (2018) Electrosense+: Empowering People to Decode the Radio Spectrum. Available: https://arxiv.org/ftp/arxiv/papers/1811/1811.12265.pdf.

**16** Voulgaris, K., Gizas, B., and Papadias, C.B. (2016) Realizing spectrum sharing through the use of a database-assisted MAC protocol. The 17th IEEE International Workshop on Signal Processing Advances in Wireless Communications (SPAWC 2016), Edinburgh, UK, July 3–6, 2016.

**17** Kassem, M.M., Marina, M.K., and Radunovic, B. (2018) DIY Model for Mobile Network Deployment: A Step Towards 5G for All, Compass '18, June 20–22, California. Available: http://homepages.inf.ed.ac.uk/mmarina/papers/compass18.pdf.

**18** Project i-Bridge with NL Government, VGGM, C2 Support (2010) Examples of an i-Bridge exercise can be seen at: http://www.youtube.com/watch?v=AY4ITXtX7PE and http://www.youtube.com/watch?v=AuMj9yeHhVQ&feature=related (audio in Dutch).

**19** Presidential Council of Advisers on Science and Technology (2012) Report to the President Realizing the Full Potential of Government-Held Spectrum to Spur Economic Growth. Available: https://obamawhitehouse.archives.gov/sites/default/files/microsites/ostp/pcast_spectrum_report_final_july_20_2012.pdf.

## Further Reading

**1** Cambridge White Spaces Consortium (2012) Cambridge TV White Spaces Trial: A Summary of the Technical Findings. Available: https://www.microsoft.com/en-us/research/uploads/prod/2016/02/spectrum-cambridge-tv-white-spaces-trial-findings.pdf.

**2** Haykin, S., Thompson, D.J., and Reed, J.H. (2009) Spectrum sensing for cognitive radio. *Proceedings of the IEEE*, 97(5): 849–877. Available: https://www.researchgate.net/profile/David_Thomson16/publication/224408386_Spectrum_Sensing_for_Cognitive_Radio/links/555625ee08aeaaff3bf5ee4d.pdf.

# 4

# Evolving Spectrum Sharing Methods, Standards and Trials: TVWS, CBRS, MulteFire and More

*Dani Anderson, K.A. Shruthi, David Crawford, and Robert W. Stewart*

*University of Strathclyde, UK*

## 4.1 Introduction

As discussed throughout this book, there are a number of techniques and technologies that can be used to deploy spectrum sharing. Although these implementations continue to evolve and improve, and are expected to contribute eventually to the operation of fifth-generation (5G) networks and beyond, the concept has been discussed for a number of years.

The University of Strathclyde's Centre for White Space Communications (CWSC) has been involved in the development of spectrum sharing through numerous projects, testbeds, and trials since 2011.

This chapter explores the evolution of the shared spectrum, including developing and emerging technologies, across multiple different operating frequency bands. Beginning with TV white space, as discussed in Chapter 3, details of several UK-based TVWS projects and trials, completed by the CWSC and various partners, will provide tangible examples of shared spectrum implementation.

Following the discussion on practical shared spectrum implementation, the remainder of the chapter will survey emerging technologies across licensed and unlicensed spectrum bands. This includes the US Citizens Broadband Radio Service (CBRS), long-term evolution (LTE) in unlicensed spectrum (LTE-U), LTE wireless local access network (WLAN) aggregation (LWA), LTE-WLAN integration with Internet protocol (IP) security tunnel (LWIP), licensed assisted access (LAA), and MulteFire.

## 4.2 TV White Space

### 4.2.1 Overview

Terrestrial, or broadcast, television transmissions have undergone a number of developments since their establishment. Most recently this has been the use of digital encoding to greatly improve spectral efficiency.

*Spectrum Sharing: The Next Frontier in Wireless Networks,* First Edition.
Edited by Constantinos B. Papadias, Tharmalingam Ratnarajah, and Dirk T.M. Slock.
© 2020 John Wiley & Sons Ltd. Published 2020 by John Wiley & Sons Ltd.

Using these techniques, multiple transmissions can occur within the same bandwidth of one analogue channel. These developments also allow for high-definition quality images and additional services such as multimedia, leading to a global digital television transition, also referred to as the digital *switchover*.

The majority of these digital television (DTV) transmissions occur within the ultrahigh frequency (UHF) band. Because of the increase in spectral efficiency, adoption of digital transmission will lead to reduced utilization of the spectrum. In addition, developments in digital signal processing (DSP) techniques, specifically in frequency shaping filters, greatly reduces the probability of causing interference to local operations in nearby spectrum.

Frequency bands that are not being used for services provided by licensed TV spectrum holders are known as white spaces. Such unoccupied spectrum slots could be licensed but geographically unused, or be assigned as guard bands for technology implementations. In areas with lower spectrum utilization, these TV band white spaces, collectively referred to as TV white space (TVWS), could be used for additional licensed-exempt connectivity services.

National and international regulatory bodies have developed a number of regulations to facilitate shared access to these vacant frequency bands. These regulations are designed to protect the transmissions of spectrum license holders, referred to as *primary* or *incumbent* users, from interference caused by licensed-exempt, or *secondary*, spectrum users. This "primary vs. secondary" categorization has dominated the spectrum sharing literature since the early days of TVWS and cognitive radio, and has only started being enriched recently with the emergence of other types of players [such as the three tiers of users defined in CBRS (see below) or the licensee users who enjoy quality of service (QoS) guarantees under the licensed shared access (LSA) paradigm (see Chapter 6)].

As the available TVWS varies geographically, the allocation of frequency bands to secondary users must also be dynamic. There are a number of techniques that could potentially be used to carry out this process, including the use of cognitive radios to perform spectrum sensing.

The most commonly implemented method involves a centralized database of primary users and their operating parameters. White space devices (WSDs) looking to access the spectrum for license-exempt transmission must coordinate with a regulator-approved white space database (WSDB) using the protocol to access white space (PAWS). This standard, developed by the Internet Engineering Task Force (IETF), defines the message format for all communication between the WSD and the WSDB [1].

On receiving a WSD request, the WSDB will calculate which transmission channels can be used and the accompanying transmit power restrictions for each. These operating parameters minimize the potential for interference to the incumbent users and allow for location-specific allocation of vacant channels to the secondary users. This process is generally referred to as dynamic spectrum allocation (DSA) through a geolocation database.

There are a number of TVWS spectrum databases, from a variety of providers, currently serving networks across the world. While the exact method required for calculating operating parameters varies according to the rules of the local spectrum regulator, the general process involves combining the supplied WSD details with known information on the spectrum license holders operating in the requested location.

In addition to complying with local regulator conditions, each WSDB should be product-agnostic to the requesting WSD. This means that the process of supplying the response should not differ among requesting hardware platforms or air interface used by the WSD (this is discussed further in section 4.2.2).

The CWSC at the University of Strathclyde has worked with many such technology providers, delivering shared spectrum projects across Scotland for a number of years. The implemented networks cover a range of different applications in both rural and urban environments, and were part of developing initial TVWS regulations for the UK. More information on the installations in Glasgow, on the Isle of Bute, and in the Orkney Islands can be found in section 4.2.3.

### 4.2.2  Operating Standards

The two most common standards implemented for TVWS device air interface were developed by the Institute of Electronic and Electrical Engineers (IEEE) and are described below. Both support methods of spectrum sharing through implementations of geolocation database requests and cognitive sensing techniques.

#### IEEE 802.22
The IEEE 802.22 standard was published in 2011, and defines media access control (MAC) and physical (PHY) layer specifications for wireless regional area (WRAN) network operation in the frequency range of 56–806 MHz. The intended topology is cellular point to multi-point (P2MP), with a single base station (BS) connected to multiple customer premise equipment (CPE) units within an operating cell.

As a WRAN, the maximum achievable coverage area is relatively large compared to a WLAN standard, such as those in the 802.11 family, even though they are intended for use in similar applications. For 802.22 implementation typical distances between BS and CPE are in the range of 17–33 km, but up to 100 km is supported [2], while a WLAN is better optimized for distances of up to 1 km, as discussed later.

The MAC and PHY layer were both designed to fully accommodate spectrum sharing. The standard adopts an orthogonal frequency division multiple access (OFDMA) transport scheme using time-division duplexing (TDD), enabling the fast channel adaptation needed for operation in shared spectrum, and supports quadrature phase shift keying (QPSK), 16 quadrature amplitude modulation (QAM) or 64-QAM modulation schemes. Bandwidths of 6, 7 or 8 MHz can be implemented. Contiguous channel bonding is supported to improve achievable bandwidth and performance, similar to the carrier aggregation techniques used in LTE systems [3]. With channel bonding the combined system must be in spectrally adjacent channels, while carrier aggregation allows for use of non-contiguous spectrum, potentially in different operating bands to form larger bandwidths [4].

The BS and CPE are capable of performing cognitive sensing both *in-band*, i.e., in the current channel being used by the device, and *out-band*, i.e., in the remaining channels. These processes occur during transmission breaks and determine local incumbent information that is then fed back to the BS. While this could impact the achievable QoS, techniques such as dynamic frequency hopping (DFH) can be used to mitigate this [5].

Two methods of spectrum sensing are implemented. *Fast*, or blind, sensing typically involves an energy detection method not particular to any signal properties [6]. The outcome of fast sensing is used to determine if further information is required. In that case, *fine*, also referred to as feature or signal-specific, sensing is used, which leads to a higher success rate in determining channel occupation but is slower and requires use of prior information [2].

Each CPE will feedback its sensing data to the BS, which will combine it with data from the geolocation database to determine spectrum information for the whole network and adjust operating channels accordingly. To ensure incumbent protection the BS cannot continually broadcast pilot signals. Unused channels are initially identified by the database then used to transmit *superframes*, which are defined structures containing a superframe control header (SCH) and potentially multiple data *frames*. At initialization stage, the CPE must first determine any channel occupancy through sensing before scanning the vacant frequencies for superframes and the information contained within the SCH regarding establishing connection to the BS.

## IEEE 802.11af

The IEEE802.11 family of standards defines the implementation of WLAN. The operations for the use of TVWS frequencies are detailed in IEEE802.11af, which defines an updated PHY layer and the necessary MAC modifications required to accommodate this.

The implemented television very high throughput (TVHT) PHY layer is OFDM based, and supports multiple modulation schemes from binary phase shift keying (BPSK) up to 256-QAM [3]. It is extremely similar to the 802.11ac implementation, down-clocked to support channel bandwidths of 6, 7 or 8 MHz in accommodation with local regulations. This design was intentional, as it would aid manufacturers of chipsets in the adaptation of their existing products [2].

Unlike IEEE 802.22, which use a cell-based architecture, IEEE 802.11af networks comprise multiple stations (STAs), or nodes, typically arranged in an *infrastructure* deployment. The 802.11af nodes are categorized as either access points (APs), which bridge connection to a wired network that is typically internet enabled, or clients, which are devices seeking access to the wired connection. Each AP might be connected to multiple client nodes, collectively referred to as its basic service set (BSS), and is responsible for routing traffic to each of them [2].

The standard allows for up to four spatial streams and simultaneous operation in up to four TVWS channels for connection between a client and an AP, which can be completed in multiple different ways [4]. Devices can choose to operate in a single channel, *TVHT_W*, or in two non-adjacent channels, *TVHT_W+W*. Alternatively, bonding can be applied to two or four adjacent channels, *TVHT_2W* and *TVHT_4W*, respectively, to form a single contiguous channel [4]. Finally, a device can use two non-contiguous pairs of bonded channels, *TVHT_2W+2W*. Overall, the standard allows for up to 35.6 Mbps per spatial stream per 8 MHz channel. Therefore, with four bonded channels and four spatial streams, the maximum peak data rate is 568.9 Mbps [2].

MAC layer functionality is based on carrier sense multiple access with collision avoidance (CSMA/CA), as is common among 802.11 standards. When a device wants to transmit, it will first listen to the desired operating channel for a set time. If there is contention, the device will wait a random amount of time before trying again. If the channel is available, transmission will begin on pre-determined slot boundaries. When a device is transmitting, it cannot simultaneously listen for others. If two devices end up operating in the same channel, a collision will occur and the resulting interference will likely cause reception of the signals to fail [2].

Transmission slot boundaries and time spent listening to channels is dependent on a number of factors. One of these is the propagation delay of the signal; the time taken for a transmission to reach the defined furthest away receiver. For IEEE802.11af the default propagation time allows for a maximum distance of roughly 900 m, giving an AP BSS operating range of 450 m. This propagation delay, combined with the CSMA/CA based scheduler, exacerbates the well-known issue referred to as the hidden node problem [2].

IEEE 802.11af doesn't specify any support for spectrum sensing-based feedback, and relies solely on information from the WSDB in order to enable shared spectrum. Any network STA transmitting in TVWS is described as a geolocation database dependent (GDD) entity, and must have its operation coordinated through a WSDB.

Client nodes are referred to as GDD-dependent stations, as their operation is completely controlled and reliant on their serving AP to allocate appropriate operating parameters for incumbent protection. An AP, referred to as a GDD-enabling STA, will consult a WSDB for every node within its BSS and has the authority to control operation of any client under its service [2].

### 4.2.3 Overview of TVWS Trials and Projects

#### Isle of Bute

In 2011 a TVWS trial network was established on the Isle of Bute, in the West of Scotland, by a six-partner consortium including BT, BBC R&D, and the University of Strathclyde. This 18-month project investigated various aspects of early TVWS implementation, with the intention of aiding and informing Ofcom's decisions in the development of UK regulations.

The project aimed to assess the overall technology performance as an alternative to copper-based infrastructure. It was also important to assess the impact of TVWS transmissions on the DTV reception in the network vicinity. This assessment included measurements of co-channel and adjacent-channel interference, as well as a validation of the UK Prediction Model (UKPM) before future use in the calculations performed by TVWS geolocation databases.

Eight trialist premises were connected to the local telephone exchange in a P2MP architecture using custom devices operating in the TV band, 470–790 MHz. One system was based on 802.11 Wi-Fi and another on the IEEE 802.16e Worldwide Interoperability for Microwave Access (WiMAX) standard, since formal TVWS standards, such as those discussed in section 4.2.2, had not been ratified at the time. A high-capacity microwave link was established as a backhaul connection from the island's exchange to BT's infrastructure on the Scottish mainland.

Following network deployment, in-field measurements were performed to ensure protection of DTV services. No co-channel or adjacent-channel interference was recorded, and no disruption was reported by trialists. At the time of the trial, the digital TV switchover had not been completed so non-interfering operating parameters were defined by project partner BBC R&D, accounting for terrain information, details on nearby DTV transmitters, and measurements taken around the island. In addition, received power levels were found to be in alignment with predictions obtained using the UKPM. These tests indicated that, even at this early development stage, TVWS technology with shared spectrum was a candidate solution for the provision of rural connectivity.

### Glasgow

After the completion of the Bute trial, the analogue-to-digital TV switchover was completed in the UK. During the same period, Ofcom published two calls for consultation as preparation for the introduction of TVWS regulations and licensed the installation and operation of 11 trial networks.

These trials were established to test the proposed frameworks, gauge stakeholder interest, and validate the use of WSDB for spectrum access as part of in-field deployments. One of the established networks was in Glasgow, on the University of Strathclyde campus, which served as an urban technology testbed and demonstrator. One of the key objectives of the project was to assess the implementation and compliance of WSD and WSDB with the Ofcom framework, and to identify areas of improvement.

The installed network was designed to function either as a P2MP topology or a collection of individual point-to-point (P2P) links. Four client stations were installed at locations around the campus: three on top of nearby university buildings and one in a communal garden area. Each unit connected over TVWS to a single base station unit installed on the roof of a central building, which had a wired connection to the university network to provide client connectivity.

Distances covered by the TVWS links ranged from 150 to 400 m with varying line-of-sight (LoS) conditions. Some client sites also had to contend with additional interference from nearby vehicular traffic.

The network was established for two urban use cases: to provide backhaul for a public external Wi-Fi AP in the communal garden, and to support real-time video-streaming from IP cameras installed at each of the client locations. This application dictated the minimum QoS provided by the network.

A combination of antenna configurations were used in the network, as requirements differed for each installation. Implemented antenna heights were varied to ensure that the received signal strength was sufficient to maintain the throughput and latency required. These parameters were verified through simulation of the P2P links prior to installation.

Following completion of the network Ofcom performed several measurements to assess the coexistence of TVWS traffic and the incumbent transmissions under their outlined framework. Ofcom's results from this and other pilot projects can be found in [7].

This project helped validate the operation of TVWS and shared spectrum in an urban environment. As with the Bute trial, devices operating in TV spectrum were able to coexist with neighboring DTV transmissions. However, unlike on Bute, device operating parameters were determined dynamically by a WSDB.

### Orkney Islands

In 2015, the CWSC and a number of project partners, including BAE Systems, the Scottish Futures Trust, and CloudNet IT Solutions, began the development of a trial network in the

Orkney Islands. The project, funded by the Scottish Government, looked to build on existing installations to create a rural TVWS testbed, and validate the use of TVWS and spectrum sharing techniques in deep rural locations.

As a remote archipelago, providing connectivity in Orkney is geographically challenging. This is exacerbated by the mountainous terrain, significant coverage area size, and long distance to the Scottish mainland, which all impact the installation of backhaul networks.

The project aimed to demonstrate the commercial viability, reliability, and overall performance of TVWS and shared spectrum as a backhaul for two applications, described below, and to assess if similar solutions could be used to support the Government's connectivity initiatives.

TVWS was used to provide broadband connectivity to passengers and crew on three ferries during voyages between islands. The same network installation was also used to simultaneously provide connectivity to six fixed premises, located in rural locations around the Orkney mainland.

The installed topology can be seen in Figure 4.1. The main network node was located at Wideford Hill, near the ferry port at Kirkwall harbor. This multi-use tower was equipped with two TVWS base station units with multiple microwave radios connecting via P2P links to other network sites and the island's point of presence (PoP) at Ayre of Cara. Every WSD in the network was connected to an Ofcom approved WSDB, provided by Fairspectrum Oy.

In addition to routing the entire backhaul for the network, the TVWS radios at Wideford also provided connectivity directly to three of the fixed premises and over the routes of the three ferries.

Four other TVWS base station sites were established at various locations around the islands. Two of these, at Sanday and Westray Pier, were used to provide additional coverage over the ferry routes, providing connectivity at distances of up to 19 miles. The remaining two sites, at Sandy Hill and Power Station, were used to supply broadband access to three additional fixed premises. Each site was established with both TVWS and microwave radios, for coverage and additional backhaul, respectively.

Each of the ferries was provided with a client unit, connected to a Wi-Fi AP in the passenger lounge. A TVWS radio and omnidirectional antenna were pole-mounted externally with a power over Ethernet (POE) connection to a network switch below deck. A similar configuration was used at each of the fixed premise sites, except since location and orientation are constant, directional antennas could be used to facilitate larger antenna gains. Unlike the generations of radio equipment devices used in previous projects, the TVWS units employed in this project were able to make use of techniques such as channel bonding to increase achievable data rates.

Following installation, information from the three Wi-Fi APs on board the ferries was used to assess demand for the services. Over a 1-month period, over 1000 unique clients were recorded on the network, with an average of 70–80 devices connected per day.

A survey of connected residents was carried out to determine the main applications. In addition to personal and recreational usage, residents took advantage of the internet connectivity to create new online businesses and increase their working efficiency through improved access to services. PoP provider Faroese Telecom was also able to use the increased bandwidth from the TVWS network to improve the radar monitoring application for its subsea fibre link.

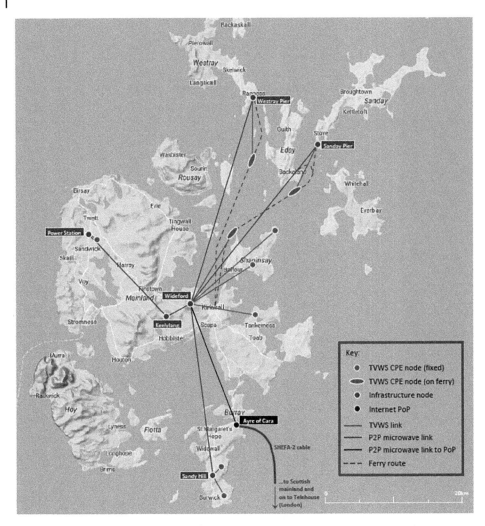

**Figure 4.1** Network diagram for the Orkney nomadic TVWS installation. The PoP for the island is located at the Ayre of Cara location.

This project demonstrates the technical and commercial viability of TVWS and shared spectrum to enhance broadband coverage in remote rural locations, and to provide connectivity to non-stationary clients in addition to rural premises.

## 4.3 Emerging Shared Spectrum Technologies

### 4.3.1 Introduction

There have been a number of standards released in recent years that either support or directly enable use of shared spectrum. Technology standards, typically based on or contributing to existing frameworks developed by the 3rd Generation Partnership Project (3GPP), have defined hardware and software operational requirements. These include systems such as Multefire, LTE-U/LAA, and LWA/LWIP. These technologies cover operations in both licensed and unlicensed spectrum.

Other standards, such as the CBRS, define regulations for the allocation of operating channels within shared frequency bands. CBRS targets the 3.5-GHz band, in the frequency range 3.55–3.76 GHz, and is described below.

### 4.3.2 CBRS

Creation of the CBRS standard was proposed initially in 2012 by the Federal Communication Commission (FCC) in the USA as a potential enabler for the deployment of shared spectrum and small cell networks. Rules were outlined initially in April 2014 before being formally adopted in 2015 [8]. CBRS uses a multi-tiered authorization framework for frequency allocation, where three levels of priority for spectrum access are defined.

The users in the highest tier, those with *incumbent access*, are protected from any interference caused by those with lower level access. In CBRS, this authorization is given to federal government transmissions and fixed satellite services.

A user (operator) with *priority access* will receive a license, referred to as a priority access license (PAL). Following a to-be-completed bidding process, a PAL will be allocated to participating operators for a portion of spectrum, which is valid for a defined licensing period. The PAL allows for protected operation within a specific geographic area, referred to as the PAL protection area (PPA). There is no guarantee of access to specific operating frequencies following the expiry of this license agreement.

The lowest spectrum access tier defined in CBRS is *general authorised access* (GAA). GAA users can operate in any portion of the 3.55–3.7-GHz band that has not been assigned to the higher tier users, but must obey specifically calculated operating parameters and receive no interference protection or QoS guarantees.

Initial CBRS regulations defined the PPA for a PAL as the size of a census tract, which include on average 4000 people, of which the USA has around 74 000. The use of Census tracts as PPA boundaries is beneficial to local Internet service providers (ISPs) since the reduced coverage area leads to lower capital- and operating expenditure (CAPEX and OPEX).

Following discussions with various stakeholders, on 23 October 2018 FCC modified the rules and settled for an intermediary solution which looked to benefit both local and national ISPs. Specifically, the license duration of PALs was extended to 10 years, from 3, and the size of a PPA was changed from census tract to that of a county area [9]. The FCC felt that dynamically managing spectrum in 3242 counties was a more practical option than the significantly larger number of census tracts.

Local ISPs believe that the increase in size of PPAs, and associated financial investment, could lead to reduced competition, resulting in slower rural and industrial deployments. However, the increase in PAL duration and probability of license renewal following expiry, rather than a new PAL auction, is consider to benefit local ISPs.

**Operation and System Architecture**

The functional architecture of the CBRS is shown in Figure 4.2. The allocation of PAL and the management of opportunistic spectrum users is carried out by a spectrum access system (SAS). The SAS accounts for information received from an FCC database of currently valid PALs, the *incumbent informer*, the environmental sensing capability (ESC) entity, and details on local terrain, among other parameters, to calculate operating conditions for a requesting Citizens Broadband Radio Service Device (CBSD).

All CBSDs must be certified and registered with a SAS before being authorized for operation in the spectrum band [10]. A SAS will store information from all registered users, including device license and certification details [11]. Although this process is similar to that used by a TVWS geolocation database, the exact procedure for the request and allocation of channels has not been fully standardized yet. In addition to protocols for communication between SAS and CBSD, there is also a protocol for SAS-to-SAS communication which enables exchange of spectrum usage data to further manage interference.

The ESC uses cognitive spectrum sensing to detect the presence of incumbents and PAL holders within the serving PPA, and communicate this information to a SAS [12] (see also Chapter 6, in particular the sensing reasoning module and dedicated sensing networks proposed for LSA in Figure 6.5). The SAS then ensures that appropriate protection is established, including sending requests to CBSDs to vacate operating channels. The ESC provides the SAS with information on the frequencies that require the highest level of protection.

Additional sensing information from these entities is fed back to the SAS to be accounted for as part of the CBSD frequency allocation. Information on the frequency bands allocated to secondary users is stored in the *incumbent informer*, and is used to monitor the network utilization and authenticate users. Even though the SAS receives information from the FCC database, for security reasons it does not include sensitive details on incumbent

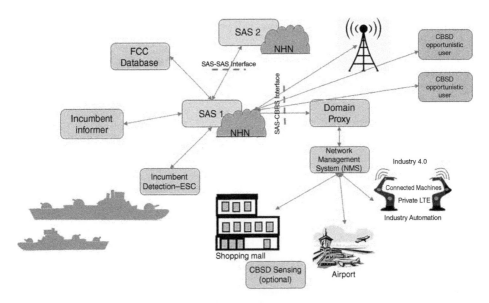

**Figure 4.2**  CBRS functional architecture [10] (modified).

transmissions. Therefore, the ESC and incumbent informer are required to ensure the accuracy of incumbent information.

As a technology designed to enhance and enable small cell deployments, the expected use cases will be for private LTE networks. These could be for providing enterprise solutions, enabling last mile connectivity via fixed wireless applications, or as supplementary networks targeting coverage "not-spots". Development of the CBRS also enables the network as a service (NaaS) provision paradigm as an alternative to fixed service traditional licensed network models.

Practical implementations of CBRS-based systems will require connection to wider networks, including those of major mobile network operators (MNO). This interoperability will most likely be handled through use of *neutral host networks* (NHN). A neutral host architecture comprises a single network infrastructure shared on an open access basis with collaborating MNOs. Deployments of this style are used to consolidate user traffic, bridge connection to larger service providers, and resolve capacity issues inside large venues or other busy locations [13].

Some of the expected benefits of implementing the CBRS in network deployments are:

1. Improved interference management for provision of indoor coverage, leading to higher capacity and better throughout.
2. Simplification of the process of coordinating multiple wireless access points.
3. Reduced cost in the deployment of radio access networks (RAN) through use of NHN. (A RAN describes the parts of a telecommunications system that enables connection between individual devices and the service provider's core network [13]).
4. Significant reduction in the financial constraint associated with access to spectrum, compared to the costs of a traditional spectrum license.

As we move towards full development, commercialization and implementation of CBRS, there are a number of challenges to be addressed:

1. Uncertainty in the overall demand for the technology.
2. Lack of practical coexistence management system for operators within the same access tier, particularly among GAA users.
3. Deployments making use of PAL and GAA have yet to produce fully proven and developed business models that take advantage of the implementation of CBRS.
4. Uncertainty on the impact that spectrum sharing and NHN will have on competition between service providers.

### The CBRS Alliance and Early Technology Trials

There have been a number of trials and research deployments of CBRS systems and technologies. The majority of these have taken place in the USA and have involved members of the CBRS Alliance, which was established in August 2016 [14].

This industry organization began with a small number of charter members, including Google, Intel, Ericsson, Nokia, Qualcomm, Ruckus Wireless, and Federated Wireless. Since then, the institution has grown to include more than 100 member organizations [14], who together support the common interest of developing the CBRS ecosystem towards the commercialization and adoption of the technology.

The consortium has adopted the OnGo name as their consumer-facing brand of LTE using CBRS [15], and have outlined a set of specifications, certifications, and standards for compliant products. Five working groups have been formed, each focusing on a separate aspect of development [14]. The most recently established deployment and operations group focuses on implementing models for deployment and operational best practices [14].

Verizon, the second largest American wireless service provider, has a long history of engaging in CBRS trials. Throughout 2018 end-to-end system tests were completed at their test facility in Irving, Texas [16, 17] and as part of a live network in Fort Lauderdale, Florida [18]. The main research outcomes for these trials were:

1. Validate the interoperability between equipment vendors.
2. Evaluate the performance and data throughput of LTE.
3. Assess the customer experience.
4. Test the mobility handoffs.
5. Verify the algorithms used by the SAS for channel allocation provide optimal results [17, 18].

Both of these tests were completed in collaboration with various CBRS Alliance members. Google and Federated Wireless both provided SAS and ESC capabilities. Base station equipment from Nokia, Ericsson, and Corning was used for both indoor and outdoor deployment testing. User equipment (UE) was supplied by Qualcomm Technologies in the form of a Snapdragon 845 test device, which used the X20 LTE modem [19].

In 2017, this same chipset was used as part of the equipment involved in a live CBRS demonstration at Qualcomm's headquarters during an event in San Diego. This demonstration was completed using an indoor small cell solution from AirSpan [20].

A private LTE network, enabled by CBRS, was deployed at the Las Vegas Motor Speedway at the beginning of 2017 [21]. Stock cars were set-up with Qualcomm modems connecting to Nokia base stations under authorization from the Google SAS to enable 4K video streaming with 360° visibility from the inside of vehicles moving in excess of 180 mph. Mobile edge computing and smart scheduler capabilities were built into the Nokia equipment to meet the low latency and high mobility requirements [22].

### 4.3.3 Other Shared Spectrum LTE Solutions

The previous section discussed the technologies and spectrum access procedures associated with the CBRS band. However, there are a number of other possible implementations of LTE using shared spectrum.

LSA offers another approach to the sharing of licensed spectrum, and is currently being researched widely in Europe (see also Chapter 2 and 6). The incumbent users in this band are typically either government or broadcasters, which tend to be using highly localized deployments. The plan is to allow secondary users, either local ISPs or MNOs, to use the band via geolocation databases as a complementary technology to their existing broadband services. LSA is discussed in depth in Chapter 6.

For deployments with a large number of simultaneous users, implementation of LTE is very appealing as it can efficiently manage high user densities and aspects of network security. However, there are significant costs associated with the requirements of traditional spectrum licenses. While an unlicensed spectrum technology, such as Wi-Fi, does not come with this additional CAPEX, the cost of managing a highly dense network with guaranteed QoS is very high. A possible solution is to develop technologies that integrate the features of LTE in unlicensed spectrum, such as in the industrial, scientific and medical (ISM) band. This section will look at some technologies designed for that purpose, including LAA/LTE-U, LWA/LWIP, and MulteFire.

The development of LTE-U/LAA and LWIP/LWA is discussed further in Chapter 14.

**MulteFire**

MulteFire is an LTE-based technology that can operate in both licensed spectrum and the 5-GHz band. It was designed to allow enterprise stakeholders to deploy private LTE networks, and supports the use of a neutral host architecture [23, 24]. It operates as a stand-alone technology in unlicensed spectrum, which means that there are no dependencies on access to licensed spectrum.

The MulteFire alliance is a consortium formed to develop and drive the MulteFire ecosystem by defining the technology specification, testing the hardware and software designs, and assessing the interoperability between various equipment manufacturers.

MulteFire can fairly coexist with other technologies operating in the same band and supports the listen before talk (LBT) protocol for deployments where there is a requirement for over-the-air spectrum contention management. Using LBT, a radio will first sense its operating environment before it starts transmitting in order to determine which channels are unused and available. This method of uncoordinated spectrum sharing is part of a feature set which also enables the use of channel aggregation techniques.

The architecture of a MulteFire supporting network is shown in Figure 4.3. This architecture is similar to that used in a typical 3GPP-defined LTE network deployment. In fact the evolved packet core (EPC), UE, the corresponding interoperability signals such as the S1 and X2 interface, and the network discovery and selection mechanisms all follow 3GPP standards [25].

The only non-3GPP defined network element is the MulteFire AP, the equivalent of an evolved NodeB (eNodeB) unit, which collectively makes up the MulteFire RAN. As the interconnectivity between MulteFire APs use the 3GPP X2 interface, there can be handover and mobility between a MulteFire RAN and a 3GPP-based RAN.

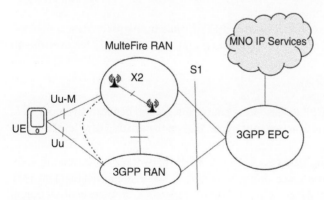

**Figure 4.3** MulteFire architecture.

Network operators can differentiate between their MulteFire RAN and 3GPP RAN deployments by assigning different tracking area code values to their operating cells [25].

The first public test of MulteFire technology was completed by MulteFire Alliance founding member Nokia, and the Saudi Telecom Company (STC) in 2016 [26, 27]. The deployment used 5-GHz-enabled FlexiZone equipment to conduct coexistence tests with surrounding Wi-Fi networks, assess the achievable range and coverage, and demonstrate neutral hosting capabilities. Results indicated harmonious coexistence with co-channel Wi-Fi deployments and significant coverage improvements [26].

### LAA/LTE-U

LAA is a 3GPP standardized LTE implementation that uses carrier aggregation techniques across a combination of licensed spectrum for the primary carrier and the unlicensed 5-GHz band for a supplementary downlink carrier [28]. The initial LAA specification was included in Release 13 with enhanced LAA (eLAA) defined in Release 14, which included additional uplink support.

Under an LAA implementation, the base station will assign a higher priority to some traffic, which is then transmitted alongside the control and signaling information using the licensed "anchor" channel. The unlicensed carrier is then used to increase the available capacity, with LBT again implemented to enable shared spectrum in the 5-GHz band. Both TDD and frequency-division duplex (FDD) implementations of carrier aggregation are supported.

With the supplementary downlink enabled, and at least 20 MHz of licensed anchor channel, an LAA deployment can deliver peak downlink speeds of 1000 Mbps gigabit LTE service.

The world's first LAA over-the-air trial was carried out by Qualcomm and Deutsche Telecom in Nurember, Germany on 20 November 2015 for both indoor and outdoor deployment scenarios [29]. Qualcomm also produced the world's first gigabit class LTE modem, the Snapdragon X16 [30], which supports both 3.5- and 5-GHz operations.

LTE-U was standardized as part of 3GPP Release 12 [28]. Similar to LAA, a licensed anchor channel is required for the control and signaling information, with carrier aggregation techniques used to combine a data channel in unlicensed spectrum. Unlike LAA, the LTE-U standard uses dynamic channel selection and carrier sense adaptive transmission to maintain coexistence in the unlicensed band.

Ericsson were involved in a number of early LTE-U trials throughout 2016, collaborating with telecommunication companies Telefónica [31] and MTN [32] for over-the-air demonstrations.

### LWA/LWIP

LWA was defined as part of 3GPP Release 13 to enable integration between LTE and Wi-Fi technologies as part of the RAN functionality, expanding on the existing core-level interoperability services.

LWA enables packet routing between LTE and Wi-Fi networks, for compliant UEs, through the packet data convergence protocol as part of the eNodeB deployment [33]. This means the activation and deactivation of LWA techniques is at the sole discretion of eNodeB. Both 2.4- and 5-GHz Wi-Fi bands are supported [34], with LBT used to enable coexistence with the shared unlicensed spectrum [28].

For the UE's uplink transmission, only LTE channels are used, while the downlink can make use of both LTE and Wi-Fi. LTE data is transported using the 802.11 MAC, which reduces the required changes to the Wi-Fi air interface [34].

The specifications for enhanced LWA were included as part of 3GPP Release 14, which enabled operation in the 60-GHz band, with 2.16 GHz of bandwidth, and the ability to perform uplink aggregation.

LWIP is also a 3GPP Release 13 feature set. Operation is similar to LWA, but aggregation between LTE and WLAN traffic is completed at the IP layer [28], with both uplink and downlink aggregation supported.

## 4.4 Conclusion

This chapter has outlined several technologies that can be used to deploy spectrum sharing: TVWS, CBRS, MulteFire, LWA/LWIP, LAA/LTE-U, and LSA. Each of these shared spectrum implementations are expected to be deployed in use-case areas such private LTE networks.

Although we have seen a number of trials for these technologies, they continue to improve and develop, and are expected to contribute to the operation of 5G networks and beyond.

## References

1 V. Chen, S. Das, L. Zhu, J. Maylar, and P. McCann, Protocol to Access White-Space (PAWS) Databases. Technical Report, Internet Engineering Task Force (IETF), May 2015.

2 D. Lekomtcev and R. Maršálek, SComparison of 802.11af and 802.22 Standards – Physical Layer and Cognitive Functionality. *Online*, pp. 12–17, 06 2012.

3 D. Vujic and M. Dukić, Comparison of TVWS Standards. In *Synthesis*, pp. 47–50, 01 2015.

4 Z. Khan, H. Ahmadi, E. Hossain, M. Coupechoux, L. A. Dasilva, and J. J. Lehtomäki, Carrier Aggregation/Channel Bonding in Next Generation Cellular Networks: Methods and Challenges. *IEEE Network*, vol. 28, pp. 34–40, Nov. 2014.

5 W. Hu, D. Willkomm, M. Abusubaih, J. Gross, G. Vlantis, M. Gerla, and A. Wolisz, Cognitive Radios For Dynamic Spectrum Access – Dynamic Frequency Hopping Communities for Efficient IEEE 802.22 Operation. *IEEE Communications Magazine*, vol. 45, pp. 80–87, May 2007.

6 S. Shellhammer, Spectrum Sensing in IEEE802.22. *IAPR Wksp. Cognitive Info. Processing*, 2008.

7 Ofcom, Implementing TV White Spaces. *Implementing TV White Spaces*, Feb. 2015.

8 FCC, 3.5 GHz Band Overview. Sept. 2019.

**9** M. Alleven, FCC adopts county-sized license areas, 10-year terms for 3.5 GHz CBRS band – FierceWireless. *Online*, Oct 2018.

**10** K. Mun, CBRS: New Shared Spectrum Enables Flexible Indoor and Outdoor Mobile Solutions and New Business Models. *Online*, March 2017.

**11** Wireless Innovation Forum, Operations for Citizens Broadband Radio Service (CBRS): Priority Access License (PAL) Database Technical Specification. *Online*, July 2017.

**12** J. Wilkins and J. Knapp, Continuing Momentum in the 3.5 GHz band – Federal Communications Commission. *Online*, May 2016.

**13** S. Weston, Is Neutral Host Infrastructure the Way Forward? *Online*, July 2018.

**14** GlobalNewsWire, CBRS Alliance Passes 100 Member Milestone, Establishes OnGo Deployment and Operations Working Group. *Online*, Aug. 2018.

**15** R. Porayath, Industry poised to ride OnGo in the CBRS innovation band. *Online*, Sept. 2018.

**16** LightReading, Verizon Tests 4G CBRS With Partners. *Online*, May 2018.

**17** C. Reichert, Verizon begins LTE trials on 3.5GHz band with Nokia, Ericsson. *Online*, April 2018.

**18** A. Szal, Verizon Announces Successful 4G Test in CBRS Band. *Online*, May 2018.

**19** Qualcomm, Snapdragon X20 LTE Modem Datasheet. 2019.

**20** C. Banke, Taking 3.5 GHz CBRS from Spec to Product with Chipset Solutions Spanning Small Cells to Mobile. *Online*, Aug. 2017.

**21** Nokia, Nokia, Alphabet's Access Group and Qualcomm showcase first live demo of a private LTE network over CBRS shared spectrum providing a 360 race car experience. *Online*, Feb. 2017.

**22** M. Allevan, Nokia, Alphabet, Qualcomm take CBRS to the Racetrack. *Online*, Feb. 2017.

**23** Y. Li and S. Xu, Traffic Offloading in Unlicensed Spectrum for 5G Cellular Network: A Two-Layer Game Approach. *Entropy*, vol. 20, no. 2, p. 88, 2018.

**24** R. Schwartz, Business Case For MulteFire Technology. *Online*, pp. 48–57, Feb. 2018.

**25** MulteFire Alliance, Release 1.0 Technical Paper. Technical Report, MulteFire Alliance, Aug. 2018.

**26** K. Kannan, Nokia and STC conduct test of MulteFire technology to bring LTE-like performance to Wi-Fi. *Online*, May 2016.

**27** MobileEurope, STC holds first live MulteFire trial with Nokia. *Online*, May 2016.

**28** GSA, LTE in Unlicensed and Shared Spectrum: Trials, Deployments and Devices. Technical Report, Global Mobile Suppliers Association, 2019.

**29** Qualcomm, World's first LTE Licensed-Assisted Access over-the-air Trial. *Online*, Feb. 2016.

**30** Qualcomm, Snapdragon X16 Modem Datasheet. 2019.

**31** MobileEurope, Telefonica holds LTE-U trial with Ericsson. *Online*, March 2016.

**32** Ericsson, Ericsson and MTN first LTE-U trial in Africa. *Online*, April 2016.

**33** H. L. Maattanen, G. Masini, M. Bergstrom, A. Ratilainen, and T. Dudda, LTE-WLAN aggregation (LWA) in 3GPP Release 13 & Release 14. In *2017 IEEE Conference on Standards for Communications and Networking, CSCN 2017*, pp. 220–226, IEEE, Sept. 2017.

**34** S. Sirotkin, LTE-WLAN Aggregation (LWA): Benefits and Deployment Considerations. Technical Report, Intel Corporation, 2017.

# 5

## Spectrum Above Radio Bands

*Abhishek K. Gupta and Adrish Banerjee*

Indian Institute of Technology (IIT), Kanpur, India

Due to the scarcity of available spectrum at traditional cellular frequencies (often known as sub-6 GHz), the use of higher frequency bands has been proposed for upcoming generations of cellular networks. These bands are collectively known as "above-6 GHz bands" and include 24–30 GHz bands and millimeter-wave (mmWave) (30–300 GHz) bands. Although mmWave bands have been used in the past for various non-communication and non-commercial communication applications, recently using mmWave as cellular and commercial communication applications has gained growing interest and push. In this chapter, we will talk about different subbands in the above-6 GHz band that have been identified for communication. It is worth noting that the communication in these bands is significantly different to communication in traditional sub-6 GHz bands owing to their sensitivity with blockages and highly directional antennas. These differences also indicate that spectrum sharing approaches and mechanisms should be different in the context of these bands. Various spectrum license sharing mechanisms such as uncoordinated, static and dynamic sharing can be seen as viable options in these bands. In comparison to the exclusive licenses, shared licenses and secondary licenses can also be considered via centralized, secondary or third-party markets.

## 5.1 Introduction and Motivation for mmWave

Up to the fourth generation, most commercial communication, including cellular services and Wi-Fi, has been limited to sub-6 GHz bands owing to their favorable channel conditions and ease of commercialization. To increase the data rate in these bands, the current generation has focused on deriving techniques for better utilization of spectrum, higher spectral efficiency, and densification [1]. However, in recent years these techniques have reached a level of maturity leading to saturation of performance. One way to increase the performance manifold, as required in the 5G standards, is to acquire larger bandwidth, which is possible by going up in the spectrum. These proposed bands are collectively known as above-6 GHz bands and include mmWave (24–300 GHz) bands. Note that (24–30 GHz) bands, which are technically centimeter-wave bands, are typically studied with mmWave

*Spectrum Sharing: The Next Frontier in Wireless Networks,* First Edition.
Edited by Constantinos B. Papadias, Tharmalingam Ratnarajah, and Dirk T.M. Slock.
© 2020 John Wiley & Sons Ltd. Published 2020 by John Wiley & Sons Ltd.

bands owing to their similarity in propagation and communication characteristics. Hence, they are included in mmWave bands for notation ease in the literature [2]. In this chapter, we will discuss mmWave bands with a focus on their characteristics and spectrum sharing options.

## 5.2 mmWave Communication: What is Different?

Recent studies have shown that communication in mmWave bands is significantly different to communication in traditional sub-6 GHz bands [3, 4] as highlighted in Figure 5.1. In this section, we will discuss the distinguishing features of communication at mmWave frequencies and their implications from the perspective of spectrum sharing.

### 5.2.1 Distinguishing Features

One of the important features of mmWave communication is its sensitivity to blocking. Measurement studies for mmWave channels [4–6] have shown that blocking affects mmWave communication severely and propagation via line-of-sight (LoS) links is significantly different than that via non-LoS (NLoS) links [7]. mmWave signals can be blocked by buildings, foliage, humans, and even the user's body, and a single blockage can lead to loss of 20–40 dB [6, 7]. It has been shown that the probability of blockage increases exponentially as the distance from the receiver increases [3]. Therefore, access points (APs) which are located at far distances are less likely to interfere, reducing the resulting interference significantly.

A second important feature of mmWave communication is its high directivity. In order to overcome the severe path loss at mmWave frequencies, it is necessary to use a large number of antennas at the transmitter and/or receiver side. Fortunately, it is possible to accommodate a large number of antennas in the same area. This is because antennas at mmWave frequencies are smaller than those at traditional frequencies due to the smaller wavelength. The use of a large number of antennas results in highly directional communication. High beam-forming gain with small beam width increases the signal strength of the serving links and reduces the average interference at receivers.

The amount of available spectrum at mmWave frequencies is very large when compared to sub-6 GHz frequencies ($\sim$ 50–100 times). As the bandwidth appears in the pre-log factor of the achievable data-rate, mmWave communication can potentially give an order of magnitude higher data rates, which is the prime reason for the popularity of mmWave for cellular and other wireless applications.

### 5.2.2 Implications

As discussed in the previous section, the inter-cell interference in mmWave communication is significantly less when compared to that in sub-6 GHz frequencies. For a typical deployment of 30 base stations per km$^2$, mmWave systems are noise-limited [3, 8] which provides transmission opportunities to other transmitters, including other operators and services. Due to high blockage losses, the range of mmWave communication is expected to be small,

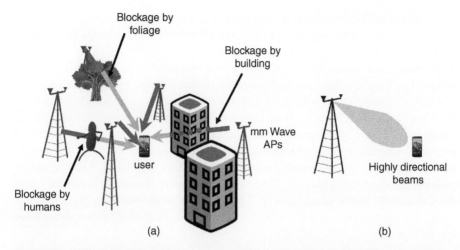

**Figure 5.1** Two distinguishing features of mmWave frequencies. (a) mmWave communication is severely impacted by blockages, including blockages by buildings, foliage, and human bodies. (b) mmWave communication is highly directional owing to the use of large antenna arrays.

with cell radius in the order of 50–150 m [3, 9]. This localization of deployment opens up the possibility of less complicated and distributed inter-cell coordination. On the one hand, directional beams help increase the serving signal strength and reduce the interference. On the other hand, they make initial access, including cell discovery, very difficult. Similarly, dynamic blocking [10, 11] caused by humans, moving vehicles, and self-body blockages can impact the reliability of connections. To tackle these problems, macro-diversity in the form of either using multiple mmWave base stations (BSs) [12] or using macro-BSs coexisting with mmWave base stations (BSs) [9] can be leveraged. It is expected that mmWave networks will be deployed with a macro-BS network to provide a reliable control channel and fail-safe mechanisms resulting in dual-band operations [7]. These differences indicate that spectrum sharing approaches and mechanisms should be different in mmWave bands from those suitable for traditional frequency bands.

### 5.2.3 Opportunity and Need for Sharing

Most spectrum bands require the operator/customer to buy a license from a competent authority, e.g., the Federal Communications Commission (FCC) in the USA, the Offce of Communications (Ofcom) in the UK or the Telecom Regulatory Authority of India (TRAI) in India. Licenses may be awarded for a specific band and may be limited to a particular geographical area or time. In conventional licensing for cellular networks, commercial operators buy exclusive licenses to get an exclusive and complete control over a spectrum band and restrict others to use or interfere in these bands. Exclusive licenses improve the reliability and quality of service (QoS) of the providers.

There are some drawbacks associated with exclusive licensing. For example, it may result in the under-utilization of the spectrum at specific locations or time. Even in conventional bands, it has been reported [13] that the spectrum remains highly under-utilized when exclusive licensing is used. Since license costs are extremely high, exclusive licensing schemes lead to a high entry barrier for new service providers, thus reducing competition and innovations. High licensing cost also promotes higher utilization of the spectrum to maximize the revenue of an operator.

In spite of the above drawbacks, most of the traditional sub-6 GHz bands have followed an exclusive licensing approach where licenses are allocated based on a spectrum auction, with the prime reason being the high level of interference at these frequencies and large overhead requirement for coordination. However, this observation may not be true for operations in mmWave bands owing to the reduced level of interference. As discussed above, mmWave communication causes less interference to neighboring BSs compared to sub-6 GHz frequencies [3]. Preliminary work has shown that even without any coordination, sharing the spectrum is beneficial compared to exclusive licenses in mmWave under specific scenarios from the perspective of the median rate achievable by users [8]. It is therefore expected that conventional exclusive licensing may not be efficient for mmWave frequencies.

Along with the opportunity of sharing, the efficient utilization of spectrum is always desired [14, 15], even for mmWave bands. This can help in meeting the increasing demand for data, making ultra-latency applications (including time-critical medical applications) a reality [16], promoting market competition and bringing benefits to both operators and users. We should remember that there are many incumbent services already present in these bands (which will be discussed in the next section). Any communication service allowed in these bands has to share the spectrum with the incumbent services. The above discussion makes it clear that there is a certain need and opportunity for spectrum sharing at mmWave frequencies and spectrum sharing is a potential way forward for mmWave systems.

## 5.3   Bands in Above-6 GHz Spectrum

Since different operators operating in the same band cause interference to each other, the transmission of data using radio waves is strictly regulated by national laws. Wireless technologies, devices architecture, target applications, and physical layer transmission techniques depend on the identified radio spectrum band. To homogenize devices across the globe, avoid interference across borders in neighboring regions or countries, and facilitate global use of any radio spectrum, different countries need to come to an agreement. To coordinate such agreements, an international body, the International Telecommunications Union (ITU), came into existence. To manage the global radio spectrum, ITU has split the world into three ITU regions:

- Region 1 covers Europe, including the UK, Russia, and African countries
- Region 2 covers America, including the USA and the Pacific islands
- Region 3 covers the Oceanian countries, including Australia, and Asian countries, including China, India, Korea, and Japan.

Along with the ITU, there are other groups and efforts such as the Radio Spectrum Policy Group (RSPG), the European Conference of Postal and Telecommunication Administration (CEPT), the World Radiocommunication Conferences (WRC), the Asia Pacific Telecommunity (APT), the Inter-American Telecommunication Commission (CITEL), and the Regional Commonwealth in the Field of Communications (RCC). These cooperate with the national regulatory bodies (such as the FCC and Ofcom) which facilitate discussions in order to arrive at a common agreement regarding spectrum usage. A number of bands in the mmWave regime are proposed for further study or use (most of these are agreed in Resolution 238 of WRC-15, which was held in 2015). The selection of these bands was dependent on many factors, including channel propagation characteristics, incumbent services, global agreement, and availability of contagious bandwidth. A description of these target mmWave bands follows.

### 5.3.1　26-GHz band: 24.25–27.5 GHz

The 24.25–27.5-GHz band, collectively known as the 26 GHz band, is identified as the pioneer band across Europe to give ultrahigh capacity for innovative new services within next-generation wireless standards [15, 17]. In ITU Region 1, this band is occupied by earth exploration satellites and space research expeditions, inter-satellites, fixed services (fixed services refer to point-to-point wireless links between specific geographic locations for various applications, including backhaul, TV broadcast distribution etc.) and fixed satellite Earth-to-space services. In the UK, this band is further divided into two parts: (i) the upper 26-GHz band (26.5–27.5 GHz), which is lightly used by the UK Ministry of Defence, and (ii) the lower 26-GHz band (24.25–26.5 GHz), which is managed by Ofcom and is currently used by fixed link services, satellite Earth station services (receiving links of Earth data, including imagery and weather data), and short-range devices [15]. In India and China, this band is being studied for future spectrum needs. However, the FCC has shown less interest in this band for mobile allocation in the USA [18].

### 5.3.2　28-GHz band: 27.5–29.5 GHz

The 27.5–29.5 GHz band has been identified as a potential band in the USA (27.5–28.35 GHz), Korea (26.5–29.5 GHz), and Japan (27.5–28.28 GHz). In the USA, the 27.5–28.35-GHz band is also known as the local multipoint distribution service (LMDS) band. There are no primary federal allocations in this band and there is a secondary allocation for an Earth-to-space fixed satellite services (FSS). The FCC has proposed allowing mobile communication in this band [18]. In Europe, this band is split into several sub-bands that allow either exclusive satellite or terrestrial communications [19].

### 5.3.3　32-GHz band: 31.8–33.4 GHz

The 32-GHz (31.8–33.4 GHz) band has been highlighted as a promising band across Europe as per recommendation of both RSPG and CEPT. In the USA, the FCC found this band challenging for mobile use. There are a few federal and non-federal allocations, a co-primary allocation, and an inter-satellite service (ISS) allocation in this band in the USA [18].

### 5.3.4　40-GHz band: 37–43.5 GHz

The 40-GHz (37–43.5 GHz) band is identified for priority study and seen as a potential band for harmonization of equipment.

#### 5.3.4.1  40-GHz lower band

The 37–38.6- and 38.6–40-GHz bands have been identified in the USA as potential bands. Mobile operations are proposed in the 38.6–40-GHz band [18]. The 37.5–40.5-GHz band is generally allocated to fixed, mobile, and fixed satellite (space-to-Earth), Earth exploration satellite (space-to-Earth and Earth-to-space), space research (space-to-Earth and Earth-to-space) and mobile satellite (space-to-Earth) services. In the USA, there are co-primary non-Federal FSS (space-to-Earth) allocations in the 37–38.6-GHz band and no Federal allocations in the 38.6–39.5-GHz band, but FSS (space-to-Earth) and mobile satellite services (space-to-Earth) are allocated in the 39.5–40-GHz band [18].

#### 5.3.4.2  40-GHz upper band

The 40-GHz upper (40.5–43.5 GHz) band has been highlighted as promising across Europe as per the recommendation of both RSPG and CEPT. Compared to the 32-GHz band, the UK believes this band to be a better candidate for global harmonization owing to its lightly loaded incumbent service status [15]. 40.5–42.5 GHz is generally used by fixed, fixed-satellite (space-to-Earth), broadcasting, and broadcasting satellite services. In ITU Region 2, 40.5–41 GHz is allocated to mobile satellite services. 42.5–43.5 GHz is allocated to fixed, fixed satellite (Earth-to-space), mobile services, and radio astronomy.

### 5.3.5  64–71-GHz band

WRC-15 has identified 66–71 GHz as a potential band for upcoming generations of mobile standards agreed by the UK and Europe. In the USA, the 64–71-GHz band is proposed to be used with unlicensed status. There are co-primary mobile allocations and ISS link authorizations on these bands as incumbent services in the USA [18].

## 5.4  Spectrum Sharing over mmWave Bands

Given the vast spectrum available in mmWave bands, reduced interference, and the plethora of spatial and time gaps in the spectrum, there is an ongoing debate on whether to allow spectrum sharing in the mmWave spectrum or go for exclusive licenses for commercial and non-commercial services. Since various bands available in the mmWave range may differ in their propagation characteristics, usage, and incumbent services, the answer 'to share or not to share' is not straightforward as it depends on many factors and deployment scenarios [20]. In this section, we will discuss factors which play a role in determining the answer to this question, while keeping mmWave in mind.

### 5.4.1  Factors Determining Sharing vs No Sharing

A wireless system's performance is primarily measured in terms of achievable data rate, which is linked to two important metrics – available spectrum and the signal-to-interference-plus-noise ratio (SINR) – via the capacity expression of the system. When two operators decide to share spectrum, available spectrum adds up, and it can naively be assumed that data rate will also increase. However, there are two negative consequences arising from spectrum sharing: (i) sharing may require some overhead, which eats away a part of the available spectrum, and (ii) the transmitters of the other operator cause additional interference, which degrades the SINR. These factors together determine whether or not spectrum sharing is beneficial for a particular network. The extent of the impact of these ramifications on a mmWave system depends on various characteristics of the system as discussed below.

### 5.4.1.1 Directionality

As discussed above, mmWave beams are narrow owing to the use of a large number of antennas [21]. This has two primary advantages: (i) the signal strength is boosted by the antenna gain as beams are steered to point towards desired users and (ii) for a typical receiver, there are fewer interferers which have their main lobes pointed towards the receiver (these interferers are termed aligned interferers) and the rest of the interferers have their side lobes (or perhaps nulls) pointing towards the receiver (these interferers are termed unaligned interferers), which reduces the average interference at this receiver. It has been shown that the beam width plays a crucial role in determining the feasibility of spectrum sharing [8, 21]. Narrowing antenna beams reduces the probability of falling into the main lobe of an interfering beam and thus is an effective way to boost the signal power without increasing interference. On the other hand, increasing the width of transmission beams can increase the inter-operator interference so severely that the gain from spectrum sharing may vanish or become negative. This is the main reason why uncoordinated spectrum sharing is not feasible in traditional frequencies. However, with a reasonable number of antennas at the transmitter (and/or at the receiver side), spectrum sharing is feasible even without coordination for mmWave communication [8]. Having said that, we should remember that the above result relies on the assumption of robust and correct beamforming towards users. If, for example, the beams point in a wrong direction, a severe loss in the strength of the serving AP's signal and/or a drastic increase in the sum interference may occur (perhaps even more than sub-6 GHz communication). Hence, the whole success relies on the ability to perform successful beamforming, which could be challenging in some scenarios, including highly dynamic environments.

### 5.4.1.2 Deployment and Blockage Density

Network deployment density plays a crucial role in determining whether or not a particular mmWave network can support spectrum sharing. Spectrum sharing with another operator effectively increases the aggregate BS density. Increasing BS density can increase the interference in the following two ways:

1. Increasing the BS density of an operator increases the serving power as well as the sum interference for users of this operator. For systems operating at traditional frequencies with single slope path loss, these two effects cancel each other out and hence the signal-to-interference ratio (SIR) is invariant with the BS density [22]. Now consider a case with multiple operators operating at the same band with closed connection-access (i.e. a user subscribed to the $i$th operator can connect to the BSs of the $i$th operator only). For users subscribed to an operator, increasing BS densities of other operators increases the sum interference with no impact on the serving power. Hence, increasing BS density will increase the interference to the users of other operators.

2. As explained above, for systems operating at traditional frequencies with single slope path loss, the SIR is invariant with the BS density. However, systems with multi-slope path loss functions are not SIR invariant with the BS density [23], which is the case with mmWave systems [7]. In mmWave systems, increasing BS density reduces the SINR of a typical user, even when no spectrum sharing is implemented.

Hence, an increase in the BS density may degrade the system's performance significantly and can even turn a noise-limited system to an interference limited system. With a practical density of 30 BS/km$^2$ (this refers to a cell size of ~100 m), mmWave networks are very much noise limited. If the effective aggregate density of all sharing operators remains within critical deployment density, spectrum sharing will work, even without any inter-operator coordination. By critical density, we mean the BS density at which mmWave systems will start becoming interference limited.

The critical density here also depends on the blockage density. In cities with higher density of blockages, including buildings, people, and foliage, a mmWave signal can be blocked with a higher probability [3]. In such environments even a denser deployment will not have significant interference and spectrum sharing will give performance gains, provided that a reasonable probability of finding a LoS-serving BS connection is maintained.

### 5.4.1.3 Traffic Characteristics

Traffic characteristics such as traffic load, time profile, and burstiness can also affect the rate gains of spectrum sharing in mmWave bands. Traffic load refers to the average load (in terms of requested data or total number of users) per BS. Spectrum sharing is shown to be effective when traffic is light, asymmetric, and inhomogeneous in time (assuming opportunistic transmission can be implemented) [24]. For example, spectrum sharing techniques such as sensing-based methods work best when there are frequent spectrum gaps. A bursty traffic consisting of large bursts and long idling can provide such conditions. In scenarios where traffic is almost constant over space and time, a static partitioning of the spectrum may be more efficient and convenient.

### 5.4.1.4 Amount of Sharing

Keeping in mind that spectrum sharing can improve per-user peak rates at the cost of degrading the edge users' performance and the system's QoS, it is natural to seek midway solutions. One way is to go for partial sharing of spectrum where licensees or operators keep some amount of spectrum exclusively for themselves and share the rest of it. It is also possible to have exclusive and shared spectrum at different frequency bands, for example the former at 28 GHz and the latter at 70 GHz.

### 5.4.1.5 Inter-operator Coordination

Inter-operator coordination refers to a variety of techniques, including centralized and distributed schemes, and dynamic, static, and semi-static approaches to reduce transmission conflicts among operators and hence inter-operator interference. Inter-operator coordination can avoid cases where there is significant interference due to other operators and help improve edge rate of users. The extent of coordination possible among operators may affect the gains from spectrum sharing.

Inter-operator coordination requires some amount of information to be exchanged among operators. Such information exchange (termed feedback overhead), if done via wireless links, occupies some part of the available spectrum. Feedback overhead increases with the extent of coordination and thus limits the gains that can be achieved by coordination. On the other hand, information exchange, if done over a fiber, increases the cost due to additional infrastructure and operating expenses. It is expected that there should be an optimal amount of coordination which can achieve the best trade-off between edge rate and median data rate. We should recall that operators typically dislike a high level of coordination, not just due to the increase in overheads, but also from a financial risk and benefit perspective.

### 5.4.1.6 Sharing of Other Resources

With ultra densification of networks, increasing density of users, growing market fluidity, and dynamic requirements, service providers are opting to share more resources day by day. Infrastructure sharing is a reality in modern wireless systems, and a continuous progression can be observed towards sharing network infrastructure such as tower sites, AP services, backhaul, and cell towers to reduce operating costs [25]. Infrastructure sharing can be either passive or active.

In passive sharing, operators choose to share the sites of APs. This is a common practice where sites are costly and owned by a third party and multiple operators just lease these sites. When APs of different operators are co-located at the same site, the interference power from each AP is equal to the serving signal power on average, resulting in the average SINR being below zero. In such cases, sharing the spectrum is not feasible in general, but in mmWave systems spectrum sharing may still be feasible while sharing the infrastructure owing to high directionality of beams [8].

In active sharing, operators choose to share the radio access network (RAN) and network core, including the connection-access to their APs. Connection-access denotes whether or not a subscriber is allowed to connect to APs of other operators. Closed connection-access refers to the case where a user can connect to APs of their own service providers. When operators with closed connection-access choose to share the spectrum, the APs of other operators will interfere, including the ones which are closer than the serving AP of the user; this can degrade the SINR severely (below 0 on average). Open connection-access refers to the case where a user can connect to the AP of any service provider. When operators with open connection-access choose to share the spectrum, interfering APs of all operators are located at a larger distance from the user than the serving AP. This creates an exclusion region around the user and ensures a decent SINR, promoting sharing [8]. Spectrum sharing with open connection-access is equivalent to a single operator system with aggregate BS density, aggregate user equipment (UE) density, and aggregate bandwidth. Due to the degradation in SINR with BS density (remember that mmWave communication has dual-slope path loss, i.e. different pathloss coefficients for LoS and NLoS links), the gain is not exactly linear but slightly less, depending on the deployment density [8]. In practical systems, it may happen that a few operators agree to allow open connection-access and some may not. In these cases, a clear scenario-specific evaluation is needed to decide whether or not to go with spectrum sharing.

### 5.4.1.7 Multi-user Communication

The use of multi-user multiple input multiple output (MIMO) techniques can increase the system throughput via spatial multiplexing, but it also increases the effective beam width of interfering BSs. Since directionality plays an important role in determining the feasibility of spectrum sharing, the use of multi-user transmission techniques impacts the decision about spectrum sharing for mmWave communication.

### 5.4.1.8 Technical vs Financial Gains

The pure technological gains offered by spectrum sharing may not be directly proportional to the net utility/revenue of the operators and users. The coverage and network services provided by cellular operators to users are considered network-based goods. When there is a network-based good, there is an extra buying incentive that is proportional to the network size for users [26]. The financial study presented in [21] shows a revenue-pricing model for both operators in the presence of a central licensing authority. In [27], it was shown that in the case of asymmetrical operators, spectrum sharing can provide technical gains to an operator but may make that operator lose its customer base, resulting in a net loss in its revenue. The interplay of technical gains, network economics, and market dynamics can lead to unexpected conclusions. Hence, it is required to include financial aspects when determining whether or not spectrum sharing is beneficial for an operator.

## 5.5 Spectrum Sharing Options for mmWave Bands

Since communication at mmWave frequencies is not a new concept, there are already many incumbent operations in these bands. These include military and research activities, fixed services, satellite to Earth communication, and unlicensed operations, as discussed previously. Hence, there is a certain need and opportunity to share mmWave spectrum among operators and service providers. Although regulatory rules for cellular services are not yet fixed for mmWave frequencies, various spectrum license sharing mechanisms such as uncoordinated, static, and dynamic sharing can be seen as viable options in these bands. In this section we will discuss multiple potential sharing options that can be considered for service providers operating at mmWave frequencies (see Figure 5.2, which summarizes these sharing options).

### 5.5.1 Exclusive Licensing

An exclusive license entitles its owner to complete control over a band of spectrum and can ensure convenient hassle-free operation with a desirable QoS. As we are heading towards more commercial deployments of mmWave cellular, it is evident that there may be a need to allow exclusive use of a spectrum band to ensure reliability, allow low latency connections and provide performance guarantees to time-critical operations. Due to the localized nature of interference in mmWave systems, it is possible to allow exclusive licensing with area specific licenses. This approach can be optimal for locations with high traffic demands such as railway or bus stations, downtown centers, and stadiums [15]. In the USA, for the 37–38.6-GHz band, the FCC has proposed implementing a hybrid licensing scheme that

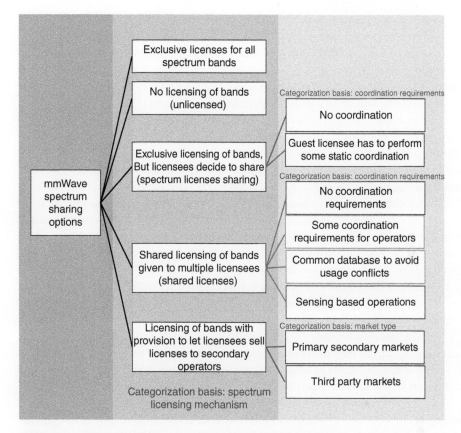

**Figure 5.2** Potential spectrum sharing options for mmWave bands.

would provide operating rights to property owners, while allowing geographic area licenses based on counties for outdoor use. Similarly, for the 27.5–28.35-GHz band, it has been proposed to allow mobile operations with county-sized geographic area exclusive licenses [18].

## 5.5.2 Unlicensed Spectrum

Unlicensed spectrum refers to a part of the spectrum that does not require operators/users to acquire any operating license from the central authority. One example of spectrum bands that are suited for unlicensed status are those that have spatially or temporally rare transmissions leading to negligible transmission conflicts among users and low-power/short-range communication. The prime purpose is to reduce the entry barrier for new or small-scale operators and to increase spectrum utilization [28]. Two of the most popular unlicensed bands at sub-sub-6 GHz frequencies are the 2.4-GHz industrial, scientific and medical (ISM) and the 5-GHz unlicensed national information infrastructure (U-NII) bands, which are well known for Wi-Fi operation. Although it has the advantages of small operating cost and ease of access, the unlicensed spectrum suffers from unpredictable interference and terrible QoS. The unlicensed spectrum exhibits good neighbor behavior which means that transmission politeness can help improve everyone's performance. Such transmission politeness can be achieved via various coordination techniques, including listen before talk (LBT), which can be implemented using cognitive spectrum sensing techniques [29–33]. One adverse consequence is the need to scan the shared medium for transmission activity from others before every transmission. One example of unlicensed access protocol proposed for LTE systems is MulteFire [34]. Multe-Fire is similar to the long-term evolution licensed assisted access (LTE-LAA) protocol and uses similar concepts to coexist with other concurrent services.

As discussed in previous sections, mmWave bands are naturally suitable for spectrum sharing and unlicensed operation can very well work in these bands. Bands above 60 GHz (specifically 59–64 GHz and 64–71 GHz) have been identified for unlicensed spectrum operation. For example, Ofcom has proposed the 66–71-GHz band to be a license-exempt band [15] whereas he FCC has proposed the 64–71-GHz band to be a license-exempt band [18]. Two of the commercial standard technologies operating over these unlicensed bands are Wireless Gigabits Alliance (WiGig) [35] and WirelessHD [36]. The WiGig standard (also known as IEEE 802.11ad) is very similar to the Wi-Fi standard and utilizes directional beams to reduce interference and improve QoS. Unlicensed mmWave spectrum is also suited to indoor deployments and systems not requiring certain QoS due to the reduced risk of inter-cell interference and insensitivity to delays for these systems [15].

### 5.5.2.1 Hybrid Spectrum Access

As unlicensed spectrum may suffer from poor QoS, there may be cases where an operator wants to own a licensed band in order to ensure reliability and a minimum level of QoS, and still wants to use an unlicensed band to increase its data rate in the favorable conditions of this spectrum. Such operation over bands consisting of at least one unlicensed and one licensed band is termed *hybrid spectrum access*. LTE-LAA, (LTE wireless-LAN aggregation (LWA) and LTE wireless-LAN aggregation using IPsec tunnel (LWIP) are key examples of hybrid spectrum access schemes in traditional frequencies. These schemes come under the umbrella name LTE-unlicensed (LTE-U) [37, 38]. For example, LTE-LAA works on the concept of carrier aggregation where licensed carrier frequencies are aggregated with the unlicensed spectrum. There is another use-case of these schemes where licensed frequencies are used for control signals and data symbols can be transmitted over either licensed or unlicensed spectrum depending on channel conditions.

Although the schemes discussed above are designed for traditional frequencies, the underlying spectrum access concepts can be extended to mmWave frequencies as well and may act as a precursor to those employed in mmWave bands. In [39], a hybrid spectrum access scheme is proposed where a mmWave band (at 28 GHz) with exclusive access and a mmWave band (at 73 GHz) with unlicensed operation (which can also be just a spectrum pool between multiple operators) are used simultaneously. The authors have compared three sharing mechanisms: (i) fully licensed when both bands have exclusive licensing, (ii) fully pooled when both bands have unlicensed operations, and (iii) the hybrid spectrum access scheme. The authors have shown the trade-off between peak, median, and edge rate by changing sharing mechanisms for the two bands, as summarized in Figure 5.3. We can observe that the achievable maximum rate is highest in the fully pooled case and lowest in the fully licensed case. The maximum data rate achievable with hybrid licensing is between the two cases. On the other hand, from the perspective of edge rates, the fully

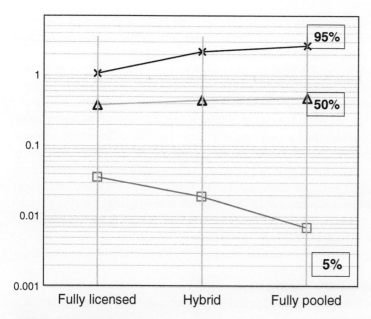

**Figure 5.3** Comparison of a hybrid spectrum access scheme with a fully licensed and fully pooled case in terms of edge rate (5% percentile), median rate (50%), and peak rate (95%) (in Gbps). This result is taken from [39]. BS density is 30 BS/km$^2$ and there are four cellular operators, each having 250 MHz spectrum at both bands.

licensed case is preferred whereas the fully pooled case performs worst, with the hybrid spectrum access scheme again performing in between the two [39].

### 5.5.3 Spectrum License Sharing

Spectrum license sharing refers to the scenario where there are multiple operators each owning an exclusive license of a different spectrum band and they decide mutually to share their licenses among each other with or without some predefined rules. Based on these rules, we can categorize the sharing of spectrum licenses further into different options.

#### 5.5.3.1 Uncoordinated Sharing of Spectrum Licenses

In uncoordinated sharing of spectrum licenses (USSL), each operator is allowed to use the other bands without any restrictions. Although it may seem that this kind of uncoordinated sharing will lead to severe interference and almost zero data rate, as expected in the sub-6-GHz bands, this is not the case with mmWave bands. In [8], the authors considered a system with coexisting mmWave cellular operators with uncoordinated spectrum license sharing. The considered setup was general enough to allow any arbitrary group granularity for license sharing and infrastructure access. By comparing license sharing with no sharing of licenses, it was shown that sharing increases the median rate available to each user. The work identified that the key reason for this feasibility of sharing is highly directional beams (5–25°). These authors also showed that there is a possibility that sharing licenses may hurt per-user edge rate in some scenarios, including dense deployments, therefore it may not provide any QoS guarantee. Further, [20] showed that uncoordinated sharing of licenses without any coordination performs better at higher mmWave frequencies (such as 70 GHz) compared to lower mmWave frequencies (such as 28 GHz).

Another important factor which degrades the gain from spectrum sharing in traditional frequencies is the possibility of infrastructure sharing among operators, for example having towers at the same location. Such infrastructure sharing helps by increasing coverage and capacity while reducing capital and operational expenditures. Owing to the highly dense deployment of mmWave BSs, the operator and site owners may be different and sites can just be leased by operators. This often leads to the co-location (at least partial) of BSs of different operators. In [8, 25], it is shown that infrastructure sharing doesn't significantly affect the gain provided by spectrum sharing and it is feasible to share the spectrum licenses and cell towers simultaneously, even without any coordination. In a similar work [40], various resource sharing methods were studied to show that a full spectrum and infrastructure sharing configuration provides significant advantages, even for the scenario in which no complicated signaling protocols are used to facilitate inter-operator information exchanges.

### 5.5.3.2  Restricted Sharing of Spectrum Licenses

In [8] a trade-off between median and edge rate is presented and it is shown that there is a possibility that sharing licenses may hurt per-user edge rate in some scenarios, including dense deployments. Figure 5.4 compares the per-user edge, median, and peak rate achieved by the exclusive licensing [8]. It can be observed that the per-user median rate and 95% percentile rate improve when spectrum is shared even without coordination. However, the per-user edge rate decreases in comparison to the exclusive licensing. Boccardi et al. [20] studied the impact of coordination and inaccurate beam-forming, and concluded that while coordination may not be generally essential for the feasibility of spectrum sharing, it may certainly help in achieving significant performance gain, especially in real-world scenarios. It is also important to remember that an operator willing to share its spectrum or other resources may still want to distinguish itself as the primary owner of the spectrum

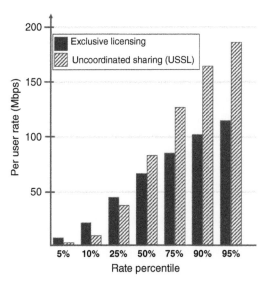

**Figure 5.4** Per-user rate achieved by uncoordinated spectrum sharing vs exclusive licensing [8]. Median rate and 95% percentile rate increases when spectrum is shared. The per-user edge rate degrades with sharing. Number of operators is two, each owning a 100-MHz band at 28 GHz frequency. BSs have directional antennas with 20° beam width while UEs are assumed to have omni-directional antennas. BS density is 30 BS/km² for each operator.

to maintain its brand value and achieve a certain level of quality of service for its own customers. In these scenarios, it is intuitive to think of a licensing scheme where the owner of the spectrum has a higher control of the owned spectrum and can impose some restrictions on the other operators while allowing them to use its spectrum. This agreement is generally mutual and the other participating operators will impose the same restrictions on this operator when it operates on bands owned by them. For a given band, let us call the owner of the band the primary licensee and the other licensees using this spectrum license the secondary licensees. One important thing when imposing such restrictions on the secondary licensee is that coordination requirements that are too strong may eat away the gains from spectrum sharing, possibly for both primary and secondary operators. This makes it important to identify a limit on coordination that should be demanded from the secondary operators. One practical solution is to demand semi-static coordination that is based only on large-scale channel or usage statistics and does not require continuous or dynamic spectrum sensing. Some possible static coordination schemes are as follows.

1. **Interference threshold restricted SSL (ITRSSL):** In this scheme, the renter operator has a restriction on the interference it can cause to the closest AP of the owner operator. Such a system was studied in [21] where it was shown that both operators benefit from restricted licensing. Figure 5.5 compares edge and peak rate trade-off achieved by USSL and ITRSSL. It can be observed that by adjusting the interference threshold carefully, both edge and median rate can be improved with ITRSSL.
2. **Power-restricted SSL:** In this scheme, the renter license has a restriction on the transmit power. A licensee can transmit with full power in the primary band, but with reduced power in the secondary bands.

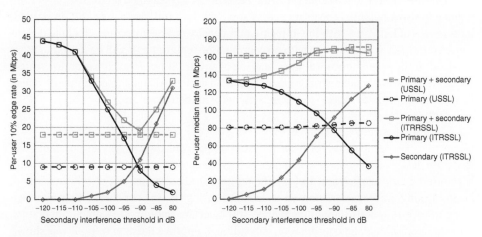

**Figure 5.5** Comparison of trade-off between edge and peak user-rate achieved by USSL and interference threshold restricted sharing of spectrum licenses (ITRSSL). By adjusting the interference threshold carefully, both edge and median rate can be improved with ITRSSL. There are two operators, each having BS density of 30 BS/km$^2$ and user density of 200 UE/km$^2$. BSs have directional antennas with 20° beam width with 20 dB mainlobe-to-sidelobe gain. Other system parameters are the same as used in [8].

3. **Restricted deployment SSL:** In this scheme, a new secondary AP can only be deployed after its impact on all the primary APs (and also secondary APs) has been investigated [15].

In [41], the authors studied uncoordinated *ad hoc* mmWave networks and showed that simple scheduling policies without any coordination can achieve similar performance to more complicated policies requiring full coordination. The authors identified that the main reason behind this conclusion is that strong interference is only present in a scattered fashion which can be handled on demand only. A similar conclusion was also reached in [42].

### 5.5.4 Shared Licenses

Shared licensing refers to the scenario where a license of a particular band is jointly bought by a group of operators primarily to reduce their individual costs and increase spectrum utilization. Such licenses are generally symmetrical, i.e. they impose the same restrictions on every operator and may require some kind of coordination (e.g., static, central, or device-to-device (D2D) communication based, or even uncoordinated). It is assumed that each entity must have a shared license for the complete spectrum. This type of access may not guarantee a fixed amount of spectrum all the time and does not give any exclusive right to any particular participant, but a general level of reliability, QoS, and fairness can be ensured given that only an exclusive group of "friendly" participants is allowed to access the spectrum [43]. Based on the coordination requirements, it can be further categorized as follows.

#### 5.5.4.1 Spectrum Pooling

The spectrum pooling term as defined in [44] refers to the scenario where a group of operators is granted access to the same spectrum band under a predefined set of rules. Spectrum pooling is also a potential spectrum access approach for mmWave bands [20], in particular for cases of networks with low transmission density or low utilization. Technical gains expected from spectrum pooling with no coordination requirements will be similar to those obtained in the case of uncoordinated sharing of spectrum licenses. However, to create a stable system, some more restrictions can be put on each participant (e.g. competitiveness, service rates, customer shares) to ensure symmetry in their revenues.

#### 5.5.4.2 Partial or Fully Coordinated

Given that the participants form an exclusive group and own a symmetric amount of spectrum, it is possible to have a moderate or high level of coordination to help get a higher edge rate for each participant. In [45], the authors studied the impact of cell association, beam-forming, and coordination on spectrum sharing. It was concluded that although using a large number of antennas reduces the need for coordination among operators, some light or on-demand coordination can still help improve the data rates. It was shown that coordination improves edge rates significantly at lower mmWave frequencies, in particular at higher density of BSs (see Figure 5.6). It can also observed that higher mmWave frequencies (e.g., 73 GHz) may not need coordination. There is still a need for extensive simulation-based studies to understand if coordination can give higher data rates under various scenarios.

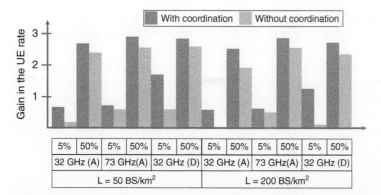

**Figure 5.6** Impact of coordination and optimal association on spectrum sharing for analog (A) and digital (D) beamforming cases [45]. The *y* axis shows the multiplicative gain in the edge (5%) and median (50%) rates offered by spectrum sharing compared to exclusive licenses. Here, BSs have 256 and UEs have 16 antennas. There are four operators with individual density of L BS/km². Coordination improves edge rates significantly at lower mmWave frequencies, in particular at higher density of BSs. On the other hand, higher mmWave frequencies may not need coordination.

### 5.5.4.3 Common Database

Since transmission conflicts between multiple licensees occur sporadically in mmWave systems, these conflicts can be resolved by the use of a common database that keeps track of the transmissions of each licensee [14]. This is a type of semi-static coordination-based approach. Although it doesn't require a continuous sensing of spectrum by each participant, it still creates feedback/transmission overheads and latencies. One example was studied in [46] for mmWave cellular systems with a central database that collects dynamic information about the interference faced by each operator on potential links and then decides which of the links cannot be scheduled simultaneously. Similarly, [47] proposed a new medium access control (MAC) layer that can jointly regulate transmissions among operators for mmWave cellular and back-haul networks.

### 5.5.4.4 Sensing/D2D Communication-based Coordination

Similar to the previous case, transmission conflicts can be resolved in a distributed manner also without requiring a central database. In [46], if an AP senses heavy interference from an AP of a different operator on a potential link it is planning to schedule, it sends a message to the interfering AP with a proposed coordination policy. The two operators can also perform sequential negotiations in order to obtain more efficient scheduling.

### 5.5.5 Secondary Licenses and Markets

Due to the highly dynamic nature of data demands in mmWave systems, it is expected that licensed spectrum may lie idle for some time or at some locations. In these scenarios, it may be beneficial for a licensee to rent its spectrum to small operators for specific locations, a specific time duration or on an opportunistic basis. Such licenses can be termed secondary licenses. As discussed earlier, a spectrum license owner would want to have privileged rights as the primary owner of a spectrum in order to maintain a QoS or get significantly better performance than the renters in order to avoid losing its customer base and brand value. Hence, it is intuitive for owners to impose restrictions on spectrum usage while ensuring a certain level of spectrum usability to secondary licensees.

Such secondary licensing can also be made mandatory to promote an efficient utilization of the spectrum by central licensing authorities. The FCC [14] proposes market-hybrid licensing where primary licensees are required to share the unused portions of geographic areas with others who are interested. Primary owners can sell licenses that allow opportunistic non-interfering use to secondary users in unused geographical areas with the understanding that renters will leave if the owner eventually decides to use that area by deploying its own network. It is also possible to allow consumers to deploy their own APs as secondary users, possibly for indoor coverage to augment the service provided by the licensee on a revenue sharing basis.

### 5.5.5.1 Primary/Secondary Markets

Secondary licenses can be provided by the central authority in a similar fashion as primary licenses are provided. However, it is also possible that primary licensees can directly sell secondary licenses to small operators. This will open up decentralized spectrum markets for secondary licenses. It may be possible to partition the bands according to area and spectrum resources and rent/sell different partitions to different secondary operators [18]. Matinmikko et al. [48] discussed other potential licensing elements on which licenses can be further partitioned into smaller chunks. This leads to *micro-licensing* of the mmWave spectrum. In particular, there is a possibility of including provisions to grant local rights to deploy and operate networks in specific places, including stadiums, downtown areas, academic areas, and malls.

### 5.5.5.2 Third-party Markets

With increasing requirements of primary/secondary markets, it is highly probable that third-party entities will come into existence. These entities can coordinate the selling and buying of secondary licenses or can even act as a middle-entity licensee such that the primary and secondary licensee don't see each other. This can guarantee a consistent revenue to the primary owner regardless of secondary owners and can guarantee a certain QoS to secondary licensees in the case when a particular primary owner doesn't wish to lease its license anymore or is causing significant interference to others.

### 5.5.6 Increasing the utilization of spectrum

Even with the vast spectrum of mmWave, the increasing requirement of data speed and exponential growth of devices necessitate an efficient use of the spectrum. To further increase the spectrum utilization and avoid idling of spectrum, it is possible to frame policies either via a central authority or on an agreement basis to bring more fluidity to the licensing, for example through secondary markets, flexibility to trade licenses, and micro-licensing. The FCC [18] proposes a policy to impose certain usage requirements on each licensee. The usage of each licensee can be monitored and its licenses can be canceled if the assigned spectrum resource is found to be under-utilized. In the case where a long-term primary license is issued, a use-or-share obligation can be enforced which forces the owner to rent its licenses (possibly for specific areas or time) to secondary users [15]. The small coverage area of mmWave makes it an optimal case for licenses based on area-specific restrictions.

## 5.6 Conclusions

Due to the reduced level of inter-operator interference, mmWave systems are naturally suitable for spectrum sharing, even without any coordination or information exchange. Spectrum sharing has been reported to provide significant gains in comparison to the exclusive licensing. Although inter-operator coordination may not be essential for spectrum sharing, some small level of coordination may help improve the per-user edge rate. In this chapter, we have discussed many potential options of spectrum sharing in mmWave bands. Important questions to ask here are how much coordination is required and how much gain can be obtained with sharing, answers to which depend on exact scenario and deployment configuration. More extensive evaluations are needed in order to answer these questions precisely for the various sharing methods discussed.

## References

1 J. Andrews, S. Buzzi, W. Choi, S. Hanly, A. Lozano, A. Soong, and J. Zhang, What will 5G be? *IEEE J. Sel. Areas Commun.*, vol. 32, no. 6, pp. 1065–1082, June 2014.

2 T. S. Rappaport, Y. Xing, G. R. MacCartney, A. F. Molisch, E. Mellios, and J. Zhang, Overview of millimeter wave communications for fifth-generation (5G) wireless networks – with a focus on propagation models. *IEEE Trans. Antennas Propag.*, vol. 65, no. 12, pp. 6213–6230, Dec. 2017.

3 T. Bai and R. W. Heath Jr., Coverage and rate analysis for millimeter wave cellular networks. *IEEE Trans. Wireless Commun.*, vol. 14, no. 2, pp. 1100–1114, Feb. 2015.

4 T. Rappaport, S. Sun, R. Mayzus, H. Zhao, Y. Azar, K. Wang, G. Wong, J. Schulz, M. Samimi, and F. Gutierrez, Millimeter wave mobile communications for 5G cellular: It will work! *IEEE Access*, vol. 1, pp. 335–349, May 2013.

5 A. R. Tharek and J. P. McGeehan, Propagation and bit error rate measurements within buildings in the millimeter wave band about 60 GHz. in *Proc. European Conference on Electrotechnics*, June 1988, pp. 318–321.

6 S. Rangan, T. Rappaport, and E. Erkip, Millimeter-wave cellular wireless networks: Potentials and challenges. *Proc. IEEE*, vol. 102, no. 3, pp. 366–385, Mar. 2014.

7 J. G. Andrews, T. Bai, M. N. Kulkarni, A. Alkhateeb, A. K. Gupta, and R. W. Heath, Modeling and analyzing millimeter wave cellular systems. *IEEE Trans. Commun.*, vol. 65, no. 1, pp. 403–430, Jan. 2017.

8 A. K. Gupta, J. G. Andrews, and R. W. Heath, On the feasibility of sharing spectrum licenses in mmWave cellular systems. *IEEE Trans. Commun.*, vol. 64, no. 9, pp. 3981–3995, Sept 2016.

9 A. Ghosh, T. Thomas, M. Cudak, R. Ratasuk, P. Moorut, F. Vook, T. Rappaport, G. MacCartney, S. Sun, and S. Nie, Millimeter-wave enhanced local area systems: A

high-data-rate approach for future wireless networks. *IEEE J. Sel. Areas Commun.*, vol. 32, no. 6, pp. 1152–1163, June 2014.

**10** I. K. Jain, R. Kumar, and S. Panwar, Can Millimeter Wave Cellular Systems provide High Reliability and Low Latency? An analysis of the impact of Mobile Blockers. *ArXiv preprint ArXiv:1807.04388*, Jul. 2018.

**11** K. Venugopal and R. W. Heath, Millimeter wave networked wearables in dense indoor environments. *IEEE Access*, vol. 4, pp. 1205–1221, 2016.

**12** A. K. Gupta, J. G. Andrews, and R. W. Heath, Macrodiversity in cellular networks with random blockages. *IEEE Trans. Wireless Commun.*, vol. 17, no. 2, pp. 996–1010, 2018.

**13** Federal Communications Commission, Spectrum policy task force. *ET Docket 02-135*, Nov. 2002.

**14** Federal Communications Commission, Notice of inquiry: FCC 14–154. Oct. 2014.

**15** Ofcom, 5G spectrum access at 26 GHz and update on bands above 30 GHz. Available: https://www.ofcom.org.uk/__data/assets/pdf_file/0014/104702/5G-spectrum-access-at-26-GHz.pdf, accessed date July 2017.

**16** R. Ford, M. Zhang, M. Mezzavilla, S. Dutta, S. Rangan, and M. Zorzi, Achieving ultra-low latency in 5G millimeter wave cellular networks. *IEEE Commun. Mag.*, vol. 55, no. 3, pp. 196–203, March 2017.

**17** M. Nekovee and R. Rudd, 5G spectrum sharing. 2017. Available: http://arxiv.org/abs/1708.03772.

**18** Federal Communications Commission, Notice of inquiry: FCC 15–138. Oct. 2015.

**19** Shared Access Terrestrial-Satellite Backhaul Network enabled by Smart Antennas Project, SANSA objectives. Available: https://sansa-h2020.eu/objectives, 2016, accessed 2019-4-1.

**20** F. Boccardi, H. Shokri-Ghadikolaei, G. Fodor, E. Erkip, C. Fischione, M. Kountouris, P. Popovski, and M. Zorzi, Spectrum pooling in mmWave networks: Opportunities, challenges, and enablers. *IEEE Commun. Mag.*, vol. 54, no. 11, pp. 33–39, Nov. 2016.

**21** A. K. Gupta, A. Alkhateeb, J. G. Andrews, and R. W. Heath, Gains of restricted secondary licensing in millimeter wave cellular systems. *IEEE J. Sel. Areas Commun.*, vol. 34, no. 11, pp. 2935–2950, Nov. 2016.

**22** J. G. Andrews, F. Baccelli, and R. K. Ganti, A tractable approach to coverage and rate in cellular networks. *IEEE Trans. Commun.*, vol. 59, no. 11, pp. 3122–3134, Nov. 2011.

**23** J. G. Andrews, X. Zhang, G. D. Durgin, and A. K. Gupta, Are we approaching the fundamental limits of wireless network densification? *IEEE Commun. Mag.*, vol. 54, pp. 184–190, Oct. 2016.

**24** A. K. Gupta, M. N. Kulkarni, E. Visotsky, F. W. Vook, A. Ghosh, J. G. Andrews, and R. W. Heath, Rate analysis and feasibility of dynamic TDD in 5G cellular systems. in *Proc. IEEE International Conference on Communications (ICC)*, May 2016, pp. 1–6.

**25** R. Jurdi, A. K. Gupta, J. G. Andrews, and R. W. Heath, Modeling infrastructure sharing in mmwave networks with shared spectrum licenses. *IEEE Transactions on Cognitive Communications and Networking*, vol. 4, no. 2, pp. 328–343, June 2018.

**26** N. Economides and C. P. Himmelberg, Critical mass and network size with application to the US fax market. NYU Stern School of Business EC-95-11, 1995.

**27** F. Fund, S. Shahsavari, S. S. Panwar, E. Erkip, and S. Rangan, Spectrum and infrastructure sharing in millimeter wave cellular networks: An economic perspective. *arXiv preprint arXiv:1605.04602*, 2016. Available: http://arxiv.org/abs/1605.04602.

**28** 5GPP, 5G empowering vertical industries: Roadmap paper. The 5G infrastructure public private partnership. The 5G infrastructure public private partnership, Brussels, 2016.

**29** S. Haykin, Cognitive radio: Brain-empowered wireless communications. *IEEE J. Sel. Areas Commun.*, vol. 23, no. 2, pp. 201–220, Feb. 2005.

**30** X. Kang, Y.-C. Liang, H. Garg, and L. Zhang, Sensing-based spectrum sharing in cognitive radio networks. *IEEE Trans. Veh. Technol.*, vol. 58, no. 8, pp. 4649–4654, Oct. 2009.

**31** X. Lin, J. Andrews, and A. Ghosh, Spectrum sharing for device-to-device communication in cellular networks. *IEEE Trans. Wireless Commun.*, vol. 13, no. 12, pp. 6727–6740, Dec. 2014.

**32** P. Karunakaran, T. Wagner, A. Scherb, and W. Gerstacker, Sensing for spectrum sharing in cognitive LTE-A cellular networks. in *Proc. IEEE WCNC*, April 2014, pp. 565–570.

**33** C. Galiotto, G. K. Papageorgiou, K. Voulgaris, M. M. Butt, N. Marchetti, and C. B. Papadias, Unlocking the deployment of spectrum sharing with a policy enforcement framework. *IEEE Access*, vol. 6, pp. 11 793–11 803, 2018.

**34** D. Chambers, Multefire lights up the path for universal wireless service. Available: https://www.thinksmallcell.com/send/3-white-papers/72-multefirelights-up-the-path-for-universal-wireless-service.html, May 2016.

**35** T. Nitsche, C. Cordeiro, A. B. Flores, E. W. Knightly, E. Perahia, and J. C. Widmer, IEEE 802.11ad: Directional 60 Ghz communication for multi-Gigabit-per-second Wi-Fi. *IEEE Commun. Mag.*, vol. 52, no. 12, pp. 132–141, Dec. 2014.

**36** WirelessHD. Available: http://www.wirelesshd.org/, 2008, accessed 2018-11-25.

**37** R. Zhang, M. Wang, L. X. Cai, Z. Zheng, X. Shen, and L. Xie, LTE-unlicensed: The future of spectrum aggregation for cellular networks. *IEEE Wireless Commun.*, vol. 22, no. 3, pp. 150–159, June 2015.

**38** Qualcomm, Extending LTE advanced to unlicensed spectrum, Qualcomm, San Diego, CA, USA. Available: https://www.qualcomm.com/media/documents/files/whitepaperextending-lte-advanced-to-unlicensed-spectrum.pdf, Dec. 2013.

**39** M. Rebato, F. Boccardi, M. Mezzavilla, S. Rangan, and M. Zorzi, Hybrid spectrum sharing in mmWave cellular networks. *IEEE Transactions on Cognitive Communications and Networking*, vol. 3, no. 2, pp. 155–168, June 2017.

**40** M. Rebato, M. Mezzavilla, S. Rangan, and M. Zorzi, Resource sharing in 5G mmWave cellular networks. in *Proc. IEEE INFOCOM Millimeter-Wave Netw. Workshop (mmNet)*, April. 2016, pp. 271–276.

**41** H. Shokri-Ghadikolaei and C. Fischione, The transitional behavior of interference in millimeter wave networks and its impact on medium access control. *IEEE Trans. Commun.*, vol. 64, no. 2, pp. 723–740, Feb. 2016.

**42** H. Shokri-Ghadikolaei, C. Fischione, G. Fodor, P. Popovski, and M. Zorzi, Millimeter wave cellular networks: A MAC layer perspective. *IEEE Trans. Commun.*, vol. 63, no. 10, pp. 3437–3458, Oct. 2015.

**43** METIS, Deliverable 5.1, intermediate description of spectrum needs and usage principles. METIS, Boston, MA, USA, Tech. Rep., 2013.

**44** T. A. Weiss and F. K. Jondral, Spectrum pooling: an innovative strategy for the enhancement of spectrum efficiency. *IEEE Commun. Mag.*, vol. 42, no. 3, pp. S8–14, March 2004.

**45** H. Shokri-Ghadikolaei, F. Boccardi, C. Fischione, G. Fodor, and M. Zorzi, Spectrum sharing in mmwave cellular networks via cell association, coordination, and beamforming for spectrum sharing in mmWave cellular networks. *IEEE J. Sel. Areas Commun.*, vol. 34, no. 11, pp. 2902–2917, Nov. 2016.

**46** G. Li, T. Irnich, and C. Shi, Coordination context-based spectrum sharing for 5G millimeter-wave networks. in *Proc. Int. Conf. Cognit. Radio Oriented Wireless Netw. Commun. (CROWNCOM)*, Jun. 2014, pp. 32–38.

**47** Y. Niu, C. Gao, Y. Li, L. Su, D. Jin, and A. V. Vasilakos, Exploiting device-to-device communications in joint scheduling of access and backhaul for mmWave small cells. *IEEE J. Sel. Areas Commun.*, vol. 33, no. 10, pp. 2052–2069, Oct. 2015.

**48** M. Matinmikko, M. Latva-aho, P. Ahokangas, and V. Seppänen, On regulations for 5G: Micro licensing for locally operated networks. *Telecommunications Policy*, vol. 42, no. 8, pp. 622–635, 2018.

# 6

# The Licensed Shared Access Approach

*António J. Morgado*

*Instituto de Telecomunicações, Aveiro, Portugal*

## 6.1 Introduction to Spectrum Management

Worldwide spectrum regulation has been a competency of the International Telecommunications Union (ITU) since 1903. Since then, even though the amount of regulated spectrum has increased significantly, the method used for regulation has not changed.

In the 1927 radio regulations the ITU decreed that several radio services should occupy different bands, separated from each other by the necessary guard bands in order to avoid interference. Should interference be noticed among users of the same service, the station power, antenna, or even frequency should be adjusted and agreed among the parties.

In 1947, the ITU concluded that by proceeding this way many spectrum resources were being wasted. To make more efficient use of the spectrum, adequate channelization schemes were proposed and frequency reuse distances were determined, i.e. several networks offering the same service were allowed to use the same band by sharing frequency and geography domains in a coordinated way. These rules became mandatory in ITU member countries.

As the number of radio services needing spectrum increased, there was a need to allow several services to be deployed in the same band (i.e., to the pioneer services such as maritime, fixed, and broadcasting communications, newer services were added, such as aeronautical, radio navigation, radar, amateur radio, radio astronomy, meteorological aids, mobile, etc.). The 1959 radio regulations therefore defined, for the first time, the concept of primary and secondary services:

- Primary services have the *highest priority* and are protected against interference.
- Secondary services *shall not cause interference to primary services*, cannot claim protection from interference from primary services, but can claim protection from interference from the same or other secondary services.

Abiding by these rules, until a few years ago spectrum management, as performed by the national regulators, consisted of issuing licenses for deploying specific primary and secondary services, using specific frequency bands in an exclusive way.

This explains why the radio spectrum between 9 kHz and 300 GHz is now highly fragmented. Fragmentation results in a significant waste of spectrum, especially in those bands

*Spectrum Sharing: The Next Frontier in Wireless Networks,* First Edition.
Edited by Constantinos B. Papadias, Tharmalingam Ratnarajah, and Dirk T.M. Slock.
© 2020 John Wiley & Sons Ltd. Published 2020 by John Wiley & Sons Ltd.

where the incumbent only needs the spectrum during very limited periods or in very limited areas. If those licensed bands could be shared by other services, particularly those which use spectrum efficiently and which already face spectrum shortages, as in the case of mobile network operators, this would allow additional useful services to be deployed without eliminating any incumbent.

## 6.2 The Dawn of Licensed Shared Access

The concept of sharing the same spectrum in the same geography at different time instants is not new (see, for example, [1]). Yet, as there was no technology to implement it efficiently until recently, this topic was erased from regulators' memory, until now.

The huge success of unlicensed Wi-Fi applications deployed in the 2.4-GHz band since the late 1990s (and later in the 5-GHz band) demonstrated that it was possible to have reliable communications even when reasonable amounts of interference were present. About two decades later, in the late 2010s, technology innovations such as spectrum agility, software-defined radio (SDR), and cognitive radio (CR) started to migrate from the research labs to the infrastructure of modern cellular networks. As the SDR and CR technologies are now mature, regulators decided it was time to look into spectrum sharing.

The first idea was to look for new uses of the TV band which was freed by switching off analogue transmissions. From 2008 to 2012, several countries around the world allowed the use of these so-called TV white spaces (TVWS) (i.e., the emptied TV bands) by unlicensed secondary users. However, as the secondary users are not protected against interference in TVWS, and for this reason it is not possible to guarantee quality of service (QoS) to the secondary users, the wireless industry, specifically the mobile network operators, were not attracted by this solution. So in 2011, two manufacturers proposed [2] a sharing scheme called authorized shared access (ASA) [2]. The main idea behind ASA was that mobile network operators (MNOs) could be allowed to use, on an exclusive basis, the international mobile telecommunications (IMT) bands that were licensed to the incumbents, as long as the MNOs knew when and where the incumbents are not using these bands. As each mobile network uses part of the available spectrum on an exclusive basis, they can benefit from predictable QoS.

In the European Union (EU), the merits of this sharing scheme were also identified by the Radio Spectrum Policy Group (RSPG), which extended the ASA concept to any licensed "secondary" user instead of being only applicable to mobile network operators [3–6] and IMT bands. This extension was renamed licensed shared access (LSA). According to [6], LSA is:

> "A regulatory approach aiming to facilitate the introduction of radiocommunication systems operated by a *limited number* of licensees under an *individual licensing regime* in a frequency band already assigned or expected to be assigned to one or more incumbent users. Under the LSA framework, the additional users are allowed to use the spectrum (or part of the spectrum) in accordance with *sharing rules* included in their rights of use of spectrum, thereby allowing *all the authorized users*, including incumbents, to provide a certain QoS."

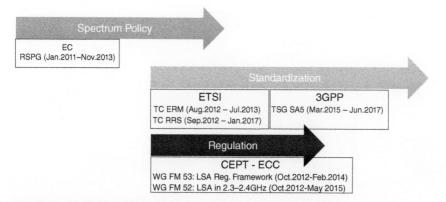

**Figure 6.1**   Roadmap of LSA in the 2300–2400 MHz band in Europe.

Unlike other spectrum sharing approaches, e.g. TVWS, the additional users that are allowed to share the band with the incumbent will be protected against interference, both from the incumbents as well as among themselves. As such, in LSA we may not consider these additional users as secondary users, and for that reason, instead of the usual primary and secondary nomenclature, in LSA we talk about incumbents and LSA licensees.

After the definition of the strategy in terms of spectrum policy, LSA development continued through its standardization and regulation paths in Europe, as depicted in Figure 6.1.

### 6.2.1   The LSA Regulatory Environment

LSA is a licensing approach that foresees the sharing of underutilized licensed bands by a limited number of new licensed users called LSA licensees. The individual rights of use obtained by LSA licensees allows them to request access to this spectrum on an exclusive basis, as long as they comply with the constraints imposed in a sharing agreement previously negotiated with the incumbent.

LSA is seen as a step to introduce new services in bands that cannot be immediately refarmed, since the LSA regulatory framework [6, 7] constrains LSA to vertical sharing situations, i.e. it imposes that the incumbent and the LSA licensees shall deploy *different radio services*. In addition, currently envisioned LSA implementations are restricted to long-term sharing arrangements that can be planned well in advance and which do not change over time.

According to the LSA regulatory framework in Europe [6, 7], LSA requires the involvement of at least three stakeholders:

- the incumbent operator, who has individual rights of use of the band
- the LSA licensee operator, who will agree to the sharing conditions with the incumbent according to which it will be allowed to use the band, with protection against interference originating from the incumbents as well as from other LSA licensees
- the national regulatory agency (NRA), which supervises the negotiation of the sharing agreements between incumbents and prospective LSA licensees, and establishes the licensing process to which prospective LSA licensees may apply.

**Figure 6.2** Baseline LSA architecture [7].

In LSA, the protection of the LSA licensees against interference originating specifically from the incumbents is ensured by the sharing agreement negotiated in advance between these two stakeholders under the supervision of the NRA. This contractual agreement should establish, among other things, the combination of channels, geographical regions, and time intervals where the incumbent promises not to cause excessive interference to the LSA licensees. The sharing agreement should also define the conditions upon which the LSA licenses may be asked to stop using the band due to an unforeseen need of the incumbent to use the band again. It should be stressed that when an NRA supervises the negotiation of the sharing agreement, it should verify if the behavior of the incumbent is predictable, i.e. whether the unforeseen needs to reclaim the use of the band are rare indeed. If this is not the case, the NRA should refuse to license that band under the LSA approach.

Regarding the protection of interference among LSA licensees, this is ensured by individual rights of use of parts of the band, i.e. when the incumbent is not using the spectrum, only one LSA licensee will be allowed to do so.

The LSA regulatory framework also establishes that at least two additional modules have to be added to the networks using bands licensed through the LSA approach, shown in Figure 6.2:

- The LSA repository, a database containing information about which part of the band will be available for use by the LSA licensees, where it will be available, and when it will be available. This database should not be owned by the LSA licensee.
- The LSA controller, who will obtain information from the LSA repository and provide to the LSA licensee information about the available spectrum. The regulatory framework does not impose any restriction regarding the ownership of the LSA controller.

Although the current LSA sharing scheme seems quite simple and easily understandable, it requires the establishment of an appropriate regulatory framework whose main distinctive features are:

- LSA consists of the vertical sharing of bands with low incumbent activity.
- A limited number of new licensed users (LSA licensees) may be allowed by the NRA to use this band when/where the incumbent is not using it.

- LSA licensees use parts of the band *exclusively*. They are protected from interference caused by neighboring LSA licensees and neighboring incumbents so they can benefit from predictable QoS.
- Implementation of LSA requires the introduction of two network modules: the LSA repository and the LSA controller.

### 6.2.2 LSA/ASA in the 2300–2400 MHz band

As mentioned before, the European Commission (EC) did not want to restrict the scope of the LSA approach to mobile operators. Nevertheless, the Commission also recognized that LSA would allow mobile operators to have a quicker access to the additional spectrum needed for deploying mobile broadband services. In other words, the ASA paradigm was unofficially recognized as the first step towards the broader LSA approach.

The 2300–2400 MHz band, which had already been identified [8] as a band where the introduction of mobile broadband was feasible, was considered as a potential ASA/LSA band since it had limited utilization across Europe [9]. Therefore, in April 2014, the EC issued a mandate to the European Conference of Postal and Telecommunications Administrations (CEPT) to develop harmonized technical conditions regarding the introduction of mobile broadband applications in the 2300–2400 MHz band under the LSA approach [10]. The CEPT answers to this mandate were published in November 2014 [11], March 2015 [12], and July 2015 [13], based on previous studies [8, 14, 15] and the LSA regulatory framework [6, 7].

The operational LSA parameters for mobile broadband applications in the 2300–2400 MHz band were first described in ECC Decision(14)02 [15] and in the cross-border coordination procedures recommended in ECC Recommendation (14)04 [14]. These documents defined the following conditions:

- The mobile broadband applications using LSA in the 2300–2400 MHz band should be based on 20 5-MHz blocks and time-division duplex (TDD) mode.
- The emissions should be below a specified block-edge mask, as indicated in Figure 6.3.
- For cross-border coordination between mobile/fixed communications network (MFCN) applications, these applications must not cause signal levels above a pre-defined threshold in the borderline between two neighboring countries. If MFCN is deployed using long-term evolution (LTE) technology, the systems on each side of the border should use different physical cell-identity (PCI) groups. When the base stations cause signal levels above the pre-defined threshold mentioned above, the regulators of the countries involved should agree on the best way to coordinate MFCN operation in the borderline.
- For cross-border coordination between MFCN and other types of application, the regulators of the countries involved should conduct specific studies to agree on the best way to coordinate the operation of those systems.

Meanwhile, within the scope of the EC's standardization mandate to develop technical specifications of reconfigurable radio systems [16], the European Telecommunications Standards Institute (ETSI) specified an LSA reference system [17], LSA system requirements [18], LSA network architecture [19], and LSA protocol for the interface between the LSA repository and the LSA controller [20].

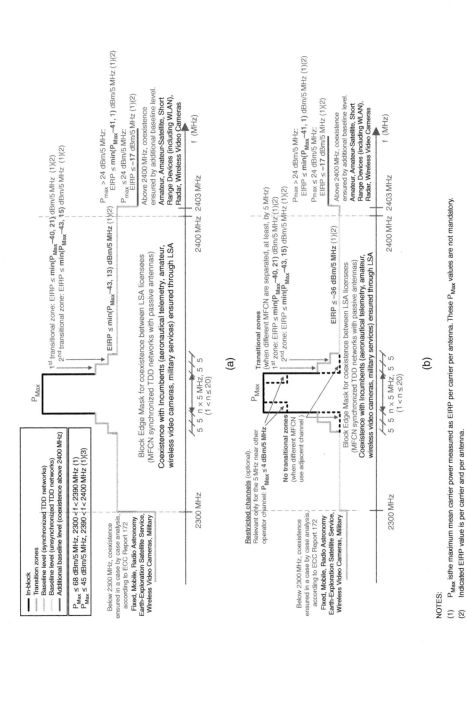

**Figure 6.3** (a) Block-edge mask for synchronized time division LTE base stations operating in the 2300−2400 MHz band. (b) Block-edge mask for unsynchronized time division LTE base stations operating in the 2300−2400 MHz band (based on [15]).

**Figure 6.4** Architecture of a baseline LSA/ASA system operating at 2300–2400 MHz.

It should be stressed that, according to the LSA regulatory framework [6, 7], if ETSI wanted to produce an LSA standard, it would have to be valid for any type of LSA licensee. However, the companies supporting LSA standardization in ETSI had the aspiration that the LSA standard would also be a facilitator of the quick and easy deployment of ASA in the 2300–2400 MHz band using LTE technology. As ETSI assumes that the LSA controller is within the mobile network domain, the standardization of the interface between the LSA controller and the LSA licensee, which was still missing, was out of scope of ETSI and had to be standardized by the 3rd Generation Partnership Project (3GPP) [21–24] (see Figure 6.4).

For completeness, it should also be mentioned that since consideration of LSA in the 2300–2400 MHz band, the adoption of LSA has also been suggested for the 3600–3800 MHz band [25].

## 6.3 An Improved LSA Network Architecture

Although LSA regulation constitutes a step forward to improve the efficiency of spectrum utilization, a weakness of the current LSA standard is that it handles protection from interference in a static way by considering pre-defined situations where the sharing may occur, as well as pre-defined methods to overcome harmful interference situations.

By confining LSA to such a static behavior, many sharing opportunities are being missed. To avoid this, the European Research Project ADEL [26] proposed that LSA support more dynamic sharing situations, where the number of active incumbents and LSA licensees, as well as their operating frequencies, may vary over time and geography, and the 'LSA system' will still react adequately to provide the predictable QoS levels envisioned by the current regulation.

The LSA architecture shown in Figure 6.5 was proposed by ADEL to perform automatic allocation of radio resources in a multitude of dynamic sharing arrangements involving multiple LSA licensees and multiple incumbents in any vertical sharing configuration. To promote a more efficient utilization of the spectrum, the ADEL LSA system contains an

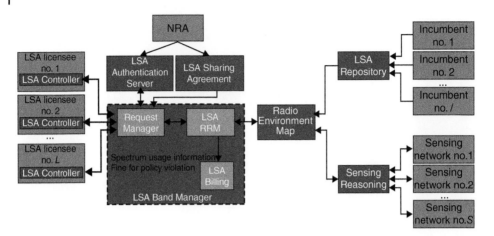

**Figure 6.5** System architecture to support the deployment of LSA in dynamic scenarios [27–30].

LSA band manager, which is a building block responsible for coordinating the access of multiple LSA licensees to the LSA band, thus avoiding the need to have a fixed band plan as presumed in the ETSI standard and in Cognitive Radio trial Environment+ (CORE+) single LSA licensee trials [31].

The basic two-node LSA architecture originally proposed by Qualcomm and Nokia is complemented in ADEL by one or several collaborative spectrum sensing networks to provide information about the radio environment, thus allowing quick detection of changes in the radio environment caused either by the incumbents or the LSA licensees.

The information provided by the LSA repository and the sensing networks is used to compute, and keep updated, a *radio coverage map* that reflects the real environment as accurately as possible. When an LSA licensee requests spectrum, the information contained in this map will be used by the LSA band manager to allocate the most adequate resources (frequency and power) to this specific LSA licensee.

As shown in Figure 6.5, the ADEL architecture also includes modules dealing with authentication, storage of the LSA sharing agreement rules, and spectrum usage accounting.

We have come up with this architecture in order to balance (i) the QoS guarantees offered to incumbent and licensee users, as per the LSA principle, and (ii) better overall spectrum utilization and control, made available through advanced radio resource management (RRM) and sensing reasoning.

The functional modules of the ADEL LSA-based system architecture are described below.

- *LSA repository*: A database that stores (and possibly updates) incumbent-specific information, that is, it stores information about each incumbent's:
  o carrier frequency and bandwidth
  o location and coverage area
  o transmitter hardware characteristics (e.g., maximum transmission power level, antenna height etc.).

In ADEL the modules in charge of processing information are decoupled from the modules responsible for storing information. Thus, unlike ETSI's LSA standard, ADEL does not include in the LSA repository the identification of the spectrum that each LSA licensee might use. In ADEL, this is a task for the LSA band manager.

- *Radio environment map (REM)*: This is a representation of the radio environment that is under the control of the LSA band manager. The information on the radio environment map may originate from (i) propagation calculations performed using terrain databases and inputs from the LSA repository/LSA band manager and (ii) measurements performed by the collaborative spectrum sensing networks. As the main objective of the map is to assist the LSA band manager to perform RRM tasks, the map sends to the manager a subset of the information it contains when requested. The radio map is also responsible for updating itself when necessary.
- *LSA sharing agreement*: This module is a database under the responsibility of the NRA that stores the rules that define the sharing agreement (e.g., LSA band, the radio service of the incumbent/LSA licensee, number of incumbents/LSA licensees, spectrum access type etc.).
- *LSA band manager*: This is the entity that implements the resource allocation procedures and it is key to guaranteeing QoS to all players. It is divided into three different functional sub-modules: the LSA request manager, the LSA RRM, and the LSA billing modules.
  - ○ The *LSA request manager* requests the authentication of the LSA licensee and performs priority management according to the LSA sharing agreement.
  - ○ The *LSA RRM* performs the computation of available resources to assign to the LSA licensee based on the LSA sharing agreement and the radio environment map. After this has been determined, it implements admission control of the LSA licensee spectrum requests. If there are available resources, the LSA RRM decides which ones are the most appropriate to assign to this LSA licensee (e.g., carrier frequency and transmitter power) and sends the information about the selected resources to the LSA licensee in question. The LSA licensee may accept or refuse the assignment. When the LSA licensee accepts the assigned resources, the LSA band manager sends information about the assigned resources to the radio environment map so the latter can update itself. Periodically, the LSA band manager analyses the radio environment map to detect potential policy violations. In such cases, it informs the LSA licensee.
  - ○ The *LSA billing module* is responsible for the financial accounting tasks.
- *LSA authentication server*: This is a module under the responsibility of the NRA that is used to store information and perform tasks related to the authentication of all the functional modules.
- *Spectrum sensing reasoning*: The functions under the responsibility of this module are:
  a) defining the sensing requirements for each sensing network
  b) detecting faulty measurements
  c) computing a sensing map (same format as the REM)
  d) updating the map (i.e., deciding which pixels of REM should be updated with the sensing results)
  e) determining which zones of the map need additional sensing. This module is connected to the REM and also assists the policy protection mechanisms.
- *LSA controller*: This module obtains the spectrum availability information from the LSA band manager and sends it to the network management system of the mobile network, where it is translated into radio transmitter configurations.

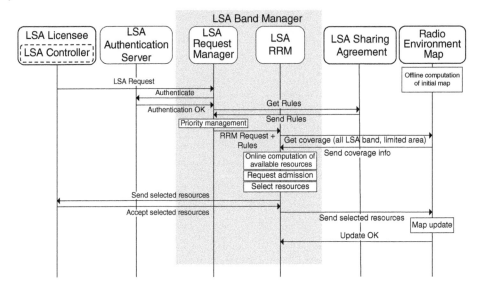

**Figure 6.6** Exchange of messages when an LSA licensee requests spectrum.

It should be stressed that the sensing networks may contain dedicated sensors or reuse the measurement capabilities already installed in 3GPP-compatible base stations and user equipment (UE). The aim of the spectrum sensing networks within the proposed architecture is four-fold: (i) to fine-tune the propagation models used in the propagation calculations within the REM, (ii) to collect information not provided by the incumbents due to privacy concerns, (iii) to detect sharing agreement violations, and (iv) to collect band usage statistics that may be provided to the band manager in order to improve future frequency allocation decisions. When there are several collaborating sensing networks, the higher number and higher diversity of sensors in the field may also allow positioning algorithms to be deployed with improved accuracy when compared with a single network.

In order to facilitate the understanding of the proposed architecture, Figure 6.6 describes the sequence of messages exchanged by the LSA system when the LSA licensee requests spectrum (the LSA repository and spectrum sensing reasoning modules are not shown due to lack of space).

## 6.4 Operation of the Improved Architecture in Dynamic LSA Use Cases

ADEL proposed a system supporting any vertical sharing situations that may arise in Europe in the future, as it considers that all radio services, i.e. aeronautical, maritime, broadcasting, military, fixed, land mobile, satellite services, etc., may be involved in spectrum sharing. Obviously, such a list of services imposes a large diversity of requirements in terms of:

- the required periodicity of assignment of resources
- the bandwidth to be assigned
- the coverage/service area.

The main challenges do not come from the different characteristics of the frequencies being assigned, but from the temporal and spatial requirements that the resource allocation algorithms must fulfil. Three use cases were selected by ADEL to exemplify those challenges:

- Scenario 1: LSA for backhaul support in railway communications
- Scenario 2: LSA for macro-cellular mobile communications
- Scenario 3: LSA for additional capacity in small cell mobile communications.

This set of scenarios allows the application of LSA to be exemplified in situations that range from very short to very long allocation periods, from very confined to very wide service areas, and are associated with different bandwidth requirements. They are described in sections 6.4.1. through 6.4.3.

## 6.4.1 Railway Scenario

In this scenario, ADEL proposed using LSA to support backhaul links between antennas located on top of moving trains and fixed base stations placed along railway tracks. Those external antennas are connected to indoor access points placed inside the train to provide mobile broadband connectivity to both staff and passenger mobile terminals using Wi-Fi or 4G mobile radio access technology.

In this scenario, LSA is used as a complement (i.e., an add-on) to the current way of implementing these backhaul links, which traditionally use other licensed spectrum. The LSA band used is the 2300–2400 MHz band. This band must be shared in time and geography with both aeronautical telemetry's ground stations and military applications that are deployed along the railway track. This means that, in this scenario, ADEL considers that the incumbents of the 2300–2400 MHz band are aeronautical telemetry and military services. Obviously, the incumbent may change when the train moves from location to location, or from country to country.

With regard to the LSA licensee role in the 2300–2400 MHz band, in this scenario ADEL considers that it is performed by a railway operator. The reader should recall that in order to be aligned with CEPT regulations, the use of LSA in the 2300–2400 MHz band by LSA licensees is constrained to multiples of 5-MHz bandwidths and TDD mode [11]. Therefore, within this scenario, the railway operator requests from the LSA system in advance (e.g., the day before or while the train is stopped in a station) access to 20 MHz of spectrum in any part of the 2300–2400 MHz band. This band is intended to support backhaul links between the moving train and the fixed infrastructure in a specific portion of the track that is expected to be crossed by the train according to a specific timetable.

Because ADEL considers moving trains with speeds of up to 360 km/h (i.e., 100 m/s), ADEL determines that LSA-capable fixed base stations should be placed at least 1 km from each other in order not to have LSA handovers more often than every 10 s. Link budget

**Table 6.1** LSA allocations in railway scenario

| LSA band | 2300–2400 MHz |
| --- | --- |
| LSA application | Backhaul link for mobile broadband applications inside moving trains |
| Incumbents | Aeronautical telemetry ground-stations and military ground stations |
| LSA licensee | Railway operator |
| LSA bandwidth | Multiples of 20 MHz |
| LSA allocation time period | Multiples of 10 s |
| LSA allocation area | 100 m wide by 1 km length along the railway track |

considerations at 2300–2400 MHz do not allow these base stations to be spaced much more than this.

The parameters relevant for LSA operation in the railway scenario are listed in Table 6.1. From the end-user perspective, this scenario should appear as follows:

1) At least 1 h before the scheduled departure of the train, the railway operator connects to the LSA band manager to request spectrum along the track that the train will cross. To deal with train delays, a 5-min safety margin should be included in the spectrum requests.

2) Some seconds afterwards, the railway operator receives a frequency profile indicating the LSA frequencies that should be used by the train when it crosses specific places of the track at specific time instants. If LSA spectrum will be not available on some parts of the track, this will be immediately known by the railway operator, who should decide how to solve the problem: will it have to warn end-users some minutes before the connection breakdown or will it rely on the legacy licensed systems? (Note: Although it will be possible that LSA spectrum might be unavailable, these situations should not be frequent. If they are frequent, this means that the band was erroneously assigned to LSA by the national regulator or that the national regulator has authorized an excessive number of LSA licensees in that band.)

3) This frequency profile may be immediately used by the railway operator to initiate the process of informing the base stations along the track that they have to use those frequencies to communicate with trains during those periods. This allows preparation for the reconfiguration procedure to start in advance.

4) The train is stopped at the departing point. It has two improved performance LTE UE modules (more transmission power, more sensitive receivers, faster processors), with 2300–2400 MHz external antennas, one placed on top of the first carriage and the other placed on top of the last carriage. These two UEs are used to connect to the fixed infrastructure and provide connectivity to the passengers through several access points distributed inside the train. While the train is at a stop, the train's UEs are connected to the fixed infrastructure, using licensed (LTE) or unlicensed (Wi-Fi) spectrum, depending on the railway operator decision.

5) An end-user arrives at the train station. He/she has a smartphone or other mobile broadband device that is powered on and registered in a mobile network.

6) The train arrives at the station. The end-user enters the train. At this instant, the end-user's mobile broadband device should detect a stronger signal coming from the train access points and should warn the end-user that a new network has been found inside the train. This is the railway operator network. The device should switch to the railway operator network automatically or manually depending on the network and device configurations.

7) The train starts its journey. As soon as it leaves the station, the train's UEs connect to the base stations placed along the track using LTE and the frequencies described in the frequency profile obtained from the LSA band manager (step 2).

8) The train position is known accurately using the railway positioning systems. As such, whenever the train approaches a location where, according to the frequency profile obtained from the LSA band manager, the trains' UEs have to switch to a different LSA carrier frequency, the UEs are required to inform the base stations, which will then require the train's UEs to handover to those frequencies. Typical LTE handover interruption times should be achieved. (Note: ADEL proposes that the switching among LSA carriers is performed in this way in order to implement this switching in the same way as regular LTE handovers, i.e. these operations should be UE-assisted network-controlled operations.)

9) Meanwhile, the LTE measurement gap patterns (6 ms periods repeated every 40 ms) configured by the base stations along the track are used by the train's UEs to perform measurements on most of the LSA carriers in the LSA band. These measurements are transmitted, like any other LTE measurement, to the base stations along the track. The base stations then forward this information to the network management system (i.e., the operations, administration, management and provisioning (OAM&P) system in 3GPP networks), which processes them and transmits the results to the LSA system. The sensing information should be used to both tune the propagation models and collect spectrum usage statistics that will assist future frequency allocation decisions. This information is also used immediately when policy violations are detected (e.g., unauthorized interference).

10) Inside the train, the use of LSA or any other type of spectrum should be entirely transparent to the end-users, which should use their mobile broadband devices as usual.

11) When the end-user leaves the train, the railway network signal distributed by the access points inside the train will become weaker. At this point, the user's device should switch to another network, automatically or manually, depending on the network and the device's configurations.

## 6.4.2 Macro-cellular Scenario

In the second scenario, ADEL proposes using the LSA band to support links between macro-cellular fixed base stations and the end-users' terminals. These terminals may be portable devices with antennas mounted on the surface of moving vehicles, handheld mobile devices, or fixed devices.

In this scenario, LSA band is seen as additional spectrum used to provide extra capacity to some macro-cellular networks in a cost-effective way. The LSA band is again 2300–2400 MHz. This band must be shared in time and geography with both aeronautical

**Table 6.2** LSA allocations in macro-cellular scenario

| LSA band | 2300–2400 MHz |
|---|---|
| LSA application | Link between end-user's devices and macro-cellular fixed base stations |
| Incumbents | Aeronautical telemetry ground-stations and military ground stations |
| LSA licensee | Mobile network operator |
| LSA bandwidth | Multiples of 5 MHz |
| LSA allocation time period | Multiples of 1 min |
| LSA allocation area | Macro cell with 1.5 km |

telemetry's ground stations and military applications that are deployed within the area that is served by macro-cellular network(s) belonging to one or more MNOs. This means that in this scenario, the incumbents of the 2300–2400 MHz band may be aeronautical telemetry and military services, while the LSA licensees are one or more MNOs. Obviously, the incumbent may change when the end-user moves from location to location within the area serviced by the macro-cellular network it is attached to. Likewise, the LSA licensees may also change from location to location, and/or from time to time, because the national regulator may have defined geographical areas where, or time periods when, some MNOs are authorized to use LSA and other MNOs are not.

As in any other LSA scenario, the LSA licensee must request in advance (e.g., in the day before), from the LSA system, the allocation of LSA spectrum. According to CEPT regulations regarding the 2300–2400 MHz band, LSA licensees should access spectrum in this LSA band in multiples of 5-MHz channels using TDD [11].

Because it is assumed that mobile terminals can move with speeds of up to 180 km/h (i.e., 50 m/s) in the aforementioned macro-cellular networks, ADEL determines that LSA-capable macro-cellular fixed base stations should be placed at least 3 km from each other in order not to have LSA handovers before 1 min. On the other hand, given that the selected LSA frequencies are higher than the traditional sub-GHz frequencies used by macro cells, the size of the cells cannot be much higher than this value.

The parameters relevant for LSA operation in the macro-cellular scenario are summarized in Table 6.2.

From the end-user perspective, this scenario is as follows:

1) When the mobile network operator predicts that it will need LSA spectrum in parts of the network during the following day, it connects, at least one day in advance, with the LSA band manager to request spectrum for the macro-cellular base stations where there is such a need.

2) The LSA band manager contacts the REM to get updated radio environment conditions, obtained both from static and sensed data. Using this information, and eventually some historical data stored internally (e.g., number of active LSA licensees on each day of the week), the LSA band manager determines the frequency profile that the LSA licensee should use to get the required QoS and minimize the probability of having to reconfigure unnecessarily. As the number of base stations to analyse might be high, the computations might take from several minutes to a couple of hours. In principle, some minutes

after placing the request, the mobile operator receives a frequency profile indicating the LSA frequencies that should be used by each of its macro-cellular base stations during the requested time interval. If LSA spectrum may not be available in any base station locations in the vicinity, this will be immediately known by the network operator, who should decide how to solve the problem, e.g. by making use of other load balancing schemes that might be available in the network. (Note: Although it will be possible that LSA spectrum might be unavailable, these situations should not be frequent. If they are frequent, this means that the band was erroneously assigned to LSA by the national regulator or that the national regulator has authorized an excessive number of LSA licensees in that band.)

3) This information may be immediately used by the network operator to initiate the process of informing the macro-cellular base stations that they can use the authorized frequencies to communicate with end-users during those periods. This allows preparation for the reconfiguration procedure to start in advance.

4) End-users have mobile broadband devices (e.g., smartphones) that are powered on and registered in a mobile network. Some of these devices are capable of operating in the 2300–2400 MHz band using TDD mode (3GPP band 40).

5) The macro-cellular base stations that have been allowed to use LSA frequencies for a specific time interval may, during that interval and as soon as they are ready, send instructions to the attached UEs commanding them to perform a reselection or a handover to the LSA frequencies indicated by the macro-cellular base station. These frequencies should be in accordance with the frequency profile obtained by the LSA band manager (step 2).

6) Whenever the end-user moves to another LSA macro-cell:

   a) If the end-user terminal is in idle mode, it should camp on the new cell during the reselection procedure as usual. The information about the target cell carrier frequencies is obtained by the broadcast channel.

   b) If the end-user terminal is in connected mode, the handover also happens in the usual way, i.e. the original base station gets information about the carrier frequencies in use on the target cell and sends this information to the mobile, commanding it to switch to that frequency. The end-user terminal then connects to the target cell and finally is disconnected from the original one.

7) Meanwhile, assuming the macro-cellular network is deploying any 3GPP standard, the measurement gap patterns defined by the base stations are used by the end-users' terminals to perform measurements on most of the LSA carriers in the LSA band. These measurements are sent to the macro-cellular base station like any other measurement. LSA measurements provided by the several end-users' terminals in the macro-cell, are concentrated on the macro-cellular base station, and sent to the mobile network management centre (OAM&P), and finally transmitted to the LSA system. This information should be used to both tune the propagation models and collect spectrum usage statistics that will assist future frequency allocation decisions. This information is also used immediately when policy violations are detected (e.g., unauthorized interference).

8) When the end-user leaves the macro-cells that are using LSA frequencies, or when the time period during which those cells are authorized to use LSA expires, the end-users' terminals should switch to other frequencies automatically, in the same way as indicated in 6.

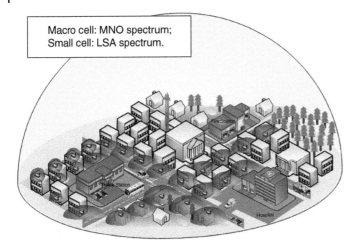

Macro cell: MNO spectrum;
Small cell: LSA spectrum.

**Figure 6.7** Indoor-to-outdoor, residential small cells.

### 6.4.3 Small Cell Scenario

ADEL proposes the deployment of small cells in highly populated areas using LSA bands in order to increase the network capacity for mobile broadband services when and where needed.

In this scenario, the links that use the LSA band are the links between the small cell fixed base stations and the end-users' mobile terminals. These terminals are typically handheld mobile devices or laptops. Naturally, before using the LSA small cell, these end-users' devices must previously register with a legacy macro/micro cell using the licensed spectrum. The network will then guide the end-user terminal to the LSA small cells when an authorized small cell is in range and when the network load requires offloading some users from the legacy network to the small cells. Although not absolutely necessary, ADEL proposes that LSA small cells deployment should always be combined with legacy licensed-spectrum micro and macro cells in order to provide the mobile terminals with initial LSA configuration, and to guarantee uninterrupted service to the users moving across the cells with low speed.

In this scenario, the LSA band is seen as additional spectrum used to provide additional capacity, in a cost-effective way, to dense networks of small cells deployed in urban environments.

We should stress that this scenario includes several implementation options, e.g. the small cell base stations may be (i) indoor, privately owned, residential LSA-capable gateways with wired connectivity, as depicted in Figure 6.7, or (ii) remote radio heads connected through fibre to a centralized baseband unit, to mention just a couple of examples. Although the resource allocation problem remains the same, we stress that these two example options correspond to different business requirements, especially in terms of initial investment.

Given the expected high demand for spectrum in these highly populated areas, it was decided to use two LSA bands in this scenario, i.e. the 2300–2400 MHz band is complemented by the 3500–3600 MHz band. Both these bands must be shared in time and geography with the respective incumbents. In this scenario, the incumbents in the

**Table 6.3** LSA allocations in small cell scenario

| LSA band | 2300–2400 and 3500–3600 MHz |
|---|---|
| LSA application | Links between end-user's devices and small cell fixed base stations |
| Incumbents | 2300–2400 MHz: PMSE operators (fixed/portable/mobile wireless cameras) |
| | 3500–3600 MHz: BWA operators (fixed/portable voice + Internet services) |
| LSA licensee | Mobile network operators with highly dynamic spectrum needs in terms of bandwidth, time, or coverage area |
| LSA bandwidth | Multiples of 5 MHz |
| LSA allocation time period | Multiples of 10 min |
| LSA allocation area | Small cell with 20–200 m radius |

2300–2400 MHz band are programme making and special events (PMSE) operators that use the band to support the operation of wireless microphones or cameras, while the incumbents in the 3500–3600 MHz band are broadband wireless access (BWA) operators using the band to provide wireless fixed/portable voice and Internet services to underpopulated areas.

LSA licensees are one or more mobile network operators willing to add capacity to their networks by using small cells in densely populated cities at reduced cost.

Obviously, we may have different incumbents from place to place and from time to time. In addition, the LSA licensee spectrum needs are also expected to change from location to location and/or from time to time.

As in any other LSA scenario, the LSA licensee must request in advance (e.g., the day before), to the LSA system, the allocation of LSA spectrum. According to CEPT regulations regarding the 2300–2400 MHz band, and as mentioned previously, LSA licensees should access spectrum in this LSA band in multiples of 5-MHz channels using TDD [11]. For the other LSA band, i.e. 3500–3600 MHz, the same access conditions are assumed. Regarding the remaining scenario parameters, it was considered that mobile terminals can move with speeds of up to 30 km/h (i.e., 8.33 m/s) between small cells having a radius from 20 to 200 m. For these parameters, link-budget considerations do not impose additional constraints.

The parameters describing LSA operation in this scenario are presented in Table 6.3.

From the end-user perspective, this scenario should appear as follows:

1) Because this scenario includes situations where small cell base stations are installed indoors by their private owners, the places where these base stations are installed might not be exactly known, but have to be estimated as accurately as possible. 3GPP requires that the mobile operator must be able to know the location of the small cell transmitter. Depending on the network operator decision, this information might be obtained in several different ways:
   a) The subscriber indicates where they will install the base station when filling out the contract.

b) The base station may be required to determine its position when it is powered on and connected to the operator network. This location might be obtained by knowing which network node was accessed by the base station when it is powered on and tries to get configuration parameters, it can be estimated through the IP address (coarse information), it can be calculated by measuring the observed time difference with which the signals from the neighbour macro cells arrive (observed time difference of arrival) at the small cell base station (intermediate accuracy), or it can be obtained through a GPS reading if possible. The Small Cell Forum [32] states that other schemes are also possible for the base station to determine its position, like determining the macro-cellular cell-ID and refining the location information using signal strength and signal quality radio measurements, or implementing multi-lateration using other signal sources whose transmitter locations are known (e.g., TV, FM or Wi-Fi).

The location of the small cell base station will have to be checked periodically by the network. Several methods are possible, e.g. asking the neighbouring macrocellular base stations to measure the uplink time difference with which the small cell signals arrive at their locations (uplink time difference of arrival) or asking the serving macro-cells to measure the receiver-transmitter time difference and combine this with angle-of-arrival measurements.

It should be stressed that ADEL's collaborative sensing networks, either composed of dedicated sensors or reusing the base station and UE measurement capabilities, can also be used to determine the position of the small cell base stations. In such cases, the collaboration of several LSA sensing networks will increase the number and diversity of sensors in the field when compared with the situation where each mobile operator performs its own measurements individually. This fact improves the accuracy, reliability, and robustness of the positioning algorithms and/or reduces the number of measurements that each individual sensing network has to perform.

2) After the location of the small cells has been determined/estimated, the interaction of the mobile network operator with the LSA system is similar to the macrocellular scenario. Therefore, when the mobile network operator predicts it will need LSA spectrum in parts of the network during the following day, it connects, at least one day in advance, with the LSA band manager to request spectrum for the small cell base stations where there is such need.

3) Some minutes afterwards, the mobile operator receives a frequency profile indicating the LSA frequencies that should be used by each of its small cell base stations during the requested time interval. Some base stations may be allocated frequencies in the 2300–2400 MHz band while others may be allocated frequencies in the 3500–3600 MHz band. If LSA spectrum is not available in any relevant base station locations, this will be immediately known by the network operator, who should decide how to solve the problem making use of other load balancing schemes that might be available in the network.

4) The frequency profile may be immediately used by the network operator to initiate the process of informing the small cell base stations that they can use the authorized 2300–2400 or 3500–3600 MHz frequencies to communicate with end-users during those periods. This allows preparation for the reconfiguration procedure to start in advance.

5) End-users have smartphones that are powered on and registered in a mobile network. Some of these phones are capable of operate in the 2300–2400 MHz and/or 3500–3600 MHz band using TDD mode (3GPP bands 40 and 42, respectively).

6) The small cell base stations that have been allowed to use LSA frequencies for a specific time interval may during that interval, and as soon as they are ready, send instructions to the attached UEs that can operate in 2300–2400 or 3500–3600 MHz bands to instruct them to perform reselection/handover to the LSA frequencies indicated by the small cell base station. These frequencies should be in accordance with the frequency profile obtained from the LSA band manager (step 3).

7) Whenever the end-user moves to another LSA small cell:
   a) If the end-user terminal is in idle mode, it should camp on the new cell during the reselection procedure as usual. Information about the target cell carrier frequencies to select is obtained from the broadcast channel.
   b) If the end-user terminal is in connected mode, the handover also happens in the usual way, i.e. the original base station gets information about the carrier frequencies in use on the target cell, then sends this information to the mobile and commands it to switch to that frequency. The end-user terminal then connects to the target cell and finally is disconnected from the original one.

8) Meanwhile, assuming the small cell network is deploying any 3GPP standard, the measurement gap patterns defined by the small cell base stations are used by the end-users' terminals to perform measurements on most of the LSA carriers in the LSA band(s). These measurements are sent to the small cell base station like any other measurement. LSA measurements provided by several end-users' terminals in the small cell are concentrated on the small cell base station, then sent to the mobile network management centre (OAM&P), and finally transmitted to the LSA system. This information should be used to both tune the propagation models and collect spectrum usage statistics that will assist future frequency allocation decisions. This information is also immediately used when policy violations are detected (e.g., unauthorized interference).

9) When the end-user leaves the small cells that are using LSA frequencies, or when the time period during which those cells are authorized to use LSA expires, the end-users' terminals should switch to other frequencies automatically, in the same way as indicated in step 7.

## 6.5 Summary

Nowadays, as more and more traffic is delivered wirelessly by mobile networks, the spectrum available to them risks becoming insufficient. In addition, inspection of the national frequency allocation tables demonstrates it is difficult to find available spectrum with desired bandwidth and appropriate propagation characteristics. Policy makers are thus returning their attention to spectrum sharing as a means of providing spectrum for the services that need it. This chapter deals with one of the solutions already encountered: LSA.

LSA is a spectrum sharing approach that targets the bands that are rarely utilized, either in time or geography, by the respective incumbent applications. Instead of refarming these bands, under LSA they are allowed to be used by a limited number of LSA licensees when and where the incumbent is not needing to use them. The conditions under which the LSA licensee may use part of the band in an exclusive way are defined in a sharing agreement contracted between the incumbent and the LSA licensee, so when using the spectrum either the incumbent or the LSA licensee will be protected against excessive interference.

According to the European regulator, the implementation of LSA requires the deployment of a LSA repository to store the information relative to the incumbent's spectrum usage and determine the spectrum available for each LSA licensee, and an LSA controller, who is in charge of obtaining this information and communicating it to the LSA licensee. The regulation does also not impose any restriction regarding the ownership of these logical entities, except that the LSA repository should not be under the jurisdiction of the LSA licensees.

These two logical entities were also considered in the LSA standard developed by the European standardization body. The goal of the standard was to define a system which would allow, as soon as possible, the use of LSA in the 2300–2400 MHz band by mobile networks. With this purpose in mind, the standard considers that the LSA controller is under the domain of the LSA licensee. As current mobile networks still have limited capabilities in terms of carrier frequency reconfiguration, first LSA deployments are expected to be based in fixed/static channel plans, being the LSA architecture simply used to determine if the pre-defined and pre-allocated channels are in use by the respective incumbents.

The European FP7 project ADEL considered that it would be desirable to support more dynamic sharing situations in which the LSA band is allocated and deallocated with a finer resolution, both in time or in geography, than what is supported by current LSA standard. To illustrate this goal, ADEL selected three scenarios where such resolutions range from 10 s to a few hours or from 20 m to 1.5 km. To support these scenarios, ADEL proposed extending the LSA architecture with several collaborative sensing networks, a radio environment map, and an entity responsible for coordinating the access of multiple LSA licensees to the band(s). These additional logical blocks would allow a faster reaction to eventual necessities to use spectrum, the detection of excessive interference or misconduct, and exploitation to the maximum extent of the sharing opportunities that arise.

Another aspect that differentiates ADEL architecture from the current LSA standard is the mapping of LSA functionalities among the logical entities. While in the LSA standard the storage of information about the incumbent's spectrum usage and the task of computing the spectrum available to each LSA licensee are aggregated in the LSA repository, in ADEL there is a decoupling between storage and processing functions that results in the distribution of the basic LSA functionality across different indivisible logical entities. This allows ADEL's LSA system to be easily adapted to different regulatory frameworks, e.g. when the regulator imposes that some functionalities are under the jurisdiction of the same, or a different, stakeholder.

## References

**1** General Secretariat of the ITU, The Selection of Frequencies and Frequency Sharing, Recommendations and Resolutions adopted by the 1947 International Radio Conference in Atlantic City, pp. 49-88. Geneva, 1949.

2 Nokia Siemens Networks, Qualcomm, Authorised Shared Access – An evolutionary spectrum authorisation scheme for sustainable economic growth and consumer benefit. Input Document FM(11)116 to the 72nd Meeting of the Working Group Frequency Management, Miesbach, Germany, 16–20 May 2011.

3 Radio Spectrum Policy Group, Report on Collective Use of Spectrum (CUS) and other spectrum sharing approaches. RSPG (11)-392 Final, November 2011.

4 European Commission, Communication from the Commission to the European Parliament, the Council, the European Economic and Social Committee and the Committee of the Regions – Promoting the shared use of radio spectrum resources in the internal market. COM(2012)478 Final, 3 September 2012.

5 Radio Spectrum Policy Group, Request for Opinion on Licensed Shared Access (LSA). RSPG12-424 Final, 8 November 2012.

6 Radio Spectrum Policy Group, RSPG Opinion on Licensed Shared Access. RSPG13-538, 12 November 2013.

7 WG FM53, ECC Report 205 – Licensed Shared Access. CEPT, February 2014.

8 WG SE7, ECC Report 172 – Broadband Wireless Systems Usage in 2300–2400 MHz. CEPT, March 2012.

9 ECO, Results of the WG FM Questionnaire to CEPT Administrations on the current and future usage of frequency band 2300–2400 MHz. Doc. FM(12)017rev1 from 74th WG FM meeting, Bern, 23–27 April 2012.

10 European Commission, Mandate to CEPT to develop harmonised technical conditions for the 2300–2400 MHz ('2.3 GHz') frequency band in the EU for the provision of wireless broadband Electronic Communications Services. 8 April 2014.

11 WG FM52, CEPT Report 55 – Report A from CEPT to the European Commission in response to the Mandate on 'Harmonised technical conditions for the 2300–2400 MHz ('2.3 GHz') frequency band in the EU for the provision of wireless broadband electronic communications services': Technical conditions for wireless broadband usage of the 2300–2400 MHz frequency band. 28 November 2014.

12 WG FM52, CEPT Report 56 – Report B1 from CEPT to the European Commission in response to the Mandate on 'Harmonised technical conditions for the 2300–2400 MHz ('2.3 GHz') frequency band in the EU for the provision of wireless broadband electronic communications services': Technological and regulatory options facilitating sharing between Wireless broadband applications (WBB) and the relevant incumbent service/application in the 2.3 GHz band. 6 March 2015.

13 WG FM52, CEPT Report 58 – Report B2 from CEPT to the European Commission in response to the Mandate on 'Harmonised technical conditions for the 2300-2400 MHz ('2.3 GHz') frequency band in the EU for the provision of wireless broadband electronic communications services': Technical sharing solutions for the shared use of the 2300-2400 MHz band for WBB and PMSE. 3 July 2015.

14 WG FM52, ECC Recommendation(14)04 – Cross-border coordination for mobile/fixed communications networks (MFCN) and between MFCN and other systems in the frequency band 2300–2400 MHz. CEPT, 30 May 2014.

15 WG FM52, ECC Decision(14)02 – Harmonised technical and regulatory conditions for the use of the band 2300–2400 MHz for Mobile/Fixed Communications Networks (MFCN). CEPT, 27 June 2014.

**16** European Commission, Standardisation Mandate M/512 to CEN, CENELEC and ETSI for Reconfigurable Radio Systems. 19 November 2012.

**17** ETSI Technical Committee Electromagnetic compatibility and Radio spectrum Matters (TC ERM), System Reference Document (SRDoc.), ETSI TR 103 113 v1.1.1 – Mobile broadband services in the 2300–2400 MHz frequency band under Licensed Shared Access regime. July 2013.

**18** ETSI Technical Committee Reconfigurable Radio Systems (TC RRS), ETSI TS 103.154 v1.1.1 – System requirements for operation of Mobile Broadband Systems in the 2300 MHz–2400 MHz band under Licensed Shared Access (LSA). October 2014.

**19** ETSI Technical Committee Reconfigurable Radio Systems (TC RRS), ETSI TS 103.235 v1.1.1 – System architecture and high level procedures for operation of Licensed Shared Access (LSA) in the 2300 MHz–2400 MHz band. October 2015.

**20** ETSI Technical Committee Reconfigurable Radio Systems (TC RRS), ETSI TS 103.379 V1.1.1 – Information elements and protocols for the interface between LSA Controller (LC) and LSA Repository (LR) for operation of Licensed Shared Access (LSA) in the 2300 MHz–2400 MHz band. January 2017.

**21** Work Group 5 of 3GPP Technical Specification Group on Services and System Aspects (TSG SA5), 3GPP TR 32.855 (Release 14) – Study on OAM support for Licensed Shared Access (LSA). March 2016.

**22** Work Group 5 of 3GPP Technical Specification Group on Services and System Aspects (TSG SA5), 3GPP TS 28.301 (Release 14) – Licensed Shared Access (LSA) Controller (LC); Integration Reference Point (IRP); Requirements. June 2017.

**23** Work Group 5 of 3GPP Technical Specification Group on Services and System Aspects (TSG SA5), 3GPP TS 28.302 (Release 14) – Licensed Shared Access (LSA) Controller (LC); Integration Reference Point (IRP); Information Service (IS). June 2017.

**24** Work Group 5 of 3GPP Technical Specification Group on Services and System Aspects (TSG SA5), 3GPP TS 28.303 (Release 14) – Licensed Shared Access (LSA) Controller (LC); Integration Reference Point (IRP); Solution Set (SS) definitions. June 2017.

**25** ECC PT1, ECC Report 254 – Operational guidelines for spectrum sharing to support the implementation of the current ECC framework in the 3600–3800 MHz range. CEPT, 18 November 2016.

**26** FP7-ICT Project ADEL (Advanced Dynamic spectrum 5G mobile networks Employing Licensed shared access) (A copy of the original www.adel.eu website is available at https://web.archive.org/web/20161023153635/http://www.fp7-adel.eu:80/).

**27** A. Morgado and A. Gomes (eds), ADEL Deliverable D3.1 – Reference scenarios, network architecture, system and user requirements and business models. December 2014.

**28** A. Morgado, A. Gomes, V. Frascolla, et al., Dynamic LSA for 5G networks – The ADEL perspective. The 2015 European Conference on Networks and Communications (EUCNC 2015), Paris, 29 June–2 July 2015.

**29** A. Morgado, A. Gomes, and V. Frascolla, ADEL: the next stop in the LSA roadmap, Special Session on Dynamic spectrum management – a building block for 5G networks. 2016 European Conference on Networks and Communications (EUCNC 2016), Athens, 27–30 June 2016.

**30** V. Frascolla, A. Morgado, and A. Gomes, Dynamic Licensed Shared Access – A new architecture and spectrum allocation techniques. 2016 IEEE 84th Vehicular Technology Conference (VTC-Fall 2016), Montréal, 18–21 September 2016.

**31** M. Palola et al., Licensed Shared Access (LSA) trial demonstration using real LTE network. 9th International Conference on Cognitive Radio Oriented Wireless Networks and Communications (CROWNCOM), 2–4 June 2014, Oulu, Finland, pp. 498–502.

**32** Small Cell Forum, Topic Brief: Femtocell synchronization and location. Document 036.05.02, December 2013.

# 7

# Collaborative Sensing Techniques

*Christian Steffens[1] and Marius Pesavento[2]*

[1]*Hyundai Mobis, Frankfurt, Germany*
[2]*Darmstadt University of Technology, Darmstadt, Germany*

Since the early days of cognitive radio and up to the advent of licensed/authorized spectrum access, spectrum sensing has been at the core of spectrum sharing technologies. The idea of licensed shared access (LSA) is to make unused spectral resources of the incumbent network, i.e. the network that owns the spectrum license, available to users of a secondary licensee network. To this end, it is required that the licensees are able to sense whether specific spectral resources are used by the incumbent network or if they are free for use of the licensee network. In the context of cognitive radio under the underlay paradigm the licensees are requested to limit their interference to the incumbent network, hence the licensees must sense the presence of an incumbent user reliably, i.e. with low probability of missed detections. On the other hand, for spectrally efficient LSA operation it is important that available spectral resources are detected with low latency and with a low false alarm rate to best utilize the available resources. Today, certain actual systems, such as IEEE 802.22, IEEE 802.11K, and Bluetooth, prescribe the application of spectrum sensing by the current regulation [1].

The stand-alone spectrum sensing approaches most commonly considered in the literature can be classified into three categories: energy detection, matched filtering, and cyclostationary detection. These three categories of spectrum sensing methods mainly differ in the knowledge of *a priori* information, detection performance, and computational complexity. Below we provide a short summary of the state-of-the-art methods and refer to the overview articles, e.g. [1–3], for more details.

Energy detection is considered the most basic approach. It does not require any information about the incumbent user signal structure and can be implemented with low complexity. On the other hand, it requires a large number of time samples and a good estimate of the receiver noise power to provide reliable detection performance [4–7]. In the matched filter approach it is assumed that part of the incumbent user's signal structure is known to the licensees, e.g. some reference or synchronization sequence, which can be used to enhance signal detection. As compared to energy detection, the matched filter approach does provide better detection performance at a lower number of time samples due to the coherent processing gain. On the other hand, matched filtering requires *a priori* signal knowledge and

*Spectrum Sharing: The Next Frontier in Wireless Networks,* First Edition.
Edited by Constantinos B. Papadias, Tharmalingam Ratnarajah, and Dirk T.M. Slock.
© 2020 John Wiley & Sons Ltd. Published 2020 by John Wiley & Sons Ltd.

synchronization to the received signal for coherent processing, hence it is more demanding in terms of hardware requirements and implementation costs [8]. Cyclostationary detection is based on some form of periodicity often encountered in the incumbent user signals, e.g. in radar and communication signals, and provides a compromise between energy detection and matched filtering in terms of prior knowledge requirements, detection performance, and system complexity. In this context various signal features may be exploited for cyclostationary detection, as, for example, the signal's modulation type, symbol rate, carrier frequency, and cyclic prefix. Depending on the knowledge of the incumbent network's signal structure, the cyclostationary detection approach can be implemented for a specific cyclic frequency or for a full bank of cyclic frequencies at the expense of complexity. In contrast to energy detection, cyclostationary detection is more resilient to noise, which usually does not exhibit cyclostationary features, and thus cyclostationary detection shows better detection performance compared to energy detection [9–12].

Most of the literature on spectrum sensing has focused on single antenna devices. In recent years multi-antenna technology has become much more mature and affordable, such that it also came into the focus of spectrum sensing applications. Besides the array gain resulting from the increased number of antennas, multi-antenna sensing devices can also exploit spatial selectivity and thus provide better spectrum sensing performance in highly dynamic environments compared to single antenna devices. Multi-antenna transceiver devices may take further advantage of the spatial domain for spectrum sharing, e.g. by estimating the direction or spatial signatures of the incumbent users and reducing the interference to them by application of appropriate transmit beamforming, as discussed in Chapter 8 in the context of cooperative communication. Motivated by these benefits, the concepts of energy detection, matched filtering, and cyclostationary detection have been extended to multi-antenna devices [13–18].

Another aspect that has received a lot of interest is the concept of collaborative spectrum sensing, where multiple licensees collaborate to jointly detect the presence of incumbent users [19, 20]. Besides improved detection performance due to the joint detection, this approach also addresses the hidden node problem, where an incumbent user is hidden from a licensee due to obstacles in the signal propagation path, such as walls, buildings, and mountains. By the joint detection approach, the hidden incumbent user can be detected and reported by other licensees, such that interference to the incumbent network is avoided. While collaborative methods show better detection performance, a general drawback of these methods lies in their increased communication and computation complexity. Different realizations of collaborative sensing with a fusion center, as well as with fully decentralized processing, have been suggested in the literature [21, 22].

The above-mentioned traditional methods of spectrum sensing have been well investigated in the signal processing literature. Comparably new in spectrum sensing are the methods developed under the paradigm of *sparse signal reconstruction (SSR)*, also referred to as sparse recovery, compressed sensing or compressive sensing. Since SSR provides many fields of application, such as parameter estimation, image processing, and machine learning, the research in this field has substantially progressed in recent years [23–30] and SSR-based methods have made their way into technical realizations [31]. SSR techniques explore prior knowledge based on which the sensed signals can be expressed by sparse models in some domain, e.g. the spectral, code, or spatial domain, in order to uniquely reconstruct a high-dimensional sparse signal vector from low-dimensional undersampled measurement vectors [25–30]. The sparsity assumption yields sensing methods that generally exhibit a number of benefits over conventional detection and estimation techniques, e.g. they avoid the requirement to define exact detection thresholds and show excellent detection performance in the low sample size regime and in scenarios with correlated signals that emerge, e.g. in multi-path propagation [32, 33]. In the context of spectrum sensing, SSR has been applied in different ways, namely for wideband sensing [34, 35], signal parameter estimation [32, 36, 37], and radio environment map construction [38]. A recent overview on SSR-based spectrum sensing techniques is provided in [39].

The focus of this chapter is on SSR methods for application in collaborative spatial spectrum sensing with multi-antenna devices. We start our discussion by considering a single- and multi-antenna sensing device, respectively, for which we establish a sparse representation model for spectral and spatial sensing. Based on this model we present the concept of SSR based on $\ell_1$ norm minimization, which provides a sparse estimate of the signal in the frequency or spatial domain. The methods designed for a single sensing device are then extended for collaborative sensing with multiple sensing devices sharing their measurements with a fusion center. In this context we exploit specific block and rank structures in the sparse signal model to jointly sense the spectrum in the fusion center using measurements that have been coherently or non-coherently recorded in different sensing devices.

The authors would like to acknowledge the financial support of the Seventh Framework Programme for Research of the European Commission under grant number ADEL-619647 and the EXPRESS project within the DFG priority program CoSIP (DFG-SPP 1798).

## 7.1 Sparse Signal Representation

In this section we introduce the general signal model for the case of a single sensing device (SD), which could, for example, be a single licensee, and multiple incumbent users (IUs) that applies to a variety of sensing problems and we show how this model applies to frequency and directional sensing applications. Both applications exhibit a model with a common structure for which we illustrate the concept of sparsity and how it can be exploited for signal detection and signal recovery. Let $\mathbf{y}(t) \in \mathbb{C}^M$, $\boldsymbol{\psi}(t) \in \mathbb{R}^L$, and $\mathbf{n}(t) \in \mathbb{C}^M$ denote the sampled measurement vector, the signal waveform vector, and the additive white Gaussian noise vector at time instant $t$, respectively. The sensing model is given by

$$\mathbf{y}(t) = \mathbf{A}(\boldsymbol{\mu})\, \boldsymbol{\psi}(t) + \mathbf{n}(t), \tag{7.1}$$

where $\mathbf{A}(\boldsymbol{\mu}) \in \mathbb{C}^{M \times L}$ denotes the signal response matrix whose $\ell$th column represents the response corresponding to the $\ell$th signal with frequency $\mu_\ell$ and $\boldsymbol{\mu} = [\mu_1, \ldots, \mu_L]^T$ is the frequency parameter vector.

### Temporal sampling and frequency domain representation

Consider the scenario of an SD with a single antenna and $L$ IUs located in the vicinity of the SD, as depicted in Figure 7.1a. Assume for simplicity of presentation that the $\ell$th IU emits the narrowband signal $\psi_l(t)$ with center frequency $f_l$ and the

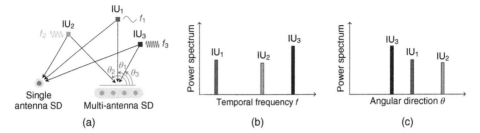

**Figure 7.1** (a) Setup with $L = 3$ IUs, one single antenna SD, and one SD with an array of $M = 4$ antennas. (b) Exemplary power spectrum over temporal frequency. (c) Exemplary power spectrum over angular direction.

frequency bands of different IUs are non-overlapping, as illustrated in Figure 7.1b. Furthermore, the received signal is sampled at Nyquist rate $1/T_s$ and $M$ samples are aggregated in a vector $\mathbf{y}(t) = [y(t), y(t + T_s), \dots, y(t + (M - 1)T_s)]^T \in \mathbb{C}^M$. The vector $\boldsymbol{\psi}^{(f)}(t) = [\psi^{(f)}(\mu_1^{(f)}), \dots, \psi^{(f)}(\mu_L^{(f)})]^T \in \mathbb{C}^L$ contains the frequency representation of the IUs' narrowband signals, where $\psi^{(f)}(\mu_\ell)$ denotes the complex amplitude of the signal received from the $\ell$th IU on frequency $\mu_\ell$. The frequency response matrix $\mathbf{A}^{(f)}(\mu^{(f)}) = [\mathbf{a}^{(f)}(\mu_1^{(f)}), \dots, \mathbf{a}^{(f)}(\mu_L^{(f)})] \in \mathbb{C}^{M \times L}$ is composed of the frequency response vectors $\mathbf{a}(\mu^{(f)}) = [1, e^{-j2\pi\mu^{(f)}}, \dots, e^{-j2\pi\mu^{(f)}(M-1)}]^T \in \mathbb{C}^M$, with $\mu_\ell^{(f)} = f_\ell T_s$ denoting the normalized signal frequency for IU $\ell$. The signal vector $\mathbf{y}(t)$ received by the SD at time instant $t$ can be expressed by model (7.1) with $\mu = \mu^{(f)}$, $\mathbf{A}(\mu) = \mathbf{A}^{(f)}(\mu^{(f)})$, and $\boldsymbol{\psi}(t) = \boldsymbol{\psi}^{(f)}(t)$, and $\mathbf{n}(t)$ containing the aggregated noise samples.

### Spatial sampling and angular domain representation

A model very similar to the previous one is obtained when signals are sampled in space rather than in time. For simplicity, we consider a uniform linear antenna array of $M$ elements with spacing $\Delta$ and assume that $L$ IUs are located in the far field, as illustrated in Figure 7.1a. In contrast to the previous section, we assume that the IUs are transmitting narrowband signals with a common center frequency, i.e. $f_\ell = f_0$, for $\ell = 1, \dots, L$, and we select the spacing $\Delta$ as half signal wavelength. The $L$ IUs are located at angular directions $\theta_1, \dots, \theta_L$, composed as the vector $\theta = [\theta_1, \dots, \theta_L]^T$, as depicted in Figure 7.1. Define the spatial frequency $\mu_\ell^{(s)}$ corresponding to the angular direction $\theta_\ell$ as $\mu_\ell^{(s)} = \cos\theta_\ell$, for $\ell = 1, \dots, L$, comprising the vector $\mu^{(s)} = [\mu_1^{(s)}, \dots, \mu_L^{(s)}]^T$. Let $\boldsymbol{\psi}^{(s)}(t) \in \mathbb{C}^L$ denote the IU signal vector and let the array steering matrix $\mathbf{A}^{(s)}(\mu^{(s)}) \in \mathbb{C}^{M \times L}$ be given by $\mathbf{A}^{(s)}(\mu^{(s)}) = [\mathbf{a}^{(s)}(\mu_1^{(s)}), \dots, \mathbf{a}^{(s)}(\mu_L^{(s)})]$, where $\mathbf{a}^{(s)}(\mu^{(s)}) = [1, e^{-j\pi\mu^{(s)}}, \dots, e^{-j\pi(M-1)\mu^{(s)}}]^T$ denotes the array steering vector for spatial frequency $\mu^{(s)}$ [40, 41]. Then the vector $\mathbf{y}(t) \in \mathbb{C}^M$ containing the sensor measurements of $M$ sensors recorded at time instant $t$ is characterized by model (7.1) with $\mu = \mu^{(s)}$, $\mathbf{A}(\mu) = \mathbf{A}^{(s)}(\mu^{(s)})$, and $\boldsymbol{\psi}(t) = \boldsymbol{\psi}^{(s)}(t)$.

For the example depicted in Figure 7.1b we observe that most of the spectrum is not used by the IUs, i.e. the IU signal spectrum is mostly zero or sparse. Similarly, from the example in Figure 7.1c we observe that the spatial spectrum for the given scenario can also be considered as sparse. The sparsity of the received signals, in either the spectral or angular domain, is the main motivation for dynamic spectrum sharing to increase the spectral efficiency. In the following we will exploit the sparsity assumption in order to devise advanced sensing techniques based on SSR. We note that model (7.1) can be extended to consider joint frequency and angular sensing, and it has also been extended to capture additional sensing domains such as the Doppler frequencies that incorporate the motion of IUs and angles of departure in the case when the IUs use multiple transmit antennas [42–44].

## 7.2 Sparse Sensing

One of the main obstacles that complicates sensing under model (7.1) follows from the nonlinearity of the model. According to the application examples considered in the previous section, the response matrix $\mathbf{A}(\mu)$ depends on the frequency parameters, i.e. the spectral or spatial frequencies $\mu_\ell$, in a highly nonlinear fashion. The idea of sparse representation techniques is to use prior knowledge regarding the sparsity of the model to avoid the nonlinearity in the model by sampling the response matrix over the fine grid of candidate frequencies and/or directions in the frequency band and/or angular field of view. In this context we define a constant overcomplete (i.e., wide) sensing matrix $\mathbf{A}(v) = [\mathbf{a}(v_1), \dots, \mathbf{a}(v_K)] \in \mathbb{C}^{M \times K}$ where the vector $v = [v_1, \dots, v_K]^T$ is obtained by sampling the frequencies in $K \gg L$ points $v_1, \dots, v_K$, as illustrated in Figure 7.2a. Assuming appropriate frequency sampling as specified in the following, the sensing model in (7.1) can equivalently be described by a sparse representation according to

$$\mathbf{y}(t) = \mathbf{A}(v)\,\check{\mathbf{x}}(t) + \mathbf{n}(t). \tag{7.2}$$

For ease of presentation we assume that the frequency grid is sufficiently fine, such that the true frequencies in $\mu$ are contained in the frequency grid $v$, i.e. $\{\mu_\ell\}_{\ell=1}^{L} \subset \{v_k\}_{k=1}^{K}$, and the sparse signal vector $\check{\mathbf{x}}(t) = [\check{x}_1(t), \dots, \check{x}_K(t)]^T$ in (7.2) contains only $L \ll K$ non-zero elements, according to

$$\check{x}_k(t) = \begin{cases} \psi_l(t) & \text{if } v_k = \mu_l \\ 0 & \text{otherwise,} \end{cases} \tag{7.3}$$

for $k = 1, \dots, K$ and $\ell = 1, \dots, L$, as illustrated in Figure 7.2b for the multiple snapshot case. The case of sparse recovery under *off-grid sources* is discussed in [45], for example.

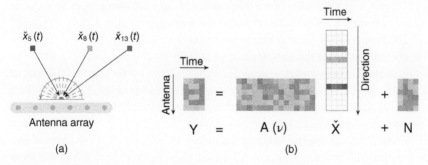

**Figure 7.2** (a) Sampling the field of view of an $M = 6$ antenna array in $K = 15$ grid points and (b) sparse representation of the corresponding sensing model for $N = 4$ snapshots.

The problem of recovering the signal vector $\check{x}(t)$ from the measurement vector $y(t)$ according to (7.2) occurs in similar form in various applications of signal processing, such as spectral analysis, direction of arrival (DOA) estimation, image processing, localization problems in geophysics, tomography and magnetic resonance imaging, and machine learning. Accordingly, the problem has received considerable research interest in the past decades and various reconstruction methods as well as theoretical reconstruction guarantees have been devised [23–30].

Considering the underdetermined system of equations in (7.2), the idea of compressed sensing is based on the assumption that the "sparsest" representation vector $x(t)$ that matches the measurements $y(t)$ corresponds to the true model (7.1). A natural approach for estimating the signal from the measurements is thus to minimize the $\ell_0$ quasi-norm of the representation vector, which counts the number of non-zero elements in $x(t)$, for all signal reconstructions under model (7.2) that match the measurement vector $y(t)$ sufficiently well. The $\ell_0$ quasi-norm is non-convex and yields a combinatorial problem that becomes computationally intractable for large problem dimensions, cf. [46], such that numerous methods to approximately solve the SSR problem have been derived [23, 24, 47–53]. The different methods vary in computational cost and reconstruction performance. Excellent overview articles on this subject are given in [46, 54].

One of the most popular strategies to approximate the $\ell_0$ minimization problem is to replace the $\ell_0$ quasi-norm by the $\ell_1$ norm, as it provides a good tradeoff between complexity and reconstruction performance. The $\ell_1$ norm of a vector $x = [x_1, \ldots, x_K]^T$ is computed as the absolute sum of the vector elements according to

$$\|x\|_1 = \sum_k |x_k|, \tag{7.4}$$

and presents a tight convex approximation of the $\ell_0$ quasi-norm that is exact if the signal in model (7.2) is sufficiently sparse and the sensing matrix $A(v)$ exhibits sufficiently different columns [25]. For the sparse sensing model in (7.2) the corresponding $\ell_1$ minimization problem can be formulated as

$$\min_x \|x\|_1 \quad \text{s.t.} \quad \|y(t) - A(v)x\|_2^2 \leq \beta, \tag{7.5}$$

where $\beta > 0$ is a regularization parameter determining the "degree of sparsity" of the minimizer $\hat{x}(t)$, i.e. larger values of $\beta$ will reduce the number of non-zero elements in $\hat{x}(t)$. Under the assumption of complex white Gaussian noise, the term $\|y(t) - A(v)\check{x}(t)\|_2^2 = \|n(t)\|_2^2$ follows a $\chi^2$ distribution with $2M$ real degrees of freedom and the regularization parameter $\beta$ can be selected such that the constraint in (7.5) is fulfilled with a desired probability $P(\|n(t)\|_2^2 \leq \beta)$, where knowledge of the noise power is required for proper computation of $\beta$. Problem (7.5) is commonly referred to as *basis pursuit denoising (BPDN)* [24] in the literature. An unconstrained version of $\ell_1$ norm minimization is given in form of the so-called *least absolute shrinkage and selection operator (LASSO)* formulation [23]. While the $\ell_1$ minimization problem (7.5) is convex and computationally tractable, the $\ell_1$ norm in the objective of (7.5) is a non-differentiable function at $x_i = 0$, which requires specific consideration in the optimization [55, 56]. Various customized optimization methods for efficiently solving the $\ell_1$ problem for large dimensions have been presented in the literature [56–63].

## Sparse sensing using multiple snapshots

The SSR approach discussed above considers parameter estimation from a single snapshot $\mathbf{y}(t)$. In the case of multiple snapshots, as common in array processing applications, the problem provides additional structure that can be exploited.

According to model (7.2), let $\mathbf{Y} = [\mathbf{y}(t_1), \ldots, \mathbf{y}(t_N)] \in \mathbb{C}^{M \times N}$ denote the matrix containing $N$ snapshots of the sensor measurements, where $[\mathbf{Y}]_{m,n}$ denotes the output at sensor $m$ in time instant $t_n$, for $m = 1, \ldots, M$ and $n = 1, \ldots, N$. The multiple sensor measurement vectors are modeled in compact notation as

$$\mathbf{Y} = \mathbf{A}(\nu)\, \check{\mathbf{X}} + \mathbf{N}, \tag{7.6}$$

where $\check{\mathbf{X}} = [\check{\mathbf{x}}(t_1), \ldots, \check{\mathbf{x}}(t_N)] \in \mathbb{C}^{K \times N}$ denotes a row-sparse (also referred to as group-sparse) signal matrix, and the sensing matrix is given by $\mathbf{A}(\nu) \in \mathbb{C}^{M \times K}$. Similar to the single snapshot case, the *on-grid assumption* is expected to hold true for ease of presentation. Correspondingly, the $K \times N$ sparse signal matrix $\check{\mathbf{X}}$ in (7.6) contains the elements

$$[\check{\mathbf{X}}]_{k,n} = \begin{cases} [\boldsymbol{\psi}(t_n)]_\ell & \text{if } \nu_k = \mu_l \\ 0 & \text{otherwise,} \end{cases} \tag{7.7}$$

for $k = 1, \ldots, K$, $\ell = 1, \ldots, L$, and $n = 1, \ldots, N$, i.e. only $L \ll K$ rows are non-zero. Thus $\check{\mathbf{X}}$ exhibits a row-sparse structure, where the elements in a row of $\check{\mathbf{X}}$ are either jointly zero or primarily non-zero, as illustrated in Figure 7.2b.

The joint SSR problem can be formulated by the $\ell_{2,1}$ mixed-norm minimization problem [32, 64–68] according to

$$\min_{\mathbf{X}} \|\mathbf{x}\|_{2,1} \quad \text{s.t.} \quad \|\mathbf{A}(\nu)\,\mathbf{X} - \mathbf{Y}\|_F^2 \leq \beta \tag{7.8}$$

where the $\ell_{2,1}$ mixed-norm is defined as

$$\|\mathbf{x}\|_{2,1} = \sum_{k=1}^{K} \|\mathbf{x}_k\|_2, \tag{7.9}$$

applying an *inner* $\ell_2$ norm on the rows $\mathbf{x}_k$, for $k = 1, \ldots, K$, in $\mathbf{X} = [\mathbf{x}_1, \ldots, \mathbf{x}_K]^T$, and an *outer* $\ell_1$ norm on the $\ell_2$ row-norms. The *inner* $\ell_2$ norm provides a nonlinear coupling among the elements in a row, which, in combination with the *outer* $\ell_1$ norm, leads to the desired row-sparse structure of the signal matrix $\mathbf{X}$. In this sense the signals are represented by a Euclidian vector norm evaluated over multiple time samples and sparsity is enforced on these norms. Another less common approach in the literature also applies a mixed $\ell_{\infty,1}$ norm in Problem (7.8).

A major drawback of the mixed-norm minimization problem in (7.8) lies in its computational cost, which increases with the number of grid points $K$ and the number of snapshots $N$, reflected in the source signal matrix $\mathbf{X}$. Different techniques to reduce the computational cost are discussed in [32, 36, 69].

## Benefits of sparse reconstruction

A general advantage of SSR methods over traditional parameter estimation methods, such as subspace-based parameter estimation techniques [70–72], is that SSR methods show good estimation performance even in the small number of snapshots regime [32, 33, 36]. Furthermore, the SSR approach gives flexibility in constructing the overcomplete sensing matrix $\mathbf{A}(v)$, which can either be provided in analytic form or in the form of a dictionary obtained from calibration measurements. While traditional subspace-based methods rely on a prior signal detection step, SSR performs signal detection and parameter estimation simultaneously, given properly chosen regularization parameters.

With regard to the computational complexity, we note that for the convex problem formulations of the SSR problem outlined above, greedy algorithms and customized descent direction algorithms have been devised that scale well with the number of sensors and the number of signals. The case of uniform sampling allows for a further simplification in the form of gridless estimation, as discussed in [36, 73–76].

## 7.3 Collaborative Sparse Sensing

In Section 7.2 we introduced SSR for temporal and spatial sensing with a single SD. If multiple SDs are available, the devices can collaborate to improve the sensing performance, where different forms of collaboration are possible. In the following we consider the application of spatial sampling and direction of arrival estimation using multiple SDs with antenna arrays, as displayed in Figure 7.3, to discuss collaboration under coherent and non-coherent sampling at the single devices.

In practice it is desired to have sensor arrays with a large aperture and a large number of sensors to improve the estimation performance in terms of resolution and number of identifiable signals [40, 41]. It is usually difficult to obtain and maintain exact calibration of such large arrays, i.e. known phase relations among signals sampled in all sensors of the array, which generally requires that all sensors have access to a common clock. A practically more feasible approach is to consider the overall array as being composed of a collection of smaller subarrays that are exactly calibrated. This setup is referred to as the partly calibrated array (PCA) [37, 77–81]. For further illustration of this concept, we assume a number of $L$ IUs located in the far-field region of a sensor array composed of $M$ antennas, with the IUs transmitting narrowband signals in overlapping frequency bands. The overall sensor array consists of $P$ SDs, each having a subarray of $M_p$ antennas, for $p = 1, \ldots, P$, such that $M = \sum_{p=1}^{P} M_p$. For ease of presentation we assume that the antennas in each subarray are arranged in a uniform linear topology with spacing given as half the wavelength of the

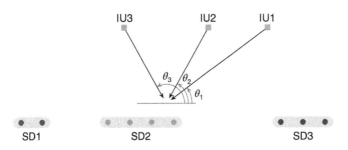

**Figure 7.3** PCA of $M = 9$ antennas partitioned in $P = 3$ subarrays and $L = 3$ IUs.

incoming signals' carrier frequency. This approach can easily be generalized to wideband signals that are commonly used in 4G and 5G networks by employing processing either jointly or disjointly on multiple subbands, cf. Section 7.5.

We consider the case when the relative sensor positions within each SD are precisely known, i.e. the subarrays are perfectly calibrated, whereas the exact positions of the SDs are unknown, resulting in direction-dependent phase offsets of the incoming signals for different SDs. Furthermore, as each SD gets its timing reference from its own local oscillators, the signals received in different devices may not be fully synchronized, resulting in a coherence loss across devices. We refer to this scenario as the non-coherent processing scheme. If a global clock reference of sufficient accuracy is available at the receivers from a common synchronization source, such as, for example, GPS signals, then the signals received in different SDs can be synchronized and coherently processed. We refer to this scenario as coherent processing.

### 7.3.1 Coherent Sparse Reconstruction

Consider the case when all subarrays of the sensing devices are synchronized in time and frequency and $N$ signal snapshots are obtained at the output of each subarray $p$, which are collected in the subarray measurement matrix $\mathbf{Y}^{(p)} = [\mathbf{y}^{(p)}(t_1), \ldots, \mathbf{y}^{(p)}(t_N)] \in \mathbb{C}^{M_p \times N}$, for $p = 1, \ldots, P$, where $[\mathbf{Y}^{(p)}]_{m,n}$ denotes the output of sensor $m$ in subarray $p$ and time instant $t_n$. The subarray measurement matrices are transmitted to a fusion center and composed in an $M \times N$ array measurement matrix $\mathbf{Y} = [\mathbf{Y}^{(1)T}, \ldots, \mathbf{Y}^{(P)T}]^T \in \mathbb{C}^{M \times N}$. According to (7.1) the measurement matrix is modeled as

$$\mathbf{Y} = \mathbf{A}(\mu, \varphi(\mu))\boldsymbol{\Psi} + \mathbf{N}, \tag{7.10}$$

where $\boldsymbol{\Psi} \in \mathbb{C}^{L \times N}$ is the source signal matrix and $\mathbf{N} \in \mathbb{C}^{M \times N}$ denotes the sensor noise matrix, defined in correspondence with the sensor measurements in $\mathbf{Y}$. The $M \times L$ matrix $\mathbf{A}(\mu, \varphi(\mu)) = [\mathbf{a}(\mu_1, \varphi(\mu_1)), \ldots, \mathbf{a}(\mu_L, \varphi(\mu_L))]$ represents the steering matrix of the entire array, where $\mathbf{a}(\mu, \varphi(\mu)) \in \mathbb{C}^M$ denotes the response of the entire array for spatial frequency $\mu$, and $\varphi(\mu) \in \mathbb{C}^P$ reflects unknown and possibly direction-dependent gain and phase offsets between the different subarrays, e.g. phase offsets in the reference clock or unknown subarray displacements of sensing devices [77, 82–84].

For further discussion, let $\mathbf{a}^{(p)}(\mu)$ denote the subarray response vector of the $p$th subarray for $p = 1, \ldots, P$, defined with the first sensor of each subarray denoting the subarray phase reference. Consider the factorization of the array response vector $\mathbf{a}(\mu)$ according to

$$\mathbf{a}(\mu, \varphi(\mu)) = \underbrace{\begin{bmatrix} \mathbf{a}^{(1)}(\mu) & 0 & 0 \\ 0 & \mathbf{a}^{(2)}(\mu) & 0 \\ 0 & 0 & \mathbf{a}^{(3)}(\mu) \end{bmatrix}}_{\mathbf{B}(\mu)} \varphi(\mu), \tag{7.11}$$

where the gray blocks illustrate exemplary vector realizations according to the PCA as illustrated in Figure 7.3. Furthermore, let $\mathbf{X} = \text{blkdiag}(\mathbf{x}^{(1)}, \ldots, \mathbf{x}^{(P)})$ denote the operator mapping the vectors $\mathbf{x}^{(1)}, \ldots, \mathbf{x}^{(P)}$ to the block-diagonal of the matrix $\mathbf{X}$, and define the

$M \times P$ subarray response matrix $\mathbf{B}(\mu) = \text{blkdiag}(\mathbf{a}^{(1)}(\mu), \ldots, \mathbf{a}^{(P)}(\mu))$. In relation to the steering matrix $\mathbf{A}(\mu, \varphi(\mu))$ of the entire array given in (7.10), define the $M \times PL$ subarray steering block matrix $\mathbf{B}(\mu) = [\mathbf{B}(\mu_1), \ldots, \mathbf{B}(\mu_L)]$ containing all the subarray response block matrices for the spatial frequencies in $\mu$, and the $PL \times L$ block-diagonal matrix $\boldsymbol{\Phi}(\mu) = \text{blkdiag}(\varphi(\mu_1), \ldots, \varphi(\mu_L))$ composed of the subarray shift vectors. Then the overall array steering matrix in (7.10) can be factorized as

$$\mathbf{A}(\mu, \varphi(\mu)) = \mathbf{B}(\mu)\,\boldsymbol{\Phi}(\mu). \tag{7.12}$$

**Sparse reconstruction problem**
Making use of the factorization (7.12), a sparse representation of the signal model in (7.10) for the PCA case is given as

$$\mathbf{Y} = \mathbf{B}(v)\,\boldsymbol{\Phi}(v)\,\check{\mathbf{X}} + \mathbf{N}, \tag{7.13}$$

where the row-sparse signal matrix $\check{\mathbf{X}}$ is defined similar to the single SD case in (7.7). The $M \times PK$ subarray sensing block matrix $\mathbf{B}(v)$ and the $PK \times K$ subarray shift matrix $\boldsymbol{\Phi}(v)$ are defined for a grid of $K \gg L$ sampled spatial frequencies $v = [v_1, \ldots, v_K]^T$.

In the PCA case, the unknown gain and phase offsets in $\boldsymbol{\Phi}(v)$ represent additional estimation variables and hence have to be appropriately included in the SSR problem. To this end, a model is introduced that couples among the variables $\check{\mathbf{x}}_k$ in the rows of $\check{\mathbf{X}} = [\check{\mathbf{x}}_1, \ldots, \check{\mathbf{x}}_K]^T$ and the subarray shifts $\varphi(v_k)$ in $\boldsymbol{\Phi}(v) = \text{blkdiag}(\varphi(v_1), \ldots, \varphi(v_K))$. Define the $KP \times N$ extended signal matrix $\check{\mathbf{Z}}$ as

$$\check{\mathbf{Z}} = \boldsymbol{\Phi}(v)\,\check{\mathbf{X}}, \tag{7.14}$$

containing the products of the subarray shifts and the signal waveforms. As compared to the total number of $K(N + P - 1)$ complex-valued unknowns in both the signal matrix $\check{\mathbf{X}}$ and the block-diagonal subarray shift matrix $\boldsymbol{\Phi}(v)$ in model (7.13), the number of unknown complex-valued signal elements in the matrix $\check{\mathbf{Z}}$, as defined in (7.14), is lifted to $KPN$. The coherent signal matrix $\check{\mathbf{Z}} = [\check{\mathbf{Z}}_1^T, \ldots, \check{\mathbf{Z}}_K^T]^T$ in (7.14) enjoys a beneficial structure as it is composed of $K$ stacked rank-one matrices

$$\check{\mathbf{Z}}_k = \varphi(v_k)\,\check{\mathbf{x}}_k^T, \tag{7.15}$$

of dimensions $P \times N$, for $k = 1, \ldots, K$, which follows from the block-diagonal structure of the subarray shift matrix $\boldsymbol{\Phi}(v) = \text{blkdiag}(\varphi(v_1), \ldots, \varphi(v_K))$. Using the coherent signal matrix in (7.14), the sparse representation for the PCA case in (7.13) is equivalently described by

$$\mathbf{Y} = \mathbf{B}(v)\check{\mathbf{Z}} + \mathbf{N}, \tag{7.16}$$

with the corresponding matrix structures illustrated in Figure 7.4.

An SSR approach to take account of the special low-rank block structure of the coherent signal matrix $\check{\mathbf{Z}}$ in (7.14) can be formulated using the *nuclear norm*, which has been successfully applied in a variety of rank minimization problems [85–88]. The definition of the nuclear norm is given as

$$\|\mathbf{Z}_k\|_* = \text{tr}((\mathbf{Z}_k^H \mathbf{Z}_k)^{1/2}) = \sum_{i=1}^{\min(P,N)} \sigma_{k,i}, \tag{7.17}$$

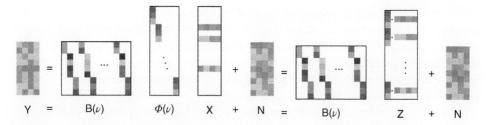

**Figure 7.4** Illustration of the matrix structure in the coherent PCA signal model.

where $\sigma_{k,i}$ is the $i$th largest singular value of $\mathbf{Z}_k \in \mathbb{C}^{P \times N}$. To gain intuition for the rank minimization properties of the nuclear norm, observe from (7.17) that the nuclear norm can be interpreted as the $\ell_1$ norm applied to the singular values of $\mathbf{Z}_k$. Thus, minimization of the nuclear norm leads to sparsity in the singular values, corresponding to a low matrix rank.

Regarding the rank minimization character of the nuclear norm, it was proposed in [89] to exploit the special structure of the signal model in (7.16) by the convex minimization problem

$$\min_{\mathbf{Z}} \|\mathbf{Z}\|_{*,1} \quad \text{s.t.} \quad \|\mathbf{B}(v)\mathbf{Z} - \mathbf{Y}\|_F^2 \leq \beta, \tag{7.18}$$

where $\|\mathbf{Z}\|_{*,1}$ denotes the $\ell_{*,1}$ mixed-norm, computed as

$$\|\mathbf{Z}\|_{*,1} = \sum_{k=1}^{K} \|\mathbf{Z}_k\|_*. \tag{7.19}$$

The formulation in (7.19) takes twofold advantage of the sparsity assumption. First, minimization of the nuclear norm terms encourages low-rank blocks $\hat{\mathbf{Z}}_1, \ldots, \hat{\mathbf{Z}}_K$ in the minimizer $\hat{\mathbf{Z}}$. Second, minimizing the sum of nuclear norms provides a block-sparse structure of $\hat{\mathbf{Z}} = [\hat{\mathbf{Z}}_1^T, \ldots, \hat{\mathbf{Z}}_K^T]^T$, i.e. the elements in each block $\hat{\mathbf{Z}}_k$, for $k = 1, \ldots, K$, are either jointly zero or primarily non-zero. Given the estimated signal matrix blocks $\hat{\mathbf{Z}}_k$, the subarray shifts $\varphi(v_k)$ and signal vectors $\mathbf{x}_k$ can be reconstructed from the principal singular vectors, as can be observed from (7.15) and explained in more detail in [37, 89]. Note that the PCA formulation for coherent processing in (7.18) reduces to the single sensing device formulation in (7.8) in the case of a single subarray, i.e. $P = 1$.

For implementation of the SSR problem in (7.18) by standard convex solvers, such as SeDuMi [90] or MOSEK [91], the semidefinite characterization of the nuclear norm discussed in [85] can be applied. Tailored implementations in form of the coordinate descent and the STELA method have been presented in [44, 89]. For means of complexity reduction the methods in [32, 37, 69] can be applied, and gridless implementations under uniform and shift invariant sampling have been proposed in [37, 80, 84].

### 7.3.2 Non-Coherent Sparse Reconstruction

We mentioned above that proper synchronization in time and frequency among the subarrays is difficult to achieve in practical applications. The resulting offsets in sampling lead to the effect that all subarrays seemingly observe the same source signal as different, independent source waveforms. We refer to this case as non-coherent processing among the subarrays, which stands in contrast to the coherent processing approach discussed in the previous section, where all subarrays observe scaled versions of the same source signals. In this context, an effect that is related to improper synchronization is that the increased size of the aperture in large PCAs might lead to a violation of the narrowband assumption, hence to a decorrelation of the signals received at the different subarrays, such that coherent processing cannot be applied.

For further discussion, consider the case of a partly calibrated array with imperfect timing synchronization among the different subarrays, where $t_1^{(p)}, \ldots, t_N^{(p)}$ denote the $N$ sampling instants of subarray $p$, for $p = 1, \ldots, P$. Similar to the previous section, the matrix $\tilde{\mathbf{Y}}^{(p)} = [\mathbf{y}^{(p)}(t_1^{(p)}), \ldots, \mathbf{y}^{(p)}(t_N^{(p)})] \in \mathbb{C}^{M_p \times N}$ contains the sensor measurements of subarray $p$ and can be modeled as [92–94]

$$\tilde{\mathbf{Y}}^{(p)} = \mathbf{A}^{(p)}(\boldsymbol{\mu}) \, \mathrm{diag}(\varphi^{(p)}(\boldsymbol{\mu})) \, \tilde{\boldsymbol{\Psi}}^{(p)} + \tilde{\mathbf{N}}^{(p)}, \tag{7.20}$$

where $\mathbf{A}^{(p)}(\boldsymbol{\mu}) = [\mathbf{a}^{(p)}(\mu_1), \ldots, \mathbf{a}^{(p)}(\mu_L)]$ denotes the steering matrix of subarray $p$, composed of the subarray steering vectors $\mathbf{a}^{(p)}(\mu)$ and $\varphi^{(p)}(\boldsymbol{\mu}) = [\varphi^{(p)}(\mu_1), \ldots, \varphi^{(p)}(\mu_L)]^T$, and contains the unknown subarray shifts $\varphi^{(p)}(\mu_l)$ due to the subarray displacement. The source signal matrix $\tilde{\boldsymbol{\Psi}}^{(p)} = [\tilde{\boldsymbol{\psi}}_1^{(p)T}, \ldots, \tilde{\boldsymbol{\psi}}_L^{(p)T}]^T$ is composed of the signal vectors $\tilde{\boldsymbol{\psi}}_l^{(p)} = [\psi_l(t_1^{(p)}), \ldots, \psi_l(t_N^{(p)})]^T$, denoting the signal transmitted by IU $\ell$ as observed and sampled by the reference sensor in subarray $p$, for $p = 1, \ldots, P$ and $\ell = 1, \ldots, L$. The matrix $\tilde{\mathbf{N}}^{(p)}$ represents the sensor noise matrix under non-coherent processing and is defined in correspondence with the subarray measurements in $\tilde{\mathbf{Y}}^{(p)}$.

Comparing the signal models for coherent processing and non-coherent processing in (7.20) and (7.10), it can be observed that the major difference in the two models lies in the representation of the source signals. In the coherent model, the same source signal matrix $\boldsymbol{\Psi}$ is used to describe the measurements in $\mathbf{Y}^{(p)}$ for the different subarrays, while for non-coherent processing a different source signal matrix $\tilde{\boldsymbol{\Psi}}^{(p)}$ is used to model the measurements $\tilde{\mathbf{Y}}^{(p)}$ for each subarray $p = 1, \ldots, P$.

**Sparse reconstruction problem**
In relation to sparse representation for a single array in (7.6), we formulate a sparse representation for the subarray signal model in (7.20) according to

$$\tilde{\mathbf{Y}}^{(p)} = \mathbf{A}^{(p)}(\boldsymbol{\mu}) \, \overset{\times}{\mathbf{X}}^{(p)} + \tilde{\mathbf{N}}^{(p)}, \tag{7.21}$$

where the sparse source signal representation under the on-grid assumption is given as

$$[\overset{\times}{\mathbf{X}}^{(p)}]_{k,n} = \begin{cases} \varphi^{(p)}(\mu_l) \, [\boldsymbol{\psi}^{(p)}(t_n^{(p)})]_\ell & \text{if } \nu_k = \mu_l \\ 0 & \text{otherwise}, \end{cases} \tag{7.22}$$

for $k = 1, \ldots, K, \ell = 1, \ldots, L, p = 1, \ldots, P$ and $n = 1, \ldots, N$. From (7.22) we observe that all subarray signal matrices $\overset{\times}{\mathbf{X}}^{(p)}$, for $p = 1, \ldots, P$, have the same support, i.e. non-zero row indices. As proposed in [93, 94], the common support can be exploited by distributed SSR techniques, e.g. by a mixed-norm minimization approach according to

$$\min_{\{\overset{\times}{\mathbf{X}}^{(p)}\}_{p=1}^{P}} \|[\overset{\times}{\mathbf{X}}^{(1)}, \ldots, \overset{\times}{\mathbf{X}}^{(P)}]\|_{2,1} \quad \text{s.t.} \quad \sum_{p=1}^{P} \|\mathbf{A}^{(p)}(\nu) \, \overset{\times}{\mathbf{X}}^{(p)} - \mathbf{Y}^{(p)}\|_F^2 \leq \beta. \tag{7.23}$$

Similar to the multiple measurement vector case in (7.8), the application of the mixed-norm in (7.23) introduces a nonlinear coupling among the rows in the subarray source signal observations $\tilde{\mathbf{X}}^{(p)}$, leading to a joint sparse solution. The sum constraint in (7.23), on the other hand, motivates a good model match.

To perform comparison of the coherent and non-coherent SSR formulations in (7.18) and (7.23), let us reformulate the subarray signal model in (7.20). In relation to the coherent signal model in (7.16), let the non-coherently sampled subarray measurements of the overall array be summarized as $\hat{\mathbf{Y}} = [\hat{\mathbf{Y}}^{(1)T}, \dots, \hat{\mathbf{Y}}^{(P)T}]^T$. We further formulate the overall signal model for the partly calibrated array under non-coherent processing as

$$\hat{\mathbf{Y}} = \mathbf{B}(\nu)\, \breve{\mathbf{Z}} + \tilde{\mathbf{N}}, \tag{7.24}$$

where the subarray sensing block matrix $\mathbf{B}(\nu)$ and the sensor noise matrix $\tilde{\mathbf{N}}$ are defined in correspondence to (7.16) and $\hat{\mathbf{Y}}$, respectively. The $KP \times N$ non-coherent source signal matrix $\breve{\mathbf{Z}} = [\breve{\mathbf{Z}}_1^T, \dots, \breve{\mathbf{Z}}_k^T]^T$ in (7.24) consists of blocks $\breve{\mathbf{Z}}_k$ containing the source signals observed by the different subarrays $p = 1, \dots, P$, for the sampled spatial frequency $\nu_k$, according to

$$[\breve{\mathbf{Z}}_k]_{p,n} = \begin{cases} \varphi^{(p)}(\mu_l) \, [\tilde{\psi}^{(p)}(t_n^{(p)})]_\ell & \text{for } \nu_k = \mu_l \\ 0 & \text{otherwise.} \end{cases} \tag{7.25}$$

The coherent and non-coherent signal models in (7.16) and (7.24) show similar structure. Both the coherent source signal matrix $\breve{\mathbf{Z}}$ under coherent processing and the non-coherent source signal matrix $\breve{\mathbf{Z}}$ exhibit a block-sparse structure, i.e. the elements in the $P \times N$ submatrices are either jointly zero or primarily non-zero. However, due to the non-coherent sampling at the different subarrays, the submatrices $\breve{\mathbf{Z}}_k$, for $k = 1, \dots, K$, of the non-coherent signal model lack the rack-one property given for the coherent source signal submatrices.

The block sparse structure of the non-coherent source signal matrix $\breve{\mathbf{Z}}$ can be taken into account by application of the Frobenius norm on the submatrices $\tilde{\mathbf{Z}}_k$, for $k = 1, \dots, K$. This results in the $\ell_{F,1}$ mixed-norm given as

$$\|\tilde{\mathbf{Z}}\|_{F,1} = \sum_{k=1}^{K} \|\tilde{\mathbf{Z}}_k\|_F = \sum_{k=1}^{K} \| \operatorname{vec}(\tilde{\mathbf{Z}}_k)\|_2. \tag{7.26}$$

From (7.26) it can be seen that the $\ell_{F,1}$ mixed-norm can be equivalently interpreted as an $\ell_{2,1}$ mixed-norm on the vectorized submatrices $\tilde{\mathbf{Z}}_k$. Using the signal model (7.24) and the $\ell_{F,1}$ mixed-norm (7.26), the non-coherent SSR problem (7.23) can equivalently be formulated as

$$\min_{\tilde{\mathbf{Z}}} \|\tilde{\mathbf{Z}}\|_{F,1} \quad \text{s.t.} \quad \|\mathbf{B}(\nu)\tilde{\mathbf{Z}} - \hat{\mathbf{Y}}\|_F^2 \le \beta. \tag{7.27}$$

Comparing the coherent and non-coherent SSR problems in (7.18) and (7.27), respectively, we observe that these mainly differ in the application of the nuclear and Frobenius norm. Note that the PCA formulation for non-coherent processing in (7.27) reduces to the single SD formulation in (7.8) in the case of a single subarray, i.e. $P = 1$. Similarly, in the case of a single snapshot, i.e. $N = 1$, the formulations (7.18) and (7.27) for coherent and non-coherent processing are identical.

## 7.4 Estimation Performance

For comparison of the estimation performance of the different SSR methods discussed in this chapter, we consider two numerical experiments regarding the angular sensing performance as well as an experiment with respect to the spatial location sensing performance. We refer to [37, 84, 89] for more rigorous experiments on the estimation performance regarding the presented SSR techniques.

For efficient implementation of the problem formulations (7.8), (7.18), and (7.23) we apply the compact equivalent formulations termed SPARROW and COBRAS with regularization parameter selection as presented in [37, 84, 89]. To admit unambiguous direction finding, the sensor spacing within the subarrays is selected as half the wavelength of the incoming signals for all simulations in this section.

### 7.4.1 Comparison of Centralized, Distributed, and Collaborative Sensing

Consider in the first experiment a uniform linear subarray of $M = 16$ sensors, equally partitioned into $P = 4$ uniform linear subarrays of $M_p = 4$ sensors per subarray, for $p = 1, \dots, 4$. Furthermore, assume $L = 2$ IUs with spatial frequencies $\mu_1 = \cos \theta_1 = 0.25$ and $\mu_2 = \cos \theta_2 = 0.5$, located in the far-field region of the array. The IUs are transmitting uncorrelated complex Gaussian signals with equal power in overlapping frequency bands, resulting in a receive signal-to-noise ratio (SNR) of 0 dB at the array. A number of $N = 50$ snapshots of the incoming signals is recorded by the array to perform sparse reconstruction on a grid of $K = 200$ spatial frequencies.

Figure 7.5 shows the spatial spectra obtained by the different SSR methods (7.8), (7.18), and (7.23). The figure also includes histograms showing the statistical frequency of the non-zero elements in the estimated sparse signals obtained for 100 Monte Carlo runs. The figure illustrates well that all SSR methods provide a sparse spectrum, as illustrated for one sample realization. Regarding the estimation performance it can be observed from the histogram that SSR using the fully calibrated array (FCA) shows the smallest variances in the estimated signal support around the true frequencies. The nuclear norm minimization approach (7.18) for coherent PCAs only shows slightly larger variances. The largest variances are obtained for the non-coherent PCA and the single subarray. The figure illustrates the basic tendency that larger array apertures and coherent processing improve the resolution of the parameter estimation in case of multiple signals, as compared to smaller apertures and non-coherent processing [37, 40, 41, 84].

To further investigate the effect of collaboration on the statistical error performance, consider a uniform linear array of $M = PM_0$ sensors, equally partitioned into a variable number of uniform linear subarrays $P$ and sensors per subarray $M_0 = M_p$, for $p = 1, \dots, P$. Two closely spaced sources with spatial frequencies $\mu_1 = \cos \theta_1 = 0.3$ and $\mu_2 = \cos \theta_2 = 0.5$ are located in the far-field region of the array, emitting equal power and uncorrelated complex Gaussian signals. The signals are impinging on the array with an SNR of 0 dB and a number of $N = 20$ snapshots is recorded. Figure 7.6 shows the resulting root mean square error (RMSE) of the estimated spatial frequencies, where the RMSE is defined according to [84].

It is observed from Figure 7.6a, that the best estimation performance is achieved when all sensors are forming a fully calibrated array. Here, the estimation performance increases

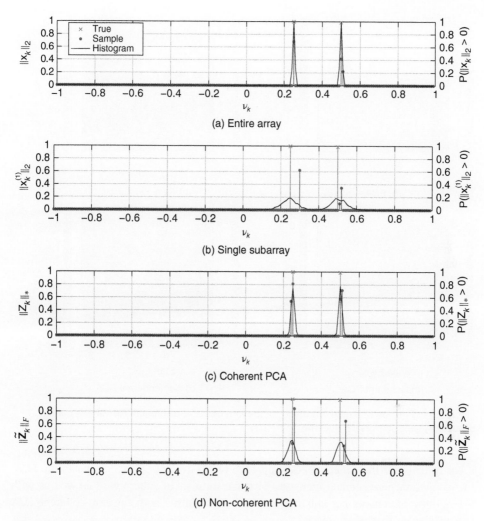

**Figure 7.5** Spatial spectrum for different SSR techniques and histogram of the non-zero elements.

with the total number of sensors available, which is given as $M = PM_0$. On the other hand, worst performance is obtained when only a single subarray is utilized, as observed from Figure 7.6b. In the case of a single subarray, the estimation performance is only affected by the number of sensors per subarray $M_0$ and not by the number of subarrays $P$.

Considering the non-coherent PCA approach (7.23). It can be observed from Figure 7.6d that for the setup under investigation, the estimation performance mainly depends on the number of sensors per subarray $M_0$, and only slightly improves with the number of subarrays $P$. This behavior can be explained by the fact, that in the case of identical subarrays, the use of non-coherent processing of data recorded with multiple subarrays is mathematically equivalent to processing multiple snapshots from a single subarray, as can be observed from (7.23). Hence, in non-coherent processing the use of multiple identical subarrays only helps to reduce noise effects. However, the effective size of the aperture, which is a key indicator for resolution performance, remains unchanged.

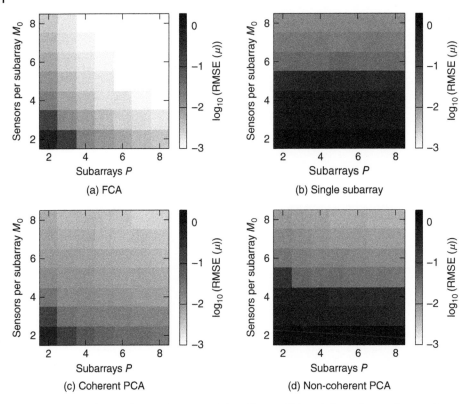

**Figure 7.6** Direction finding performance of the different SSR techniques for varying number of subarrays and sensors per subarray.

Lastly, consider the coherent PCA approach (7.18) in Figure 7.6c. Similar to the FCA, the estimation performance increases with the number of subarrays $P$ as well as with the number of sensors per subarray $M_0$ and outperforms both the non-coherent PCA and the single subarray approach.

### 7.4.2 Source Localization

As proposed in [92, 95, 96], distributed partly calibrated sensor arrays can also be used for direct source position estimation, where the signal phase shift information resulting at the subarrays for different source positions is exploited for estimation in a single step, in contrast to the more common approach of first estimating the directions of arrival at the single subarrays and subsequently performing triangulation. The corresponding system model can similarly be utilized for coherent and non-coherent sparse reconstruction according to (7.18) and (7.27). To this end, in the third numerical experiment a variable number $P$ of subarrays of two sensors in the $x$–$y$ plane and with different orientations, as indicated in Figure 7.7, is considered. Furthermore, a variable number of sources in the $x$–$y$ plane, are transmitting narrowband signals in the same frequency band. For the experiment pathloss effects are neglected, and the SNR is set to 20 dB while the number

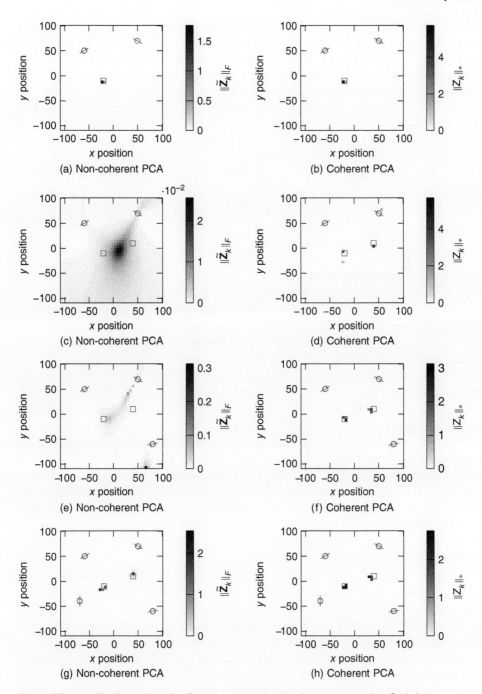

**Figure 7.7** Localization estimation for non-coherent and coherent processing. Red circle markers ⌀ indicate subarray positions and orientations, while red square markers □ indicate source positions.

of snapshots is set to $N = 50$, providing a good signal quality in the experiment such that noise effects can be neglected in the signal reconstruction performance. A uniform rectangular gird of $K = 40 \times 40$ positions is used for the two SSR methods.

As seen from Figure 7.7a,b, both the non-coherent and coherent reconstruction methods can well estimate a single source position if $P = 2$ subarrays are used. In the case of $L = 2$ source signals, the non-coherent method fails to resolve the two source positions using $P = 2$ and $P = 3$ subarrays, as displayed in Figure 7.7c,e. Only for the case of $P = 4$ subarrays can the non-coherent method resolve the source signals.

The coherent method, in turn, can well resolve the two signals independent of the number of subarrays considered in the setup, as can be observed from Figure 7.7d,f,g. The experiment illustrates that for the multiple source case and non-coherent processing the estimation problem may become non-identifiable if the number of distributed subarrays is not sufficiently large, resulting in ambiguous position estimates.

## 7.5 Concluding Remarks

In this chapter we discussed different techniques for collaborative sensing in frequency and angular direction that are based on sparse optimization. Sparse representation based sensing approaches have the benefit that they are computationally efficient as the underlying optimization problems are convex and can be solved using efficient parallel algorithms. In the collaborative approaches we have considered the scenario that different sensing devices share their measurements, whether coherently or non-coherently recorded, with a fusion center, in which the sparse optimization is performed and the network wide sensing solution is computed centrally. The fusion center could, for example, be an intelligent edge node of a 5G communication network to which the sensing devices are reporting. In distributed sensing networks a fusion center may not exist. In this case the parallel optimization algorithm can be implemented in a fully distributed fashion over the network, where each SD in the network is exchanging gradient information only locally with its neighbors and updates are performed locally using consensus and diffusion type algorithms [59, 63, 97]. These distributed implementations based on in-network processing exhibit a low communication overhead and have the benefit that communication bottlenecks at the fusion center are avoided.

The sparse sensing approaches discussed in the chapter combine high-resolution sensing and detection of available spectral resources. Unlike conventional detection approaches based on hypothesis testing, in which some test metric is compared to a detection threshold, sparsity based sensing approaches avoid the requirement of defining a threshold, which is often difficult in practice. Instead in SSR methods, a regularization parameter is balancing between the model mismatch and the sparsity of the solution, and thus it inherently defines the detection thresholds. Different formulations of the compressed sensing problem exist. We considered the basis pursuit problem formulation of type (7.5) and (7.8) in which sparsity is induced in the objective function and the model is matched to the measurements in the constraint. A closely related variant of this problem formulation is the LASSO formulation in which the constraint is relaxed into the Lagrangian [23].

Finally, we remark that for ease of presentation, in this chapter the narrowband far-field assumption under the line-of-sight model has been used and the simple uniform linear sensor geometry has been considered. While these assumptions widely simplify the description of the sensing techniques, we remark that the sparse representation based spectrum sensing techniques are flexible and can be applied in a much wider class of sensing configurations. This includes near-field sensing for which the sensing matrix is, for example, obtained by sampling range and direction of arrival (cf. the numerical examples in Section 7.4) and sensing with arrays that do not exhibit an analytic description of the array manifold, e.g. arrays characterized by calibration tables. Furthermore, the sparse representation based sensing techniques described in this chapter show excellent performance also in the generally difficult scenario of correlated source signals, such that the methods naturally extend to spatial sensing under multi-path propagation [44]. We would also like to remark that, with respect to the multiple measurements model, we considered the static source case in which the support of the sparse representation vector does not change over time. Motion models can be incorporated in the row sparsity based approaches outlined in this chapter using, for example, the concept of group sparsity with overlapping groups to account for the motion of the incumbent user's transmitters [98]. Finally, for simplicity in this chapter we only considered one-dimensional sensing, i.e. collaborative wideband frequency spectrum sensing with a single sensor or spatial spectrum sensing of narrowband signal with multiple sensors but not joint spatial and frequency sensing of wideband signals. The sensing techniques can, however, be extended to joint sensing in multiple dimensions [44, 99, 100]. Joint sensing yields improved sensing performance as signals can be resolved in two or multiple domains, e.g. frequency, direction of arrival, Doppler. The improved resolution performance comes at the expense of computational complexity.

# References

**1** T. Yucek and H. Arslan, "A survey of spectrum sensing algorithms for cognitive radio applications," *IEEE Communications Surveys Tutorials*, vol. 11, no. 1, pp. 116–130, Jan. 2009.

**2** D. Cabric, S. M. Mishra, and R. W. Brodersen, "Implementation issues in spectrum sensing for cognitive radios," in *Conference Record of the Thirty-Eighth Asilomar Conference on Signals, Systems and Computers, 2004.*, vol. 1, Nov. 2004, pp. 772–776.

**3** E. Axell, G. Leus, E. G. Larsson, and H. V. Poor, "Spectrum sensing for cognitive radio: State-of-the-art and recent advances," *IEEE Signal Processing Magazine*, vol. 29, no. 3, pp. 101–116, May 2012.

**4** H. Urkowitz, "Energy detection of unknown deterministic signals," *Proceedings of the IEEE*, vol. 55, no. 4, pp. 523–531, Apr. 1967.

**5** R. Tandra and A. Sahai, "Fundamental limits on detection in low SNR under noise uncertainty," in *2005 International Conference on Wireless Networks, Communications and Mobile Computing*, vol. 1, Jun. 2005, pp. 464–469.

**6** F. F. Digham, M. Alouini, and M. K. Simon, "On the energy detection of unknown signals over fading channels," *IEEE Transactions on Communications*, vol. 55, no. 1, pp. 21–24, Jan. 2007.

**7** R. Tandra and A. Sahai, "SNR walls for signal detection," *IEEE Journal of Selected Topics in Signal Processing*, vol. 2, no. 1, pp. 4–17, Feb. 2008.

**8** S. Kapoor, S. Rao, and G. Singh, "Opportunistic spectrum sensing by employing matched filter in cognitive radio network," in *2011 International Conference on Communication Systems and Network Technologies*, Jun. 2011, pp. 580–583.

**9** W. A. Gardner, A. Napolitano, and L. Paura, "Cyclostationarity: Half a century of research," *Signal processing*, vol. 86, no. 4, pp. 639–697, 2006.

**10** A. V. Dandawate and G. B. Giannakis, "Statistical tests for presence of cyclostationarity," *IEEE Transactions on Signal Processing*, vol. 42, no. 9, pp. 2355–2369, Sep. 1994.

**11** J. Lunden, V. Koivunen, A. Huttunen, and H. V. Poor, "Spectrum sensing in cognitive radios based on multiple cyclic frequencies," in *2007 2nd International Conference on Cognitive Radio Oriented Wireless Networks and Communications*, Aug. 2007, pp. 37–43.

**12** V. Turunen, M. Kosunen, A. Huttunen, S. Kallioinen, P. Ikonen, A. Parssinen, and J. Ryynanen, "Implementation of cyclostationary feature detector for cognitive radios," in *2009 4th International Conference on Cognitive Radio Oriented Wireless Networks and Communications*, Jun. 2009, pp. 1–4.

**13** M. Wax and T. Kailath, "Detection of signals by information theoretic criteria," *IEEE Transactions on Acoustics, Speech, and Signal Processing*, vol. 33, no. 2, pp. 387–392, Apr. 1985.

**14** A. Taherpour, M. Nasiri-Kenari, and S. Gazor, "Multiple antenna spectrum sensing in cognitive radios," *IEEE Transactions on Wireless Communications*, vol. 9, no. 2, pp. 814–823, Feb. 2010.

**15** P. Wang, J. Fang, N. Han, and H. Li, "Multiantenna-assisted spectrum sensing for cognitive radio," *IEEE Transactions on Vehicular Technology*, vol. 59, no. 4, pp. 1791–1800, May 2010.

**16** P. Bianchi, M. Debbah, M. Maida, and J. Najim, "Performance of statistical tests for single-source detection using random matrix theory," *IEEE Transactions on Information Theory*, vol. 57, no. 4, pp. 2400–2419, Apr. 2011.

**17** P. Urriza, E. Rebeiz, and D. Cabric, "Multiple antenna cyclostationary spectrum sensing based on the cyclic correlation significance test," *IEEE Journal on Selected Areas in Communications*, vol. 31, no. 11, pp. 2185–2195, Nov. 2013.

**18** S. Sedighi, A. Taherpour, T. Khattab, and M. O. Hasna, "Multiple antenna cyclostationary-based detection of primary users with multiple cyclic frequency in cognitive radios," in *2014 IEEE Global Communications Conference*, Dec. 2014, pp. 799–804.

**19** S. M. Mishra, A. Sahai, and R. W. Brodersen, "Cooperative sensing among cognitive radios," in *2006 IEEE International Conference on Communications*, vol. 4, Jun. 2006, pp. 1658–1663.

**20** J. Lunden, V. Koivunen, A. Huttunen, and H. V. Poor, "Collaborative cyclostationary spectrum sensing for cognitive radio systems," *IEEE Transactions on Signal Processing*, vol. 57, no. 11, pp. 4182–4195, Nov. 2009.

**21** Z. Li, F. R. Yu, and M. Huang, "A distributed consensus-based cooperative spectrum-sensing scheme in cognitive radios," *IEEE Transactions on Vehicular Technology*, vol. 59, no. 1, pp. 383–393, Jan. 2010.

**22** F. Penna and S. Stańczak, "Decentralized largest eigenvalue test for multi-sensor signal detection," in *2012 IEEE Global Communications Conference (GLOBECOM)*, Dec. 2012, pp. 3893–3898.

**23** R. Tibshirani, "Regression shrinkage and selection via the LASSO," *Journal of the Royal Statistical Society. Series B (Methodological)*, vol. 58, pp. 267–288, 1996.

**24** S. S. Chen, D. L. Donoho, and M. A. Saunders, "Atomic decomposition by basis pursuit," *SIAM Journal On Scientific Computing*, vol. 20, pp. 33–61, 1998.

**25** D. L. Donoho and M. Elad, "Optimally sparse representation in general (nonorthogonal) dictionaries via $\ell^1$ minimization," in *Proceedings of the National Academy of Sciences*, vol. 100, no. 5, 2003, pp. 2197–2202.

**26** E. Candès and T. Tao, "Decoding by linear programming," *IEEE Transactions on Information Theory*, vol. 51, no. 12, pp. 4203–4215, Dec. 2005.

**27** D. Donoho, "Compressed sensing," *IEEE Transactions on Information Theory*, vol. 52, no. 4, pp. 1289–1306, Apr. 2006.

**28** E. Candès, J. Romberg, and T. Tao, "Robust uncertainty principles: Exact signal reconstruction from highly incomplete frequency information," *IEEE Transactions on Information Theory*, vol. 52, no. 2, pp. 489–509, Feb. 2006.

**29** E. J. Candès, J. K. Romberg, and T. Tao, "Stable signal recovery from incomplete and inaccurate measurements," *Communications on pure and applied mathematics*, vol. 59, no. 8, pp. 1207–1223, Aug. 2006.

**30** E. J. Candès and J. Romberg, "Quantitative robust uncertainty principles and optimally sparse decompositions," vol. 6, no. 2, pp. 227–254, 2006.

**31** T. Chernyakova and Y. C. Eldar, "Fourier-domain beamforming: the path to compressed ultrasound imaging," *IEEE Transactions on Ultrasonics, Ferroelectrics, and Frequency Control*, vol. 61, no. 8, pp. 1252–1267, Aug. 2014.

**32** D. Malioutov, M. Çetin, and A. Willsky, "A sparse signal reconstruction perspective for source localization with sensor arrays," *IEEE Transactions on Signal Processing*, vol. 53, no. 8, pp. 3010–3022, 2005.

**33** S. Fortunati, R. Grasso, F. Gini, and M. S. Greco, "Single snapshot DOA estimation using compressed sensing," in *2014 IEEE International Conference on Acoustics, Speech and Signal Processing (ICASSP)*, May 2014, pp. 2297–2301.

**34** D. D. Ariananda and G. Leus, "Compressive wideband power spectrum estimation," *IEEE Transactions on Signal Processing*, vol. 60, no. 9, pp. 4775–4789, Sep. 2012.

**35** D. Cohen and Y. C. Eldar, "Sub-nyquist sampling for power spectrum sensing in cognitive radios: A unified approach," *IEEE Transactions on Signal Processing*, vol. 62, no. 15, pp. 3897–3910, Aug 2014.

**36** C. Steffens, M. Pesavento, and M. E. Pfetsch, "A compact formulation for the $\ell_{2,1}$ mixed-norm minimization problem," *IEEE Transactions on Signal Processing*, vol. 66, no. 6, pp. 1483–1497, Mar. 2018.

**37** C. Steffens and M. Pesavento, "Block- and rank-sparse recovery for direction finding in partly calibrated arrays," *IEEE Transactions on Signal Processing*, vol. 66, no. 2, pp. 384–399, Jan. 2018.

**38** J. A. Bazerque and G. B. Giannakis, "Distributed spectrum sensing for cognitive radio networks by exploiting sparsity," *IEEE Transactions on Signal Processing*, vol. 58, no. 3, pp. 1847–1862, March 2010.

**39** S. K. Sharma, E. Lagunas, S. Chatzinotas, and B. Ottersten, "Application of compressive sensing in cognitive radio communications: A survey," *IEEE Communications Surveys Tutorials*, vol. 18, no. 3, pp. 1838–1860, thirdquarter 2016.

**40** H. Krim and M. Viberg, "Two decades of array signal processing research: The parametric approach," *IEEE Signal Processing Magazine*, vol. 13, no. 4, pp. 67–94, Jul. 1996.

**41** H. L. van Trees, Optimum Array Processing: Part IV of Detection, Estimation, and Modulation Theory. New York: John Wiley & Sons, Inc., 2002.

**42** M. Pesavento, C. F. Mecklenbräuker, and J. F. Böhme, "Multidimensional rank reduction estimator for parametric mimo channel models," *EURASIP Journal on Applied Signal Processing*, vol. 2004, pp. 1354–1363, 2004.

**43** S. Byun, W. Seong, and S. Kim, "Sparse underwater acoustic channel parameter estimation using a wideband receiver array," *IEEE Journal of Oceanic Engineering*, vol. 38, no. 4, pp. 718–729, Oct. 2013.

**44** C. Steffens, Y. Yang, and M. Pesavento, "Multidimensional sparse recovery for MIMO channel parameter estimation," in *Proceedings of the European Signal Processing Conference (EUSIPCO)*, Budapest, Hungary, Sep. 2016, pp. 66–70.

**45** Y. Chi, L. Scharf, A. Pezeshki, and A. Calderbank, "Sensitivity to basis mismatch in compressed sensing," *IEEE Transactions on Signal Processing*, vol. 59, no. 5, pp. 2182–2195, May 2011.

**46** S. Foucart and H. Rauhut, *A mathematical introduction to compressive sensing*. Basel: Birkhäuser, 2013.

**47** S. Mallat and Z. Zhang, "Matching pursuits with time-frequency dictionaries," *IEEE Transactions on Signal Processing*, vol. 41, no. 12, pp. 3397–3415, Dec. 1993.

**48** T. Blumensath and M. E. Davies, "Iterative thresholding for sparse approximations," *Journal of Fourier Analysis and Applications*, vol. 14, no. 5-6, pp. 629–654, Dec. 2008.

**49** H. Zou, "The adaptive LASSO and its oracle properties," *Journal of the American statistical association*, vol. 101, no. 476, pp. 1418–1429, 2006.

**50** E. J. Candès, M. B. Wakin, and S. P. Boyd, "Enhancing sparsity by reweighted $\ell_1$ minimization," *Journal of Fourier analysis and applications*, vol. 14, no. 5, pp. 877–905, 2008.

**51** D. L. Donoho, A. Maleki, and A. Montanari, "Message-passing algorithms for compressed sensing," *Proceedings of the National Academy of Sciences*, vol. 106, no. 45, pp. 18 914–18 919, 2009.

**52** M. E. Tipping, "Sparse Bayesian learning and the relevance vector machine," *Journal of machine learning research*, vol. 1, pp. 211–244, Jun. 2001.

**53** D. P. Wipf and B. D. Rao, "Sparse Bayesian learning for basis selection," *IEEE Transactions on Signal processing*, vol. 52, no. 8, pp. 2153–2164, 2004.

**54** J. A. Tropp and S. J. Wright, "Computational methods for sparse solution of linear inverse problems," *Proceedings of the IEEE*, vol. 98, no. 6, pp. 948–958, 2010.

**55** R. T. Rockafellar, Convex Analysis. Princeton University Press, 1970.

**56** D. Bertsekas, *Nonlinear Programming*, 2nd ed. Belmont: Athena Scientific, 1999.

**57** B. Efron, T. Hastie, I. Johnstone, and R. Tibshirani, "Least angle regression," *Annals of Statistics*, vol. 32, pp. 407–499, 2004.

**58** J. Friedman, T. Hastie, H. Höfling, and R. Tibshirani, "Pathwise coordinate optimization," *The Annals of Applied Statistics*, vol. 1, pp. 302–332, 2007.

**59** Y. Yang and M. Pesavento, "A unified successive pseudoconvex approximation framework," *IEEE Transactions on Signal Processing*, vol. 65, no. 13, pp. 3313–3328, Jul. 2017.

**60** K. Koh, S.-J. Kim, and S. Boyd, "An interior-point method for large-scale $\ell_1$-regularized logistic regression," *Journal of Machine learning research*, vol. 8, no. Jul., pp. 1519–1555, 2007.

**61** P. L. Combettes and V. R. Wajs, "Signal recovery by proximal forward-backward splitting," *Multiscale Modeling & Simulation*, vol. 4, no. 4, pp. 1168–1200, 2005.

**62** N. Parikh, S. Boyd *et al.*, "Proximal algorithms," *Foundations and Trends® in Optimization*, vol. 1, no. 3, pp. 127–239, 2014.

**63** S. Boyd, N. Parikh, E. Chu, B. Peleato, and J. Eckstein, "Distributed optimization and statistical learning via the alternating direction method of multipliers," *Foundations and Trends® in Machine Learning*, vol. 3, no. 1, pp. 1–122, 2011.

**64** B. A. Turlach, W. N. Venables, and S. J. Wright, "Simultaneous variable selection," *Technometrics*, vol. 47, no. 3, pp. 349–363, 2005.

**65** M. Yuan and Y. Lin, "Model selection and estimation in regression with grouped variables," *Journal of the Royal Statistical Society. Series B (Statistical Methodology)*, vol. 68, no. 1, pp. 49–67, 2006.

**66** J. A. Tropp, "Algorithms for simultaneous sparse approximation. Part II: Convex relaxation," *Signal Processing*, vol. 86, no. 3, pp. 589–602, 2006.

**67** M. Kowalski, "Sparse regression using mixed norms," *Applied and Computational Harmonic Analysis*, vol. 27, no. 3, pp. 303–324, 2009.

**68** M. M. Hyder and K. Mahata, "Direction-of-arrival estimation using a mixed $\ell_{2,0}$ norm approximation," *IEEE Transactions on Signal Processing*, vol. 58, no. 9, pp. 4646–4655, Sep. 2010.

**69** Z. Yang, J. Li, P. Stoica, and L. Xie, "Sparse methods for direction-of-arrival estimation," in Academic Press Library in Signal Processing - Array, Radar and Communications Engineering, 1st ed., S. Theodoridis and R. Chellappa, Eds. Academic Press, Oct. 2017, vol. 7, ch. 11.

**70** A. B. Gershman, M. Rübsamen, and M. Pesavento, "One- and two-dimensional direction-of-arrival estimation: An overview of search-free techniques," *Signal Processing*, vol. 90, no. 5, pp. 1338–1349, 2010.

**71** A. Barabell, "Improving the resolution performance of eigenstructure-based direction-finding algorithms," in *Proceedings of the IEEE International Conference on Acoustics, Speech, and Signal Processing (ICASSP)*, vol. 8, Apr. 1983, pp. 336–339.

**72** R. Schmidt, "Multiple emitter location and signal parameter estimation," *IEEE Transactions on Antennas and Propagation*, vol. 34, no. 3, pp. 276–280, Mar. 1986.

**73** E. J. Candès and C. Fernandez-Granda, "Towards a mathematical theory of super-resolution," *Communications on Pure and Applied Mathematics*, vol. 67, no. 6, pp. 906–956, 2014.

**74** ——, "Super-resolution from noisy data," *Journal of Fourier Analysis and Applications*, vol. 19, no. 6, pp. 1229–1254, 2013.

**75** G. Tang, B. Bhaskar, P. Shah, and B. Recht, "Compressed sensing off the grid," *IEEE Transactions on Information Theory*, vol. 59, no. 11, pp. 7465–7490, Nov. 2013.

**76** Z. Yang and L. Xie, "Continuous compressed sensing with a single or multiple measurement vectors," in *Proceedings of the IEEE Workshop on Statistical Signal Processing (SSP)*, Jun. 2014, pp. 288–291.

**77** M. Pesavento, A. Gershman, and K. M. Wong, "Direction finding in partly calibrated sensor arrays composed of multiple subarrays," *IEEE Transactions on Signal Processing*, vol. 50, no. 9, pp. 2103–2115, 2002.

**78** P. Parvazi and M. Pesavento, "A new direction-of-arrival estimation and calibration method for arrays composed of multiple identical subarrays," in *Proceedings of the IEEE International Workshop on Signal Processing Advances in Wireless Communications (SPAWC)*, Jun. 2011, pp. 171–175.

**79** W. Suleiman, M. Pesavento, and A. M. Zoubir, "Performance analysis of the decentralized eigendecomposition and ESPRIT algorithm," *IEEE Transactions on Signal Processing*, vol. 64, no. 9, pp. 2375–2386, May 2016.

**80** C. Steffens, W. Suleiman, A. Sorg, and M. Pesavento, "Gridless compressed sensing under shift-invariant sampling," in *Proceedings of the IEEE International Conference on Acoustics, Speech and Signal Processing (ICASSP)*, Mar. 2017, pp. 4735–4739.

**81** W. Suleiman, P. Parvazi, M. Pesavento, and A. M. Zoubir, "Non-coherent direction-of-arrival estimation using partly calibrated arrays," *IEEE Transactions on Signal Processing*, vol. 66, no. 21, pp. 5776–5788, Nov 2018.

**82** C. See and A. Gershman, "Direction-of-arrival estimation in partly calibrated subarray-based sensor arrays," *IEEE Transactions on Signal Processing*, vol. 52, no. 2, pp. 329–338, 2004.

**83** P. Parvazi, "Sensor array processing in difficult and non-idealistic conditions," Ph.D. dissertation, Technische Universität Darmstadt, 2012.

**84** C. Steffens, "Compact formulations for sparse reconstruction in fully and partly calibrated arrays," Ph.D. dissertation, Technische Universität Darmstadt, 2018.

**85** M. Fazel, H. Hindi, and S. Boyd, "A rank minimization heuristic with application to minimum order system approximation," in *Proceedings of the American Control Conference (ACC)*, vol. 6, 2001, pp. 4734–4739.

**86** J. Cai, E. Candès, and Z. Shen, "A singular value thresholding algorithm for matrix completion," *SIAM Journal on Optimization*, vol. 20, pp. 1956–1982, 2010.

**87** B. Recht, M. Fazel, and P. A. Parrilo, "Guaranteed minimum-rank solutions of linear matrix equations via nuclear norm minimization," *SIAM Review*, vol. 52, no. 3, pp. 471–501, 2010.

**88** E. Candès and B. Recht, "Exact matrix completion via convex optimization," *Communications of the ACM*, vol. 55, no. 6, pp. 111–119, 2012.

89 C. Steffens, P. Parvazi, and M. Pesavento, "Direction finding and array calibration based on sparse reconstruction in partly calibrated arrays," in *IEEE Sensor Array and Multichannel Signal Processing Workshop (SAM)*, Jun. 2014, pp. 21–24.

90 J. Sturm, "Using SeDuMi 1.02, a MATLAB toolbox for optimization over symmetric cones," *Optimization Methods and Software*, vol. 11–12, pp. 625–653, 1999.

91 *The MOSEK optimization toolbox for MATLAB manual. Version 7.1 (Revision 28)*, MOSEK ApS, 2015. [Online]. Available: http://docs.mosek.com/7.1/toolbox/index .html7.

92 M. Wax and T. Kailath, "Decentralized processing in sensor arrays," *IEEE Transactions on Acoustics, Speech, and Signal Processing*, vol. 33, no. 5, pp. 1123–1129, 1985.

93 M. F. Duarte, S. Sarvotham, D. Baron, M. B. Wakin, and R. G. Baraniuk, "Distributed compressed sensing of jointly sparse signals," in *Conference Record of the Asilomar Conference on Signals, Systems and Computers*, Oct. 2005, pp. 1537–1541.

94 Z. Lu, R. Ying, S. Jiang, P. Liu, and W. Yu, "Distributed compressed sensing off the grid," *IEEE Signal Processing Letters*, vol. 22, no. 1, pp. 105–109, Jan. 2015.

95 A. J. Weiss and A. Amar, "Direct position determination of multiple radio signals," *EURASIP Journal on Applied Signal Processing*, vol. 2005, pp. 37–49, 2005.

96 D. Xie, J. Huang, and H. Ge, "Localization of near-field sources with partly calibrated subarray-based array," in *2010 5th IEEE Conference on Industrial Electronics and Applications*, Jun. 2010, pp. 1758–1761.

97 M. Mardani, G. Mateos, and G. B. Giannakis, "Decentralized sparsity-regularized rank minimization: Algorithms and applications," *IEEE Transactions on Signal Processing*, vol. 61, no. 21, pp. 5374–5388, Nov. 2013.

98 N. Rao, R. Nowak, C. Cox, and T. Rogers, "Classification with the sparse group lasso," *IEEE Transactions on Signal Processing*, vol. 64, no. 2, pp. 448–463, Jan. 2016.

99 Z. Yang, L. Xie, and P. Stoica, "Vandermonde decomposition of multilevel Toeplitz matrices with application to multidimensional super-resolution," *IEEE Transactions on Information Theory*, vol. 62, no. 6, pp. 3685–3701, Jun. 2016.

100 Z. Tian, Z. Zhang, and Y. Wang, "Low-complexity optimization for two-dimensional direction-of-arrival estimation via decoupled atomic norm minimization," in *Proceedings of the IEEE International Conference on Acoustics, Speech and Signal Processing (ICASSP)*, Mar. 2017, pp. 3071–3075.

# 8

# Cooperative Communication Techniques for Spectrum Sharing

*Faheem Khan[1], Miltiades C. Filippou[2], and Mathini Sellathurai[3]*

[1]*University of Huddersfield, UK*
[2]*Intel Deutschland GmbH, Germany*
[3]*Heriot-Watt University, Edinburgh, UK*

## 8.1  Introduction

Cognitive radio (CR) has received significant attention in recent years as a key enabling technology to implement dynamic spectrum access (DSA) to achieve efficient spectrum utilization [11, 12, 15, 19, 20]. The early research on CR mainly dealt with the opportunistic spectrum access (OSA) capability of CR, an interference avoidance approach, wherein the CR ensures that it only transmits when the primary user does not transmit and stops as soon as it detects any primary user's transmission. This approach is also referred to as "interweave" in the CR literature. IEEE 802.22 is the first worldwide standard that is based on OSA CR, targeting wireless regional area networks (WRAN). It operates in TV white spaces from 54 to 862 MHz on a non-interfering basis with primary users (TV broadcasters) and provides rural wireless broadband access.

The key challenge in OSA CR is the detection of primary user activity over a wide range of frequencies. The CR network is required to continuously monitor the used spectrum to detect the presence of any primary user. Spectrum sensing thus plays a critical role in OSA CR networks. Most of the OSA CR research has focused on developing highly accurate spectrum sensing algorithms to reliably detect extremely weak primary signals. Several single-user spectrum sensing and later cooperative spectrum sensing techniques were developed to improve the detection capability of the CR networks [19].

While significant efforts in CR research were made to achieve spectrum sharing between primary and secondary devices in TV bands on a non-interfering basis, such efforts were mainly limited to the use of white spaces in TV bands based on OSA CR, i.e. on a "coexistence avoidance" basis. Recently, the focus of CR research has shifted to the coexistence of primary and secondary systems in more generalized settings. Such an approach, known as spectrum sharing CR, allows primary and cognitive users to transmit concurrently without interfering with each other and can be one of two types: underlay and overlay. In underlay CR, the CR operates under an interference constraint to the primary users so that a threshold primary performance is guaranteed. On the other hand, overlay CR facilitates the primary transmission, besides its own transmission, with the availability of the primary

*Spectrum Sharing: The Next Frontier in Wireless Networks,* First Edition.
Edited by Constantinos B. Papadias, Tharmalingam Ratnarajah, and Dirk T.M. Slock.
© 2020 John Wiley & Sons Ltd. Published 2020 by John Wiley & Sons Ltd.

user's message in a non-causal or causal manner at the CR transmitter [11, 15–17, 20]. This CR paradigm, referred to as spectrum sharing CR, involves underlay and overlay CR that are differentiated by the nature and use of side information available at the CR transmitter. In underlay CR, the side information at the CR transmitter is in the form of knowledge of the maximum tolerable interference at the primary receiver (also termed *interference temperature*) so that the CR can adjust its transmit power in a way that does not degrade the primary performance. In contrast to underlay CR, overlay CR makes use of the side information about the primary's codebooks and/or message knowledge at the CR transmitter and uses advanced coding and signal processing techniques to mitigate the interference to primary and secondary users.

Cooperative communication has recently emerged as a powerful technique to enhance performance in a wireless environment, leading to significant improvement in capacity, reliability, and area of coverage [18]. As an example, cooperation can be implemented by exploiting intermediate relay nodes between a source and a destination to provide spatial diversity, thus leading to significant performance improvements in wireless networks. Motivated by the promise of CR and cooperative communication to achieve efficient spectrum utilization and reliable communication, cooperative communication techniques have recently been used in interweave, underlay, and overlay CR setups to improve the overall spectrum efficiency as well as performance tradeoffs for both primary and secondary users. For instance, considering interweave CR, as elaborated in Chapter 7, collaborative spectrum sensing techniques have resulted in highly accurate spectrum sensing while solving the hidden terminal problem. On the other hand, cooperative communication has been employed in underlay and overlay CR to improve the primary and secondary user throughput by exploiting cooperation between primary and secondary users, as well as among secondary users.

In this chapter, we consider different aspects of coordination within a spectrum sharing environment, with the existence of imperfect and possibly even distributed channel knowledge across the system. Concentrating on the design of enhanced performance transmission (reception) schemes for downlink (uplink) communication, such partial channel knowledge at the transmitter (receiver) side is represented by a realistic model. According to this model, assuming a CR system composed of a primary and a secondary transmitter–receiver pair, the direct links are assumed to be instantaneously known at the transmitter (or receiver), whereas the interference links are merely statistically known, i.e. by means of channel covariance information, which is exchanged between the primary and the secondary system, possibly together with statistical primary traffic information. Following such an assumption on statistical cooperation, with regards to channel information and the activity profile of the primary service, the notion of *service coexistence* is emphasized through formulating problems, the aim of which is to optimize the data transmission/reception and/or the spectrum sensing parameters. The goal of the considered optimized system design is to efficiently multiplex two different services instantiated at a given primary and secondary transmitter–receiver pair, respectively, by means of achieving an enhanced performance at the secondary receiver, constrained by a performance requirement at the primary receiver. Considering such a service prioritization framework, in section 8.2, two enhanced mobile broadband (eMBB) services [1] are efficiently multiplexed, whereas in section 8.3, by exploiting the advantages of a hybrid interweave/underlay CR system design, a secondary service in need of high data rates effectively coexists in space and in time with a high reliability service running at the primary system side. For further details on the solution frameworks, interested readers may refer to [8–10].

## 8.2 Distributed Precoding Exploiting Commonly Available Statistical CSIT for Efficient Coordination

In this section, we focus on cooperative communication between transmitters in underlay CR networks operating in the downlink. The system comprises a multiple input single output (MISO) primary transmitter–receiver pair, which offers a spectrum sharing opportunity to a MISO secondary transmitter–receiver pair. Assuming that both receivers aim to run high data rate (i.e., eMBB) services, where the average experienced data rate is the dominant QoS metric, the objective is to design a cooperative transmission scheme, towards maximizing the average data rate of the secondary user, subject to an average data rate constraint imposed by the primary service for its own user. In contrast to most prior works on underlay CR systems, the two transmitters here cooperate under a realistic channel state information (CSI) scenario where each transmitter has access to the instantaneous direct channel of its associated terminal and only statistical information is available regarding all involved links. Such a CSI knowledge setting brings about a formulation based on the theory of *team decisions* [3, 5, 13, 24, 30], whereby the transmitters aim at optimizing a common objective given the same constraint set on the basis of locally available channel information.

In further detail, the spectrum sharing system is composed of a MISO primary system, comprising a transmitter, TX$p$, equipped with $M_p$ antennas, along with its assigned single-antenna terminal, RX$p$. Focusing on downlink communication, the primary system is willing to share its resources with a MISO secondary system. The latter system consists of a multiple antenna transmitter, TX$s$, equipped with $M_s$ antennas, as well as of a secondary user, RX$s$, associated with TX$s$. Considering the involved channels, spatially correlated Rayleigh fading is assumed for both the direct and the interfering channel links. As a result, for the channel between TX$j$ and RX$i$, we have $\mathbf{h}_{i,j} \sim \mathcal{CN}(\mathbf{0}_{M_j}, \mathbf{R}_{i,j}), i,j \in \{p,s\}$, where, $\mathbf{R}_{i,j}$ denotes the covariance matrix of channel $\mathbf{h}_{i,j}$. The system setup is depicted in Figure 8.1

Focusing on the described system model, a realistic channel state information at the transmitter (CSIT) assumption that can be made is that TX $i, i \in \{p,s\}$, has both instantaneous and statistical knowledge of its direct link $\mathbf{h}_{i,i}$, whereas the interference cross-links are only statistically known via the knowledge of their covariance matrices. The second-order statistics of all involved channels constitute slow-varying information that can be realistically collected and exchanged between the two transmitters even through low capacity/high delay backhaul links.

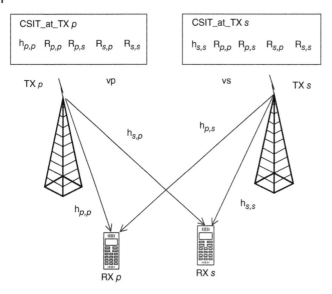

**Figure 8.1** System setup, along with the available CSIT at each transmitter.

### 8.2.1 Problem Formulation

Capitalizing on the available CSIT at TX $i$, $i \in \{p, s\}$, the optimization problem of maximizing the average rate of the secondary user, subject to an average rate constraint for the primary user, can be formulated as a *functional optimization problem*, with functional dependencies related to the available CSI. Hence, the service coexistence optimization problem can be described as follows:

$$
\begin{cases}
(\mathbf{v}_p^*, \mathbf{v}_s^*) = \arg\max \; \mathbb{E}(R_s(\mathbf{v}_p(\mathbf{h}_{p,p}), \mathbf{v}_s(\mathbf{h}_{s,s}))) \\
\text{subject to } \mathbb{E}(R_p(\mathbf{v}_p(\mathbf{h}_{p,p}), \mathbf{v}_s(\mathbf{h}_{s,s}))) \geq \tau_p > 0, \\
0 \leq \| \mathbf{v}_p(\mathbf{h}_{p,p}) \|^2 \leq P_p^{\max}, \; 0 \leq \| \mathbf{v}_s(\mathbf{h}_{s,s}) \|^2 \leq P_s^{\max},
\end{cases}
\tag{8.1}
$$

where $R_i$ denotes the instantaneous data rate of user $i$, $\tau_p$ stands for the quality of service (QoS) demand of RX $p$ in terms of average rate, and $\mathbf{v}_i$ denotes the transmit beamforming vector at TX $i$, $i \in \{p, s\}$. It is assumed that $\mathbf{v}_i = \sqrt{P_i}\mathbf{w}_i$, with $P_i \leq P_i^{\max}$ and $\| \mathbf{w}_i \| = 1$, where $P_i^{\max}$ is the maximum instantaneous power level at TX $i$, $i \in \{p, s\}$.

With the aim of deriving a practical solution, slow power control depending on the long-term statistical channel information is assumed. Hence, instantaneous power levels $P_p$ and $P_s$ can be replaced by slow power allocation levels $\overline{P}_p$ and $\overline{P}_s$, where $0 \leq \overline{P}_i \leq P_i^{\max}$, $i \in \{p, s\}$. Nevertheless, the optimization problem is still a challenging one due to the consideration of interference when computing the expected user rates. To relax the problem's complexity, an approximated problem can be designed instead by properly applying Jensen's inequality, thanks to the fact that the precoders $\mathbf{v}_p$ and $\mathbf{v}_s$ are dependent, apart from the instantaneous direct links $\mathbf{h}_{p,p}$ and $\mathbf{h}_{s,s}$, respectively, only on the statistics of the cross-links. For any RX $i$, $i \in \{p, s\}$, and assuming a bandwidth of 1 Hz, we thus obtain

$$
\begin{aligned}
\mathbb{E}(R_i) &= \mathbb{E}_{\mathbf{h}_{i,i}, \mathbf{h}_{\bar{i},i}} \left( \log_2 \left( 1 + \frac{\overline{P}_i |\mathbf{h}_{i,i}^H \mathbf{w}_i|^2}{N_0 + \overline{P}_{\bar{i}} |\mathbf{h}_{\bar{i},i}^H \mathbf{w}_{\bar{i}}|^2} \right) \right) \\
&\geq \mathbb{E}_{\mathbf{h}_{i,i}} \left( \log_2 \left( 1 + \frac{\overline{P}_i |\mathbf{h}_{i,i}^H \mathbf{w}_i|^2}{N_0 + \mathbb{E}_{\mathbf{h}_{\bar{i},i}} \left( \overline{P}_{\bar{i}} |\mathbf{h}_{\bar{i},i}^H \mathbf{w}_{\bar{i}}|^2 \right)} \right) \right) \\
&= \mathbb{E}_{\mathbf{h}_{i,i}} \left( \log_2 \left( 1 + \frac{\overline{P}_i |\mathbf{h}_{i,i}^H \mathbf{w}_i|^2}{N_0 + \overline{P}_{\bar{i}} \mathbf{w}_{\bar{i}}^H \mathbf{R}_{\bar{i},i} \mathbf{w}_{\bar{i}}} \right) \right) \\
&\triangleq \mathbb{E}(\tilde{R}_i(\mathbf{w}_i, \mathbf{w}_{\bar{i}})),
\end{aligned}
\tag{8.2}
$$

where $\bar{i}$ denotes the complementary index of $i$. It should be noted that the tightness of the derived lower bound depends on the structure of the covariance matrices of the cross-links and the applied unit norm precoders $\mathbf{w}_i$ $i \in \{p, s\}$. However, in any case, although the derivation of a lower bound of each user's rate may reduce the average rate potential of the secondary user, it will guarantee the feasibility of the constraint set (with regards to the average rate of the primary user), as will be numerically shown. Hence, the equivalent approximated problem to be solved is

$$
\begin{cases}
(\overline{P}_p^*, \mathbf{w}_p^*, \overline{P}_s^*, \mathbf{w}_s^*) = \arg\max \ \mathbb{E} \left( \tilde{R}_s(\overline{P}_p, \mathbf{w}_p, \overline{P}_s, \mathbf{w}_s) \right) \\
\quad \text{subject to } \mathbb{E} \left( \tilde{R}_p(\overline{P}_p, \mathbf{w}_p, \overline{P}_s, \mathbf{w}_s) \right) \geq \tau_p, \\
\quad 0 \leq \overline{P}_p \leq P_p^{\max}, \ 0 \leq \overline{P}_s \leq P_s^{\max}, \\
\quad \| \mathbf{w}_p \|^2 = 1, \ \| \mathbf{w}_s \|^2 = 1,
\end{cases}
\tag{8.3}
$$

where quantities $\mathbb{E}(\tilde{R}_p(\overline{P}_p, \mathbf{w}_p, \overline{P}_s, \mathbf{w}_s))$ and $\mathbb{E}(\tilde{R}_s(\overline{P}_p, \mathbf{w}_p, \overline{P}_s, \mathbf{w}_s))$ are the obtained lower bounds of the true expected rates of the primary and secondary users, respectively.

### 8.2.2 Distributed Statistically Coordinated Precoding

Given that the derivation of closed-form expressions for the optimal downlink precoders is hardly tractable due to the functional nature of optimization problem 8.3, which requires optimizing over an infinite-dimensional space, the functional space of the applicable precoding solutions can be restricted to a set of transmission strategies. Such a search space compression and discretization allows for every transmission strategy (i.e., joint precoding scheme) to be evaluated in terms of both feasibility and performance, hence it provides a simple and practical method for coordinating the transmitters.

The constructed set of transmission strategies is obtained by following these two steps:

- First, design beamforming schemes (i.e., choose unit norm vectors $\mathbf{w}_i$) that can be potentially applicable to each of the two transmitters. Although any beamforming scheme could be chosen in theory, a good heuristic choice is key to the tractability and efficiency of the approach. Here, we restrict our analysis to the maximum ratio transmission

(MRT), also known as matched filter (MF) and statistical ZF (sZF) strategies, as they reflect the nature of the available CSIT and also represent a tradeoff between maximizing the received power of direct communication and minimizing the leaked interference to the receiver of the other communication pair.

- Power control is a key ingredient to ensure that the average rate constraint for the primary user is not violated. As shown in Filippou et al. [9, Proposition 1], in optimality the average rate constraint for primary communication is fulfilled with equality and, on top of that, one of the two transmitters transmits with full power, as per Filippou et al. [9, Proposition 2]. Consequently, the other transmitter will need to regulate its transmission power to comply with the imposed QoS constraint. Therefore, we denote by $P_p$ the joint power policy where TX $p$ transmits with full power and by $P_s$ the joint power policy where TX $s$ transmits with full power.

As a result of the above, considering the potential applicability of the two power control policies for each joint beamforming solution, where each transmitter chooses to apply a unit norm precoding vector, corresponding to either MF or sZF, such a formulation leads to a *joint transmission strategy set*, $S$, which consists of eight possible joint transmission schemes, i.e. $S = \{\text{MF-MF} - P_i, \text{MF-sZF} - P_i, \text{sZF-MF} - P_i, \text{sZF-sZF} - P_i, i \in \{p, s\}\}$, where the first acronym of each transmission scheme refers to the precoding strategy applied by TX $p$, the second one refers to the precoding strategy applied by TX $s$, while the last one indicates the transmitter transmitting with full power. However, the primary user data rate constraint is only fulfilled with a probability of one for some of these strategies and has to be verified otherwise. It is hence necessary, for each of these eight joint transmission schemes to verify that a power transmission level for the other transmitter can be found within its transmit power limits, such that the ergodic rate at RX $p$ satisfies the imposed data rate requirement, $\tau_p$, at least with equality. Once the feasible transmission schemes are obtained, the best solution, in terms of the average data rate achieved at the secondary user, is directly obtained. Algorithm 1 explains the steps of the statistically coordinated precoding procedure towards solving problem 8.3.

---

**Algorithm 1**  Distributed statistically coordinated precoding

---

**Input available at both transmitters:** $\tau_p, M_p, M_s, P_p^{\max}, P_s^{\max}, \mathbf{R}_{p,p}, \mathbf{R}_{p,s}, \mathbf{R}_{s,p}, \mathbf{R}_{s,s}$
**Input exclusively available at TX p:** $\mathbf{h}_{p,p}$
**Input exclusively available at TX s:** $\mathbf{h}_{s,s}$
**Output:** Joint transmission scheme $D \in S$ which solves problem 8.3

1  For each element of joint transmission strategy set $S$ determine whether the transmitter, able to regulate its transmit power, can apply a transmit power level that is enough to satisfy the average rate constraint of problem 8.3.
2  Formulate the set of feasible joint transmission schemes, $F \subseteq S$.
3  Select the element of feasible set $F$ that maximizes the average rate of the secondary user, i.e. $D = \arg\max_{F_i \in F} \mathbb{E}\left(\tilde{R}_s(F_i)\right)$, where $\mathbb{E}\left(\tilde{R}_s(F_i)\right)$ denotes the calculated lower bound of the average rate of the secondary user when joint transmission scheme $F_i \in F$, $i \in \{1, \cdots, |F|\}$, is applied.

---

### 8.2.3 Performance Evaluation

With the aim of evaluating the performance of the proposed statistically coordinated precoding scheme, extensive Monte Carlo simulations have been performed. We choose $M_p = M_s = M = 4$ antennas at each transmitter and consider a classical exponential channel correlation model [21] where the $(m, n)$th entry of channel covariance matrix $\mathbf{R}_{i,j}$, $i, j \in \{p, s\}$, is equal to $\beta_{i,j}\rho^{|n-m|}$, $m, n \in \{1, \cdots, M\}$, where $\beta_{i,j}$ symbolizes the distance-based pathloss and $\rho$ stands for the antenna correlation factor. In our example, we assume that $\rho = 0.25$ and $\beta_{i,j} = 1$, when $i = j$, otherwise it is equal to 0.3. Furthermore, we consider unit noise variance and a QoS threshold $\tau_p = 1.75$ bps/Hz.

To effectively evaluate the potential of the solution, we compare its performance to the performance of two reference precoding schemes: the first one, denoted as "interference temperature-based" precoding, is an adaptation of the approaches in the literature [[9], Section VI.A], where the average interference temperature is derived based on the rate demand of RX $p$. Intuitively, it corresponds to the conventional, *uncoordinated* underlay CR paradigm, where the secondary transmitter merely adapts its transmission strategy in order for the interference received by the primary user to be below a given threshold [2]. The second reference scheme constitutes a *coordination benchmark* and it is *a priori* not practically attainable as it is assumed that each transmitter can achieve simultaneously both goals, i.e. maximize the direct signal power of its own assigned user and minimize the interference received by the other user. Hence, it represents a performance upper bound that illustrates the sub-optimality and coordination limitations of the state-of-the-art and proposed approaches [9, Section VI.B]. In Figures 8.2 and 8.3, the average data rates of the secondary and primary users, respectively, are depicted as a function of the system's transmit SNR $= \frac{P_p^{\max}}{N_0} = \frac{P_s^{\max}}{N_0}$, where $N_0 = 0$ dB stands for the noise variance. The three curves

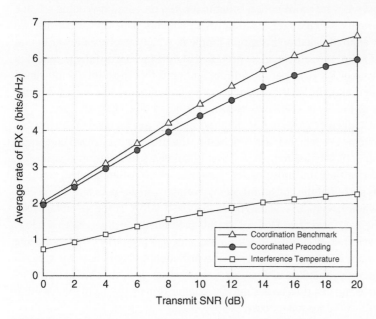

**Figure 8.2** Average data rate of RX $s$ versus transmit SNR, when $\tau_p = 1.75$ bps/Hz.

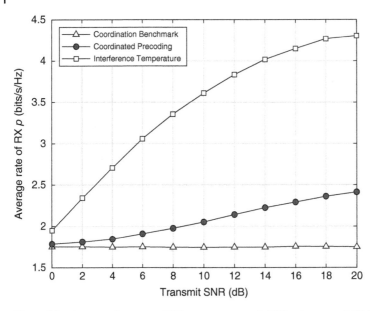

**Figure 8.3** Average data rate of RX $p$ versus transmit SNR, when $\tau_p = 1.75$ bps/Hz.

represent the throughput performance achieved by the proposed statistically coordinated precoding scheme, the interference temperature-based (i.e., non-coordinated) precoding scheme, as well as the considered coordination benchmark. Focusing on RX $s$, the coordination benchmark outperforms both the proposed precoding scheme as well as the interference temperature-based scheme, as expected. Also, by observing Figure 8.3 it should be noted that, in contrast with the coordination benchmark, the proposed precoding scheme fails to satisfy the primary average rate constraint with equality. This occurs because we resort to tackling an approximated optimization problem, which involves a lower bound of the average rate of RX $p$. Nevertheless, the proposed algorithm successfully manages to control the average rate of RX $p$, as per its requirement, and this capability is translated to a significant throughput gain for the secondary user, as compared to the performance achieved by the state-of-the-art interference temperature-based precoding scheme.

Finally, in Figure 8.4, the average data rate of the secondary user for each of the feasible joint transmission schemes is depicted as a function of the primary user's QoS constraint $\tau_p$, when the transmit SNR is equal to 8 dB. It should be noted that the term "feasible" is used here to characterize the joint transmission schemes, which, when applied, result in transmit power levels such that the constraint on the average primary user rate is satisfied. It is observed that, when $\tau_p \in [0.5\ 3]$ bits/s/Hz, strategy MF-MF-$\mathcal{P}_s$ is the secondary user rate-optimal one, whereas, for stricter QoS constraints posed by the primary user, strategy MF-MF-$\mathcal{P}_p$ has to be selected, exactly because the system focuses primarily on guaranteeing the feasibility of the optimization problem. It is also worth mentioning that, as the value of threshold $\tau_p$ increases, only the subset of the most "primary user-protective" joint transmission schemes is feasible and can thus be put into the comparison by means of the resulting achievable average rate at the secondary user.

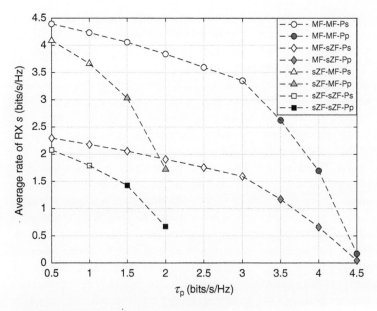

**Figure 8.4** Average data rate of RX $s$ versus $\tau_p$, SNR = 8 dB.

## 8.3 A Statistical Channel and Primary Traffic-aware Cooperation Framework for Optimal Service Coexistence

Interweave and underlay are the two of most important CR system approaches that have been proposed in recent years. Considering the performance analysis of the aforementioned CR system approaches, substantial work has been carried out, e.g., considering the calculation of the achievable average rate of the secondary user [6, 7, 14, 28, 29]. Nonetheless, these two approaches are characterized by some drawbacks that may critically affect system performance, especially when it comes to the coexistence of dissimilar services running on a primary and a secondary system. More specifically, the interweave CR approach, when implemented with high quality spectrum sensing, successfully protects the primary receiver from secondary interference. However, reliable sensing comes at the expense of increased sensing time and energy resources that could lead to degraded throughput or reliability performance for the secondary system. Also, as discussed in Chapter 7, stand-alone (i.e., non-collaborative/distributed) spectrum sensing is often unable to cope with the "hidden node" problem. Moreover, as per the interweave approach, the time intervals during which primary activity is found to occur, as a result of spectrum sensing, remain totally unexploited by the secondary system. On the other hand, focusing on the underlay CR approach, the same secondary system transmission policy is applied both at times when the primary system is silent (equivalently, in frequency slots where it is absent) and at times (equivalently, frequencies) of primary system activity, as the only criterion is for the primary receiver to not receive interference overcoming a tolerable level (i.e., the interference temperature). This way, the imposed interference temperature constraint restricts secondary transmissions, even when the primary system is idle, leading to degraded throughput performance for the secondary receiver.

As a result of the above, the notion of a hybrid interweave/underlay CR system approach has been developed in the academic literature, motivated by the need to better exploit the implementation advantages of the two "classical" CR system approaches. According to this concept, the secondary system implements a modified version of the interweave CR approach, according to which, when the primary channel is found to be busy, as a result of spectrum sensing, the secondary transmitter, instead of keeping silent, transmits by applying a power (and, possibly, precoding/antenna combining) policy driven by the primary QoS requirement. Based on this idea, several recent works have considered combined interweave/underlay CR approaches, such as [23, 26, 27, 31], and [25]. Although in these works multiple solutions have been provided, referring to various system setups, several simplifications and non-realistic assumptions seem to have been made (e.g., perfect spectrum sensing, single input single output (SISO) communications or multiple antenna communications, but assuming uncorrelated antennas, and special focus on eMBB services), which call for new design proposals through the formulation of problems concentrating on system parameter optimization. Moving a step forward, one would argue that it is meaningful to *jointly* design the spectrum sensing and transmission/reception parameters of a hybrid CR system that encapsulates the advantages of both conventional CR approaches and then evaluate the performance of the optimized system with regards to different QoS metrics tailored to wireless services of different types.

To this end, in the remainder of this section, although the level of primary/secondary system cooperation, focusing on channel knowledge, is of a similar nature as before (i.e., statistical CSIT/channel state information at the receiver (CSIR) exchanged between the primary and the secondary system), we aim at multiplexing a primary high-reliability service with a secondary eMBB service exploiting the additional design degrees of freedom offered by the primary traffic statistics (i.e., average ON/OFF time duration of the service) that are provided to the secondary system. To achieve this we focus on the hybrid interweave/underlay CR approach and expand the optimization parameter set, i.e. we aim to include the spectrum sensing parameters as well. Then, to measure the performance of the optimized system, we evaluate the system behavior over a range of reliability requirements for the service instantiated at the primary user.

### 8.3.1 Joint Design of Spectrum Sensing and Reception for a SIMO Hybrid CR System

In this section, the problem of jointly designing spectrum sensing and receive antenna combining, with reference to the uplink of a CR [equivalently, a licensed shared access (LSA)] system, is considered. The aim of the proposed design is the maximization of the achievable average uplink rate of a secondary CR (equivalently, the LSA licensee) user, subject to an outage-based, communication reliability constraint for primary (equivalently, the LSA incumbent) communication. A hybrid CR system approach is studied, according to which the system either operates as an interweave or as an underlay CR system, depending on the results of the performed spectrum sensing procedure at the beginning of each medium access control (MAC) frame of the secondary user.

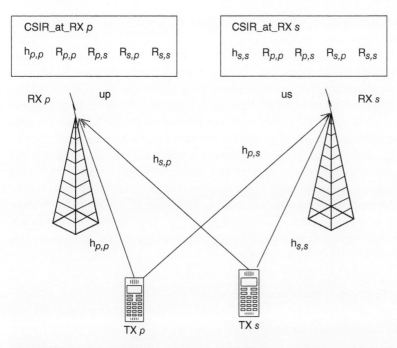

**Figure 8.5** The investigated system scenario, along with the available CSIR at each receiver.

More specifically, as illustrated in Figure 8.5, the uplink of a CR system is considered, which is composed of a single-antenna primary user, TX $p$, transmitting with a fixed power level, $P_p$, that communicates with a multiple-antenna primary receiver (i.e., base station), RX $p$, along with a single-antenna secondary user, TX $s$, communicating with a multiple antenna secondary receiver, RX $s$. In what follows, it is assumed that RX $p$ and RX $s$ are equipped with $N$ antennas each. Regarding the channel model, it is assumed that both the SIMO channels $\mathbf{h}_{ij}$ between TX $i$ and RX $j$, $i, j \in \{p, s\}$, and the SISO channel between TX $p$ and TX $s$ undergo Rayleigh fading. Also, it is assumed that the elements of any given SIMO channel, $\mathbf{h}_{ij}, i, j \in \{p, s\}$, are spatially correlated. Regarding the availability of CSIR, a practical scenario is considered, according to which RX $i$, $i \in \{p, s\}$, is aware of both the instantaneous direct channel $\mathbf{h}_{ii}$ as well as its covariance matrix $\mathbf{R}_{ii}$, while it only has statistical knowledge of the other links in the form of covariance information. Such a CSIR formulation is chosen because standard releases for 4G wireless systems require that a given user equipment is allowed to report instantaneous CSI to its home base station, but it cannot report such information to interfering base stations [4].

Since spectrum sensing constitutes an essential feature of the investigated hybrid CR system, focusing on secondary communication, each MAC frame of the secondary user that has a duration of $T$ time units consists of (i) a spectrum sensing subframe, the duration of which is $\tau$ time units, followed by (ii) a data tranmission subframe, which lasts for the remaining $T - \tau$ time units. Concerning spectrum sensing, energy detection (ED) is chosen to be applied since it is characterized by low implementation complexity and analytical expressions for false alarm and detection probabilities. Also importantly, it is assumed that the length of each MAC frame is such that the involved wireless channels remain fixed for the duration of it.

Concentrating now on the previously introduced hybrid interweave/underlay CR system approach, and focusing on a specific data transmission subframe, one needs to discriminate between two secondary user transmission policies, depending on the results of the ED procedure carried out by the secondary transmitter during the corresponding spectrum sensing subframe: (i) Absence of primary user transmissions is detected. We denote this event as $\hat{\mathcal{H}}_0$. Whenever such an event occurs, the secondary user transmits using a power level $P_s = P_0 = P_{peak}$, where $P_{peak}$ is a peak power constraint at the secondary user. On the other hand, RX $s$ employs a unit-norm receive beamforming vector $\mathbf{u}_s = \mathbf{u}_0(\mathbf{h}_{ss}) \in \mathbb{C}^{N \times 1}$ for the detection of the signal transmitted by the secondary user. (ii) Presence of primary user transmission is detected. We denote this event as $\hat{\mathcal{H}}_1$. Whenever $\hat{\mathcal{H}}_1$ occurs, the secondary user transmits using a power level, $P_s = P_1$, $0 < P_1 \leq P_{peak}$. In addition, RX $s$ employs a unit-norm receive vector, $\mathbf{u}_s = \mathbf{u}_1(\mathbf{h}_{ss}) \in \mathbb{C}^{N \times 1}$ that is designed taking into account the fact that primary user activity has been detected.

### 8.3.1.1 Problem Formulation and Solution Framework

Under such system operation and CSIR assumptions, and also considering that RX $p$ is assumed to apply maximal ratio combining (MRC), i.e. $\mathbf{u}_p = \frac{\mathbf{h}_{pp}}{\|\mathbf{h}_{pp}\|}$, a lower bound for the average rate of secondary communication, as well as a closed form approximation for the outage probability of primary communication have been derived in [10], where an outage event is declared by the primary system when the received SINR is below a threshold value, $\gamma_0$. These expressions are been shown to be useful towards formulating and then solving a joint spectrum sensing and multi-antenna reception optimization problem, the objective of which is the effective coexistence of a data rate dominant service instantiated by the secondary user, subject to an outage constraint on primary communication. The optimization problem is

$$
\begin{cases}
\underset{\mathbf{u}_1 \in \mathbb{C}^{N \times 1}, \tau, \varepsilon, P_1}{\text{maximize}} & \mathbb{E}_{|\mathbf{h}_{ss}}\{\mathcal{R}\} \\
\text{subject to} & \mathcal{P}_{out} \leq \tilde{\mathcal{P}}_{out}, \ \mathcal{P}_d = \tilde{\mathcal{P}}_d, \ \|\mathbf{u}_1\| = 1, \\
& 0 < P_1 \leq P_{peak}, \ 0 < \tau \leq T, \ \varepsilon \geq 0,
\end{cases}
\tag{8.4}
$$

where $\mathbb{E}_{|\mathbf{h}_{ss}}\{\mathcal{R}\}$ is the expected data rate of the secondary user, given the knowledge of channel $\mathbf{h}_{ss}$, $\tilde{\mathcal{P}}_{out}$ is the predetermined constraint of outage probability $\mathcal{P}_{out}$, imposed by the primary service, $\varepsilon$ stands for the ED threshold, and $\tilde{\mathcal{P}}_d$ is a targeted value of the average detection probability $\mathcal{P}_d$ for the implementation of ED-based spectrum sensing.

As shown in [10], the original complex *stochastic optimization* problem 8.4 can be approximated by a better tractable problem,

$$
\begin{cases}
\underset{\mathbf{u}_1 \in \mathbb{C}^{N \times 1}, \tau, \varepsilon, P_1}{\text{maximize}} & \mathcal{C} \\
\text{subject to} & \mathcal{P}_{out} \leq \tilde{\mathcal{P}}_{out}, \ \mathcal{P}_d = \tilde{\mathcal{P}}_d, \ \|\mathbf{u}_1\| = 1, \\
& 0 < P_1 \leq P_{peak}, \ 0 < \tau \leq T, \ \varepsilon \geq 0,
\end{cases}
\tag{8.5}
$$

where $C$ is a lower bound of the expected secondary user rate, given the knowledge of channel $\mathbf{h}_{ss}$, which can be described in closed form. The above optimization problem can be successfully decomposed into two subproblems, each of which focuses on the optimization of either ED parameters $\tau$ and $\varepsilon$ or the optimization of transmission/reception parameters $P_1$ and $\mathbf{u}_1$, respectively. Having this solution framework in hand, an iterative algorithm can be implemented as follows.

---

**Algorithm 2** Jointly optimizing antenna combining vector, $\mathbf{u}_1$, and sensing parameters $\tau$ and $\varepsilon$

---

1 Initialization ($n = 0$). Fix the antenna combining scheme such that $\mathbf{u}_1 = \mathbf{u}_1^{(0)}$ and increase counter by one.
2 For the $n$th iteration, solve the resulting spectrum sensing optimization problem with $\mathbf{u}_1 = \mathbf{u}_1^{(n-1)}$ and find values $\tau_n$ and $\varepsilon_n$.
3 Utilizing the obtained values $\tau_n$ and $\varepsilon_n$, solve the resulting antenna combining optimization problem and determine antenna combining vector $\mathbf{u}_1^{(n)}$.
4 Compute the value of the objective function $C_n(\mathbf{u}_1^{(n)}, \tau_n, \varepsilon_n)$.
5 Increase the counter by one and if $|C_n - C_{n-1}| < \zeta$, where $n \geq 2$ and $\zeta > 0, \zeta \in \mathbb{R}$ is an arbitrary small number, stop, otherwise go to Step 2.

---

**Remark 8.1** The solution framework falls within the category of *block coordinate ascent* optimization. Considering the existence of two blocks of optimization variables, as has been described above, i.e. one for the spectrum sensing parameters $\tau$ and $\epsilon$ and one for the antenna combining vector, $\mathbf{u}_1$, and observing that, for each block of variables, the equivalent optimization problem consists of a concave objective function and a convex constraint, then iterative Algorithm 2 converges to a stationary point $(\mathbf{u}_1^*, \tau^*, \epsilon^*)$ [22].

### 8.3.1.2 Performance Evaluation

To evaluate the service coexistence efficiency of the obtained solution, the performance of the designed hybrid interweave/underlay CR system is compared by means of extensive Monte Carlo numerical simulations to the equivalent (i.e., average secondary user rate optimized) interweave and underlay systems, assuming the same reliability requirement on primary user communication (i.e., a required SINR of $\gamma_0 = 4$ dB for correct message decoding) imposed on an interference-limited SIMO system, where each receiver is equipped with $N = 4$ antennas. The MAC frame size is equal to 100 ms, the sampling frequency for ED is equal to 5 MHz, the noise variance is equal to 0 dB, and the variance of the primary user–secondary user side link is equal to −3 dB. We adopt the exponential antenna correlation model [21] and assume that the antenna correlation factor, $\rho$, is equal to 0.4. Furthermore, both the peak power level at the secondary user and the (fixed) power level at the primary user are equal to 15 dB and the targeted detection probability is $\tilde{P}_d = 0.98$. It should be noted that the values of these parameters remain fixed in the remainder of this section unless otherwise stated.

In Figure 8.6 the achievable average rate of RX $s$ is depicted for the three investigated systems, i.e. the optimized hybrid CR system and the two optimized interweave and underlay CR systems, as a function of the activity profile of the primary system when the outage

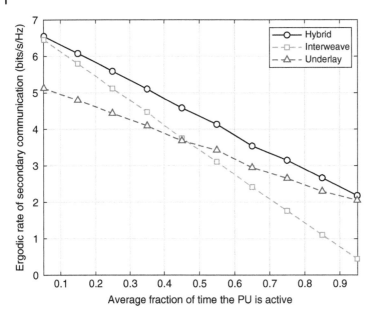

**Figure 8.6** Ergodic rate of RX $s$ versus primary user activity profile when $\tilde{P}_{out} = 10^{-2}$.

probability requirement of the primary user is fixed to 1%. One can observe that the average throughput of RX $s$ regarding the hybrid system balances between two "extremes" with respect to the activity profile of the primary user. More specifically, the hybrid CR system behaves similarly to the interweave one, when the primary user is idle for most of the time, whereas it approaches the throughput performance of the underlay system, when the primary user is active for most of the time. Also importantly, all three curves are decreasing. This occurs because, when the primary system is busy for an increased fraction of time, more interference will be received by RX $s$ over time, on average.

The impact of the number of receive antennas, $N$, and the spatial correlation factor, $\rho$, on the optimized performance of the hybrid CR system is shown in Figure 8.7, where the average rate of secondary communication is depicted for different primary user activity profiles when the primary user outage probability constraint is equal to 1%. It is observed that when $N = 8$ antennas are used at either receiver, the average rate at RX $s$, in the existence of strongly correlated Rayleigh fading, overcomes the performance obtained when the $N$ branches of each SIMO channel are close to being independent and identically distributed (i.i.d.) and this performance gap increases as the primary user becomes active more often. This behavior occurs because since we focus on an interference-limited system, and given the assumptions made on CSIR knowledge, it is more critical for RX $s$ to zero-force the incoming interference from the primary user when the latter becomes active more frequently than to exploit a receive antenna diversity gain. Also interestingly, one observes that when the antenna correlations are low, the performance gain by applying an excess of receive antennas, as compared to a smaller antenna number, vanishes when the primary user is in transmission mode for more than 70% of the time, on average.

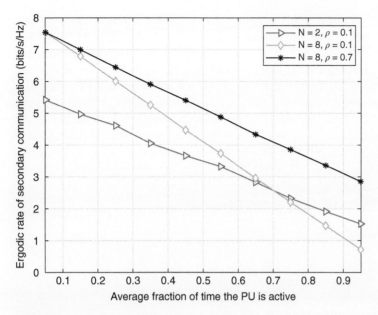

**Figure 8.7** Ergodic rate of RX $s$ versus primary user activity profile (optimized hybrid CR system) for different values of receive antenna number, $N$, and receive antenna correlation factor, $\rho$, when $\tilde{P}_{out} = 10^{-2}$.

### 8.3.2 Throughput Performance of Sensing-optimized Hybrid MIMO CR Systems

In this section we analytically examine the secondary user throughput performance of a hybrid MIMO interweave/underlay CR system, subject to a reliability constraint posed by the primary system, where the secondary user operates as in the previous section, but this time focusing on downlink communication and assuming that an energy-based antenna selection scheme is performed at the user equipment side.

#### 8.3.2.1 Problem Formulation and Solution Framework

Regarding the system and channel model, the downlink of a CR system is considered that is composed of a primary transmitter, i.e. base station $TX$ $p$, and its assigned primary user RX $p$, and a secondary transmitter, i.e. base station $TX$ $s$, with its assigned secondary user RX $s$. It is assumed that the transmitters are equipped with $M$ antennas each, while each of the receivers is equipped with $N$ antennas. We denote the MISO channel between $TX$ $i$ and the $n$th antenna of RX $j$ as $\mathbf{h}_{ij,n} \in \mathbb{C}^{1 \times M}$, $i,j \in \{p,s\}, n = 1, \cdots, N$, and the channel between $TX$ $p$ and the $m$th antenna of $TX$ $s$ as $\mathbf{h}_{00,m} \in \mathbb{C}^{1 \times M}, m = 1, \cdots, M$. As far as the channel model is concerned, Rayleigh fading is assumed, i.e. it holds that $\mathbf{h}_{ij,n} \sim \mathcal{CN}(\mathbf{0}, \sigma_{ij,n}^2 \mathbf{I}_M)$ and $\mathbf{h}_{00,m} \sim \mathcal{CN}(\mathbf{0}, \sigma_{00,m}^2 \mathbf{I}_M)$, $i,j \in \{p,s\}, n = 1, \cdots, N, m = 1, \cdots, M$. For simplicity, we assume that $\{\sigma_{ij,n}^2\}_{n=1}^N = \sigma_{ij}^2$, $i,j \in \{p,s\}$, and $\{\sigma_{00,m}^2\}_{m=1}^M = \sigma_{00}^2$.

Always concentrating on a cooperative, however imperfect, CSIT knowledge framework, it is assumed that $TX$ $i$ has perfect knowledge of the $N \times M$ MIMO channel matrix

$\mathbf{H}_{ii} = [\mathbf{h}_{ii,1}^T, \cdots, \mathbf{h}_{ii,N}^T]^T, i \in \{p, s\}$, which justifies the use of MRT transmissions. Perfect instantaneous knowledge of the channel matrix $\mathbf{H}_{sp} = [\mathbf{h}_{sp,1}^T, \cdots, \mathbf{h}_{sp,N}^T]^T$ at both *TX s* and RX *s* is also considered. On the other hand, it is assumed that the MIMO interference link $\mathbf{H}_{ps} = [\mathbf{h}_{ps,1}^T, \cdots, \mathbf{h}_{ps,N}^T]^T$ as well as the $M \times M$ channel matrix $\mathbf{H}_{00} = [\mathbf{h}_{00,1}^T, \cdots, \mathbf{h}_{00,M}^T]^T$ are statistically known to the TXs. While data transmission is based on an MRT policy, data reception at the receivers is based on a maximum instantaneous direct channel energy criterion. It should be noted that the MAC frame is decomposed, as described in the previous section, into a spectrum sensing (ED) subframe followed by a data transmission subframe.

Taking this model into consideration, the average secondary user rate-optimal values of the sensing time $\tau^*$ and the ED threshold $\epsilon^*$ for a targeted value, $\mathcal{P}_t$, of the primary user outage probability, $\mathcal{P}_{\text{out}}^{\text{hyb}}$, can be found by solving the problem

$$
\begin{cases}
(\epsilon^*, \tau^*) = \underset{\epsilon, \tau}{\arg\max} \; \mathbb{E}(\mathcal{R}_{\text{hyb}}) \\
\text{subject to } \mathcal{P}_{\text{out}}^{\text{hyb}} = \mathcal{P}_t, \; 0 < \tau \le T, \; \epsilon \ge 0,
\end{cases}
\tag{8.6}
$$

where $\mathbb{E}(\mathcal{R}_{\text{hyb}})$ denotes the expected rate of the secondary user. Expressions describing this key performance quantity, as well as primary user outage probability $\mathcal{P}_{\text{out}}^{\text{hyb}}$, have been analytically derived in [8], where it is also shown that by applying the second derivative criterion for the objective function $\mathbb{E}\{\mathcal{R}_{hyb}(\tau, \epsilon(\tau))\}$ it can be proved that the latter is a concave function of $\tau$ for $\tau \in (0, T]$, when as a result of applying ED for spectrum sensing the CR system operates as an interweave one. As a result, any convex optimization algorithm can be applied in order to find optimal values $\tau^*$ and thus $\epsilon^*$ as well.

### 8.3.2.2 Performance Evaluation

With the aim of cross-validating the theoretical expressions derived in [8], as well as with the objective of comparing the performance of the sensing-optimized hybrid CR system with those of the optimized interweave and underlay CR systems, extensive Monte Carlo simulations have been performed. According to the simulation scenario, each transmitter is equipped with $M = 4$ antennas, while each receiver is equipped with $N = 2$ antennas. The variances of the involved channel links are $\sigma_{pp}^2 = \sigma_{ss}^2 = 10\,\text{dB}$, $\sigma_{sp}^2 = 8.75\,\text{dB}$, $\sigma_{ps}^2 = 9\,\text{dB}$, and $\sigma_{00}^2 = 7.92\,\text{dB}$. Additionally, we set the sampling frequency for spectrum sensing to 5 MHz and the length of a secondary user MAC frame to $T = 100\,\text{ms}$, while unit variance noise is considered. The SINR threshold below which an outage event at the primary user is declared is $\gamma_0 = 8\,\text{dB}$.

In Figure 8.8 the achievable average secondary user rate is depicted as a function of the outage probability of primary communication, $\mathcal{P}_t$, for a system setup characterized by relatively low primary user activity (i.e., $\mathbb{P}(\text{"busy primary user"}) = 0.25$). Both experimental and theoretical curves are illustrated with respect to the optimized hybrid CR system as well as for the equivalent optimized interweave and underlay CR systems. Considering the two conventional CR approaches, the secondary user rate-optimal design parameters are found by solving problems equivalent to the ones in [7]. One can observe that the empirical curves closely converge to those corresponding to the derived approximate expressions for the average secondary user rate, which confirms the validity of the derived expressions. Second, it

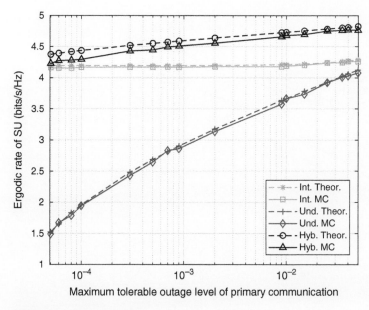

**Figure 8.8**  Ergodic secondary user rate versus primary user outage probability requirement, $\mathcal{P}_t$, when $\mathbb{P}(\text{"busy primary user"}) = 0.25$.

is evident that the hybrid CR system outperforms the two conventional CR approaches for the whole examined range of primary user outage constraints. This happens because, in contrast with the underlay CR approach, where only a fraction of the total power of $TX$ $s$ is constantly transmitted, regardless of the activity profile of the primary system, under the hybrid CR system there are times where, as a result of sensing decisions upon the absence of the primary user, full transmit power is transmitted by $TX$ $s$. On the other hand, unlike the interweave CR approach, where when primary user activity is detected by sensing $TX$ $s$ becomes silent during the data transmission subframe, according to the hybrid CR system secondary transmission is still carried out by switching to the underlay CR mode. As a result, the existence of more transmission opportunities for the secondary system is translated into better exploitation of the shared frequency resources, given the assumed level of CSIT knowledge, which is cooperatively exchanged between the two systems.

In Figure 8.9 the same performance curves are illustrated, this time for a system scenario according to which the primary system is busy for 75% of the time. In this case, the hybrid CR system still outperforms the two conventional CR approaches, but the performance of the underlay CR system now partially overcomes the one achieved by the interweave CR system. This is due to the fact that intensive primary user activity results in reducing the time slots during which the frequency resources are free for use by the interweave CR system (i.e., fewer transmission opportunities for the secondary system). Thus, the achievable throughput of the interweave CR system is reduced. It should be also noted that, as in Figure 8.8, for all CR system approaches the average secondary user rate increases as a function of reliability threshold $\mathcal{P}_t$. This occurs because as the targeted primary user outage probability increases, the equivalent maximum tolerated interference at the primary user also increases, hence for all examined approaches $TX$ $s$ allocates its available resources

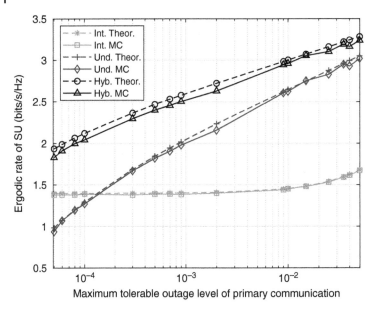

**Figure 8.9** Ergodic secondary user rate versus primary user outage probability requirement, $\mathcal{P}_t$, when $\mathbb{P}($"busy primary user"$) = 0.75$.

primarily with the aim of maximizing the secondary user rate. Finally, fixing a value for $\mathcal{P}_t$, the average secondary user rate for all approaches will be lower for highly active primary systems. This is explained by the fact that when $\mathbb{P}($"busy primary user"$)$ converges to one, more interference by primary transmissions is experienced by the secondary user, on average.

The above numerical evaluation clearly shows that the hybrid CR system outperforms both conventional ones for the investigated primary user activity profiles and communication reliability regimes. Interesting extensions can be thought of, e.g., in terms of investigating the existence of multiple primary and secondary user equipment, where the latter will conduct collaborative spectrum sensing.

## 8.4 Summary

In this chapter we discussed cooperative communication techniques for spectrum sharing in the CR system. Both conventional underlay and new hybrid interweave/underlay approaches were investigated. In the first part of the chapter we considered cooperative statistical channel information exchange between a primary and a secondary transmitter in an underlay CR network. The objective of this investigation was to design a statistically coordinated transmission scheme towards maximizing the average data rate of the secondary user, subject to an average data rate constraint imposed by the primary service. The transmitters cooperate under a realistic CSI scenario where each transmitter has sole instantaneous access to the direct channel of its associated terminal, while it has a full statistical view of the global channel. A formulation based on the theory of team decisions whereby the transmitters aim at optimizing a common objective given the same constraint set on the basis of locally available channel information was considered. Efficient coordination is ensured by exploiting the commonly available statistical CSI of all involved (i.e., direct and interference) links.

In the second part of the chapter we investigated a hybrid CR system approach where the CR operates as either an interweave or an underlay CR system, depending on the results of a spectrum sensing procedure performed at the beginning of each MAC frame of the secondary user. Both SIMO and MIMO system setups were investigated, focusing on uplink and downlink communication, respectively. The problem of jointly designing spectrum sensing and receive antenna combining was considered. The throughput performance of the designed hybrid CR system was evaluated and compared to the throughput performance achieved by the equivalent standard interweave and underlay CR systems, assuming the same level of channel and primary traffic information exchange. Further research challenges include the extension of research to more generalized settings, i.e. multiple primary users and secondary users with multiple antennas, as well as the coexistence of further emerging wireless services.

# References

**1** NGMN Alliance. *Recommendations for NGMN KPIs and Requirements for 5G*. Frankfurt, June 2016.

**2** E. Biglieri, A. J. Goldsmith, L.J. Greenstein, N. Mandayam, and H.V. Poor. *Principles of Cognitive Radio*. Cambridge University Press, 2012.

**3** P. de Kerret and D. Gesbert. CSI sharing strategies for transmitter cooperation in wireless networks. *IEEE Wireless Communications*, 20(1): 43–49, Feb. 2013.

**4** M.C. Filippou, D. Gesbert, and G.A. Ropokis. Optimal combining of instantaneous and statistical CSI in the SIMO interference channel. In *2013 IEEE 77th Vehicular Technology Conference (VTC Spring)*, pages 1–5, Jun. 2013.

**5** M.C. Filippou, G.A. Ropokis, and D. Gesbert. A team decisional beamforming approach for underlay cognitive radio networks. In *IEEE 24th International Symposium on Personal, Indoor and Mobile Radio Communications (PIMRC), 2013*, pages 575–579, Sep. 2013.

**6** M.C. Filippou, D. Gesbert, and G.A. Ropokis. Underlay versus interweaved cognitive radio networks: A performance comparison study. In *2014 9th International Conference on Cognitive Radio Oriented Wireless Networks and Communications (CROWNCOM)*, pages 226–231, Jun. 2014.

**7** M.C. Filippou, D. Gesbert, and G.A. Ropokis. A comparative performance analysis of interweave and underlay multi-antenna cognitive radio networks. *IEEE Transactions on Wireless Communications*, 14(5): 2911–2925, May 2015.

**8** M.C. Filippou, G.A. Ropokis, D. Gesbert, and T. Ratnarajah. Performance analysis and optimization of hybrid MIMO cognitive radio systems. In *2015 IEEE International Conference on Communication Workshop (ICCW)*, pages 555–561, Jun. 2015.

**9** M.C. Filippou, P. de Kerret, D. Gesbert, T. Ratnarajah, A. Pastore, and G.A. Ropokis. Coordinated shared spectrum precoding with distributed CSIT. *IEEE Transactions on Wireless Communications*, 15(8):5182–5192, Aug. 2016.

**10** M.C. Filippou, G.A. Ropokis, D. Gesbert, and T. Ratnarajah. Joint sensing and reception design of SIMO hybrid cognitive radio systems. *IEEE Transactions on Wireless Communications*, 15(9):6327–6341, Sep. 2016.

**11** A. Goldsmith, S. A. Jafar, I. Maric, and S. Srinivasa. Breaking spectrum gridlock with cognitive radios: An information theoretic perspective. *Proceedings of the IEEE*, 97(5):894–914, May 2009.

**12** S. Haykin. Cognitive radio: brain-empowered wireless communications. *IEEE Journal on Selected Areas in Communications*, 23(2):201–220, Feb. 2005.

**13** Y.-C. Ho. Team decision theory and information structures. *Proceedings of the IEEE*, 68(6):644–654, 1980.

**14** X. Kang, Y.-C. Liang, A. Nallanathan, H.K. Garg, and R. Zhang. Optimal power allocation for fading channels in cognitive radio networks: Ergodic capacity and outage capacity. *IEEE Transactions on Wireless Communications*, 8(2):940–950, Feb. 2009.

**15** F.A. Khan, C. Masouros, and T. Ratnarajah. Interference-driven linear precoding in multiuser MISO downlink cognitive radio network. *IEEE Transactions on Vehicular Technology*, 61(6):2531–2543, Jul. 2012.

**16** F.A. Khan, C. Masouros, and T. Ratnarajah. Enhanced outage performance with adaptive linear precoding in cognitive radio downlink. In *2012 IEEE International Conference on Communications (ICC)*, pages 1859–1863, Jun. 2012.

**17** F.A. Khan, T. Ratnarajah, and Z. Ding. Outage performance of cognitive radio wireless network with secondary relaying. In *2012 International Conference on Computer Systems and Industrial Informatics*, pages 1–5, Dec. 2012.

**18** J.N. Laneman, D.N.C. Tse, and G.W. Wornell. Cooperative diversity in wireless networks: Efficient protocols and outage behavior. *IEEE Transactions on Information Theory*, 50(12):3062–3080, Dec. 2004.

**19** K.B. Letaief and W. Zhang. Cooperative communications for cognitive radio networks. *Proceedings of the IEEE*, 97(5):878–893, May 2009.

**20** L. Li, F. Khan, M. Pesavento, T. Ratnarajah, and S. Prakriya. Sequential search based power allocation and beamforming design in overlay cognitive radio networks. *Signal Processing*, 97:221–231, Apr. 2014.

**21** S.L. Loyka. Channel capacity of MIMO architecture using the exponential correlation matrix. *IEEE Communications Letters*, 5(9):369–371, Sep. 2001.

**22** D.G. Luenberger and Y. Ye. *Linear and nonlinear programming*, volume 2. Springer, 1984.

**23** J. Oh and W. Choi. A hybrid cognitive radio system: A combination of underlay and overlay approaches. In *2010 IEEE 72nd Vehicular Technology Conference Fall (VTC 2010-Fall)*, pages 1–5, Sep. 2010.

**24** R. Radner. Team decision problems. *The Annals of Mathematical Statistics*, 33(3):857–881, 1962.

**25** G.A. Ropokis, C.G. Tsinos, M.C. Filippou, K. Berberidis, D. Gesbert, and T. Ratnarajah. Joint power and sensing optimization for hybrid cognitive radios with limited CSIT. In *Proceedings of European Wireless 2015, 21st European Wireless Conference*, May 2015.

**26** S. Senthuran, A. Anpalagan, and O. Das. Throughput analysis of opportunistic access strategies in hybrid underlay-overlay cognitive radio networks. *IEEE Transactions on Wireless Communications*, 11(6): 2024–2035, Jun. 2012.

**27** H. Song, J.-P. Hong, and W. Choi. On the optimal switching probability for a hybrid cognitive radio system. *IEEE Transactions on Wireless Communications*, 12(4):1594–1605, Apr. 2013.

**28** H.A. Suraweera, P.J. Smith, and M. Shafi. Capacity limits and performance analysis of cognitive radio with imperfect channel knowledge. *IEEE Transactions on Vehicular Technology*, 59(4):1811–1822, 2010.

**29** K. Tourki, F.A. Khan, K.A. Qaraqe, Hong-Chuan Y., and M.-S. Alouini. Exact performance analysis of MIMO cognitive radio systems using transmit antenna selection. *IEEE Journal on Selected Areas in Communications*, 32(3):425–438, Mar. 2014.

**30** R. Zakhour and D. Gesbert. Team decision for the cooperative MIMO channel with imperfect CSIT sharing. In *Information Theory and Applications Workshop (ITA), 2010*, pages 1–6, Jan. 2010.

**31** J. Zou, H. Xiong, D. Wang, and C.W. Chen. Optimal power allocation for hybrid overlay/underlay spectrum sharing in multiband cognitive radio networks. *IEEE Transactions on Vehicular Technology*, 62(4):1827–1837, May 2013.

# 9

## Reciprocity-Based Beamforming Techniques for Spectrum Sharing in MIMO Networks

*Kalyana Gopala and Dirk T.M. Slock*

*Institut Eurecom, France*

The efficiency of several spectrum sharing techniques, especially those used for horizontal sharing (i.e., between users of the same type, in the same band, at the same time), relies heavily on the accurate and timely knowledge of the involved user and interference channels. Powerful spectrum sharing techniques are possible with beamforming (BF) over multiple antennas to separate users by exploiting the spatial dimension. However, these BF techniques require channel state information at the transmitter (CSIT). CSIT requirements become crucial in massive multiple input multiple output (MaMIMO) systems, which are a key ingredient of 5G and are also well suited for licensed shared access (LSA) as they offer higher spatial resolution and multiplexing. However, MaMIMO is harder to implement in frequency division duplex (FDD) systems due to the complications imposed by the feedback channel. An alternative approach is to consider time division duplex (TDD) systems, where, in theory at least, the forward and reverse channels are equal, hence do not require a feedback channel.

The goal of this chapter is to discuss two approaches for handling CSIT that lead to limited overhead and are applicable to LSA. The approaches consider either *pathwise* or *instantaneous CSIT reciprocity*. Both approaches extract CSIT information from the uplink (UL) to be used for downlink (DL) transmission. While the second approach is geared towards TDD systems, the first approach is based on exploiting the multipath structure and would be applicable to FDD systems or uncoordinated TDD systems. For the case of instantaneous reciprocity, this chapter deals with the actual TDD scenario in which reciprocity is only obtained after calibration of the radio frequency (RF) parts in the transmit and receive chains.

## 9.1 Multi-antenna Cognitive Radio Paradigms

Before delving into multi-antenna LSA, we shall consider the extension of a number of standard cognitive radio (CR) paradigms to the multi-antenna case. These extensions are not as straightforward and unambiguous as it may seem at first. Here we discuss some possible multi-antenna extensions for these paradigms. These proposals were first put forward in a

*Spectrum Sharing: The Next Frontier in Wireless Networks*, First Edition.
Edited by Constantinos B. Papadias, Tharmalingam Ratnarajah, and Dirk T.M. Slock.
© 2020 John Wiley & Sons Ltd. Published 2020 by John Wiley & Sons Ltd.

CR Panel Session organized by the European Union FET project CROWN-233843 consortium at the CogART conference in Barcelona, Spain, in October 2011 (see also [18]).

### 9.1.1 Spatial Overlay: MISO/MIMO Interference Channel

In the overlay paradigm, Primary and secondary systems collaborate (see Figure 9.1). This collaboration could be interpreted at multiple levels, at the level of an exchange of transmit signals [as in network multiple input multiple output (MIMO) or cooperative multi-point (CoMP)], or just at the level of CSIT, which in the single antenna case translates to coordinated power control. In the case of multiple antennas, if we limit cooperation to CSIT, this would lead to the exploitation of the multiple antennas for coordinated beamforming to achieve parallel interference-free channels. Coordinated beamforming applies to multiple antennas at the transmit side [multiple input single output (MISO) interference channel]. In the case of multiple antennas at the receivers, we can have coordinated receivers. The case of coordination of the multiple antennas on both sides corresponds to the (noisy) MIMO interference channel and spatial interference alignment (IA)

The authorized shared access (ASA) proposal by Qualcomm and Nokia [22] fits in this realm of overlay cognitive radio.

### 9.1.2 Spatial Underlay

In the underlay paradigm, interference caused by a secondary transmitter to a primary receiver is acceptable as long as the interference remains under a maximum tolerance level. One possible definition of spatial underlay then would be that a primary receiver equipped with multiple antennas allows primary interference as long as it has enough antennas to handle it. Hence the primary receiver needs to be active. So, the primary receiver allows an interference subspace of maximum dimension equal to the excess of its number of antennas over the number of primary streams it needs to receive. The primary system is secondary aware. Of course, the secondary transmitters need to align the interference caused to primaries in subspaces of limited dimension.

### 9.1.3 Spatial Interweave

In the interweave paradigm, the primary system should not be disturbed at all, and is not required to exhibit any cooperation with the secondary systems. So in a spatial interweave version with multiple primary receive antennas, the secondary systems need to zero-force to all primary receive antennas individually. In this case there is still room for secondary transmission if a secondary transmitter has more antennas than the combined primary receivers. The spatial interweave paradigm requires significant CSIT and can be reciprocity based in TDD or a location based on the case of line-of-sight (LoS) secondary–primary cross channels. In the LOS case, the number of primary receive antennas becomes irrelevant (assuming they are in the far field of the secondary) because the MIMO cross channel becomes rank one [15]. In the case of non-LoS (NLoS) and a pathwise CSIT-based approach, the secondary transmitter needs to have more antennas than the number of propagation paths to all primary receivers.

## 9.2 From Multi-antenna Underlay to LSA Coordinated Beamforming

We consider exploiting multiple antennas for a much more dynamic form of LSA in the form of coordinated beamforming (CoBF) between incumbent cells and licensee cells, see Figure 9.1. The BF is based on a combined form of partial CSIT, comprising both channel estimates (mean CSIT) and covariance CSIT (of which pathwise CSIT is one particular form). In particular, multipath induced structured low rank covariances are considered that arise in MaMIMO and millimeter-wave (mmWave) settings. For the beamforming optimization, we first revisit the weighted sum rate (WSR) maximization with perfect CSIT [1]. We then turn to the partial CSIT case where we consider expected WSR (EWSR) maximization. We also establish an explicit link between underlay cognitive radio and coordinated beamforming: the optimal choice for the Lagrange multipliers for the interference temperature constraints in underlay cognitive radio in fact corresponds to the ratio of the rate weight (in the WSR) for the stream interfered with over the mean squared error (MSE) attained for that stream. The work considered there is applicable to both macrocellular and small cell or even heterogeneous scenarios.

### 9.2.1 CoBF and CSIT Discussion

Interference is the main limiting factor in wireless transmission. Base stations (BSs) disposing of multiple antennas are able to serve multiple user equipments (UEs) simultaneously,

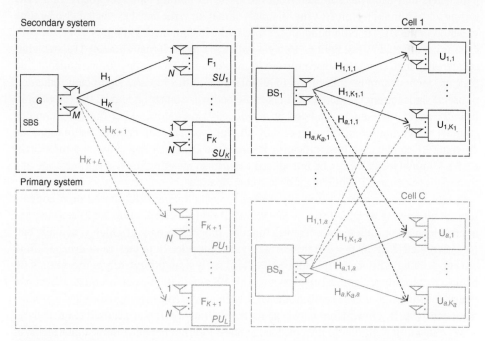

**Figure 9.1** Traditional underlay cognitive radio systems (left) vs. coordinated beamforming in multiple cells (right).

which is called spatial division multiple access (SDMA) or multi-user (MU) MIMO. However, MU systems have precise requirements for CSIT, which is more difficult to acquire than CSI at the Rx (CSIR). Hence we focus here on the more challenging DL. The main difficulty in realizing linear IA for MU MIMO in multi-cell settings (interference (broadcast) channel (IC/IBC)) is that the design of any BS Tx filter depends on all Rx filters whereas in turn each Rx filter depends on all Tx filters [19]. As a result, all Tx/Rx filters are globally coupled and their design requires global CSIT. To carry out this Tx/Rx design in a distributed fashion, global CSIT is required at all BS [21]. The overhead required for this global distributed CSIT is substantial, even if done optimally, leading to substantially reduced net degrees of freedom (DoFs) [14]. We refer to [16] for a further discussion of the state of the art. Recent works focus on intercell exchange of only scalar quantities, at fast fading rate, and on two-stage approaches in which the intercell interference is zero forced (ZF). The recent development of MaMIMO [13] opens up new possibilities for increased system capacity while at the same time simplifying system design. We refer to [1] for a further discussion of the state of the art. Note that MaMIMO in most works refers actually to MU MISO.

Whereas the exploitation of covariance CSIT may be beneficial, in a MaMIMO context it may quickly lead to high computational complexity and estimation accuracy issues. Computational complexity may be reduced (and the benefit of exploiting covariance CSIT enhanced) in the case of low rank covariance structure, such as can be expected in MaMIMO and mmWave settings, but the use and tracking of subspaces may still be cumbersome. In the pathwise approach, these subspaces are very parsimoniously parameterized in terms of directions of arrival or departure (DoAs/DoDs). In an FDD setting, these parameters may even be estimated from the UL by exploiting path reciprocity. In a TDD setting with channel reciprocity, the channel estimation error may be affected also by time variation in the UL/DL ping-pong. In contrast to the instantaneous channel CSIT, the path CSIT is not affected by fast fading. Whereas path CSIT by itself may allow ZF [15], which is of interest at high signal-to-noise ratio (SNR), maximum WSR designs accounting for finite SNR are more desirable. Indeed, ZF of all interfering links leads to significant reduction of useful signal strength. MaMIMO makes the pathwise approach viable: the (cross-link) beamformers (BFs) can be updated at a reduced (slow fading) rate, the parsimonious channel representation facilitates not only UL but especially DL channel estimation, the crosslink BF can be used to significantly improve the DL direct link channel estimates (in FDD). Minimal feedback can be introduced to perform meaningful WSR optimization at a finite SNR (whereas ZF requires much less coordination).

In [30] we review some recent approaches for maximizing WSR, based on a connection to minimizing weighted sum MSE (WSMSE) [2] [see also Section 9.4.2) and another approach based on difference of convex function programming [9] (which is actually better interpreted as an instance of minorization maximization)]. In fact, the Karush–Kuhn–Tucker (KKT) conditions for maximum WSR lead fairly straightforwardly to the optimal BF being a maximum generalized eigenvector. This BF turns out to be an optimized form of a heuristic solution that has been introduced to maximize a so-called signal-to-leakage-plus-noise-ratio (SLNR), in which the leakage represents the interference caused to other users when transmitting, similar to the familiar signal-to-interference-plus-noise-ratio (SINR) at a receiver. The SLNR optimizing BF is also a generalized eigenvector, but of unweighted quantities. The minorization of the WSR cost function by a concave approximation also leads to an optimization of the stream powers, in a fashion that can be interpreted as "interference leakage aware water filling".

Also, in the case of partial CSIT, the WSR criterion needs to be modified. We consider the EWSR, also called the ergodic WSR. We consider various approaches for maximizing EWSR to handle partial CSIT. The existing expected WSMSE (EWSMSE) approach [20] (also termed "use and forget lower bound" in [17]) improves over naive EWSR (NEWSR) by accounting for covariance CSIT in the interference. This can have significant impact, even on the sum rate prelog (DoF) if the instantaneous channel CSIT quality does not scale with SNR. A further improvement is proposed in [30] in the expected signal and interference power WSR (ESIP-WSR) approach, which represents a better approximation of the EWSR. In a MaMIMO setting (in which case ESIP-WSR becomes an upper bound of EWSR), the way mean and covariance CSIT are combined in the EWSMSE or ESIP-WSR approaches as the interference terms become equally optimal as in the EWSR for a large number of users. However, ESIP-WSR represents an improvement over EWSMSE by capturing the signal power in the covariance CSIT (matched filtering and diversity aspects) and only leads to a finite gain (in SNR), but its remaining approximation error over EWSR may be quite limited [3]. Strictly speaking, in a large number of user settings, EWSMSE $\leq$ EWSR $\leq$ ESIP-WSR. The step from EWSMSE to ESIP-WSR also deals with the following question. Covariance CSIT can be used to improve the channel estimate from a basic deterministic estimate to a Bayesian estimate. The question then arises: is that enough? The answer is no and the question has been settled in [11], at least for ESIP-WSR, in which the posterior expectation of a quadratic expression in the channel is required. This can be constructed from the minimum mean squared error (MMSE) channel estimate and corresponding error covariance matrix. Hence, beyond the MMSE channel estimate, it is important to also exploit the posterior (error) covariance, which exhibits the same subspace structure as the prior channel covariance.

For covariance CSIT in the form of pathwise CSIT, in [29] we introduced a heuristic to design the Tx separately using pathwise CSIT only. It turns out that this heuristic is recovered by the ESIP-WSR approach proposed in [12], which furthermore provides expressions for a number of auxiliary quantities that are needed and allows the combination of channel estimate and pathwise CSIT. In these works we typically considered multiple cells (incumbent and licensee) with multi-antenna BSs serving possibly multiple users with single or multiple antennas.

### 9.2.2 Some LoS Results

We now discuss the extension of the perfect CSIT approach to optimization of the EWSR combining both channel estimates and covariance CSIT. The goal here is multi-fold. A naive approach would just perform (possibly regularized) zero forcing (R-ZF) BF. We derived algorithms that maximize the WSR at any finite SNR, exploiting not only channel estimate information, accounting for the channel estimation error level, but also exploiting channel covariance information. The resulting approach even works when only channel covariance information is available (e.g., possibly for the intercell channels). This could arise when the user location information is translated into a channel covariance based on LoS propagation. In the simulations below, we consider just a single cell MIMO system with two users, with

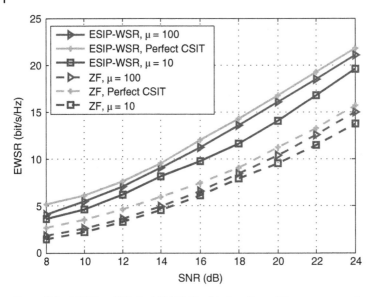

**Figure 9.2** EWSR vs SNR for MU MIMO with four Tx and Rx antennas and two users.

CSIT based on the MIMO Ricean channel model [hence the CSIT comprises the (downlink) Tx side LoS antenna array response and the Rice factor $\mu$, both of which can be estimated from the uplink channel]. This may be a simple way to account for unmodeled multipath components. In Figure 9.2 the expected sum rate is plotted versus SNR for for Tx and Rx antennas and two users. For the Tx design, we consider either ZF on the LoS component, with uniform power loading, or an optimized design based on the minorization approach to ESIP-WSR. For each design, three cases of Rice factor are considered: $\mu = 10, 100$ or $\infty$ (this last case is labeled "Perfect CSIT" in Figure 9.2). The expected sum rate is obtained by averaging over channel realizations, according to the Ricean distribution, with one of the three possible values for $\mu$. The optimized approach which accounts for both CSIT imperfections and finite SNR clearly improves over naive (LoS based) ZF.

### 9.2.3 Noncoherent Multi-user MIMO Communications using Covariance CSIT

We briefly allude again to the general case of Gaussian partial CSIT, in which the combined availability of channel estimates (mean CSIT) and covariance CSIT can be exploited. Such general partial CSIT scenario can, for example, be particularized as in [28] to the case of perfect instantaneous CSIT for intracell channels and pathwise CSIT for intercell channels. This leads to two-stage BF expressions, similar to hybrid beamforming. The slow stage handles intercell interference and is frequency-flat. It can be exploited also to separate the cells for channel estimation purposes. In [12] we consider in more detail pathwise CSIT for all channels (both intercell and intracell; see Figure 9.3. The ESIP-WSR approach mentioned earlier leads to a loss of all (narrowband) frequency-selectivity in the channel and also leaves no utility for space-time coding (as far as EWSR is concerned), though this can be expected to bring some benefits (e.g., for outage). The exploitation of pathwise CSIT in a (especially massive) MIMO setting leads to non-Kronecker MIMO channel covariance structures, as opposed to the widely assumed Kronecker MIMO model. The non-Kronecker

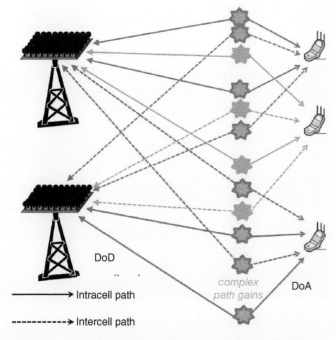

DoD

Intracell path

Intercell path

*complex path gains*

DoA

**Figure 9.3** Pathwise multi-user heterogeneous network scenario.

MIMO channel pathwise CSIT leads to a split between the roles of transmitters and receivers in MIMO systems. For the BF optimization, we consider in [12] a minorization approach applied to ESIP-WSR. Simulations indicate that the pathwise CSIT-based designs may lead to limited spectral efficiency loss compared to instantaneous CSIT based designs, while trading fast fading CSIT for slow fading CSIT. We also point out in [12] that the pathwise approach may lead to distributed designs requiring only local pathwise CSIT, and analyze the sum rates for instantaneous and pathwise CSIT in the low and high SNR limits.

## 9.3 TDD Reciprocity Calibration

Whereas the pathwise CSIT approaches considered above can be applied to FDD or uncoordinated TDD approaches, we now turn our attention to coordinated TDD systems. We first provide an overview of the state of the art in reciprocity calibration techniques, with an emphasis on internal calibration usable in MaMIMO. Then a number of promising reciprocity-based techniques are presented for the design of transmit precoders for spectrum sharing between incumbents and licensees. In particular, we present the concept of naive UL/DL duality, which allows further reduction of the additional information exchange required for utility optimization or to deal with non-cooperative nodes.

### 9.3.1 Fundamentals

Figure 9.4 shows a detailed break up of the end-to-end channel between a set of antennas **A** communicating with another set of antennas **B** clearly demarcating the propagation

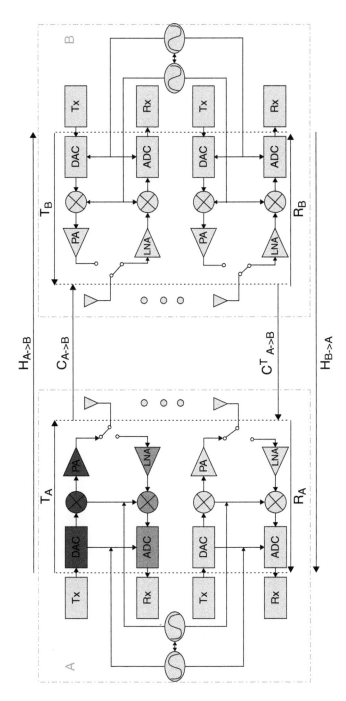

**Figure 9.4** Reciprocity model in TDD.

channel and the RF chains. $\mathbf{C}$ refers to the propagation channel, which is reciprocal. The $(i, j)$th entry of $\mathbf{C}$ corresponds to the propagation channel between the antennas $i$ and $j$ and is reciprocal: $\mathbf{C}_{i,j} = \mathbf{C}_{j,i}$. The part of the channel covered by $\mathbf{C}$ actually starts from where the antenna feeding line connects to the Tx and Rx RF chains at side $A$ to a similar point at side $B$. The overall DL and UL channels observed in the digital domain are denoted by $\mathbf{H}_{A\to B}$ and $\mathbf{H}_{B\to A}$. It can be clearly seen from Figure 9.4 that paths traversed by the Tx and Rx signals are different at the RF level, which results in non-reciprocity of the overall digital channel. In the frequency domain, over a narrow frequency band, we get:

$$\begin{cases} \mathbf{H}_{A\to B} = \mathbf{R}_B \mathbf{C}\, \mathbf{T}_A, \\ \mathbf{H}_{B\to A} = \mathbf{R}_A \mathbf{C}^T \mathbf{T}_B. \end{cases} \tag{9.1}$$

Matrices $\mathbf{T}_A$, $\mathbf{R}_A$, $\mathbf{T}_B$, and $\mathbf{R}_B$ model the response of the transmit and receive RF front-ends and are called the absolute calibration factors. It is assumed that the impact of the Tx and Rx chains over a narrow frequency band may be modeled as a complex scaling factor. This has been validated in several real implementations, such as [27] and [25]. The diagonal elements in these matrices represent the linear effects attributable to the attenuation or amplification and phase shift in the Tx and Rx parts of the RF front-end, whereas the off-diagonal elements correspond to RF cross-talk. Thus, the DL channel $\mathbf{H}_{A\to B}$ may be derived from the UL channel $\mathbf{H}_{B\to A}$ by eliminating $\mathbf{C}$ between both equations in (9.1) as follows:

$$\mathbf{H}_{A\to B} = \mathbf{R}_B (\mathbf{R}_A^{-1} \mathbf{H}_{B\to A} \mathbf{T}_B^{-1})^T \mathbf{T}_A = \underbrace{\mathbf{R}_B \mathbf{T}_B^{-T}}_{\mathbf{F}_B^{-T}} \mathbf{H}_{B\to A}^T \underbrace{\mathbf{R}_A^{-T} \mathbf{T}_A}_{\mathbf{F}_A} = \mathbf{F}_B^{-T} \mathbf{H}_{B\to A}^T \mathbf{F}_A. \tag{9.2}$$

Thus, the lack of reciprocity at the level of the RF chains brings in a need for reciprocity calibration factors $\mathbf{F}_A$ and $\mathbf{F}_B$. If we drop the indication of the sets $A$ and $B$, we have

$$\mathbf{F} = \mathbf{R}^{-T} \mathbf{T}. \tag{9.3}$$

$\mathbf{F}$ is called a relative calibration factor as it is obtained as a ratio of the absolute calibration factors $\mathbf{T}$, $\mathbf{R}$. It is important to note that for the purpose of DL channel estimation, there is no need to estimate the absolute calibration factors (which would be useful for, for example, antenna array response determination). Instead, we only need the relative calibration factors.

From the UL/DL relation (9.2), it can be seen that the relative calibration factors $\mathbf{F}_A$, $\mathbf{F}_B$ can only be determined up to a common complex scale factor. Indeed, replacing $\mathbf{F}_A$, $\mathbf{F}_B$ by $\alpha \mathbf{F}_A$, $\alpha \mathbf{F}_B$ for some scalar $\alpha$ does not modify the relation (9.2). Note that despite this scale ambiguity in $F$, $\mathbf{H}_{A\to B}$ can be uniquely determined from $\mathbf{H}_{B\to A}$. Another remark concerns the common MISO case where typically side $A$ is a multi-antenna BS whereas side $B$ would be a single-antenna UE. In this case $\mathbf{F}_B$ is a scalar. Based on the previous scale ambiguity remark, it is clear that we can choose $\mathbf{F}_B = 1$ so that $\mathbf{F}_A$ becomes uniquely identifiable. So, in the MISO case, only a one-sided (BS side) calibration is required. It is tempting to continue to calibrate only the BS side even in the MIMO case, since the internal calibration procedure to be discussed below is geared towards one-sided calibration. Doing so will lead to a predicted DL channel $\mathbf{H}_{A\to B}$, which is off at the UE side by a (matrix) factor $\mathbf{F}_B$.

Such discrepancy would be unimportant if the BS side beamforming would perform zero forcing to all UE antennas. It will lead to suboptimality though in a WSR or other utility optimization design at finite SNR or with a reduced number of streams compared to the number of Rx antennas, except perhaps in a ping-pong design between Tx and Rx optimization by BS(s) and UE(s) alternatingly, in which a UE optimizes its Rx based on the estimated actual DL channel. In any case, such one-sided calibration has been considered also in 3GPP standardization work [23, 24].

A TDD reciprocity-based MIMO system normally has two phases for its operation. First, during the initialization of the system or the training phase, the reciprocity calibration process is activated, which consists of estimating $\mathbf{F}_A$ and $\mathbf{F}_B$. Then, during the data transmission phase, these calibration coefficients are used together with instantaneous measured UL channel $\widehat{\mathbf{H}}_{B \to A}$ to estimate the CSIT $\mathbf{H}_{A \to B}$, based on which advanced beamforming algorithms can be performed. Since the calibration coefficients remain stable for quite a long time [27] (in the order of hours), the calibration process does not have to be performed very frequently.

### 9.3.2 Diagonality of the Calibration Matrix

In the rest of this chapter we assume that the calibration matrix $\mathbf{F}$ is diagonal. This was validated experimentally in [6], where the off-diagonal elements of $\mathbf{F}$ were found to be less than 30 dB compared to the diagonal elements. Note, however, that this does not imply that there would be no mutual coupling between antennas of an array on a given side, or cross-talk between their feeding lines. Indeed, if $\mathcal{M}_A$, $\mathcal{M}_B$ represent such (reciprocal) non-diagonal matrices that encapsulate the antenna mutual coupling and cross-talk, we get

$$\begin{cases} \mathbf{H}_{A \to B} = \mathbf{R}_B(\mathcal{M}_B \mathbf{C} \mathcal{M}_A)\mathbf{T}_A, \\ \mathbf{H}_{B \to A} = \mathbf{R}_A(\mathcal{M}_A \mathbf{C}^T \mathcal{M}_B)\mathbf{T}_B. \end{cases} \tag{9.4}$$

Hence, by treating the reciprocal mutual coupling and cross-talk as part of the propagation channel, by renaming $\mathcal{M}_B \mathbf{C} \mathcal{M}_A$ as $\mathbf{C}$, we get back the diagonal calibration factors $\mathbf{R}$, $\mathbf{T}$, and hence $\mathbf{F}$.

### 9.3.3 Coherent and Non-coherent Calibration Scheme

The calibration parameters of the antenna may be considered to remain constant in the order of several hours. However, the variation of the physical propagation channel is typically much faster. This leads to two ways of approaching the estimation of the relative calibration parameters. We could complete the entire estimation of these parameters in a short time span where the propagation channel stays a constant. Such a time duration would be called a coherent time slot. When the estimation happens within one coherent time slot, it is called a coherent calibration scheme. Alternatively, the problem may well be formulated over several different coherent time slots (during which the calibration parameters themselves are assumed constant), and in this case it is called non-coherent calibration. This is illustrated in Figure 9.5.

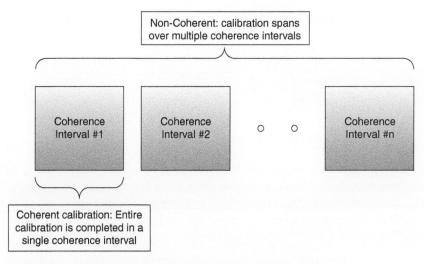

**Figure 9.5** Illustration of coherent and non-coherent calibration.

### 9.3.4 UE-aided vs Internal Calibration

There are two main approaches to reciprocity calibration based on whether or not a UE is involved in its determination.

1. In *UE-aided calibration*, explicit channel feedback from a UE during the calibration phase is used to estimate the calibration parameters. Hence, during a training phase, explicit pilots are exchanged between the BS and UE over the air. Based on these pilots, the UE feeds back its estimate of the DL channel to the BS which, together with its estimate of the UL channel, derives the calibration parameters.

2. It was noted in [10] that the BS side calibration factor $\mathbf{F}_A$ is independent of which is considered as side $B$ (see Figure 9.4). This was exploited in [10] to calibrate the channel from a secondary BS to a primary UE without primary cooperation by performing calibration between secondary BS and UE. This was then pushed further in [27] to replace the cooperative UE by one BS antenna. This led to an approach called *internal calibration* or *self-calibration*, in which the calibration is performed entirely between the antennas of the BS (as already suggested in [23, 24] in fact). Internal calibration only estimates the $\mathbf{F}_A$ up to a scale factor and does not estimate the $\mathbf{F}_B$ at all. An important advantage of this kind of calibration is that it ensures tight time and frequency synchronization amongst the antennas that are being calibrated in the case of co-located MaMIMO.

### 9.3.5 Group Calibration System Model

The existing literature on reciprocity calibration typically considers transmission from a single antenna at a time for purposes of calibration. A more general system model was proposed in [7] that allows the grouping of multiple antennas during transmission. This model falls back to the single antenna transmission scenario when each group has only one antenna.

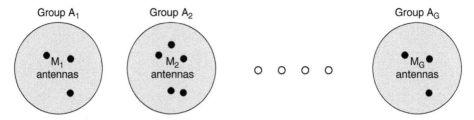

**Figure 9.6** Illustration of the group calibration system model.

Here, as shown in Figure 9.6, the total of $M$ (BS) antennas is partitioned into $G$ groups with $M_i$ antennas each. Each group $A_i$ transmits pilots $\mathbf{P}_i$ for $L_i$ time instants (or channel uses). Let $\mathbf{Y}_{i \to j}$ be the received signal at antenna $j$ on transmission of pilot $\mathbf{P}_i$ from antenna $i$. Then for every pair of transmission between antennas $i$ and $j$ (bi-directional Tx), we obtain

$$\text{bi-directional Tx} \begin{cases} \mathbf{Y}_{i \to j} = \underbrace{\mathbf{R}_j}_{M_j \times L_i} \underbrace{\mathbf{C}_{i \to j}}_{M_j \times M_j} \underbrace{\mathbf{T}_i}_{M_j \times M_i} \underbrace{\mathbf{P}_i}_{M_i \times M_i} + \mathbf{N}_{i \to j}, \\[2mm] \mathbf{Y}_{j \to i} = \mathbf{R}_i \mathbf{C}_{i \to j}^T \mathbf{T}_j \mathbf{P}_j + \mathbf{N}_{j \to i}. \end{cases} \qquad (9.5)$$

$\mathbf{N}_{i \to j}$ represents the noise seen at antenna $j$ when antenna $i$ is transmitting. Equation (9.5) also shows the dimensions of the matrices involved for clarity. It is important to note that the channel is assumed to be constant during this bi-directional Tx. Eliminating the propagation channel $\mathbf{C}_{i \to j}$, we get

$$\mathbf{P}_i^T \mathbf{F}_i^T \mathbf{Y}_{j \to i} - \mathbf{Y}_{i \to j}^T \mathbf{F}_j \mathbf{P}_j = \mathbf{P}_i^T \mathbf{F}_i^T \mathbf{N}_{j \to i} - \mathbf{N}_{i \to j}^T \mathbf{F}_j \mathbf{P}_j, \qquad (9.6)$$

where $\mathbf{F}_i = \mathbf{R}_i^{-T} \mathbf{T}_i$ and $\mathbf{F}_j = \mathbf{R}_j^{-T} \mathbf{T}_j$ are the calibration matrices for groups $i$ and $j$. Using the vec operator (stacking consecutive columns of a matrix into a tall vector) and its properties, equation (9.6) may be rewritten as

$$\begin{aligned} \text{vec}(\mathbf{P}_i^T \mathbf{F}_i^T \mathbf{Y}_{j \to i}) &= \text{vec}(\mathbf{Y}_{i \to j}^T \mathbf{F}_j \mathbf{P}_j) + \text{vec}(\mathbf{P}_i^T \mathbf{F}_i^T \mathbf{N}_{j \to i} - \mathbf{N}_{i \to j}^T \mathbf{F}_j \mathbf{P}_j), \\ (\mathbf{Y}_{j \to i}^T \otimes \mathbf{P}_i^T) \, \text{vec}(\mathbf{F}_i^T) &= (\mathbf{P}_j^T \otimes \mathbf{Y}_{i \to j}^T) \, \text{vec}(\mathbf{F}_j) + \tilde{\mathbf{N}}_{ij}. \end{aligned} \qquad (9.7)$$

Here, we have used the property that for any matrices $\mathbf{X}_1$, $\mathbf{X}_2$, and $\mathbf{X}_3$,

$$\text{vec}(\mathbf{X}_1 \mathbf{X}_2 \mathbf{X}_3) = (\mathbf{X}_3^T \otimes \mathbf{X}_1) \, \text{vec}(\mathbf{X}_2), \qquad (9.8)$$

where $\otimes$ denotes the Kronecker product. $\tilde{\mathbf{N}}_{ij} = \text{vec}(\mathbf{P}_i^T \mathbf{F}_i^T \mathbf{N}_{j \to i} - \mathbf{N}_{i \to j}^T \mathbf{F}_j \mathbf{P}_j)$. In addition, as the matrices $\mathbf{F}_i$ are diagonal, all the columns in matrices such as $\mathbf{Y}_{j \to i}^T \otimes \mathbf{P}_i^T$ corresponding to the zero entries in $\text{vec}(\mathbf{F}_i^T)$ can be eliminated. Hence, equation (9.7) may be further rewritten as

$$(\mathbf{Y}_{j \to i}^T * \mathbf{P}_i^T)\mathbf{f}_i - (\mathbf{P}_j^T * \mathbf{Y}_{i \to j}^T)\mathbf{f}_j = \tilde{\mathbf{N}}_{ij}, \qquad (9.9)$$

where $*$ denotes the Khatri–Rao product [8] (or column-wise Kronecker product) and vector $\mathbf{f}_i$ contains the diagonal elements of $\mathbf{F}_i$. With matrices $\mathbf{A}$ and $\mathbf{B}$ partitioned into columns $A = [\mathbf{a}_1 \, \mathbf{a}_2 \dots \mathbf{a}_M]$ and $B = [\mathbf{b}_1 \, \mathbf{b}_2 \dots \mathbf{b}_M]$ where $\mathbf{a}_i$ and $\mathbf{b}_i$ are column vectors, $A * B = [\mathbf{a}_1 \otimes \mathbf{b}_1 \; \mathbf{a}_2 \otimes \mathbf{b}_2 \; \cdots \; \mathbf{a}_M \otimes \mathbf{b}_M]$. Here, we have used the identity $\text{vec}(\mathbf{X}_1 \, \text{diag}(\mathbf{x}) \, \mathbf{X}_3) = (\mathbf{X}_3^T * \mathbf{X}_1)\,\mathbf{x}$.

### 9.3.6 Least-squares Solution

Collecting all these bi-directional transmissions, we arrive at a least-squares (LS) formulation to solve for the relative calibration factors $\mathbf{f}$:

$$\widehat{\mathbf{f}} = \arg\min_{\mathbf{f}} \sum_{i,j\in\mathcal{G}} ||(\mathbf{Y}_{j\to i}^T * \mathbf{P}_i^T)\mathbf{f}_i - (\mathbf{P}_j^T * \mathbf{Y}_{i\to j}^T)\mathbf{f}_j||^2, \tag{9.10}$$

where $\mathcal{G}$ defines the set of all bi-directional transmissions. Of course, this needs to be augmented with a constraint,

$$C(\widehat{\mathbf{f}}, \mathbf{f}) = 0, \tag{9.11}$$

in order to exclude the trivial solution $\widehat{\mathbf{f}} = \mathbf{0}$ in equation (9.10). The constraint on $\widehat{\mathbf{f}}$ may depend on the true parameters $\mathbf{f}$. As we shall see further this constraint needs to be complex valued (which represents two real constraints). Typical choices for the constraint are

1) Norm plus phase constraint (NPC):

$$\text{norm: } \mathrm{Re}\{C(\widehat{\mathbf{f}}, \mathbf{f})\} = ||\widehat{\mathbf{f}}||^2 - c\,, \ c = ||\mathbf{f}||^2, \tag{9.12}$$

$$\text{phase: } \mathrm{Im}\{C(\widehat{\mathbf{f}}, \mathbf{f})\} = \mathrm{Im}\{\widehat{\mathbf{f}}^H\mathbf{f}\} = 0. \tag{9.13}$$

2) Linear constraint:

$$C(\widehat{\mathbf{f}}, \mathbf{f}) = \widehat{\mathbf{f}}^H\mathbf{g} - c = 0. \tag{9.14}$$

If we choose the vector $\mathbf{g} = \mathbf{f}$ and $c = ||\mathbf{f}||^2$, then the $\mathrm{Im}\{.\}$ part of equation (9.14) corresponds to equation (9.13). The most popular linear constraint is the first coefficient constraint (FCC), which is equation (9.14) with $\mathbf{g} = \mathbf{e}_1, c = 1$.

The LS solution presented here reduces to the algorithm presented in [25, 26] for the special case of a single antenna per group. With a single antenna per group, for a coherent calibration estimation scheme, the minimum number of transmissions required to obtain the calibration parameters is $M$. However, it is shown in [7] that with an optimal grouping of antennas, this can be reduced to the order of $\sqrt{M}$.

### 9.3.7 A Bilinear Model

To obtain more insight into the performance limits for the calibration parameter estimation, we can rewrite the received signal as follows:

$$\mathbf{Y}_{i\to j} = \underbrace{\mathbf{R}_j\mathbf{C}_{i\to j}\mathbf{R}_i^T}_{\mathcal{H}_{i\to j}}\mathbf{F}_i\mathbf{P}_i + \mathbf{N}_{i\to j}. \tag{9.15}$$

We define $\mathcal{H}_{i\to j} = \mathbf{R}_j\mathbf{C}_{i\to j}\mathbf{R}_i^T$ to be an auxiliary internal channel (not corresponding to any physically measurable quantity) that appears as a nuisance parameter in the estimation of the calibration parameters. Note that the auxiliary channel $\mathcal{H}_{i\to j}$ inherits the reciprocity from the propagation channel $\mathbf{C}_{i\to j}$: $\mathcal{H}_{i\to j} = \mathcal{H}_{j\to i}^T$. Thus, the received signal in (9.15) takes on a bilinear form (separately linear in each of the parameters $\mathcal{H}$ and $\mathbf{F}$). This is exploited in [7] and [5] to derive the Cramer–Rao bound for the general calibration framework. It is also shown that the maximum likelihood solution for the calibration parameter estimation may be viewed as a weighted least-squares, where the weights are dependent on the calibration parameters.

## 9.4 MIMO IBC Beamformer Design

Here we discuss the beamformer design problem for an IBC (multi-cell MU DL), as illustrated in Figure 9.7. We shall introduce the concept of naive UL/DL duality as a low complexity practical BF design. To set this in context, we first review proper UL/DL duality.

### 9.4.1 System Model

Consider an IBC with $C$ cells and a total of $K$ UEs. Consider a system-wide UE numbering. UE $k$ has $N_k$ antennas and is served by BS $b_k$ which has $M_{b_k}$ antennas. Only one stream is transmitted per UE. The $N_k$ length received signal vector in the DL at user $k$ in cell $b_k$ is

$$\mathbf{y}_k = \underbrace{\mathbf{H}_{k,b_k} \mathbf{g}_k s_k}_{\text{signal}} + \underbrace{\sum_{\substack{i \neq k \\ b_i = b_k}} \mathbf{H}_{k,b_k} \mathbf{g}_i s_i}_{\text{intracell interf.}} + \underbrace{\sum_{j \neq b_k} \sum_{i : b_i = j} \mathbf{H}_{k,j} \mathbf{g}_i s_i}_{\text{intercell interf.}} + \mathbf{v}_k. \tag{9.16}$$

Here $s_k$ is the intended (white, unit variance) signal and $\mathbf{H}_{k,b_k}$ is the $N_k \times M_{b_k}$ channel from BS $b_k$ to user $k$. BS $b_k$ serves $K_{b_k} = \sum_{i : b_i = b_k} 1$ UEs. The noise $\mathbf{v}_k \sim \mathcal{CN}(0, \mathbf{I})$. The $M_{b_k} \times 1$ spatial Tx filter or BF is $\mathbf{g}_k$.

### 9.4.2 WSR Optimization via WSMSE

We now look at the weighted sum rate optimization when complete CSIT is available. Assuming Gaussian signaling, the WSR [2] for the IBC scenario would be

$$\text{WSR}(\mathbf{g}) = \sum_k u_k \ln \det(1 + \mathbf{g}_k^H \mathbf{H}_{k,b_k}^H \mathbf{R}_{\bar{k}}^{-1} \mathbf{H}_{k,b_k} \mathbf{g}_k) = \sum_k u_k \ln \frac{1}{e_{k,\text{MMSE}}}. \tag{9.17}$$

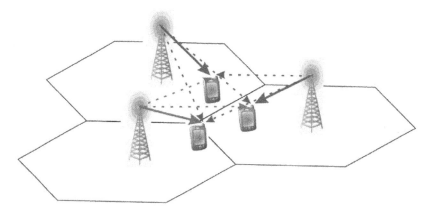

**Figure 9.7** Illustration of a MIMO IBC scenario.

Here $\mathbf{R}_{\bar{k}} = \sum_{i \neq k} \mathbf{H}_{k,b_i} \mathbf{g}_i \mathbf{g}_i^H \mathbf{H}_{k,b_i}^H + \mathbf{I}_{N_k}$ is the received interference plus noise covariance matrix, $u_k$ are rate weights, and $e_{k,\text{MMSE}}$ refers to the MSE at the output of a linear minimum MSE (LMMSE) Rx filter. The WSR is to be optimized under the power constraint $\sum_{k:b_k=j} ||\mathbf{g}_k||^2 \leq P_j$, where $P_j$ is the total output power of BS $j$. This inspired the following WSMSE approach to WSR maximization [2] [or see [20] for the EWSMSE extension]. Introduce a linear Rx filter $\mathbf{f}_k$ with output

$$\hat{s}_k = \mathbf{f}_k^H \mathbf{y}_k = \mathbf{f}_k^H \mathbf{H}_{k,b_k} \mathbf{g}_k s_k + \sum_{i \neq k} \mathbf{f}_k^H \mathbf{H}_{k,b_i} \mathbf{g}_i s_i + \mathbf{f}_k^H \mathbf{v}_k. \tag{9.18}$$

The MSE $\mathbb{E}|s_k - \hat{s}_k|^2$ may be obtained as

$$e_k(\mathbf{f}_k, \mathbf{g}_k) = 1 - \mathbf{f}_k^H \mathbf{H}_{k,b_k} \mathbf{g}_k - \mathbf{g}_k^H \mathbf{H}_{k,b_k} \mathbf{f}_k + \sum_i \mathbf{f}_k^H \mathbf{H}_{k,b_i} \mathbf{g}_i \mathbf{g}_i^H \mathbf{H}_{k,b_i}^H \mathbf{f}_k + ||\mathbf{f}_k||^2.$$

$$\tag{9.19}$$

Let $\mathbf{g}, \mathbf{f}$ represent the collection of Tx/Rx BFs $\mathbf{g}_k$ and $\mathbf{f}_k$, respectively. Now introduce the WSMSE cost function (subject to same Tx power constraints),

$$\text{WSMSE}(\mathbf{g}, \mathbf{f}, \mathbf{w}) = \sum_k u_k (w_k \, e_k(\mathbf{f}_k, \mathbf{g}) - 1 - \ln w_k). \tag{9.20}$$

$\mathbf{w}$ corresponds to a collection of scalar weights $w_k \geq 0$. It can be shown that

$$\min_{\mathbf{f},\mathbf{w}} \text{WSMSE}(\mathbf{g}, \mathbf{f}, \mathbf{w}) = -\text{WSR}(\mathbf{g}). \tag{9.21}$$

Hence, an optimization over the additional parameters $\mathbf{f}, \mathbf{w}$ gives us back the original WSR cost function. This suggests an alternating optimization over $\mathbf{f}, \mathbf{w}$ and $\mathbf{g}$ to optimize the original WSR metric. The attractiveness of the WSMSE alternative compared to optimizing the WSR metric directly is that every step of the alternating optimization only involves simple quadratic or convex metrics. The overall algorithm for determining the BFs is to perform the alternating optimization

$$\min_{w_k} \text{WSMSE} \Rightarrow \quad w_k = 1/e_k$$

$$\min_{\mathbf{f}_k} \text{WSMSE} \Rightarrow \quad \mathbf{f}_k = \left( \sum_i \mathbf{H}_{k,b_i} \mathbf{g}_i \mathbf{g}_i^H \mathbf{H}_{k,b_i}^H + \mathbf{I}_{N_k} \right)^{-1} \mathbf{H}_{k,b_k} \mathbf{g}_k \tag{9.22}$$

$$\min_{\mathbf{g}_k} \text{WSMSE} \Rightarrow \quad \mathbf{g}_k = \left( \sum_i u_i w_i \mathbf{H}_{i,b_k}^H \mathbf{f}_i \mathbf{f}_i^H \mathbf{H}_{i,b_k} + \lambda_{b_k} \mathbf{I}_M \right)^{-1} \mathbf{H}_{k,b_k}^H \mathbf{f}_k u_k w_k$$

Here, $\lambda_{b_k}$ corresponds to the Lagrange multiplier for the transmit power constraint at BS $b_k$. We remark here that a key interpretation of the WSMSE solution is that, whereas the optimal Rx $\mathbf{f}_k$ is a LMMSE filter in the DL, the optimal transmit BF $\mathbf{g}_k$ has the form of a linear MMSE receiver for the dual UL. Note that duality leads to $\mathbf{H}_{i,b_k}^H$ whereas physical reciprocity corresponds to $\mathbf{H}_{i,b_k}^T$.

### 9.4.3 Naive UL/DL Duality-based Beamformer Exploiting Reciprocity

We present a BF design based on the WSMSE approach for a specific case of single antenna UEs ($N_k = 1$) [4]. This assumption is not too restrictive as single antenna UEs are more common. The BF is designed based on a naive UL/DL duality inspired by the WSMSE alternating minimization equations. The relations between Rx $\mathbf{f}_k$ and Tx $\mathbf{g}_k$ in equation (9.22) represent a proper UL/DL duality as one can observe that the optimal DL BF $\mathbf{g}_k$ corresponds to an LMMSE Rx in a dual UL in which the UL channels would be $\mathbf{H}_{i,b_k}^H$, the UL Tx filters are $\mathbf{f}_i$, the UL stream powers are $u_i w_i$, and the white noise variance at the BS is $\lambda_{b_k}$. These dual UL quantities are obviously different from corresponding actual UL transmission quantities. However, in order to largely simplify BF design and reduce signaling overhead, we propose a naive duality BF design in which we use the actual UL LMMSE Rx as DL BF. Note that one difference between actual and dual UL is a complex conjugation on the channel responses. Also, in the case of $N_k = 1$, we can ignore the UE side BF $f_k$. Note, however, that the resulting naive UL/DL duality BF design will converge to a matched filter at low SNR and to a ZF filter at high SNR. Hence the naive duality-based BF does give optimal results at both low and high SNR. Finally, we replace statistical averaging by temporal averaging.

The (actual) UL received signal $\check{\mathbf{y}}_k$ at BS $b_k$ may be written as

$$\check{\mathbf{y}}_{b_k} = \check{\mathbf{h}}_{k,b_k} \check{s}_k + \check{\mathbf{v}}_{b_k}. \tag{9.23}$$

Here, $\check{\mathbf{v}}_{b_k}$ includes the AWGN (Additive White Gaussian Noise) channel noise as well as the received signal from all other users, both intracell and intercell. Note that we use the ˘ to indicate a quantity in the UL. $\check{\mathbf{h}}_{k,b_k}$ denotes the UL channel from the user $k$ to BS $b_k$ and $\check{s}_k$ is the signal transmitted by the $k$th UE. Let $\mathbf{R}_{\check{y}_{b_k}\check{y}_{b_k}}$ be the uplink Rx covariance matrix. Then the UL MMSE filter is given by

$$\check{\mathbf{g}}_{MMSE,k} = \mathbf{R}_{\check{y}_{b_k}\check{y}_{b_k}}^{-1} \check{\mathbf{h}}_{k,b_k}. \tag{9.24}$$

Using reciprocity in TDD and accounting for the calibration factor $\mathbf{F}$, we get from (9.2) that $\mathbf{H}_{DL} = \mathbf{H}_{UL}^T \mathbf{F}$, hence that $\mathbf{h}_{i,b_k} = \check{\mathbf{h}}_{i,b_k}^T \mathbf{F}_{b_k}$. This leads to the DL MMSE filter

$$\begin{aligned} \mathbf{g}_{MMSE,k} &= \mathbf{R}_{y_{b_k}y_{b_k}}^{-1} \mathbf{h}_{k,b_k}^H = (\mathbf{F}_{b_k}^H \mathbf{R}_{\check{y}_{b_k}\check{y}_{b_k}}^* \mathbf{F}_{b_k})^{-1} \mathbf{F}_{b_k}^H \check{\mathbf{h}}_{k,b_k}^* \\ &= \mathbf{F}_{b_k}^{-1} (\mathbf{R}_{\check{y}_{b_k}\check{y}_{b_k}}^{-1} \check{\mathbf{h}}_{k,b_k})^* = \mathbf{F}_{b_k}^{-1} \check{\mathbf{g}}_{MMSE,k}^*. \end{aligned} \tag{9.25}$$

Here, ()* denotes the complex conjugate operation. The UL covariance matrix can be estimated as a sample covariance:

$$\mathbf{R}_{\check{y}_{b_k}\check{y}_{b_k}} = \frac{1}{L} \sum_{i=1}^{L} \check{\mathbf{y}}_{b_k,i} \check{\mathbf{y}}_{b_k,i}^H. \tag{9.26}$$

A known issue with this approach is signal cancellation, which can occur due to the mismatch between the estimated channel of the desired UE and the implicit contribution of the true desired channel present in the sample covariance matrix [31]. A known solution in this context is the subtraction of the desired signal from $\check{\mathbf{y}}_{b_k}$ before computing the covariance matrix. This requires an iterative receiver for joint detection and channel estimation so that the BS can subtract out the contribution from its own UE(s) before computing the sample covariance matrix.

## 9.5 Experimental Validation

Figure 9.8 shows an image of the MaMIMO prototype that is a part of the Eurecom OpenAir-Interface platform. The Eurecom MaMIMO array is constructed with several microstrip

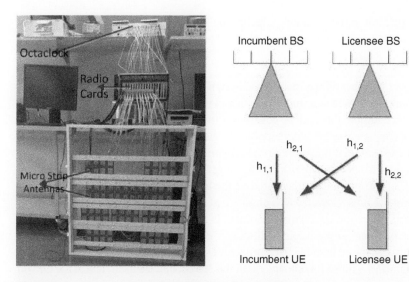

**Figure 9.8** Eurecom MaMIMO prototype and demo set up.

antenna cards, 12 of which are used in the current validation. Each such microstrip card, in turn, has four antennas. The 48 antennas are driven by 12 radio cards, where each radio card has four transceiver units. The synchronization between the radio units is achieved using an Octoclock clock distribution module. The transmission happens at carrier frequency $f_c = 2.66$ GHz and the sampling frequency $f_s = 7.68$ MHz corresponding to a 5-MHz long-term evolution (LTE) orthogonal frequency division multiplexing (OFDM) transmission that uses 300 occupied subcarriers. The beamformer design is computed and applied individually on every subcarrier. Figure 9.8 also illustrates the demo scenario. The two BS units consist of 23 antennas each and the two UEs have one antenna each. Thus, the 48 antennas of the MaMIMO antenna array are used to mimic the two BS as well as the two single antenna UEs.

The demo exploits channel reciprocity to derive the DL beamformer weights based on the UL channel/covariance estimates. Hence, when the prototype is initialized, we perform a reciprocity calibration and store the reciprocity calibration parameters $\mathbf{F}$ in a file. Subsequently, this file is read to derive the DL beamformer using instantaneously estimated UL channel information, as was given in equation (9.25). The instantaneous UL channel estimation is based on UL pilots. In the experiment, we assume all the useful subcarriers as pilots in the UL. The quality of the channel estimates is further improved by exploiting the limited delay spread in time domain. Our DL LMMSE design assumes no (explicit) knowledge of the cross-links between the BS of one cell and UE of another. However, to serve as a reference, we also consider a ZF receiver that has full knowledge of all cross-links. In this case, let the UL channel matrix be

$$\check{\mathcal{H}} = \begin{bmatrix} \check{\mathbf{h}}_{1,1} & \check{\mathbf{h}}_{2,1} \end{bmatrix}. \tag{9.27}$$

Then, the DL channel $\mathcal{H} = \check{\mathcal{H}}^T \mathbf{F}$ and

$$\mathbf{g}_{ZF} = \mathcal{H}^H (\mathcal{H}\mathcal{H}^H)^{-1} \mathbf{e}_1. \tag{9.28}$$

Here $\mathbf{e}_1 = [1 \quad 0]^T$. The other popular Tx technique in a MaMIMO scenario is maximum ratio transmission (MRT), which in this case would be

$$\mathbf{g}_{MRT} = \mathbf{h}_{1,1}^H = \mathbf{F}^* \check{\mathbf{h}}_{1,1}^*. \tag{9.29}$$

The estimation of the covariance matrix needs significant averaging, particularly as the number of BS antennas increases. In our prototype, we exploit the low delay spread of the environment and compute the average covariance matrix across all the subcarriers. However, the limited frequency selectivity increases the risk of signal cancellation and the need for signal of interest subtraction before sample covariance computation.

Figure 9.9 shows the need for calibration by taking the example of MRT in a single BS single UE setting. The performance is measured on the basis of the ratio between the received signal power and the noise power (SNR) observed at the UE. The curve labeled *ideal* here refers to the case where the DL channel estimate is available and estimated directly. The curve labeled *calib* refers to the implementation of equation (9.29) and the curve labeled *no_calib* directly uses the estimated UL channel for DL beamforming without applying any reciprocity calibration. The SNR is shown for all the 300 occupied subcarriers of the OFDM symbol.

Figure 9.10 shows the relative gains of MRT and ZF beamformers compared to no beamforming (omnidirectional antenna) by measuring the SINR at incumbent UE as a result of using the different beamformer techniques. It is remarkable that the performance of the ZF beamformer is far superior to that of the MRT, which is the most widespread beamforming technique used for MaMIMO.

In Figure 9.11, the covariance matrix is estimated for the interfering links in the UL and the DL MMSE BF is derived based on the UL covariance estimates and the reciprocity calibration parameters. The curve *ZF* serves as a reference where the UL channels of the interfering links are known (estimated) so that the DL ZF beamforming can be done with the

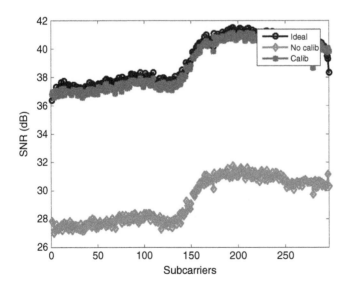

**Figure 9.9** Performance of MRT with and without calibration for a 23-antenna BS with a single antenna UE.

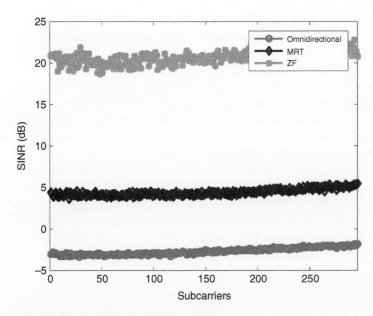

**Figure 9.10** Performance of MRT and ZF beamformers compared to no beamforming.

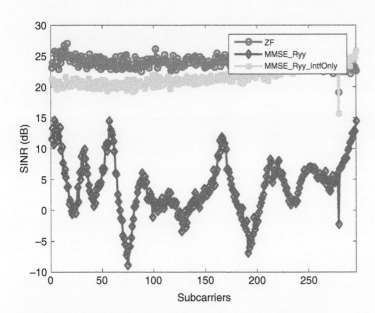

**Figure 9.11** Comparison of the performance of a naive LMMSE beamformer with that of ZF which requires full information of cross-links.

help of reciprocity calibration, as shown in equation (9.28). The curve *MMSE_Ryy* is the scenario in which the BS computes the covariance based on the total received signal from both its own UE and the interfering UE. We are limited here by the accuracy of the channel estimation and the averaging required for the covariance estimation. For the massive MIMO BS configuration, the averaging requirement for the covariance matrix estimation is very stringent as the dimension of the covariance matrix grows proportionally to the square of the number of BS antennas. Due to the inaccuracy in channel estimation, signal cancellation occurs between the channel estimate (in the matched filter factor of the LMMSE expression) and the true channel contribution implicit in $\mathbf{R}_{\bar{y}\bar{y}}$. The curve *MMSE_Ryy_IntfOnly* corresponds to the scenario in which the covariance computation does not include the contribution from the desired UE (desired signal subtraction). This approach avoids the signal cancellation issue. Hence, we observe that the performance of *MMSE_Ryy_IntfOnly* is much better compared to that of the curve *MMSE_Ryy* for the MaMIMO BS. In fact, the performance of the curve *MMSE_Ryy_IntfOnly* is quite close to that of the ZF which has explicit knowledge of the interfering links as well.

## 9.6 Conclusions

In this chapter, we have reviewed two types of reciprocity that can be exploited to reduce the signaling overhead required to acquire CSIT. One is pathwise reciprocity leading to pathwise CSIT, which is a parametric form of covariance CSIT and can be used in FDD or unsynchronized TDD systems. The other is instantaneous channel reciprocity in TDD systems. We have reviewed the key ingredients of relative RF calibration to extend the reciprocity of propagation channels to the estimated channels in the digital domain. We have proposed a DL beamformer design for a MaMIMO IBC scenario with no cooperation assumed across the different BSs. Inspired by the structure of the WSMSE alternating minimization operations, a naive UL/DL duality-based BF is proposed. Using the naive duality, and incorporating the concepts of TDD channel reciprocity, the actual UL LMMSE receiver was transformed into a DL BF. The UL covariance matrix is estimated while avoiding the signal cancellation issue as part of the overall beamformer design. The resulting beamformer was then validated on Eurecom's MaMIMO OpenAirInterface platform and shows a performance close to that of a ZF beamformer that needs information for the cross-links. Further work is required to combine channel estimate and sample covariance information.

## References

**1** M. Bashar and D. Slock. Cognitive Multi-User MIMO Downlink with Mixed Feedback/Location based Gaussian CSIT. In *Proceedings of the IEEE Workshop on Signal Processing Advances in Wireless Communications (SPAWC)*, Toronto, Canada, June 2014.

**2** S.S. Christensen, R. Agarwal, E. de Carvalho, and J.M. Cioffi. Weighted sum-rate maximization using weighted MMSE for MIMO-BC beamforming design. *IEEE Transactions on Wireless Communications*, December 2008.

**3** K. Gopala and D. Slock. A refined analysis of the gap between expected rate for partial CSIT and the massive MIMO rate limit. *ICASSP, IEEE International Conference on Acoustics, Speech and Signal Processing, Calgary, Canada*, April 2018.

**4** K. Gopala and D. Slock. Robust LMMSE beamformer design by naive UL/DL duality and validation for non-cooperative massive MIMO. *VTC Fall, Vehicular Technology Conference, Chicago, USA*, August 2018.

**5** K. Gopala and D. Slock. Optimal algorithms and CRB for reciprocity calibration in Massive MIMO. *IEEE International Conference on Acoustics, Speech and Signal Processing (ICASSP)*, April 2018.

**6** X. Jiang, M. Čirkić, F. Kaltenberger, et al. MIMO-TDD reciprocity and hardware imbalances: experimental results. *IEEE International Conference on Communication (ICC)*, June 2015.

**7** X. Jiang, A. Decurninge, K. Gopala, et al. A Framework for Over-the-air Reciprocity Calibration for TDD Massive MIMO Systems. *IEEE Transactions on Wireless Communications*, September 2018.

**8** C.G. Khatri and C.R. Rao. Solutions to some functional equations and their applications to characterization of probability distributions. *Sankhyā: The Indian Journal of Statistics, Series A*, 1968.

**9** S.-J. Kim and G.B. Giannakis. Optimal Resource Allocation for MIMO Ad Hoc Cognitive Radio Networks. *IEEE Transactions on Information Theory*, May 2011.

**10** B. Kouassi, B. Zayen, R. Knopp, F. Kaltenberger, D. Slock, I. Ghauri, F. Negro, and L. Deneire. Design and Implementation of Spatial Interweave LTE-TDD Cognitive Radio Communication on an Experimental Platform. *IEEE Wireless Communications Magazine*, April 2013.

**11** C. Kurisummoottil Thomas and D. Slock. Large Multi-Antenna Stochastic Geometry. In *Proceedings of the IEEE Workshop on Information Theory and Applications (ITA)*, San Diego, CA, USA, February 2018.

**12** C. Kurisummoottil Thomas, W. Tabikh, D. Slock, and Y. Yuan-Wu. Noncoherent Multi-User MIMO Communications using Covariance CSIT. In *Proceedings of the Asilomar Conference on Signals, Systems and Computers*, Pacific Grove, CA, USA, November 2017.

**13** E.G. Larsson, O. Edfors, F. Tufvesson, and T.L. Marzetta. Massive MIMO for Next Generation Wireless Systems. *IEEE Communications Magazine*, February 2014.

**14** Y. Lejosne, D. Slock, and Y. Yuan-Wu. Foresighted Delayed CSIT Feedback for Finite Rate of Innovation Channel Models and Attainable NetDoFs of the MIMO Interference Channel. In *Proceedings on Wireless Days*, Valencia, Spain, November 2013.

**15** Y. Lejosne, M. Bashar, D. Slock, and Y. Yuan-Wu. Decoupled, Rank Reduced, Massive and Frequency-Selective Aspects in MIMO Interfering Broadcast Channels. In *Proceedings of the IEEE International Symposium on Communications, Control and Signal Processing (ISCCSP)*, Athens, Greece, May 2014.

**16** Y. Lejosne, M. Bashar, D. Slock, and Y. Yuan-Wu. From MU Massive MISO to Pathwise MU Massive MIMO. In *Proceedings of the IEEE Workshop on Signal Processing Advances in Wireless Communications (SPAWC)*, Toronto, Canada, June 2014.

**17** T.L. Marzetta, E.G. Larssson, H. Yang, and H.Q. Ngo. Fundamentals of Massive MIMO). Cambridge University Press, 2016.

**18** F. Negro, I. Ghauri, and D. Slock. Spatial Interweave for a MIMO Secondary Interference Channel with Multiple Primary Users. In *4th International Conference on Cognitive Radio and Advanced Spectrum Management (CogART)*, Barcelona, Spain, October 2011.

**19** F. Negro, I. Ghauri, and D.. Slock. Deterministic Annealing Design and Analysis of the Noisy MIMO Interference Channel. In *Proceedings of the IEEE Information Theory and Applications workshop (ITA)*, San Diego, CA, USA, February 2011.

**20** F. Negro, I. Ghauri, and D. Slock. Sum Rate Maximization in the Noisy MIMO Interfering Broadcast Channel with Partial CSIT via the Expected Weighted MSE. In *IEEE International Symposium on Wireless Communication Systems (ISWCS)*, Paris, France, August 2012.

**21** F. Negro, D. Slock, and I. Ghauri. On the Noisy MIMO Interference Channel with CSI through Analog Feedback. In *International Symposium on Communications, Control and Signal Processing (ISCCSP)*, Rome, Italy, May 2012.

**22** Qualcomm. Authorized Shared Access (ASA) – A New Licensed Model to Access Underutilized Spectrum, January 2014. URL https://www.fcc.gov/file/14784/download.

**23** 3GPP. Report R1-091752: Performance study on Tx/Rx mismatch in LTE TDD dual-layer beamforming. Nokia, Nokia Siemens Networks, CATT, ZTE, 3GPP RAN1 meeting #57, May 2009.

**24** 3GPP. Report R1-091794: Hardware calibration requirement for dual layer beamforming. Huawei, 3GPP RAN1 meeting #57, May 2009.

**25** R. Rogalin, O.Y. Bursalioglu, H.C. Papadopoulos, et al. Hardware-Impairment Compensation for Enabling Distributed Large-Scale MIMO. In *Information Theory and Applications (ITA) Workshop*, San Diego, CA, USA, February 2013.

**26** R. Rogalin, O.Y. Bursalioglu, H.C. Papadopoulos, et al. Scalable synchronization and reciprocity calibration for distributed multiuser MIMO. *IEEE Transactions on Wireless Communications*, April 2014.

**27** C. Shepard, H. Yu, N. Anand, E. Li, et al. Argos: Practical Many-Antenna Base Stations. In *Proceedings of the 18th Annual International Conference on Mobile Computing and Networking (Mobicom)*, Istanbul, Turkey, August 2012.

**28** W. Tabikh, D. Slock, and Y. Yuan-Wu. The Pathwise MIMO Interfering Broadcast Channel. In *Proceedings of the IEEE Workshop on Information Theory and Applications (ITA)*, San Diego, CA, USA, February 2015.

**29** W. Tabikh, D. Slock, and Y. Yuan-Wu. The Pathwise MIMO Interfering Broadcast Channel. In *IEEE 16th Workshop on Signal Processing Advances in Wireless Communications (SPAWC)*, 2015.

**30** W. Tabikh, K. Gopala, Y. Yuan-Wu, and D. Slock. From Multi-Antenna Underlay to LSA Coordinated Beamforming. In *Proceedings of the SAS5G Workshop, ISWCS*, Poznan, Poland, September 2016.

**31** B. Widrow, K. Duvall, R. Gooch, and W. Newman. Signal cancellation phenomena in adaptive antennas: Causes and cures. *IEEE Transactions on Antennas and Propagation*, May 1982.

# 10

## Spectrum Sharing with Full Duplex

*Sudip Biswas[1], Ali Cagatay Cirik[2], Miltiades C. Filippou[3], and Tharmalingam Ratnarajah[4]*

[1] *Indian Institute of Information Technology Guwahati, India*
[2] *Ofinno Technologies, USA*
[3] *Intel Deutschland GmbH, Germany*
[4] *University of Edinburgh, UK*

## 10.1 Introduction

Among the various spectrum sharing techniques, cognitive radio (CR) is a promising technology to enhance spectrum utilization efficiency by allowing unlicensed secondary users (SUs) to operate within the service area of licensed primary users (PUs). Traditionally, the secondary network is deployed in half-duplex (HD) mode, whereby transmission and reception happen orthogonally in time or frequency [26]. However, the recent proliferation in demand for wireless communication services has motivated the industry to strive to finding contemporary solutions, among which full duplex (FD) communications is one of the emerging technologies for next-generation wireless networks. Theoretically, communicating in FD mode can potentially double the throughput of wireless communication systems due to the ability of an FD transceiver to transmit and receive at the same time and the same frequency, thus utilizing the spectrum fully. As such, CR can be deployed in the FD mode, whereby an FD CR can concurrently transmit and sense the transmission status of other nodes [2], thus providing improved sensing efficiency and secondary throughput. Furthermore, FD communication can also combat several problems faced by CR at the medium access control (MAC) layer, such as hidden terminals, large delays, and congestion [15, 23]. Accordingly, several research problems related to interweave FD CR systems have been investigated in the literature [19, 26]. In particular, these works employ FD communications at the SUs to simultaneously perform spectrum sensing and data transmission towards significantly improving sensing performance, while also increasing data transmission efficiency. Another emerging research trend concerns underlay cooperative systems [5, 6, 9, 10, 16], where FD CRs are employed. In underlay CR systems, a set of unlicensed SUs operate within the service range of licensed PUs, where the amount of interference from SUs to PUs is constrained to meet the quality-of-service (QoS) requirements of the latter.

*Spectrum Sharing: The Next Frontier in Wireless Networks,* First Edition.
Edited by Constantinos B. Papadias, Tharmalingam Ratnarajah, and Dirk T.M. Slock.
© 2020 John Wiley & Sons Ltd. Published 2020 by John Wiley & Sons Ltd.

However, the performance of an FD system is hindered by several predicaments, the most important of which is the inherent interference from its own transmit signal, also known as self-interference (SI). Fortunately, many feasible solutions, including antenna, analog, and digital cancellation methods, have been demonstrated experimentally to mitigate the overwhelming SI [4, 12], which have made FD communication more practical in recent years. Nevertheless, the performance of an FD system is still limited by the residual SI (RSI) that is induced by the imperfection of the transmit and receive front-end chains [17]. In addition to the SI, co-channel interference (CCI) from uplink (UL) users to downlink (DL) users is another challenge in FD networks that needs to be overcome to fully exploit the multi-access nature of the wireless medium in conjunction with fully utilizing the spectrum. In this respect, the application of beamforming techniques is known to be effective to optimize the system performance, where the impact of the CCI as well as the SI are jointly taken into account [7–10, 20, 25].

In light of the above discussion, this chapter focuses on two underlay CR scenarios, a *CR cellular system* (involving an FD secondary cellular network) and (ii) a *CR interference channel* (involving an ad hoc FD secondary Internet-of-Things (IoT) network), and analyzes the performance of both systems with respect to baseline HD CR systems. Potential transceiver design algorithms to enable FD underlay CR operation and to mitigate the residual SI and CCI at the digital domain are provided based on data stream decoding mean-squared error (MSE) minimization, the rationale for which is the good performance and significantly reduced complexity of the MSE metric. When a minimum MSE (MMSE) receiver is used, MSE minimization problems are equivalent to signal-to-interference-plus-noise ratio (SINR)-based optimization problems, since they are related as $MSE = 1/(1 + SINR)$ [21]. Therefore, rate-based optimization using $\log_2(1 + SINR)$ can be conveniently transformed into MSE-based optimization, $-\log_2(MSE)$. Furthermore, as mentioned in [14], the user-wise MSE can be used to approximate the achievable rate of the users when they jointly decode their streams.

## 10.2 Transceiver Design for an FD MIMO CR Cellular Network

In this section, we discuss the transceiver design problem for an FD cognitive cellular system in which a secondary FD BS communicates with HD mode UL and DL SUs simultaneously within the service range of PUs.

### 10.2.1 System Model

As illustrated in Figure 10.1, the FD BS, equipped with $M_0$ transmit and $N_0$ receive antennas, serves $K$ UL and $J$ DL users, simultaneously. The numbers of antennas of the $k$th UL and the $j$th DL user are denoted by $M_k$ and $N_j$, respectively.

#### 10.2.1.1 Signal and Channel Model

The channels $\mathbf{H}_k^{UL} \in \mathbb{C}^{N_0 \times M_k}$ and $\mathbf{H}_j^{DL} \in \mathbb{C}^{N_j \times M_0}$ represent the $k$th UL and the $j$th DL channels, respectively. Similarly, $\mathbf{H}_0 \in \mathbb{C}^{N_0 \times M_0}$ is the SI channel from the transmitter to the

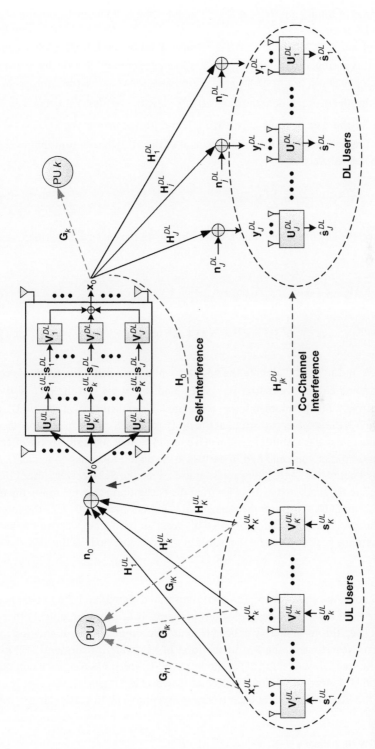

**Figure 10.1** An illustration of an FD multi-user MIMO CR cellular system.

receiver antennas of the BS and $\mathbf{H}_{jk}^{DU} \in \mathbb{C}^{N_j \times M_k}$ denotes the CCI channel from the $k$th UL user to the $j$th DL user.

The vector of source symbols of length $d_k^{UL}$ transmitted by the $k$th UL user is denoted $\mathbf{s}_k^{UL} \in \mathbb{C}^{d_k^{UL} \times 1}$. It is assumed that the symbols are independent and identically distributed (i.i.d.) with unit power, i.e. $\mathbb{E}[\mathbf{s}_k^{UL}(\mathbf{s}_k^{UL})^H] = \mathbf{I}_{d_k^{UL}}$. Similarly, the vector of transmit symbols of length $d_j^{DL}$ for the $j$th DL user is denoted by $\mathbf{s}_j^{DL} \in \mathbb{C}^{d_j^{DL} \times 1}$, with $\mathbb{E}[\mathbf{s}_j^{DL}(\mathbf{s}_j^{DL})^H] = \mathbf{I}_{d_j^{DL}}$. Denoting the precoders for the data streams of the $k$th UL and $j$th DL user as $\mathbf{V}_k^{UL} \in \mathbb{C}^{M_k \times d_k^{UL}}$ and $\mathbf{V}_j^{DL} \in \mathbb{C}^{M_0 \times d_j^{DL}}$, respectively, the transmitted signal of the $k$th UL user and that of the BS can be written, respectively, as

$$\mathbf{x}_k^{UL} = \mathbf{V}_k^{UL}\mathbf{s}_k^{UL}, \qquad \mathbf{x}_0 = \sum_{j=1}^{J} \mathbf{V}_j^{DL}\mathbf{s}_j^{DL}. \tag{10.1}$$

The signal received by the BS and that received by the $j$th DL user can be written, respectively, as

$$\mathbf{y}_0 = \sum_{k=1}^{K} \mathbf{H}_k^{UL}(\mathbf{x}_k^{UL} + \mathbf{c}_k^{UL}) + \mathbf{H}_0(\mathbf{x}_0 + \mathbf{c}_0) + \mathbf{e}_0 + \mathbf{n}_0, \tag{10.2}$$

$$\mathbf{y}_j^{DL} = \mathbf{H}_j^{DL}(\mathbf{x}_0 + \mathbf{c}_0) + \sum_{k=1}^{K} \mathbf{H}_{jk}^{DU}(\mathbf{x}_k^{UL} + \mathbf{c}_k^{UL}) + \mathbf{e}_j^{DL} + \mathbf{n}_j^{DL}, \tag{10.3}$$

where $\mathbf{n}_0 \in \mathbb{C}^{N_0}$ and $\mathbf{n}_j^{DL} \in \mathbb{C}^{N_j}$ denote the additive white Gaussian noise (AWGN) vector with zero mean and covariance matrix $\mathbf{R}_0 = \sigma_0^2 \mathbf{I}_{N_0}$ and $\mathbf{R}_j^{DL} = \sigma_j^2 \mathbf{I}_{N_j}$ at the BS and the $j$th DL user, respectively. Furthermore, $\mathbf{c}_k^{UL}$ ($\mathbf{c}_0$) is the transmitter distortion at the $k$th UL user (BS) and $\mathbf{e}_j^{DL}$ ($\mathbf{e}_0$) is the receiver distortion at the $j$th DL user (BS). They model the effect of limited dynamic range of transmitters and receivers, and closely approximate the effects of additive power-amplifier noise and non-linearities in the digital-to-analog converter (DAC), analog-to digital converter (ADC), and phase noise. Since the SU receiver cannot distinguish the interference generated by the PUs from the background thermal noise, the noise vectors in (10.2) and (10.3) capture the background thermal noise as well as the interference generated by the PUs, possibly after prewhitening. In particular, we assume that the PU sum-interference is estimated and measured at the receiving node of the SUs.

### 10.2.1.2 SI Cancellation

Since the BS knows the codeword $\mathbf{x}_0$ (its own transmitted signal) and the SI channel $\mathbf{H}_0$ in practice, the term $\mathbf{H}_0\mathbf{x}_0$ can be canceled out in Figure 10.2. However, unless otherwise stated, we will keep this term merely to be able to use the simplification of notation in the next subsection. Nevertheless, the RSI in the form of hardware impairments still exists. We adopt the limited DR model in [11] to model the RSI, which has also been commonly used in [18] and [17]. Accordingly, the covariance matrix of $\mathbf{c}_k^{UL}$ is given by $\kappa$ ($\kappa \ll 1$) times the energy of the intended signal at each transmit antenna [11]. In particular $\mathbf{c}_k^{UL}$ can be modeled as

$$\mathbf{c}_k^{UL} \sim \mathcal{CN}(\mathbf{0}, \kappa \, \text{diag}(\mathbf{V}_k^{UL}(\mathbf{V}_k^{UL})^H)), \quad \mathbf{c}_k^{UL} \perp \mathbf{x}_k^{UL}. \tag{10.4}$$

Similarly, the covariance matrix of $\mathbf{e}_j^{DL}$ is given by $\beta$ ($\beta \ll 1$) times the energy of the undistorted received signal at each receive antenna [11]. In particular, $\mathbf{e}_j^{DL}$ can be modeled as

$$\mathbf{e}_j^{DL} \sim \mathcal{CN}(\mathbf{0}, \beta \mathrm{diag}(\boldsymbol{\Phi}_j^{DL})), \mathbf{e}_j^{DL} \perp \mathbf{u}_j^{DL}, \tag{10.5}$$

where $\boldsymbol{\Phi}_j^{DL} = \mathrm{Cov}\{\mathbf{u}_j^{DL}\}$ and $\mathbf{u}_j^{DL}$ is the undistorted received vector at the $j$th DL user, i.e. $\mathbf{u}_j^{DL} = \mathbf{y}_j^{DL} - \mathbf{e}_j^{DL}$. The discussion on the transmitter/receiver distortion model holds for $\mathbf{c}_0$ and $\mathbf{e}_0$ as well.

### 10.2.1.3 MSE of the Received Data Stream

The received signals are processed by linear decoders, denoted $\mathbf{U}_k^{UL} \in \mathbb{C}^{N_0 \times d_k^{UL}}$ and $\mathbf{U}_j^{DL} \in \mathbb{C}^{N_j \times d_j^{DL}}$, by the BS and $j$th DL user, respectively. Therefore the estimate of data streams of the $k$th UL user at the BS is given as $\hat{\mathbf{s}}_k^{UL} = (\mathbf{U}_k^{UL})^H \mathbf{y}_0$, and similarly the estimate of data streams of the $j$th DL user is $\hat{\mathbf{s}}_j^{DL} = (\mathbf{U}_j^{DL})^H \mathbf{y}_j^{DL}$. Using these estimates, the MSE of the $k$th UL and $j$th DL users can be respectively given as [10]

$$\mathbf{MSE}_k^{UL} = \mathbb{E}\{(\hat{\mathbf{s}}_k^{UL} - \mathbf{s}_k^{UL})(\hat{\mathbf{s}}_k^{UL} - \mathbf{s}_k^{UL})^H\}$$

$$= ((\mathbf{U}_k^{UL})^H \mathbf{H}_k^{UL} \mathbf{V}_k^{UL} - \mathbf{I}_{d_k^{UL}})((\mathbf{U}_k^{UL})^H \mathbf{H}_k^{UL} \mathbf{V}_k^{UL} - \mathbf{I}_{d_k^{UL}})^H + (\mathbf{U}_k^{UL})^H \overset{UL}{\underset{k}{\boldsymbol{\Sigma}}} \mathbf{U}_k^{UL}, \tag{10.6}$$

$$\mathbf{MSE}_j^{DL} = \mathbb{E}\{(\hat{\mathbf{s}}_j^{DL} - \mathbf{s}_j^{DL})(\hat{\mathbf{s}}_j^{DL} - \mathbf{s}_j^{DL})^H\}$$

$$= ((\mathbf{U}_j^{DL})^H \mathbf{H}_j^{DL} \mathbf{V}_j^{DL} - \mathbf{I}_{d_j^{DL}})((\mathbf{U}_j^{DL})^H \mathbf{H}_j^{DL} \mathbf{V}_j^{DL} - \mathbf{I}_{d_j^{DL}})^H + (\mathbf{U}_j^{DL})^H \overset{DL}{\underset{j}{\boldsymbol{\Sigma}}} \mathbf{U}_j^{DL}. \tag{10.7}$$

In (10.6) and (10.7), $\boldsymbol{\Sigma}_k^{UL}$ and $\boldsymbol{\Sigma}_j^{DL}$ are the approximated aggregate interference-plus-noise terms[1] at the $k$th UL and $j$th DL users, respectively, and are expressed as [11]

$$\overset{UL}{\underset{k}{\boldsymbol{\Sigma}}} \approx \sum_{j \neq k}^{K} \mathbf{H}_j^{UL} \mathbf{V}_j^{UL} (\mathbf{V}_j^{UL})^H (\mathbf{H}_j^{UL})^H + \kappa \sum_{j=1}^{K} \mathbf{H}_j^{UL} \mathrm{diag}(\mathbf{V}_j^{UL} (\mathbf{V}_j^{UL})^H)$$

$$\times (\mathbf{H}_j^{UL})^H + \sum_{j=1}^{J} \mathbf{H}_0 (\mathbf{V}_j^{DL} (\mathbf{V}_j^{DL})^H + \kappa \mathrm{diag}(\mathbf{V}_j^{DL} (\mathbf{V}_j^{DL})^H)) \mathbf{H}_0^H$$

$$+ \beta \sum_{j=1}^{K} \mathrm{diag}(\mathbf{H}_j^{UL} \mathbf{V}_j^{UL} (\mathbf{V}_j^{UL})^H (\mathbf{H}_j^{UL})^H) + \beta \sum_{j=1}^{J} \mathrm{diag}(\mathbf{H}_0 \mathbf{V}_j^{DL} (\mathbf{V}_j^{DL})^H$$

$$\times \mathbf{H}_0^H) + \sigma_0^2 \mathbf{I}_{N_0}, \tag{10.8}$$

$$\overset{DL}{\underset{j}{\boldsymbol{\Sigma}}} \approx \sum_{i \neq j}^{J} \mathbf{H}_j^{DL} \mathbf{V}_i^{DL} (\mathbf{V}_i^{DL})^H (\mathbf{H}_j^{DL})^H + \kappa \sum_{i=1}^{J} \mathbf{H}_j^{DL} \mathrm{diag}(\mathbf{V}_i^{DL} (\mathbf{V}_i^{DL})^H)$$

$$\times (\mathbf{H}_j^{DL})^H + \sum_{k=1}^{K} \mathbf{H}_{jk}^{DU} (\mathbf{V}_k^{UL} (\mathbf{V}_k^{UL})^H + \kappa \mathrm{diag}(\mathbf{V}_k^{UL} (\mathbf{V}_k^{UL})^H)) (\mathbf{H}_{jk}^{DU})^H$$

$$+ \beta \sum_{i=1}^{J} \mathrm{diag}(\mathbf{H}_j^{DL} \mathbf{V}_i^{DL} (\mathbf{V}_i^{DL})^H (\mathbf{H}_j^{DL})^H) + \beta \sum_{k=1}^{K} \mathrm{diag}(\mathbf{H}_{jk}^{DU} \mathbf{V}_k^{UL} (\mathbf{V}_k^{UL})^H$$

$$\times (\mathbf{H}_{jk}^{DU})^H) + \sigma_j^2 \mathbf{I}_{N_j}. \tag{10.9}$$

---

1 Note that $\boldsymbol{\Sigma}_k^{UL}$ and $\boldsymbol{\Sigma}_j^{DL}$ are approximated under $\kappa \ll 1$ and $\beta \ll 1$, which is a practical assumption [4, 11]. Therefore, the terms including the multiplication of $\kappa$ and $\beta$ are negligible and have been ignored in the approximation.

Without loss of generality, we assume that there is only DL transmission over the considered frequency band in the primary network. Therefore, the power of the interference resulting from the secondary UL users and BS at the $l$th PU equipped with $T_l$ receive antennas can be written as

$$I_l^{PU} = \sum_{k=1}^{K} \text{tr} \left\{ \mathbf{G}_{lk} (\mathbf{V}_k^{UL} (\mathbf{V}_k^{UL})^H + \kappa \text{diag}(\mathbf{V}_k^{UL} (\mathbf{V}_k^{UL})^H)) \mathbf{G}_{lk}^H \right\}$$

$$+ \sum_{j=1}^{J} \text{tr} \left\{ \mathbf{G}_l (\mathbf{V}_j^{DL} (\mathbf{V}_j^{DL})^H + \kappa \text{diag}(\mathbf{V}_j^{DL} (\mathbf{V}_j^{DL})^H)) \mathbf{G}_l^H \right\}, \tag{10.10}$$

where $\mathbf{G}_{lk} \in \mathbb{C}^{T_l \times M_k}$ ($\mathbf{G}_l \in \mathbb{C}^{T_l \times M_0}$) is the channel between the $l$th PU and $k$th UL users ($l$th PU and the BS).

### 10.2.2 Joint Transceiver Design

As previously stated, we tackle the transceiver design problem as a sum-MSE minimization problem, which is formulated as

$$\min_{\mathbf{V}, \mathbf{U}} \quad \sum_{k=1}^{K} \text{tr} \left\{ \mathbf{MSE}_k^{UL} \right\} + \sum_{j=1}^{J} \text{tr} \left\{ \mathbf{MSE}_j^{DL} \right\} \tag{10.11}$$

$$\text{s.t.} \quad \text{tr} \left\{ \mathbf{V}_k^{UL} (\mathbf{V}_k^{UL})^H \right\} \le P_k, \ k = 1, \dots, K, \tag{10.12}$$

$$\sum_{j=1}^{J} \text{tr} \left\{ \mathbf{V}_j^{DL} (\mathbf{V}_j^{DL})^H \right\} \le P_0, \tag{10.13}$$

$$I_l^{PU} \le \lambda_l, \ l = 1, \dots, L, \tag{10.14}$$

where $P_k$ is the transmit power constraint at the $k$th UL user, $P_0$ is the total power constraint at the BS, and $\lambda_l$ is the threshold of allowed interference temperature at the $l$th PU. Here, $\mathbf{V} = \{\mathbf{V}_k^{UL}, k = 1, \dots, K, \mathbf{V}_j^{DL}, j = 1, \dots, J\}$ and $\mathbf{U} = \{\mathbf{U}_k^{UL}, k = 1, \dots, K, \mathbf{U}_j^{DL}, j = 1, \dots, J\}$ are the sets of all transmit and receive transceiver matrices, respectively. Next, to simplify the notations, we will combine UL and DL channels, similar to [10]. Let $S^{UL}$ and $S^{DL}$ represent the set of $K$ UL and $J$ DL channels, respectively. Denoting $\mathbf{H}_{ij}$, $\mathbf{G}_{lj}$, $\mathbf{n}_i$, and receive (transmit) antenna numbers $\tilde{N}_i(\tilde{M}_i)$ as

$$\mathbf{H}_{ij} = \begin{cases} \mathbf{H}_j^{UL}, & i \in S^{UL}, j \in S^{UL}, \\ \mathbf{H}_0, & i \in S^{UL}, j \in S^{DL}, \\ \mathbf{H}_{ij}^{DU}, & i \in S^{DL}, j \in S^{UL}, \\ \mathbf{H}_i^{DL}, & i \in S^{DL}, j \in S^{DL}, \end{cases} \quad \mathbf{G}_{lj} = \begin{cases} \mathbf{G}_{lj}, & j \in S^{UL}, \\ \mathbf{G}_l, & j \in S^{DL}, \end{cases}$$

$$\mathbf{n}_i = \begin{cases} \mathbf{n}_0, & i \in S^{UL}, \\ \mathbf{n}_i^{DL}, & i \in S^{DL}, \end{cases} \quad \tilde{N}_i(\tilde{M}_i) = \begin{cases} N_0(M_i), & i \in S^{UL}, \\ N_i(M_0), & i \in S^{DL}, \end{cases}$$

and referring to $\mathbf{V}_i^X$, $\mathbf{U}_i^X$, $d_i^X$, and $\Sigma_i^X$, $X \in \{UL, DL\}$ as $\mathbf{V}_i$, $\mathbf{U}_i$, $d_i$, and $\Sigma_i$, respectively, the MSE of the $i$th link, $i \in S \triangleq S^{UL} \bigcup S^{DL}$ can be written as

$$\mathbf{MSE}_i = (\mathbf{U}_i^H \mathbf{H}_{ii} \mathbf{V}_i - \mathbf{I}_{d_i})(\mathbf{U}_i^H \mathbf{H}_{ii} \mathbf{V}_i - \mathbf{I}_{d_i})^H + \mathbf{U}_i^H \sum_i \mathbf{U}_i, \tag{10.15}$$

where

$$\underset{i}{\Sigma} = \sum_{j \in S, j \neq i} \mathbf{H}_{ij} \mathbf{V}_j \mathbf{V}_j^H \mathbf{H}_{ij}^H$$

$$+ \kappa \sum_{j \in S} \mathbf{H}_{ij} \, \text{diag}(\mathbf{V}_j \mathbf{V}_j^H) \mathbf{H}_{ij}^H + \beta \sum_{j \in S} \text{diag}(\mathbf{H}_{ij} \mathbf{V}_j \mathbf{V}_j^H \mathbf{H}_{ij}^H) + \sigma_i^2 \mathbf{I}_{\tilde{N}_i}, \tag{10.16}$$

and the interference power at the $l$th PU, $I_l^{PU}$ in (10.10), can be rewritten as

$$I_l^{PU} = \sum_{j \in S} \text{tr}\{\mathbf{G}_{lj}(\mathbf{V}_j \mathbf{V}_j^H + \kappa \text{diag}(\mathbf{V}_j \mathbf{V}_j^H))\mathbf{G}_{lj}^H\}. \tag{10.17}$$

Using the simplified notation, the problem (10.11)–(10.14) can be rewritten as

$$\underset{\mathbf{V},\mathbf{U}}{\min} \sum_{i \in S} \text{tr}\{\mathbf{MSE}_i\} \tag{10.18}$$

$$\text{s.t.} \ \ \text{tr}\{\mathbf{V}_i \mathbf{V}_i^H\} \leq P_i, \ i \in S^{UL}, \tag{10.19}$$

$$\sum_{i \in S^{DL}} \text{tr}\{\mathbf{V}_i \mathbf{V}_i^H\} \leq P_0, \tag{10.20}$$

$$I_l^{PU} \leq \lambda_l, \ l = 1, \dots, L. \tag{10.21}$$

## 10.2.3 Imperfect CSI and Robust Design

### 10.2.3.1 CSI Acquisition

We assume that the secondary BS has knowledge of the nominal channels and the radii of uncertainty regions. We undertake a centralized approach where the secondary BS coordinates the calibration of channel matrices, collects all channel matrices, computes the transceiver matrices based on the imperfect channel state information (CSI), and then distributes them to the SUs. The estimation of CSI matrices in the secondary network follows a similar strategy to that of traditional systems, as the secondary nodes cooperate with the secondary BS. This is performed via the exchange of training sequences and feedback, and the application of the usual CSI estimation methods [22]. On the other hand, it is more challenging to obtain an accurate estimate for the CSI between the secondary and primary networks, as the primary network is usually not willing to cooperate with the secondary network. In this regard, a few methods have been suggested to address this problem. First, if the primary system adopts the TDD scheme, the secondary network can obtain the CSI to the primary nodes by taking advantage of the channel reciprocity and overhearing the transmissions from the primary network [22]. Second, a partial CSI can be obtained via blind environmental learning [13]. Third, an estimate of CSI can be obtained via the realization of a *band manager* with the ability to exchange the CSI between the secondary and primary networks [9]. Finally, if possible, the primary system can cooperate with the secondary network to exchange the channel estimates [22]. Of course, since the primary and secondary systems are not fully coordinated, the quality of these channel estimates will be degraded. Hence, we choose to model these imperfections by considering norm-bounded estimation errors for the links between the secondary transmitters and primary receivers.

### 10.2.3.2 CSI Modeling

The imperfect CSI is modeled using a deterministic norm-bounded error model [26], which is expressed as

$$\mathbf{H}_{ij} \in \mathcal{H}_{ij} = \{\tilde{\mathbf{H}}_{ij} + \mathbf{\Delta}_i : \|\mathbf{\Delta}_i\|_F \leq \delta_i, \, j \in S\}, \tag{10.22}$$

$$\mathbf{G}_{lj} \in \mathcal{G}_{lj} = \{\tilde{\mathbf{G}}_{lj} + \mathbf{\Lambda}_l : \|\mathbf{\Lambda}_l\|_F \leq \theta_l, \, j \in S\}, \tag{10.23}$$

where $\tilde{\mathbf{H}}_{ij}$, $\tilde{\mathbf{G}}_{lj}$, and $\delta_i$, $\theta_l$ denote the nominal value of the CSI and uncertainty bounds, respectively. Under channel uncertainties, the optimization problem (10.18)–(10.21) can be rewritten as

$$\min_{\mathbf{V},\mathbf{U}} \, \max_{\forall \mathbf{H}_{ij} \in \mathcal{H}_{ij}} \, \sum_{i \in S} \text{tr}\{\mathbf{MSE}_i\} \tag{10.24}$$

$$\text{s.t.} \quad \text{tr}\{\mathbf{V}_i \mathbf{V}_i^H\} \leq P_i, \, i \in S^{UL}, \tag{10.25}$$

$$\sum_{i \in S^{DL}} \text{tr}\{\mathbf{V}_i \mathbf{V}_i^H\} \leq P_0, \tag{10.26}$$

$$I_l^{PU} \leq \lambda_l, \, \forall \mathbf{G}_{lj} \in \mathcal{G}_{lj}, \, l = 1, \ldots, L. \tag{10.27}$$

Due to the constraint (10.27), the problem (10.24)–(10.27) is a semi-infinite program, and we will derive an equivalent constraint in linear matrix inequality (LMI) form in section 10.2.3.3 so that the problem (10.24) will turn into an equivalent semi-definite programming (SDP) problem, which can be efficiently solved by standard interior point methods.

### 10.2.3.3 Robust Transceiver Design

Since the problem (10.24)–(10.27) is an intractable semi-infinite optimization problem [3], in the following we turn it into a tractable form. Using epigraph form and introducing slack variables $\tau_i$, the minimax problem can be equivalently rewritten as

$$\min_{\mathbf{V},\mathbf{U},\tau} \, \sum_{i \in S} \tau_i \tag{10.28}$$

$$\text{s.t.} \quad \text{tr}\{\mathbf{MSE}_i\} \leq \tau_i, \, \forall \mathbf{H}_{ij} \in \mathcal{H}_{ij}, \, i \in S, \tag{10.29}$$

$$\text{tr}\{\mathbf{V}_i \mathbf{V}_i^H\} \leq P_i, \, i \in S^{UL}, \tag{10.30}$$

$$\sum_{i \in S^{DL}} \text{tr}\{\mathbf{V}_i \mathbf{V}_i^H\} \leq P_0, \tag{10.31}$$

$$I_l^{PU} \leq \lambda_l, \, \forall \mathbf{G}_{lj} \in \mathcal{G}_{lj}, \, l = 1, \ldots, L, \tag{10.32}$$

where $\tau$ is a stacked vector composed of $\tau_i$, $i \in S$. The problem (10.28)–(10.32) can be formulated as a standard SDP problem, which is defined as minimizing a linear objective under LMI constraints. Writing $\text{tr}\{\mathbf{MSE}_i\}$ and $I_l^{PU}$ in vector forms and utilizing Lemma 10.1, the SDP formulation of problem (10.28)–(10.32) is expressed as

$$\min_{\mathbf{V},\mathbf{U},\tau,\epsilon_i \geq 0, \eta_l \geq 0} \, \sum_{i \in S} \tau_i \tag{10.33}$$

$$\text{s.t.} \quad \begin{bmatrix} \tau_i - \epsilon_i & \tilde{\boldsymbol{\mu}}_i^H & \mathbf{0}_{1 \times \tilde{N}_i \tilde{M}} \\ \tilde{\boldsymbol{\mu}}_i & \mathbf{I}_{A_i} & -\delta_i \mathbf{D}_{\Delta_i} \\ \mathbf{0}_{\tilde{N}_i \tilde{M} \times 1} & -\delta_i \mathbf{D}_{\Delta_i}^H & \epsilon_i \mathbf{I}_{\tilde{N}_i \tilde{M}} \end{bmatrix} \succeq 0, \ i \in S, \tag{10.34}$$

$$\|\text{vec}(\mathbf{V}_i)\|_2^2 \le P_i, \ i \in S^{UL}, \tag{10.35}$$

$$\| \lfloor \text{vec}(\mathbf{V}_i) \rfloor_{i \in S^{DL}} \|_2^2 \le P_0, \tag{10.36}$$

$$\begin{bmatrix} \lambda_l - \eta_l & \tilde{\boldsymbol{\iota}}_l^H & \mathbf{0}_{1 \times T_l \tilde{M}} \\ \tilde{\boldsymbol{\iota}}_l & \mathbf{I}_{B_l} & -\theta_l \mathbf{E}_{\Lambda_l} \\ \mathbf{0}_{T_l \tilde{M} \times 1} & -\theta_l \mathbf{E}_{\Lambda_l}^H & \eta_l \mathbf{I}_{T_l \tilde{M}} \end{bmatrix} \succeq 0, \ l = 1, \ldots, L. \tag{10.37}$$

The variables $A_i$, $B_l$, $\tilde{\boldsymbol{\mu}}_i$, $\mathbf{D}_{\Delta_i}$, $\tilde{\boldsymbol{\iota}}_l$, and $\mathbf{E}_{\Lambda_l}$ are defined as follows:

$$A_i = d_i \left( \sum_{j \in S} (d_j + \tilde{M}_j) + \tilde{N}_i \right) + \tilde{N}_i \sum_{j \in S} d_j, \tag{10.38}$$

$$B_l = T_l \sum_{j \in S} (d_j + \tilde{M}_j), \tag{10.39}$$

$$\tilde{\boldsymbol{\mu}}_i = \begin{bmatrix} (\mathbf{V}_i^T \otimes \mathbf{U}_i^H) \text{vec}(\tilde{\mathbf{H}}_{ii}) - \text{vec}(\mathbf{I}_{d_i}) \\ \lfloor (\mathbf{V}_j^T \otimes \mathbf{U}_i^H) \text{vec}(\tilde{\mathbf{H}}_{ij}) \rfloor_{j \in S, j \neq i} \\ \lfloor \lfloor \sqrt{\kappa}((\boldsymbol{\Gamma}_\ell \mathbf{V}_j)^T \otimes \mathbf{U}_i^H) \text{vec}(\tilde{\mathbf{H}}_{ij}) \rfloor_{\ell \in D_j^{(T)}} \rfloor_{j \in S} \\ \lfloor \lfloor \sqrt{\beta}(\mathbf{V}_j^T \otimes (\mathbf{U}_i^H \boldsymbol{\Gamma}_\ell)) \text{vec}(\tilde{\mathbf{H}}_{ij}) \rfloor_{\ell \in D_i^{(R)}} \rfloor_{j \in S} \\ \sigma_i \text{vec}(\mathbf{U}_i) \end{bmatrix}, \tag{10.40}$$

$$\boldsymbol{\mu}_{\Delta_i} = \underbrace{\begin{bmatrix} (\mathbf{V}_i^T \otimes \mathbf{U}_i^H) \\ \lfloor (\mathbf{V}_j^T \otimes \mathbf{U}_i^H) \rfloor_{j \in S, j \neq i} \\ \lfloor \lfloor \sqrt{\kappa}((\boldsymbol{\Gamma}_\ell \mathbf{V}_j)^T \otimes \mathbf{U}_i^H) \rfloor_{\ell \in D_j^{(T)}} \rfloor_{j \in S} \\ \lfloor \lfloor \sqrt{\beta}(\mathbf{V}_j^T \otimes (\mathbf{U}_i^H \boldsymbol{\Gamma}_\ell)) \rfloor_{\ell \in D_i^{(R)}} \rfloor_{j \in S} \\ \mathbf{0}_{d_i \tilde{N}_i \times \tilde{N}_i \tilde{M}} \end{bmatrix}}_{\mathbf{D}_{\Delta_i}} \text{vec}(\boldsymbol{\Delta}_i), \tag{10.41}$$

$$\tilde{\boldsymbol{\iota}}_l = \begin{bmatrix} \lfloor (\mathbf{V}_j^T \otimes \mathbf{I}_{T_l}) \text{vec}(\tilde{\mathbf{G}}_{lj}) \rfloor_{j \in S} \\ \sqrt{\kappa} \lfloor \lfloor ((\boldsymbol{\Gamma}_\ell \mathbf{V}_j)^T \otimes \mathbf{I}_{T_l}) \text{vec}(\tilde{\mathbf{G}}_{lj}) \rfloor_{\ell \in D_j^{(T)}} \rfloor_{j \in S} \end{bmatrix}, \tag{10.42}$$

$$\boldsymbol{\iota}_{\Lambda_l} = \underbrace{\begin{bmatrix} \lfloor (\mathbf{V}_j^T \otimes \mathbf{I}_{T_l}) \rfloor_{j \in S} \\ \sqrt{\kappa} \lfloor \lfloor ((\boldsymbol{\Gamma}_\ell \mathbf{V}_j)^T \otimes \mathbf{I}_{T_l}) \rfloor_{\ell \in D_j^{(T)}} \rfloor_{j \in S} \end{bmatrix}}_{\mathbf{E}_{\Lambda_l}} \text{vec}(\boldsymbol{\Lambda}_l). \tag{10.43}$$

The problem 10.33–10.37 does not hold a jointly convex structure over the optimization variables. Nevertheless it is a separately convex optimization problem over the transmit beamforming matrices $\mathbf{V}$ and the receiving beamforming matrices $\mathbf{U}$ once the other variables are fixed. This facilitates an alternating optimization algorithm (see Table 10.1) where in each iteration the solution to (10.33)–(10.37) is calculated as a convex optimization problem, assuming an alternatively fixed $\mathbf{V}$ or $\mathbf{U}$. The iterations continue until a stationary point is obtained or a pre-defined number of iterations is reached.

**Table 10.1** Sum-MSE minimization algorithm for FD cellular CRN.

---

1) Set the iteration number $n = 0$ and initialize $\mathbf{V}^{[n]}$.

2) $n \leftarrow n + 1$. Update $\mathbf{U}_i^{[n]}$, $i \in S$ by solving the SDP problem (10.33)–(10.37) under fixed $\mathbf{V}^{[n-1]}$.

3) Update $\mathbf{V}_i^{[n]}$, $i \in S$ by solving the SDP problem (10.33)–(10.37) under fixed $\mathbf{U}^{[n]}$.

4) Repeat steps 2 and 3 until convergence.

---

### 10.2.4 Numerical Results

We now numerically investigate the robust sum-MSE minimization algorithm for an FD MIMO CR cellular system. Note that smart channel assignment prior to precoder/decoder design is essential for an FD setup as the CCI can be reduced by assigning the users with weaker interference paths into the same channel. In order to incorporate the effect of channel assignment, we assume an attenuation coefficient, namely $v$, on the CCI channels, which represents the degree of isolation among UL and DL users. The tolerance (the difference between the MSEs of two consecutive iterations) of the proposed iterative algorithm is set to $10^{-4}$, the maximum number of iterations is set to 50, and the results are averaged over 100 independent channel realizations.

We consider small cell deployments and compare the FD system with HD ones under 3rd Generation Partnership Project Long-Term Evolution (3GPP LTE) specifications [1]. A small cell is considered to be suitable for deployment of FD technology due to its low transmit power, short transmission distances, and low mobility [20]. We consider a single hexagonal cell consisting of a BS in the center with $M_0$ transmit and $N_0$ receive antennas. $K = 2$ UL and $J = 2$ DL users equipped with $N$ antennas are randomly distributed in the cell. For simplicity, we assume $M_0 = N_0 = N = \tilde{N}$. The primary system has $L = 2$ PUs, with the same maximum allowed interfering power (i.e., $\lambda_l = 0$dB). The channels between BS and users (both SUs and PUs) are assumed to experience a pathloss model for line-of-sight (LoS) communications, and the channels between UL and DL users are assumed to experience a pathloss model for non-line-of-sight (NLoS) communications. Detailed simulation parameters are shown in Table 10.2. The estimated channel gain between the BS to the $k$th UL user is given by $\tilde{\mathbf{H}}_k^{UL} = \sqrt{\kappa_k^{UL}} \hat{\mathbf{H}}_k^{UL}$, where $\hat{\mathbf{H}}_k^{UL}$ denotes the small-scale fading following a complex Gaussian distribution with zero mean and unit variance, and $\kappa_k^{UL} = 10^{(-X/10)}$, $X \in \{\text{LOS}, \text{NLOS}\}$ represents the large-scale fading consisting of path loss and shadowing, where LoS and NLoS are calculated from a specific path-loss model given in Table 10.2. The channels between BS and DL users, between UL users and DL users, between BS and PUs, and between UL users and PUs are defined similarly. We adopt the Rician model in [10], where the SI channel is distributed as $\tilde{\mathbf{H}}_0 \sim \mathcal{CN}\left(\sqrt{\frac{K_R}{1+K_R}}\hat{\mathbf{H}}_0, \frac{1}{1+K_R}\mathbf{I}_{N_0} \otimes \mathbf{I}_{M_0}\right)$, with $K_R$ being the Rician factor and $\hat{\mathbf{H}}_0$ a deterministic matrix.[2] Unless otherwise stated, we consider $\tilde{N} = 2$, $\kappa = \beta = -70$dB, $v = 0.5$, and $\delta = \theta = 0.1$.

---

2 Similar to [20], without loss of generality, we set $K_R = 1$ and $\hat{\mathbf{H}}_0$ to be the matrix of all ones for all experiments.

**Table 10.2** Simulation parameters.

| Parameter | Settings |
|---|---|
| Cell radius | 40 m |
| Carrier frequency | 2 GHz |
| Bandwidth | 10 MHz |
| Thermal noise density | −174 dBm/Hz |
| Noise figure | BS: 13 dB, user: 9 dB |
| Path loss (dB) between BS and users | $103.8 + 20.9 \log_{10} d$ ($d$ in km) |
| Path loss (dB) between users ($d$ in km) | $145.4 + 37.5 \log_{10} d$ |
| Shadowing standard deviation | LoS: 3 dB, NLoS: 4 dB |

**Figure 10.2** Convergence behavior of the proposed algorithm.

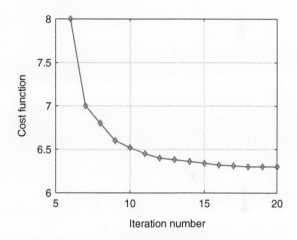

We begin by showing the evolution of the proposed algorithm (Table 10.1) in Figure 10.2. The monotonic decrease of the cost function (sum-MSE) verifies the convergence of the algorithm.

Next, the complementary cumulative distribution (CCD) of the total interference power from the secondary users to the primary users ($\mathbb{P}[I^{PU} \geq \lambda]$, where $I^{PU} = \sum_{l=1}^{L} I_{l}^{PU}$ and $\lambda = \sum_{l=1}^{L} \lambda_{l}$), is shown. It can be seen from Figure 10.3 that the probability of total interference power from the secondary network to the PUs is zero when it is higher than $\lambda = 3$dB, which is the maximum allowed total interfering power (considering two PUs, with each allowing 0 dB interference). This is in conjunction to constraint (10.32), which ensures that the interference to the PUs is always kept below or equal to the maximum allowed total interfering power. While achieving the equality condition in (10.32) will ensure the maximum sum rate for the SUs, the proposed algorithm mainly operates below the maximum allowed interfering power to protect the PUs, but still satisfying the required quality of service of the SUs. Moreover, the area under the CCD function curve can be contemplated as the region under which the proposed algorithm is feasible.

Hereinafter, FD with HD systems will be compared in terms of sum-rate performance as a function of RSI cancellation strength (interpreted here in terms of $\kappa = \beta$

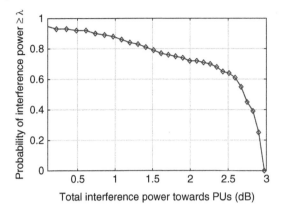

**Figure 10.3** Probability of interference power from secondary to primary network.

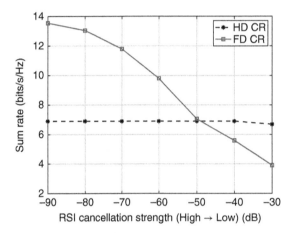

**Figure 10.4** Sum-rate comparison of FD and HD systems with respect to RSI cancellation strength.

values). In particular, the sum-rate of a MIMO FD cellular system can be expressed as $I_{sum} = \sum_{i \in S} \sum_{k=1}^{d_i} \log_2(1 + \text{SINR}_{i_k})$, where $\text{SINR}_{i_k}$ is the SINR of the $k$th stream of user $i$ and can be obtained from the MSE expression given in section 10.1 As seen in Figure 10.4, the performance of the HD system is invariant to the strength of RSI cancellation, and at high RSI cancellation levels the FD system achieves around 1.6 times more sum-rate than that of HD. However, at low RSI cancellation levels (below around −55dB) the distortion is magnified with the increasing number of antennas and the performance of the FD system drops below that of the HD system.

In Figure 10.5, the importance of the smart channel assignment at a stage prior to the precoder/decoder design is depicted. The CCI attenuation represents the provided isolation among UL and DL users. It is seen that as the suppression level of CCI increases, the FD system starts outperforming the HD system, and thus isolation among UL and DL users is essential for a successful coexistence of UL and DL users in an FD setup.

Finally, in Figure 10.6, sum-rate performance of FD and HD systems is compared as a function of channel uncertainty factor (interpreted here as $\delta = \theta$ values). It can be seen that the performance of both systems degrades as the size of the uncertainty region increases.

**Figure 10.5** Sum-rate comparison of FD and HD systems with respect to the CCI attenuation factor.

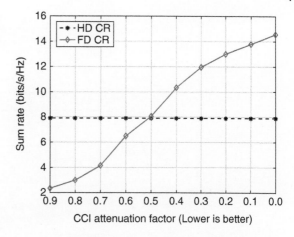

**Figure 10.6** Sum-rate comparison of FD and HD systems with respect to the channel uncertainty factor.

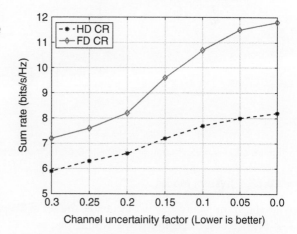

However, the FD system suffers more and as a result of that the gap between the FD and HD system decreases. However, if the channel uncertainty is nominal along with a low distortion level (−70 dB in this case), FD systems achieve around 1.4 times more sum-rate than HD systems. This degradation in performance of the FD system is explained as follows. Since there are more interference channels (SI and CCI) in FD systems, with increasing channel uncertainty the degradation in performance of the FD system is accelerated. This indicates that channel estimation is a critical factor for successful deployment of FD systems.

## 10.3 Transceiver Design for an FD MIMO IoT Network

In this section, we discuss the transceiver design problem for an FD MIMO CR IoT system in which $K$ pairs of FD SUs (IoT nodes, e.g. pairs of sensors and actuators/controllers when considering an industrial IoT setup) communicate simultaneously within the service range of $L$ PUs.

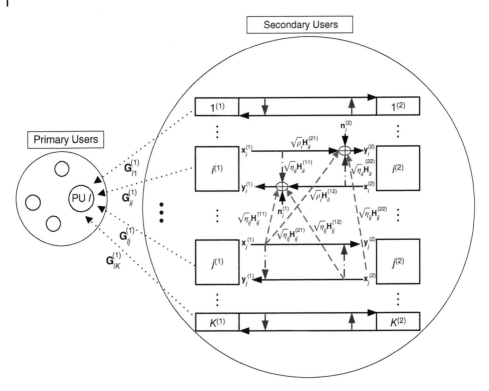

**Figure 10.7** An illustration of an FD MIMO CR IoT network.

### 10.3.1 System Model

As shown in Figure 10.7, let us denote the set of SU pairs and PUs by $\mathcal{K} \triangleq \{1, \dots, K\}$ and $\mathcal{L} \triangleq \{1, \dots, L\}$, respectively. We assume that each SU node that belongs to the $i$th pair is equipped with $N_i$ and $M_i$ transmit and receive antennas, respectively.

#### 10.3.1.1 Signal and Channel Model

In what follows, $i^{(a)}$ denotes SU $a \in \{1, 2\}$ belonging to pair $i \in \mathcal{K}$. The SU $i^{(a)}$, $i \in \mathcal{K}$, $a \in \{1, 2\}$ receives signals from all the SU transmitters in the system via MIMO channels. $\mathbf{H}_{ii}^{(ab)} \in \mathbb{C}^{M_i \times N_i}$ is the desired channel between the transmitter of node $b$, where $b \in \{1, 2\}, b \neq a$, and the receiver of node $a$, when both nodes (SUs) belong to the $i$th pair. $\mathbf{H}_{ii}^{(aa)} \in \mathbb{C}^{M_i \times N_i}, a \in \{1, 2\}$ denotes the SI channel of the SU $i^{(a)}$. Also, $\mathbf{H}_{ij}^{(ac)} \in \mathbb{C}^{M_i \times N_j}, (a, c) \in \{1, 2\}$ denotes the CCI channel from the transmit antennas of the SU $c$ in the $j$th SU pair to the receive antennas of SU $a$ in the $i$th pair, $(i, j) \in \mathcal{K}$ and $j \neq i$. All the channel matrices are assumed to be mutually independent, and the entries of each matrix are circular complex Gaussian variables with zero mean, independent real and imaginary parts, each with variance $1/2$.

The transmitted data streams of size $d_i$ at the SU $i^{(a)}$ are denoted as $\mathbf{d}_i^{(a)} \in \mathbb{C}^{d_i}$, $i \in \mathcal{K}$, $a \in \{1, 2\}$, and are assumed to be complex, zero mean, i.i.d. with unit variance. The $N_i \times 1$ signal vector transmitted by the SU $i^{(a)}$ is given by $\mathbf{x}_i^{(a)} = \mathbf{V}_i^{(a)} \mathbf{d}_i^{(a)}$, $i \in \mathcal{K}$, $a \in \{1, 2\}$, where $\mathbf{V}_i^{(a)} \in \mathbb{C}^{N_i \times d_i}$ represents the transmit beamforming matrix applied at the node $i^{(a)}$. According to the investigated system model, we consider an FD MIMO interference channel between SUs that suffers from SI and CCI from other pairs. Thus, the SU $i^{(a)}$ receives a combination of the signals transmitted by all the transmitters along with additive noise. The $M_i \times 1$ received signal at the SU $i^{(a)}$ is written as

$$\mathbf{y}_i^{(a)} = \sqrt{\rho_i}\mathbf{H}_{ii}^{(ab)}(\mathbf{x}_i^{(b)} + \mathbf{c}_i^{(b)}) + \underbrace{\sqrt{\eta_{ii}}\mathbf{H}_{ii}^{(aa)}(\mathbf{x}_i^{(a)} + \mathbf{c}_i^{(a)})}_{SI}$$

$$+ \underbrace{\sum_{j \neq i}^{K}\sum_{c=1}^{2}\sqrt{\eta_{ij}^{(ac)}}\ \mathbf{H}_{ij}^{(ac)}\ (\mathbf{x}_j^{(c)} + \mathbf{c}_j^{(c)}) + \mathbf{e}_i^{(a)}}_{CCI}$$

$$+ \mathbf{n}_i^{(a)},\ i \in \mathcal{K},\ (a,b) \in \{1,2\}\text{ and }a \neq b. \tag{10.44}$$

Here, $\mathbf{n}_i^{(a)} \in \mathbb{C}^{M_i}$ is the AWGN vector at SU $i^{(a)}$ with zero mean and covariance matrix $\mathbf{I}_{M_i}$, and it is uncorrelated to all the transmitted signals. In (10.44), $\rho_i$ denotes the average power gain of the $i$th SU transmitter–receiver pair, $\eta_{ii}$ denotes the average power gain of the SI channel at the $i$th SU pair, and $\eta_{ij}^{(ac)}$ denotes the average power gain of the CCI channel between the nodes at the $i^{(a)}$th and $j^{(c)}$th SU pairs.[3] Like before, in (10.44), $\mathbf{c}_i^{(a)} \in \mathbb{C}^{N_i}$, $i \in \mathcal{K}$, $a \in \{1,2\}$ is the noise at the transmit antennas of SU $i^{(a)}$, which models the effect of limited transmitter DR, and its covariance matrix is given as $\mathbf{c}_i^{(a)} \sim \mathcal{CN}(\ \mathbf{0}_{N_i}, \kappa\ \text{diag}\ (\mathbf{V}_i^{(a)}(\mathbf{V}_i^{(a)})^H)\ )$, $\mathbf{c}_i^{(a)} \perp \mathbf{x}_i^{(a)}$. Similarly, $\mathbf{e}_i^{(a)} \in \mathbb{C}^{M_i}$, $i \in \mathcal{K}$, $a \in \{1,2\}$ is the additive receiver distortion at the receive antennas of the SU $i^{(a)}$, which models the effect of limited receiver DR, the covariance matrix of which is given as $\mathbf{e}_i^{(a)} \sim \mathcal{CN}(\mathbf{0}_{M_i}, \beta\text{diag}(\mathbf{\Phi}_i^{(a)}))$, $\mathbf{e}_i^{(a)} \perp \mathbf{u}_i^{(a)}$, where $\mathbf{\Phi}_i^{(a)} = \text{Cov}\{\mathbf{u}_i^{(a)}\}$ and $\mathbf{u}_i^{(a)}$ is the undistorted received signal vector at SU $i^{(a)}$, i.e. $\mathbf{u}_i^{(a)} = \mathbf{y}_i^{(a)} - \mathbf{e}_i^{(a)}$.

### 10.3.1.2 SI Cancellation

We assume that SU $i^{(a)}$ knows the self-interfering codewords $\mathbf{x}_i^{(a)}$ and its SI channel $\mathbf{H}_{ii}^{(aa)}$. So, the SI term $\sqrt{\eta_{ii}}\mathbf{H}_{ii}^{(aa)}\mathbf{x}_i^{(a)}$ is known and thus can be cancelled [11], giving the received signal

$$\tilde{\mathbf{y}}_i^{(a)} = \mathbf{y}_i^{(a)} - \sqrt{\eta_{ii}}\mathbf{H}_{ii}^{(aa)}\mathbf{x}_i^{(a)} = \sqrt{\rho_i}\mathbf{H}_{ii}^{(ab)}\mathbf{x}_i^{(b)} + \tilde{\mathbf{n}}_i^{(a)}, \tag{10.45}$$

where $\tilde{\mathbf{n}}_i^{(a)} \in \mathbb{C}^{M_i \times 1}$ is the RSI component of (10.45) after SI cancellation and is given by

$$\tilde{\mathbf{n}}_i^{(a)} = \sqrt{\rho_i}\mathbf{H}_{ii}^{(ab)}\mathbf{c}_i^{(b)} + \sqrt{\eta_{ii}}\mathbf{H}_{ii}^{(aa)}\mathbf{c}_i^{(a)} + \mathbf{e}_i^{(a)} + \mathbf{n}_i^{(a)}$$

$$+ \sum_{j \neq i}^{K}\sum_{c=1}^{2}\sqrt{\eta_{ij}^{(ac)}}\ \mathbf{H}_{ij}^{(ac)}(\mathbf{x}_j^{(c)} + \mathbf{c}_j^{(c)}). \tag{10.46}$$

---

3 Note that in (10.44), the power gains $\rho$ and $\eta$ correspond to the large-scale fading factors, which are distance-based, therefore they are assumed to be constant from time-slot to time-slot as mobility is not taken into account in the studied scenario. On the contrary, channels $\mathbf{H}$ are considered to model the fast fading phenomena.

Similar to the case of the cellular system model, the covariance matrix of $\tilde{\mathbf{n}}_i^{(a)}$ can be approximated as

$$
\sum_i^{(a)} \approx \rho_i \kappa \mathbf{H}_{ii}^{(ab)} \operatorname{diag}(\mathbf{V}_i^{(b)}(\mathbf{V}_i^{(b)})^H) (\mathbf{H}_{ii}^{(ab)})^H + \eta_{ii} \kappa \mathbf{H}_{ii}^{(aa)} \operatorname{diag}(\mathbf{V}_i^{(a)}(\mathbf{V}_i^{(a)})^H)(\mathbf{H}_{ii}^{(aa)})^H
$$
$$
+ \beta \rho_i \operatorname{diag}(\mathbf{H}_{ii}^{(ab)}\mathbf{V}_i^{(b)}(\mathbf{V}_i^{(b)})^H (\mathbf{H}_{ii}^{(ab)})^H) + \beta \eta_{ii} \operatorname{diag}(\mathbf{H}_{ii}^{(aa)}\mathbf{V}_i^{(a)}(\mathbf{V}_i^{(a)})^H(\mathbf{H}_{ii}^{(aa)})^H)
$$
$$
+ \sum_{j \neq i}^{K} \sum_{c=1}^{2} \eta_{ij}^{(ac)} [\mathbf{H}_{ij}^{(ac)}(\mathbf{V}_j^{(c)}(\mathbf{V}_j^{(c)})^H + \kappa \operatorname{diag}(\mathbf{V}_j^{(c)}(\mathbf{V}_j^{(c)})^H))(\mathbf{H}_{ij}^{(ac)})^H]
$$
$$
+ \sum_{j \neq i}^{K} \sum_{c=1}^{2} \beta \eta_{ij}^{(ac)} \operatorname{diag}(\mathbf{H}_{ij}^{(ac)}\mathbf{V}_j^{(c)}(\mathbf{V}_j^{(c)})^H(\mathbf{H}_{ij}^{(ac)})^H) + \mathbf{I}_{M_i}. \tag{10.47}
$$

Now, assuming that the SU $i^{(a)}$ applies a linear receiver $\mathbf{R}_i^{(a)} \in \mathbb{C}^{d_i \times M_i}$ to estimate the signal transmitted from SU $i^{(b)}$, i.e. $\mathbf{d}_i^{(b)}$, we have

$$
\hat{\mathbf{d}}_i^{(b)} = \mathbf{R}_i^{(a)}\tilde{\mathbf{y}}_i^{(a)} = \sqrt{\rho_i}\mathbf{R}_i^{(a)}\mathbf{H}_{ii}^{(ab)}\mathbf{V}_i^{(b)}\mathbf{d}_i^{(b)} + \mathbf{R}_i^{(a)}\tilde{\mathbf{n}}_i^{(a)}. \tag{10.48}
$$

### 10.3.1.3 MSE of the Received Data Stream

Using (10.48), the MSE matrix of the SU $i^{(a)}$ can be written as

$$
\mathbf{MSE}_i^{(a)} = \mathbb{E}\{(\hat{\mathbf{d}}_i^{(b)} - \mathbf{d}_i^{(b)})(\hat{\mathbf{d}}_i^{(b)} - \mathbf{d}_i^{(b)})^H\}
$$
$$
= (\sqrt{\rho_i}\mathbf{R}_i^{(a)}\mathbf{H}_{ii}^{(ab)}\mathbf{V}_i^{(b)} - \mathbf{I}_{d_i}) (\sqrt{\rho_i}\mathbf{R}_i^{(a)}\mathbf{H}_{ii}^{(ab)}\mathbf{V}_i^{(b)} - \mathbf{I}_{d_i})^H + \mathbf{R}_i^{(a)}\sum_i^{(a)}(\mathbf{R}_i^{(a)})^H. \tag{10.49}
$$

As mentioned before, the SUs are located within the service range of $L$ PUs, for which the SUs should provide protection according to a QoS-based criterion. We assume that the PUs are equipped with $N$ receive antennas. The received interference signal at the $l$th PU from SU $i^{(b)}$ is expressed as

$$
\mathbf{z}_{i,l}^{(b)} = \sqrt{\mu_{i,l}^{(b)}}\mathbf{G}_{i,l}^{(b)} (\mathbf{x}_i^{(b)} + \mathbf{c}_i^{(b)}), \ i \in \mathcal{K}, \ b = 1, 2, \ l \in \mathcal{L}, \tag{10.50}
$$

where $\mathbf{G}_{i,l}^{(b)} \in \mathbb{C}^{N \times N_i}$ is the channel between the $l$th PU and $i^{(b)}$th SU, which is modeled similar to $\mathbf{H}_{ij}^{(ab)}$, discussed in section 10.3.1.1, and $\mu_{i,l}^{(b)}$ is the average power gain of $\mathbf{G}_{i,l}^{(b)}$. Using (10.50), the power of the interference resulting from the $i^{(b)}$th SU at the $l$th PU can be written as

$$
I_{i,l}^{(b)}(\mathbf{V}_i^{(b)}) = \mu_{i,l}^{(b)}\operatorname{tr}\{\mathbf{G}_{i,l}^{(b)}(\mathbf{V}_i^{(b)}(\mathbf{V}_i^{(b)})^H + \kappa \operatorname{diag}(\mathbf{V}_i^{(b)}(\mathbf{V}_i^{(b)})^H))(\mathbf{G}_{i,l}^{(b)})^H\}. \tag{10.51}
$$

### 10.3.2 Joint Transceiver Design

Similar to the cellular system scenario, we take sum-MSE as the performance metric to design the transceivers under a transmit power constraint imposed on the SUs and an interference power constraint at the $l$th PU, which can be formulated as follows

$$
\min_{\mathbf{V},\mathbf{R}} \sum_{i=1}^{K} \sum_{a=1}^{2} \operatorname{tr}\{\mathbf{MSE}_i^{(a)}\} \tag{10.52}
$$
$$
\text{s.t.} \quad \operatorname{tr}\{\mathbf{V}_i^{(b)}(\mathbf{V}_i^{(b)})^H\} \leq P_i^{(b)}, \ i \in \mathcal{K}, \ b = 1, 2, \tag{10.53}
$$
$$
I_{i,l}^{(b)}(\mathbf{V}_i^{(b)}) \leq \lambda_{i,l}^{(b)}, \ i \in \mathcal{K}, \ b = 1, 2, \ l \in \mathcal{L}. \tag{10.54}
$$

Here, $P_i^{(b)}$ is the power constraint at the $i^{(b)}$th SU transmitter, $\lambda_{i,l}^{(b)}$ is the maximum allowed interference temperature at the $l$th PU receiver [24], and $\mathbf{V}(\mathbf{R}) = \{\mathbf{V}_i^{(b)}(\mathbf{R}_i^{(b)}) : \forall(i, b)\}$ is the set of all transmitting (receiving) beamforming matrices. Now, fixing the transmit beamforming matrix, the optimal receive beamforming matrices at the SU $i^{(a)}$ is the MMSE receive filter, which can be expressed as

$$
\mathbf{R}_i^{(a)*} = \arg\min_{\mathbf{R}_i^{(a)}} \, \mathrm{tr}\{\mathbf{MSE}_i^{(a)}\} = \sqrt{\rho_i}(\mathbf{V}_i^{(b)})^H(\mathbf{H}_{ii}^{(ab)})^H
$$

$$
\times \, (\rho_i \mathbf{H}_{ii}^{(ab)}\mathbf{V}_i^{(b)}(\mathbf{V}_i^{(b)})^H(\mathbf{H}_{ii}^{(ab)})^H + \overset{(a)}{\Sigma}_i)^{-1}. \tag{10.55}
$$

Substituting (10.55) for the objective function $\mathbf{MSE}_i^{(a)}$ in (10.52) gives $\mathbf{C}_i^{(a)}(\mathbf{V})$, which is the error matrix for the node $i^{(a)}$ given that the MMSE receive filter is applied. The error matrix can be written as

$$
\mathbf{C}_i^{(a)}(\mathbf{V}) = \mathbf{I}_{d_i} - \rho_i(\mathbf{V}_i^{(b)})^H(\mathbf{H}_{ii}^{(ab)})^H
$$

$$
\times \, (\, \rho_i \mathbf{H}_{ii}^{(ab)}\mathbf{V}_i^{(b)}(\mathbf{V}_i^{(b)})^H \, (\mathbf{H}_{ii}^{(ab)})^H + \overset{(a)}{\Sigma}_i)^{-1} \times \mathbf{H}_{ii}^{(ab)}\mathbf{V}_i^{(b)}. \tag{10.56}
$$

Since the first term, i.e. the identity matrix in (10.56), has no effect in the optimization problem, we only consider the second term in (10.56), and the negative sign in front of the second term in (10.56) changes the minimization problem (10.52) to a maximization problem (10.57). Accordingly, substituting $\mathbf{C}_i^{(a)}(\mathbf{V})$ into the objective function (10.52), and writing $\mathbf{Q}_i^{(b)} = \mathbf{V}_i^{(b)}(\mathbf{V}_i^{(b)})^H$, the problem of determining the optimum transmit beamforming matrices under fixed receiver matrices can be rewritten as

$$
\max_{\mathbf{Q}} \quad \sum_{i=1}^{K} \sum_{a=1}^{2} \mathrm{tr}\{\mathbf{A}_i^{(a)}(\mathbf{Q})\} \tag{10.57}
$$

$$
\text{s.t.} \quad \mathrm{tr}\{\mathbf{Q}_i^{(b)}\} \le P_i^{(b)}, \; i \in \mathcal{K}, \; b = 1, 2, \tag{10.58}
$$

$$
I_{i,l}^{(b)}(\mathbf{Q}_i^{(b)}) \le \lambda_{i,l}^{(b)}, \; i \in \mathcal{K}, \; b = 1, 2, \; l \in \mathcal{L}, \tag{10.59}
$$

$$
\mathbf{Q}_i^{(b)} \succeq \mathbf{0}, \; i \in \mathcal{K}, \; b = 1, 2, \tag{10.60}
$$

where $\mathbf{Q} = \{\mathbf{Q}_i^{(b)} : \forall(i, b)\}$ and the matrix $\mathbf{A}_i^{(a)}(\mathbf{Q})$ is defined as

$$
\mathbf{A}_i^{(a)}(\mathbf{Q}) = \rho_i \mathbf{H}_{ii}^{(ab)}\mathbf{Q}_i^{(b)}(\mathbf{H}_{ii}^{(ab)})^H \, (\rho_i \mathbf{H}_{ii}^{(ab)}\mathbf{Q}_i^{(b)}(\mathbf{H}_{ii}^{(ab)})^H + \tilde{\Sigma}_i^{(a)})^{-1}. \tag{10.61}
$$

Here, $\tilde{\Sigma}_i^{(a)}$ in (10.61) and $I_{i,l}^{(b)}(\mathbf{Q}_i^{(b)})$ in (10.59) are obtained by replacing $\mathbf{V}_i^{(b)}(\mathbf{V}_i^{(b)})^H$ in (10.47) and (10.51) with $\mathbf{Q}_i^{(b)}$, respectively.

### 10.3.3 Imperfect CSI and Robust Design

Similar to the previous scenario, the imperfect CSI is modeled using the deterministic norm-bounded error model [26], expressed as

$$
\mathbf{G}_{i,l}^{(b)} \in \mathcal{G}_{i,l}^{(b)} = \{\tilde{\mathbf{G}}_{i,l}^{(b)} + \mathbf{\Lambda}_{i,l}^{(b)} : \|\mathbf{\Lambda}_{i,l}^{(b)}\|_F \le \theta_{i,l}^{(b)}\}, \; \forall(i, b, l). \tag{10.62}
$$

In the above equation, $\tilde{\mathbf{G}}_{i,l}^{(b)}$, $\mathbf{\Lambda}_{i,l}^{(b)}$, and $\theta_{i,l}^{(b)}$ denote the nominal value of the CSI, the error matrix, and the uncertainty bounds, respectively. With the imperfect CSI, the optimization problem in (10.57)–(10.60) can be rewritten as

$$\max_{\mathbf{Q}} \ \sum_{i=1}^{K} \sum_{a=1}^{2} \mathrm{tr}\{\mathbf{A}_i^{(a)}(\mathbf{Q})\} \tag{10.63}$$

$$\text{s.t.} \quad \mathrm{tr}\{\mathbf{Q}_i^{(b)}\} \leq P_i^{(b)}, \ i \in \mathcal{K}, \ b = 1, 2, \tag{10.64}$$

$$I_{i,l}^{(b)}(\mathbf{Q}_i^{(b)}) \leq \lambda_{i,l}^{(b)}, \ \forall \mathbf{G}_{i,l}^{(b)} \in \mathcal{G}_{i,l}^{(b)}, \ \forall (i, b, l), \tag{10.65}$$

$$\mathbf{Q}_i^{(b)} \geq \mathbf{0}, \ i \in \mathcal{K}, \ b = 1, 2. \tag{10.66}$$

The above problem can be solved in a way similar to the previous scenario in polynomial time through standard interior point methods by converting it into an SDP problem.

### 10.3.4 Numerical Results

To numerically investigate this scenario, a robust sum-MSE minimization algorithm for the MIMO CR IoT network is formulated similar to the previous scenario. We set the tolerance level to $10^{-4}$ and the maximum number of iterations to 100, and the results are averaged over numerous independent channel realizations. The distance between the desired links is set to $d_i = 30$ m. The PU receiver is located at a distance from the SUs that is uniformly distributed over $70 - 100$ m. For brevity, we assume that the maximum transmit power for all SUs is the same, i.e. $P = P_i^{(b)}$, $\forall(i, b)$. The path loss obeys the model $d^{-\varsigma}$, where $d$ is the distance between nodes and $\varsigma = 3.5$ is the path-loss exponent. The maximum transmit powers are set so that the (maximum) signal-to-noise ratio (SNR) is defined as SNR $= Pd_i^{-\varsigma} = 15$ dB. The total interference threshold is set to $4 \times 10^{-7}$W and for simplicity it is equally split among the SUs. Unless otherwise stated, the channel uncertainty is set to $\theta_i^{(b)} = s\|\tilde{\mathbf{G}}_i^{(b)}\|_F$, with $s \in (0, 1]$, the transmitter/receiver distortion parameters are chosen as $\kappa = \beta = -70$ dB, and we set the same number of transmit and receive antennas at each node, i.e. $M_i = N_i = N$, $i = 1, \dots, K$.

In Figure 10.8, the sum-rate performance of an FD system is compared with corresponding HD systems for different RSI cancellation strengths (interpreted here as $\kappa = \beta$ values). Note that for a HD system, transmission is carried out in two time-slots, i.e. in the first time slot all the SUs on the left-hand side in Figure 10.7 transmit to their peers on the right, whereas in the second time slot these roles are reversed. As a result, although SI does not exist, CCI is present and the sum-rate should be divided by 2 because of the two time slot transmissions. As can be seen in the figure, the performance of the HD systems is not affected by RSI cancellation values, and at high RSI cancellation levels the FD system achieves around 1.6 times more sum-rate than that of the corresponding HD system. It is also worth mentioning that the performance of the FD system starts to drop below that of the HD systems in and around $-55$ dB, which is similar to the results obtained for the cellular system model.

Finally, the robust FD precoding scheme is compared with the robust HD one for different values of the channel uncertainty parameters. It can be seen from Figure 10.9 that as the

**Figure 10.8** Sum-rate comparison of FD and HD systems with respect to RSI cancellation strength. Here, $K = 4$, $s = 0.2$, $N = 2$.

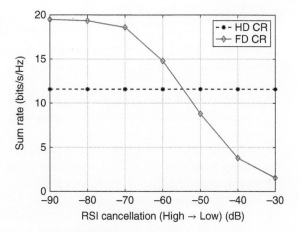

**Figure 10.9** Sum-rate comparison of FD and HD systems with respect to channel uncertainty factors. Here, $K = 2$, $N = 2$.

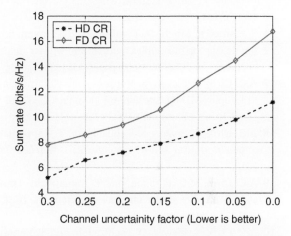

size of the uncertainty region increases, the performance of the FD system degrades, and the performance gap between the considered FD and HD systems decreases. This degradation in performance of the FD system can be explained in a similar way to the cellular system model.

## 10.4 Summary

In this chapter we studied robust MSE-based transceiver design problems for an FD MIMO CR cellular system and an FD MIMO CR IoT network that suffer from SI and CCI under limited DR at the transmitters and receivers, and norm-bounded channel uncertainties. Since globally optimal solutions for both scenarios are difficult to obtain due to the non-convex nature of the problems, alternating SDP-based algorithms that iteratively optimize the transmit and receiving beamforming matrices in a block coordinate descent fashion are given. Sum-rate performance gains are observed for FD systems, as compared to their HD counterparts, which are driven by numerical results under reasonable RSI cancellation and/or CCI attenuation values.

## References

**1** TR 36.828 3GPP. Further enhancements to LTE time division duplex (TDD) for downlink-uplink (DL-UL) interference management and traffic adaptation (release 11), 2012.

**2** W. Afifi and M. Krunz. Exploiting self-interference suppression for improved spectrum awareness/efficiency in cognitive radio systems. In *Proceedings of the IEEE INFOCOM*, pages 1258–1266, Turin, Italy, Apr. 2013.

**3** A. Ben-Tal and A. Nemirovski. *Lectures on modern convex optimization: analysis, algorithms, and engineering applications*, volume 2. SIAM, Philadelphia, PA, USA, 2001.

**4** D. Bharadia and S. Katti. Full duplex MIMO radios. In *Proceedings of the USENIX NSDI*, pages 359–372, Seattle, WA, Apr. 2014.

**5** S. Biswas, K. Singh, O. Taghizadeh, and T. Ratnarajah. Coexistence of MIMO radar and FD MIMO cellular systems with QoS considerations. *IEEE Trans. Wireless Commun.*, 17(11):7281–7294, Nov. 2018.

**6** S. Biswas, K. Singh, O. Taghizadeh, T. Ratnarajah, and M. Sellathurai. Beamforming design for full-duplex cellular and MIMO radar coexistence: A rate maximization approach. In *IEEE ICASSP*, pages 3384–3388, Apr. 2018.

**7** A.C. Cirik, S. Biswas, S. Vuppala, and T. Ratnarajah. Beamforming design for full-duplex MIMO interference channels: QoS and energy-efficiency considerations. *IEEE Trans. Commun.*, 64(11):4635–4651, Nov. 2016.

**8** A.C. Cirik, S. Biswas, S. Vuppala, and T. Ratnarajah. Robust transceiver design for full duplex multiuser MIMO systems. *IEEE Wireless Commun. Lett.*, 5(3):260–263, Jun. 2016.

**9** A.C. Cirik, M.C. Filippou, and T. Ratnarajaht. Transceiver design in full-duplex MIMO cognitive radios under channel uncertainties. *IEEE Trans. Cog. Commun. Netw.*, 2(1):1–14, Mar. 2016.

**10** A.C. Cirik, S. Biswas, O. Taghizadeh, and T. Ratnarajah. Robust transceiver design in full-duplex MIMO cognitive radios. *IEEE Trans. Veh. Technol.*, 67(2):1313–1330, Feb. 2018.

**11** B.P. Day, A.R. Margetts, D.W. Bliss, and P. Schniter. Full-duplex bidirectional MIMO: Achievable rates under limited dynamic range. *IEEE Trans. Signal Process.*, 60(7):3702–3713, Jul. 2012.

**12** M. Duarte, C. Dick, and A. Sabharwal. Experiment-driven characterization of full-duplex wireless systems. *IEEE Trans. Wireless Commun.*, 11(12):4296–4307, Dec. 2012.

**13** F. Gao, R. Zhang, Y. Liang, and Xiaodong Wang. Multi-antenna cognitive radio systems: Environmental learning and channel training. In *Proceedings of the IEEE ICASSP*, pages 2329–2332, Taipei, Taiwan, Apr. 2009.

**14** R. Hunger, M. Joham, and W. Utschick. On the MSE-duality of the broadcast channel and the multiple access channel. *IEEE Trans. Signal Process.*, 57(2):698–713, Feb. 2009.

**15** M. Jain, J. Il Choi, T. Kim, D. Bharadia, S. Seth, K. Srinivasan, P. Levis, S. Katti, and P. Sinha. Practical, real-time, full duplex wireless. In *Proceedings of the ACM MobiCom*, pages 301–312, Las Vegas, NV, Sep. 2011.

**16** H. Kim, S. Lim, H. Wang, and D. Hong. Optimal power allocation and outage analysis for cognitive full duplex relay systems. *IEEE Trans. Wireless Commun.*, 11(10):3754–3765, Oct. 2012.

**17** T.M. Kim, H.J. Yang, and A.J. Paulraj. Distributed sum-rate optimization for full-duplex MIMO system under limited dynamic range. *IEEE Signal Process. Lett.*, 20(6):555–558, Jun. 2013.

**18** W. Li, J. Lilleberg, and K. Rikkinen. On rate region analysis of half- and full-duplex OFDM communication links. *IEEE J. Sel. Areas Commun.*, 32 (9):1688–1698, Sep. 2014.

**19** Y. Liao, L. Song, Z. Han, and Y. Li. Full duplex cognitive radio: a new design paradigm for enhancing spectrum usage. *IEEE Commun. Mag.*, 53 (5):138–145, May 2015.

**20** D. Nguyen, L. Tran, P. Pirinen, and M. Latva-aho. On the spectral efficiency of full-duplex small cell wireless systems. *IEEE Trans. Wireless Commun.*, 13(9):4896–4910, Sep. 2014.

**21** D.P. Palomar, J.M. Cioffi, and M.A. Lagunas. Joint Tx-Rx beamforming design for multicarrier MIMO channels: a unified framework for convex optimization. *IEEE Trans. Signal Process.*, 51(9):2381–2401, Sep. 2003.

**22** K.T. Phan, S.A. Vorobyov, N.D. Sidiropoulos, and C. Tellambura. Spectrum sharing in wireless networks via QoS-aware secondary multicast beamforming. *IEEE Trans. Signal Process.*, 57(6):2323–2335, Jun. 2009.

**23** N. Singh, D. Gunawardena, A. Proutiere, B. Radunovi, H.V. Balan, and P. Key. Efficient and fair MAC for wireless networks with self-interference cancellation. In *Proc. Int. Symp. WiOpt*, pages 94–101, Princeton, NJ, May 2011.

**24** J. Wang, G. Scutari, and D.P. Palomar. Robust MIMO cognitive radio via game theory. *IEEE Trans. Signal Process.*, 59(3):1183–1201, Mar. 2011.

**25** J. Xue, S. Biswas, A.C. Cirik, H. Du, Y. Yang, T. Ratnarajah, and M. Sellathurai. Transceiver design of optimum wirelessly powered full-duplex MIMO IoT devices. *IEEE Trans. Commun.*, 66(5):1955–1969, May 2018.

**26** L. Zhang, Y. Liang, Y. Xin, and H.V. Poor. Robust cognitive beamforming with partial channel state information. *IEEE Trans. Wireless Commun.*, 8(8):4143–4153, Aug. 2009.

# Appendix for Chapter 10

## 10.A.1 Useful lemmas

**Lemma 1** Given matrices $\mathbf{P}$, $\mathbf{Q}$, $\mathbf{A}$ with $\mathbf{A} = \mathbf{A}^H$, the semi-infinite LMI of the form of

$$\mathbf{A} \geq \mathbf{P}^H \mathbf{X} \mathbf{Q} + \mathbf{Q}^H \mathbf{X}^H \mathbf{P}, \qquad \forall \mathbf{X} : \|\mathbf{X}\|_F \leq \rho, \tag{10.A.1}$$

holds if and only if $\exists \epsilon \geq 0$ such that

$$\begin{bmatrix} \mathbf{A} - \epsilon \mathbf{Q}^H \mathbf{Q} & -\rho \mathbf{P}^H \\ -\rho \mathbf{P} & \epsilon \mathbf{I} \end{bmatrix} \geq 0. \tag{10.A.2}$$

**Lemma 2**   Given $N \times N$ Hermitian matrices $\mathbf{D}$, $\mathbf{A}$; $N \times 1$ vector $\mathbf{b}$, and the scalars $c, e$, there exists an $\bar{\mathbf{x}}$ satisfying $\bar{\mathbf{x}}^H \mathbf{D} \bar{\mathbf{x}} < e$. Then the inequality

$$\mathbf{x}^H \mathbf{A} \mathbf{x} + 2\Re\{\mathbf{b}^H \mathbf{x}\} + c \geq 0, \ \forall \mathbf{x}^H \mathbf{D} \mathbf{x} \leq e \tag{10.A.3}$$

holds if and only if $\exists \epsilon \geq 0$ such that

$$\begin{bmatrix} \epsilon \mathbf{D} + \mathbf{A} & \mathbf{b} \\ \mathbf{b}^H & c - e\epsilon \end{bmatrix} \geq \mathbf{0}. \tag{10.A.4}$$

# 11

## Communication and Radar Systems: Spectral Coexistence and Beyond

*Fan Liu and Christos Masouros*

University College London, UK

## 11.1 Background and Applications

Having developed for almost a century, radar systems are now deployed in various frequency bands worldwide, with extensive usage in environment sensing, navigation, surveillance and localization, etc. Below 6 GHz, radar applications are allocated primary use of a significant portion of the spectrum at the time of writing, e.g. airborne navigation radars close to the 3.4 GHz band, shipborne and vessel traffic service (VTS) radar at 5.6 GHz. It is worth noting that these bands have seen increasing cohabitation with commercial wireless systems such as long-term evolution (LTE) and Wi-Fi [1]. With the allocation of the available spectrum to newer communication technologies, the interference in radar bands is on the rise and has raised concerns from governmental and military organizations on the safeguarding of critical radar operations. Accordingly, there is a rising interest in reliable solutions to enable the spectral coexistence of communication and radar transmission. As an emerging research topic, communication and radar spectrum sharing (CRSS) not only presents the advantage of enabling the efficient usage of the spectrum, but also provides a new way to designing novel systems that can benefit from the cooperation of the two functionalities. Below we briefly overview the application scenarios of CRSS from both the civilian and military aspects.

### 11.1.1 Civilian Applications

As one of its original motivations, a direct application of CRSS is the coexistence of the L-band air traffic control (ATC) radar and the frequency division duplex (FDD)-LTE base station (BS), both of which are deployed in the 1.3 GHz band [2]. A similar case holds for the coexistence in the S-band of ATC radars and 802.11 wireless local area network (WLAN) networks [3]. It is expected that by the shared use of the above bands spectral congestion can be significantly eased and interference can be reduced. More recently, the rapidly growing vehicle-to-everything (V2X) network calls for the design of joint communication and sensing techniques, via which vehicles can communicate with each other and sense the traffic environment simultaneously. Typically, such systems are required to operate in

*Spectrum Sharing: The Next Frontier in Wireless Networks,* First Edition.
Edited by Constantinos B. Papadias, Tharmalingam Ratnarajah, and Dirk T.M. Slock.
© 2020 John Wiley & Sons Ltd. Published 2020 by John Wiley & Sons Ltd.

the millimeter-wave (mmWave) band. Indeed, existing research investigates the feasibility for generating radar probing waveforms based on the IEEE 802.11ad protocol, which is a WLAN standard operating at 60 GHz [4]. The idea is in fact not new, since similar techniques have been well-studied in the area of WLAN-based indoor positioning in light of the 802.11n and 802.11ac standards [5]. As another important application scenario, radio frequency identification (RFID) technology, has to some extent integrated remote sensing and backscatter communication into the system design [6]. Besides this research, it is also interesting to highlight that the CRSS might even find use in the medical field. The deeply embedded bio-sensors only support low-power sensing, and thus the measured data needs to be transmitted to external devices for further processing, where joint sensing and communication techniques are naturally required [7].

### 11.1.2 Military Applications

The biggest support of CRSS technologies to date has come from the US Armed Forces, who have launched several projects for the corresponding investigations. These projects aim at not only maintaining military radar coverage (e.g., the shipborne radar AN/SPN-43C) with the presence of coexisting civilian wireless systems (e.g., the 3.5-GHz LTE systems), but also the co-design of novel systems from the ground up, which motivates the study of the military dual-functional radar communication (DFRC) platform [8]. To realize this, one can either implement the target detection/estimation functions on a communication system (comm-centric DFRC) or, conversely, transmit useful information by a radar (radar-centric DFRC). Note that the two design philosophies might lead to completely different applications. For instance, it is possible to transform cellular BSs into low-power radars, which can monitor ground traffic and unmanned aerial vehicles (UAVs) while offering wireless communication services to the user equipment (UE) [9]. By such modifications, the future ultra-dense network (UDN) with a large number of cooperative micro BSs (see Chapter 8) can be exploited as the urban air defense system, which performs early warning and surveillance of incoming UAVs and threats. Given the high transmission power and strong directionality of military radar, there are also a number of interesting applications of radar-centric DFRC, e.g. using radar as a communication relay [10] or to embed confidential information into the radar probing waveforms, which allows for low-probability-of-intercept (LPI) communications [11]. Finally, we note that from a broader viewpoint the passive radar that exploits cellular/TV signals as probing waveforms can be also regarded as a type of DFRC system.

## 11.2 Radar Basics

Complementary to the communication basics detailed in previous chapters, let us briefly revisit the essential concepts for radar systems, which will be useful for the discussion of the CRSS methods.

Depending on the signaling strategies used, radar systems can be generally classified into pulsed radar and continuous wave (CW) radar. In this chapter, we focus on pulsed radar without loss of generality. As shown in Figure 11.1, in its simplest form the radar transmits a probing pulse that is known *a priori* and receives the echo wave reflected by the target. A

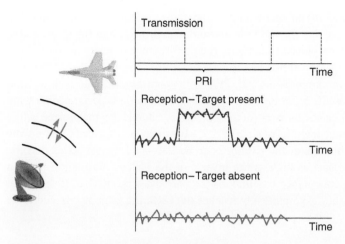

**Figure 11.1** Basic operations for a pulsed radar.

single transmission, together with the corresponding reception form the operation period of the radar called the *pulse repetition interval (PRI)*, where the ratio between the pulse length and the PRI is called the *duty cycle*. The received echo is then analyzed to extract the target information. To gain an intuitive impression, let us consider the example of a multi-antenna radar detecting a point-like target located in the far-field. The echo signal vector received by the radar at the $l$th epoch can be expressed as

$$\mathbf{y}_R[l] = \alpha_0 e^{j2\pi(l-1)f_d} \mathbf{v}_r(\theta_0) \mathbf{v}_t^T(\theta_0) \mathbf{s}_R[l - l_0] + \mathbf{y}_I[l] + \mathbf{n}_R[l], \forall l, \tag{11.1}$$

where $\mathbf{s}_R[l]$ is the radar transmitted signal, which is also called the $l$th *snapshot*, $\alpha_0$ is the complex path loss including the propagation loss and the radar cross-section (RCS) of the target, $l_0$ represents the time delay determined by the relative range from the target to the radar, $f_d$ denotes the target normalized Doppler frequency associated with its relative velocity, $\theta_0$ stands for the azimuth angle of the target with $\mathbf{v}_t(\theta)$ and $\mathbf{v}_r(\theta)$ being the transmit and receive steering vectors of the radar, respectively, and finally $\mathbf{y}_I[l]$ and $\mathbf{n}_R[l]$ are the signal-dependent clutter and the noise, where the clutter is typically composed of the reflections from obstacles and false targets from other directions. By processing the echo in (11.1), the radar aims to obtain the accurate estimates of the three motion parameters of the target, i.e. range, velocity, and azimuth angle. Nevertheless, it is possible for the radar to receive nothing but interference plus noise when it is listening to the target return, in which case the estimates will be ineffective. To resolve this issue, the radar needs to detect whether a target is present. This is equivalent to the following binary hypothesis testing (HT) problem:

$$\mathbf{y}_R[l] = \begin{cases} \mathcal{H}_0 : \mathbf{y}_I[l] + \mathbf{n}_R[l], \forall l, \\ \mathcal{H}_1 : \alpha_0 e^{j2\pi(l-1)f_d} \mathbf{v}_r(\theta_0) \mathbf{v}_t^T(\theta_0) \mathbf{s}_R[l - l_0] + \mathbf{y}_I[l] + \mathbf{n}_R[l], \forall l, \end{cases} \tag{11.2}$$

To solve this HT problem, a detector $T(\cdot)$ is designed to map the received signal to a real number, which is then compared with a given threshold $\gamma$ to determine which hypothesis to choose. This is expressed as

$$T(\mathbf{y}_R) \underset{\mathcal{H}_0}{\overset{\mathcal{H}_1}{\gtrless}} \gamma. \tag{11.3}$$

There are a number of figures of merit for the radar being defined by the academia, wherein the *detection probability* $P_D$, the false-alarm probability $P_{FA}$ and the *Cramér–Rao bound (CRB)* are typically employed to measure the performance of the detector and the estimator. To be specific, $P_D$ is defined as the probability that the radar detects a target while $\mathcal{H}_1$ holds true, i.e. the target is present. $P_{FA}$, on the contrary, is the probability that the radar detects a target while $\mathcal{H}_0$ holds true. Typically, the detector is designed following the criterion of constant false-alarm rate (CFAR), i.e. $P_D$ should be maximized while maintaining a constant and low $P_{FA}$ (e.g., $10^{-5}$)[1]. CRB expresses a lower bound on the variance of unbiased estimators of the parameter to be estimated, which states that the estimation variance is at least as high as the inverse of the Fisher information [12]. In practice, CRB acts as a performance baseline for the parameter estimation. If the mean squared error (MSE) reaches/asymptotically reaches the CRB, the estimator is said to be optimal/asymptotically optimal.

## 11.3 Radar Communication Coexistence

### 11.3.1 Opportunistic Access

From the perspective of the cognitive radio (CR), a naive coexistence scheme is the so-called *opportunistic spectrum sharing*. In such methods, the radar is regarded as the primary user (PU) in the band of interest, whereas the communication system, acting as the secondary user (SU), performs spectrum sensing to detect whether the radar is active and in the mean-time checks if the communication power exceeds the tolerable interference threshold of the radar. By doing so, a communication opportunity is gained whenever the spectrum is unoccupied [13]. Another option is to avoid the mutual interference by physically separating radar and LTE/Wi-Fi systems through large distances [14, 15]. Although fairly easily implemented in realistic scenarios, these schemes are unable to support a highly efficient use of the spectrum as they require either temporal or spatial isolations between the radar and the communication operations. In view of the above, and given the multi-antenna nature of modern radar and communication systems, a more powerful approach from a spectrum efficiency viewpoint would be to cancel the mutual interference by precoding designs.

### 11.3.2 Precoding Designs

#### 11.3.2.1 Interfering Channel Estimation

Before designing a precoder, most of the precoding techniques require interfering channel state information (ICSI), in our case between the radar and the communication systems, i.e. the channel through which the mutual interfering signals propagate. Conventionally, the radar and the communication systems periodically cooperate by transmitting training symbols to estimate the ICSI [16]. Nevertheless, this inevitably occupies extra computational and signaling resources of the radar. Moreover, since it is the cellular operator

---

1 Note that a false alarm may inflict more damage to the radar compared to losing a target, as it leads to unnecessary waste of radar's computational and energy resources.

who exploits the spectrum of the radar, it is the performance of the latter that should be primarily guaranteed, i.e. the radar resources should be allocated to target detection rather than obtaining the ICSI.

Towards this direction, a more practical line of work involves interfering channel estimation in the coexistence of a multiple input multiple output (MIMO) BS and a MIMO radar performing "search and track" [17]. As can be seen in Figure 11.2, in such working modes the radar randomly transmits searching or tracking waveforms during each PRI, and is assumed to be agnostic to the interference or even the operation of the BS while the latter is attempting to acquire the ICSI with limited information from the radar. Note that the communication Tx remains silent while it is listening to the radar transmission, and hence does not interfere with the radar Rx. Let us suppose that an $N$-antenna BS is receiving the interference from an $M$-antenna MIMO radar. The received signal matrix can be accordingly expressed as

$$\mathbf{Y} = \mathbf{G}\mathbf{S}_R + \mathbf{N}_C, \tag{11.4}$$

where $\mathbf{G} \in \mathbb{C}^{N \times M}$ is the interfering channel to be estimated, $\mathbf{S}_R \in \mathbb{C}^{M \times L}$ is the radar probing waveform matrix with $L$ being its length, and finally $\mathbf{N}_C \in \mathbb{C}^{N \times L}$ is the Gaussian noise matrix. Before estimating the channel, the BS needs to decide whether the radar is searching or tracking based on the received $\mathbf{Y}$. Denote the searching waveform matrix as $\mathbf{S}_0 \in \mathbb{C}^{M \times L}$. According to the MIMO radar literature, $\mathbf{S}_0$ is spatially orthogonal, which leads to omnidirectional transmission, and thus its sampled covariance matrix satisfies

$$\mathbf{R}_S = \frac{1}{L}\mathbf{S}_0\mathbf{S}_0{}^H = \frac{P_R}{M}\mathbf{I}_M, \tag{11.5}$$

where $P_R$ is the transmit power of the radar. As an omnidirectional searching waveform, there is no reason for $\mathbf{S}_0$ to vary. Indeed, in some cases the radar may only use a single fixed waveform for omni-searching. On the other hand, the tracking waveform, denoted as $\mathbf{S}_1 \in \mathbb{C}^{M \times L}$, forms a directional transmission that points to the angles of interest under the uniform linear array (ULA) geometry, and is likely to vary from pulse to pulse following the movement of the target. Therefore $\mathbf{S}_1$ is not spatially orthogonal anymore, i.e. its covariance matrix does not satisfy (11.5). Based on the above, it is reasonable to assume that the searching waveform $\mathbf{S}_0$ is known to the BS, which can be done by information exchange once off-line. On the other hand, it is not possible for the BS to know $\mathbf{S}_1$ a priori.

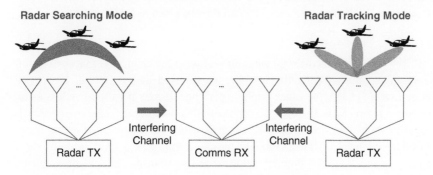

**Figure 11.2** "Search and track" MIMO radar coexists with the BS.

Consequently, the recognition for the radar waveforms is equivalent to the following HT problem:

$$\begin{aligned}\mathcal{H}_0 &: \mathbf{S}_R = \mathbf{S}_0, \mathbf{G}, \\ \mathcal{H}_1 &: \mathbf{S}_R \neq \mathbf{S}_0, \mathbf{G},\end{aligned} \tag{11.6}$$

where $\mathbf{G}$ is the so-called nuisance parameter [12]. This HT problem can be solved via the *Rao test*. According to [12, 17], the Rao detector is given by

$$T_R(\mathbf{Y}) = \frac{2}{\sigma^2} \mathrm{tr}\left(\left(\mathbf{I}_L - \frac{M}{LP_R}\mathbf{S}_0^H\mathbf{S}_0\right)\mathbf{Y}^H\mathbf{Y}\mathbf{S}_0^H(\mathbf{S}_0\mathbf{Y}^H\mathbf{Y}\mathbf{S}_0^H)^{-1}\mathbf{S}_0\mathbf{Y}^H\mathbf{Y}\right) \overset{\mathcal{H}_1}{\underset{\mathcal{H}_0}{\gtrless}} \gamma, \tag{11.7}$$

where $\gamma$ is a preset threshold. By performing the Rao test on $\mathbf{Y}$, the BS is able to determine if the radar is operating in search mode, i.e. whether $\mathbf{S}_0$ is transmitted in the current radar PRI. In that case, the BS could obtain the channel by the following least-squares estimator

$$\hat{\mathbf{G}}_0 = \mathbf{Y}\mathbf{S}_0^H(\mathbf{S}_0\mathbf{S}_0^H)^{-1} = \frac{M}{LP_R}\mathbf{Y}\mathbf{S}_0^H. \tag{11.8}$$

Otherwise, the BS is required to wait until an orthogonal waveform is transmitted by the radar.

In the case where the channel $\mathbf{G}$ is a rank-1 line-of-sight (LoS) channel, the Rao test is no longer effective. Nevertheless, it is possible for the BS to employ a simpler *energy detection* method to identify the radar working mode, given the differences in the radar's searching and tracking beampatterns. More details on this topic are provided in [17].

### 11.3.2.2 Closed-form Precoding

After the channel matrix is estimated, the precoding strategy can be designed either at the radar or the communication's side, where a simple idea to cancel the mutual interference is zero-forcing (ZF), or *null-space projection (NSP)* as it is called in [16]. Typically, such a design requires the ICSI at the radar, which needs to control the power of the interference generated to the BSs. The received interference signal at the BS is

$$\mathbf{C}_I = \mathbf{G}\mathbf{W}_R\mathbf{S}_R, \tag{11.9}$$

where $\mathbf{G}$ and $\mathbf{S}_R$ are defined as the interfering channel matrix and the radar probing waveform matrix as in (11.4), and $\mathbf{W}_R \in \mathbb{C}^{M \times M}$ denotes the radar precoder to be designed. To ensure zero-interference to the BS, it is necessary to have $\mathbf{G}\mathbf{W}_R\mathbf{S}_R = \mathbf{0}$, which suggests that each column of $\mathbf{W}_R\mathbf{S}_R$ falls into the null-space of channel $\mathbf{G}$. Let us denote the right singular matrix of $\mathbf{G}$ as $\mathbf{V}_G \in \mathbb{C}^{M \times M}$. The NSP precoder can thus be given as [18]

$$\mathbf{W}_R = \overline{\mathbf{V}}_G(\overline{\mathbf{V}}_G^H\overline{\mathbf{V}}_G)^{-1}\overline{\mathbf{V}}_G^H, \tag{11.10}$$

where $\overline{\mathbf{V}}_G \in \mathbb{C}^{M \times (M-N)}$ contains the right singular vectors of the channel $\mathbf{G}$ that correspond to its zero singular values. While the interference received at the BS will be strictly zero-forced by use of (11.10), the radar might experience serious performance loss as its probing waveform matrix $\mathbf{S}_R$ is distorted. To cope with this issue, a more effective approach [18] is to set a threshold $\lambda$ for the singular values of $\mathbf{G}$, which corresponds to the minimum tolerable interference level of the BS, and formulate a matrix $\overline{\mathbf{V}}_{G,\lambda}$ that contains right singular vectors associated with singular values that less than $\lambda$. This precoder can be obtained as [18]

$$\mathbf{W}_R = \overline{\mathbf{V}}_{G,\lambda}(\overline{\mathbf{V}}_{G,\lambda}^H\overline{\mathbf{V}}_{G,\lambda})^{-1}\overline{\mathbf{V}}_{G,\lambda}^H. \tag{11.11}$$

It is easy to see that when $\lambda \to 0$, the interference imposed on the BS will be zero. Conversely, if $\lambda \to \infty$, we have $\overline{\mathbf{V}}_{G,\lambda} \to \mathbf{V}_G$ and hence $\mathbf{W}_R \to \mathbf{I}_M$ as $\mathbf{V}_G$ is a unitary square matrix, which guarantees the best performance of the radar. By using the precoder in (11.11), a flexible performance tradeoff can be achieved between the radar and the BS.

### 11.3.2.3 Optimization-based Precoding

In order to optimize the system performance under controllable constraints, convex optimization techniques have been exploited for the coexistence of radar and communications. Recent research has addressed the scenario that a MIMO radar shares its spectrum with a multi-user MIMO (MU-MIMO) downlink system [19]. As depicted in Figure 11.3, let us suppose an $N$-antenna BS is communicating with $K$ single-antenna UEs, while an $M$-antenna MIMO radar is operating at the same frequency band. The radar received signal model in a given range-Doppler bin can be expressed as

$$\mathbf{y}_R[l] = \alpha_0 \mathbf{v}_r(\theta_0) \mathbf{v}_t^T(\theta_0) \mathbf{s}_R[l] + \mathbf{G}_1 \mathbf{x}_C[l] + \mathbf{n}_R[l], \forall l, \tag{11.12}$$

where the definitions of all the parameters follow those in previous sections, and $\mathbf{G}_1 \in \mathbb{C}^{M \times N}$ represents the channel matrix from communication Tx to the radar. The second term of (11.12) is the interference generated by the BS, which is assumed to be the only interference received by radar. The radar probing signals $\mathbf{s}_R[l], \forall l$ are unprecoded and are spatially orthogonal with each other, i.e. $\frac{1}{L} \sum_{l=1}^{L} \mathbf{s}_R[l] \mathbf{s}_R^H[l] = \frac{P_R}{M} \mathbf{I}_M$. Therefore, the precoding design will be performed only at the communication's side to overcome radar's interference. The received symbol for the $i$th UE at the $l$th time epoch is given by

$$y_{C,i}[l] = \mathbf{h}_i^T \sum_{i=1}^{K} \mathbf{w}_i s_{C,i}[l] + \mathbf{g}_{2,i}^T \mathbf{s}_R[l] + n_{C,i}[l]$$

$$= \mathbf{h}_i^T \mathbf{w}_i s_{C,i}[l] + \mathbf{h}_i^T \sum_{\substack{k=1, \\ k \neq i}}^{K} \mathbf{w}_k s_{C,k}[l] + \mathbf{g}_{2,i}^T \mathbf{s}_R[l] + n_{C,i}[l], \forall i, \forall l, \tag{11.13}$$

where $\mathbf{g}_{2,i}$, $\mathbf{w}_i$, $s_{C,i}$, and $n_{C,i} \sim \mathcal{CN}(0, \sigma_C^2)$ stand for the interfering channel vector, the precoding vector, the transmitted symbol, and the received noise of the $i$th UE, respectively. It

**Figure 11.3** MIMO radar coexists with MU-MIMO downlink.

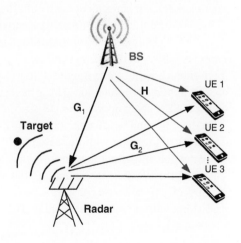

can be seen from (11.13) that the SINR of the UE is impacted not only by the multi-user interference (MUI) term, but also the interference from the radar. The SINR of the $i$th UE is therefore obtained as

$$\gamma_i = \frac{|\mathbf{h}_i^T \mathbf{w}_i|^2}{\sum_{k=1, k \neq i}^{K} |\mathbf{h}_i^T \mathbf{w}_k|^2 + \frac{P_R}{M} \|\mathbf{g}_{2,i}\|^2 + \sigma_C^2}, \forall i, \tag{11.14}$$

where $P_R$ is the radar transmit power as defined in (11.5), and the first term in the denominator represents the MUI. Based on (11.12)–(11.14), the detection probability $P_D$ of the MIMO radar can be maximized under the individual SINR requirements of the UEs and the transmit power budget $P_C$ of the BS, which leads to the following optimization problem [19]:

$$\mathcal{P}_1 : \max_{\mathbf{w}_i} P_D \text{ s.t. } \gamma_i \geq \Gamma_i, \forall i, \sum_{i=1}^{K} \text{tr}(\mathbf{w}_i \mathbf{w}_i^H) \leq P_C. \tag{11.15}$$

The objective function of (11.15) is non-concave, but can be relaxed as a convex lower bound. An approximated solution of (11.15) can then be obtained via the classic semi-definite relaxation (SDR) method [19].

By taking a closer look at (11.14), we see that the MUI term is regarded as harmful to all the UEs in the SINR constraint. To improve the performance of the BS in the above scenario, further research has employed the concept of *constructive interference* (CI) in the design of the precoders [19], where the CI is defined as the interference that pushes the received symbol away from the decision thresholds [20]. In contrast to conventional designs, the CI concept allows the BS to utilize the known MUI as a useful signal source, which benefits the symbol demodulation at the UEs. For phase shift keying (PSK), this involves replacing the SINR constraint of (11.15) by the following [19]

$$\left| \text{Im} \left( \mathbf{h}_i^T \sum_{k=1}^{K} \mathbf{w}_k e^{j(\phi_k - \phi_i)} \right) \right|$$
$$\leq \left( \text{Re} \left( \mathbf{h}_i^T \sum_{k=1}^{K} \mathbf{w}_k e^{j(\phi_k - \phi_i)} \right) - \Gamma_i \left( \frac{P_R}{M} \|\mathbf{g}_{2,i}\|^2 + \sigma_C^2 \right) \right) \tan \frac{\pi}{M_P}, \forall i, \tag{11.16}$$

where $\phi_k$ is the phase of the desired PSK symbol for the $k$th user, and $M_P$ is the PSK modulation order. For convenience, we omit the time index $l$. Note that (11.16) is nothing but a linear constraint, which is much easier to tackle compared to the quadratic constraints in (11.15) and yields a convex optimization problem that can be readily solved via numerical tools [19]. By taking advantage of the CI power, the BS is able to guarantee the downlink performance while generating less interference to the radar. As a consequence, both the detection and the estimation performances of the radar are significantly improved compared to SDR-based approaches [19].

In the following we demonstrate some numerical results in Figure 11.4, where the performance of the radar with increased SINR requirement at UEs is provided. In particular, the detection and estimation performance metrics are chosen as the $P_D$ and the root mean squared error (RMSE) for estimating the azimuth angle of the target, respectively. Without loss of generality, we consider a 10-antenna BS serving five single-antenna users while coexisting with a five-antenna radar. The three channels $\mathbf{H}$, $\mathbf{G}_1$, and $\mathbf{G}_2$ are assumed to

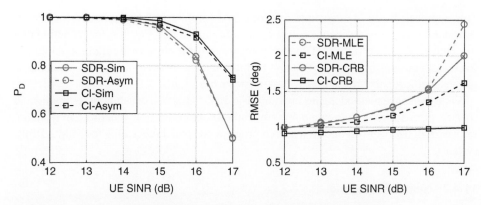

**Figure 11.4** Performance tradeoff between radar and communication systems.

be flat Rayleigh fading. Accordingly, the associated three large-scale fading factors are set as $\kappa_H = 1$ and $\kappa_{G,1} = \kappa_{G,2} = 10^{-3}$, respectively. The noise variance at UEs is assumed to be $\sigma_C^2 = 10^{-4}$. The MIMO radar transmits at a relative high power, e.g. 10 kW, while the BS power budget is 40 dBm. The receive SNR for the MIMO radar is fixed at 8 dB. For comparison, we employ both the SDR approach in (11.15) and its CI counterpart to design the communication precoder. For notational convenience, we use "Asym" and "Sim" for the theoretically asymptotic and the simulated $P_D$, and use "MLE" and "CRB" for the RMSE of the maximum likelihood estimator (MLE) and the associated Cramér–Rao bound. We see that for lower UE SINR requirements an improved radar performance can be achieved. As expected, by exploiting the known interference at the BS, the CI technique brings benefits to both radar target detection and estimation compared to the SDR approach.

## 11.4 Dual-functional Radar Communication Systems

### 11.4.1 Temporal and Spectral Processing

The earliest research on designing the DFRC waveform can be traced to the 1960s, when communication bits were modulated on radar pulses by classic pulse interval modulation (PIM) [21]. Such an integration can be also extended to the combination of the chirp signal, a waveform that is typically used for radar probing, and PSK modulations, where the quasi-orthogonality between the up and down chirp waveforms is exploited to differentiate 0 and 1 in the data sequence [22]. Recently, a simpler approach has been proposed based on the time-division (TD) framework, where the system operation period is divided between radar and communication time slots, respectively [23]. Such a DFRC system adopts up and down chirp signals for radar target detection, while allowing arbitrary modulation formats to be used for communication as this will not affect the radar operation.

In addition to designing a novel waveform from the ground up, an alternative approach would be to employ the existing communication signal as the radar probing waveform. The widely used orthogonal frequency division multiplexing (OFDM) signal is considered as a promising candidate for the DFRC, where the fast Fourier transform (FFT) and inverse FFT (IFFT) can be used to estimate the velocity and the range of the target, respectively [24].

This approach is able to decouple temporal and Doppler processing, and hence offers considerably better performance compared to other methodologies. It is worth noting that the sinusoidal carrier in the OFDM can be replaced by the chirp carrier [25]. Accordingly, fractional Fourier transform (FrFT) and its inverse are employed for signal processing as they use orthogonal chirp signals as the transform basis [26].

While the above methodologies provide basic dual-functional capabilities, temporal and spectral processing on their own offer only limited communication data rates. In what follows we will overview the recent research progress on spatial processing for MIMO DFRC systems.

### 11.4.2 Spatial Processing

In contrast to the conventional phased-array radar that transmits the phase-shifted versions of a benchmark signal on each element of the antenna array, MIMO radar transmits individual waveforms on each antenna, which offers the advantage of *waveform diversity*, allowing more degrees of freedom (DoFs) to be exploited for the system design. This is similar to the concept of *spatial multiplexing* for MIMO communications [27]. Given such a property, it is possible to embed communication bits into multiple radar waveforms, where a straightforward way is to transmit useful information by controlling the side lobes of the MIMO radar beampattern [28]. Suppose that an $N$-antenna communication RX is located at angle $\theta$ relative to an $M$-antenna MIMO DFRC Tx. Upon letting $\psi(t) = [\psi_1(t); \psi_2(t)...; \psi_Q(t)]$ be $Q$ orthogonal waveforms, the receive signal model can be written as

$$\mathbf{y}_C(t) = \beta \mathbf{b}(\theta)\, \mathbf{a}^T(\theta)\, \mathbf{W}_R \psi(t) + \mathbf{n}_C(t), \tag{11.17}$$

where $\mathbf{W}_R = [B_1 \mathbf{w}_1 + (1 - B_1)\mathbf{w}_0, ..., B_Q \mathbf{w}_1 + (1 - B_Q)\mathbf{w}_0] \in \mathbb{C}^{M \times Q}$ is the DFRC beamforming matrix with $B_q$ being the $q$th information bit, $\mathbf{w}_i, i = 0, 1$ the beamforming vectors associated with 0 and 1 data, $\beta$ is the path-loss factor, $\mathbf{a}(\theta)$ and $\mathbf{b}(\theta)$ are the transmit and receive steering vectors, and finally $\mathbf{n}_C(t)$ is the noise signal. By matched-filtering the received signal with the $q$th waveform, the resulting signal vector is

$$\mathbf{y}_q = \begin{cases} \beta \mathbf{b}(\theta)\, \mathbf{a}^T(\theta)\, \mathbf{w}_0 + \mathbf{z}_q, & B_q = 0, \\ \beta \mathbf{b}(\theta)\, \mathbf{a}^T(\theta)\, \mathbf{w}_1 + \mathbf{z}_q, & B_q = 1, \end{cases} \tag{11.18}$$

where $\mathbf{z}_q$ is the output noise at the $q$th matched filter. The beamforming vectors $\mathbf{w}_0$ and $\mathbf{w}_1$ are optimized by use of convex optimization methods, such that a given radar beampattern can be approximated with controllable sidelobe power at the angle $\theta$. After matched-filtering, a simple receive beamformer $\mathbf{b}^H(\theta)$ is applied to $\mathbf{y}_q$. By doing this, the $q$th information bit is interpreted as

$$\hat{B}_q = \begin{cases} 0, & \text{if } |\mathbf{b}^H(\theta)\mathbf{y}_q| \leq \gamma, \\ 1, & \text{if } |\mathbf{b}^H(\theta)\mathbf{y}_q| \geq \gamma, \end{cases} \tag{11.19}$$

where $\gamma$ is a preset threshold.

It is worth highlighting that the above technique relies on the assumption of a LoS channel between the DFRC and the communication users, which unfortunately restricts the use of the designed waveforms as a rich multi-path communication channel is more commonly seen in practical scenarios. Moreover, it should be noted that in these approaches one communication symbol is represented by one or several radar pulses, which leads to a low data

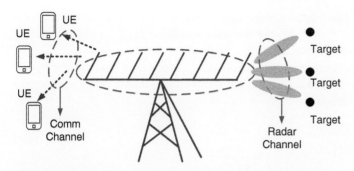

**Figure 11.5** MIMO dual-functional radar communication system.

rate in the order of the pulse repetition frequency (PRF) of the radar, e.g. kbits per second. To address these drawbacks, several waveform designs have been proposed for supporting simultaneous target detection and MU-MIMO downlink transmission [9]. Let us consider the scenario shown in Figure 11.5, in which an $M$-antenna MIMO DFRC system is detecting targets while serving $K$ single-antenna communication users. The communication channel is denoted as $\mathbf{H} \in \mathbb{C}^{K \times M}$, which can be LoS or NLoS. Given the desired constellation symbol matrix $\mathbf{D} \in \mathbb{C}^{K \times L}$ for the downlink users, the received signal matrix at the users is

$$\mathbf{Y} = \mathbf{D} + \underbrace{(\mathbf{HX} - \mathbf{D})}_{\text{MUI}} + \mathbf{N}_C, \tag{11.20}$$

where $\mathbf{X} = [\mathbf{x}_1, \mathbf{x}_2, ..., \mathbf{x}_L] \in \mathbb{C}^{M \times L}$ is the transmitted dual-functional signal matrix, with $L$ being the length of the communication frame/radar pulse, and $\mathbf{N}_C$ is the noise matrix. The second term in (11.20) represents the MUI signals. The total MUI energy can be measured as

$$P_{\text{MUI}} = \|\mathbf{HX} - \mathbf{D}\|_F^2. \tag{11.21}$$

It is understood that the sum-rate of the downlink users can be significantly improved by minimizing the MUI energy above [9]. Furthermore, note that the transmit beampattern of the MIMO radar can be expressed as

$$P_d(\theta) = \mathbf{a}^H(\theta) \mathbf{R}_X \mathbf{a}(\theta), \tag{11.22}$$

where $\mathbf{a}(\theta)$ is the ULA steering vector of the antenna array and $\mathbf{R}_X = \frac{1}{L}\mathbf{XX}^H$ is the spatial covariance matrix of $\mathbf{X}$. To formulate a desired radar beampattern while minimizing the communication MUI term, the following DFRC waveform design problem has been considered

$$\mathcal{P}_2 : \min_{\mathbf{X}} \|\mathbf{HX} - \mathbf{D}\|_F^2$$
$$s.t. \ \frac{1}{L}\mathbf{XX}^H = \mathbf{R}_d, \tag{11.23}$$

where $\mathbf{R}_d$ is a given covariance matrix that corresponds to a well-designed radar beampattern. In particular, when $\mathbf{R}_d = \frac{P_T}{M}\mathbf{I}_M$ with $P_T$ the total transmit power budget of the DFRC system, the designed waveform tends to be spatially orthogonal, which yields an omnidirectional beampattern. While the problem (11.23) is non-convex, it can be optimally solved in closed form by use of SVD. By contrast to the methods in [28], here each symbol of $\mathbf{X}$ is represented by a radar snapshot. Noting the fact that one radar pulse is typically composed of a number of snapshots, the designed waveform in (11.23) is expected to achieve a far better sum-rate performance.

To achieve a favorable performance tradeoff between radar and communication, one may also incorporate the radar-related constraints in the objective function. Let us consider the weighted optimization problem

$$\mathcal{P}_3 : \min_{\mathbf{X}} \rho \, \|\mathbf{HX} - \mathbf{D}\|_F^2 + (1 - \rho)\|\mathbf{X} - \mathbf{X}_0\|_F^2$$
$$\text{s.t.} \ \tfrac{1}{L}\|\mathbf{X}\|_F^2 = P_T, \tag{11.24}$$

where $0 \le \rho \le 1$ is a weighting factor that determines the weights for radar and communication performances and $\mathbf{X}_0$ is a benchmark radar signal matrix with desired properties, e.g. the optimal solution of (11.23). By imposing the equality power constraint, problem (11.24) is a non-convex quadratically constrained quadratical programming (QCQP). Fortunately, a globally optimal solution can be found via low-complexity algorithms [9].

To show the performance of the above waveform designs, we consider a 20-antenna DFRC system, which serves six single-antenna UEs while detecting three targets. The communication channel is subject to flat Rayleigh fading while the radar targets are located at $-60°, 0°$, and $60°$. The DFRC first transmits omnidirectional waveform to search potential targets, then formulates three beams pointing to the angles of interest by transmitting the directional waveform. It can be seen in the right half of Figure 11.6 that by solving the strict equality constrained problem (11.23) we obtain exactly the desired beam patterns, which are shown by solid lines. Nevertheless, the resulting communication sum-rates, shown as solid lines on the left-hand side, are relatively low. We then apply a small weighting factor $\rho = 0.2$ at the communication side in the tradeoff design (11.24). By doing this, the communication sum-rates increase significantly by approaching the zero MUI performance bound. In the meantime, the obtained radar beampatterns only experience slight performance loss, which is shown as dashed lines in the right half of the figure. The above results suggest that in the tradeoff waveform design the communication performance can be considerably improved by allowing small mismatches in the radar beampattern design.

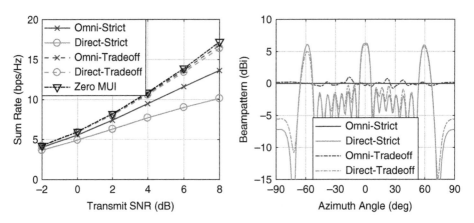

**Figure 11.6** Performance of the designed DFRC waveforms.

So far we have focused on designing a DFRC waveform without considering the *constant-modulus constraint (CMC)*, which is an important requirement for the radar system. This is because the radar typically transmits at its maximum available power, which requires the power amplifiers (PAs) to operate at the saturation region, i.e. the nonlinear region. The transmission of the constant-modulus waveform avoids signal distortion and thus leads to energy-efficient transmission.[2] As an example, we consider the following DFRC waveform optimization by imposing the CMC,

$$\mathcal{P}_4 : \min_{\mathbf{X}} \|\mathbf{HX} - \mathbf{D}\|_F^2$$
$$s.t \ \ \|\text{vec}\,(\mathbf{X} - \mathbf{X}_0)\|_\infty \le \eta,$$
$$|x_{i,l}| = \sqrt{\tfrac{P_T}{M}}, \forall i, l, \tag{11.25}$$

where $\mathbf{X}_0 \in \mathbb{C}^{M \times L}$ is a known benchmark radar signal matrix that has constant-modulus entries, e.g. chirp signals, vec $(\cdot)$ denotes the vectorization of a matrix, and $x_{i,l}$ is the $(i, l)$th entry of $\mathbf{X}$. The first constraint in (11.25) is called the *similarity constraint (SC)* and controls the difference between the designed waveform and the benchmark, with $\eta$ being the tolerable difference. Again, problem (11.25) is challenging due to the second non-convex CMC. Nevertheless, by employing a specifically tailored branch-and-bound (BnB) method, the globally optimal solution of (11.25) can be efficiently obtained [9].

## 11.5 Summary and Open Problems

This chapter has reviewed the latest developments in CRSS. While a number of contributions have been made towards both radar-communication coexistence and DFRC systems, the topic remains to be explored within a broader range of constraints and scenarios. Accordingly, there are numerous open problems in the area.

**Machine learning (ML) approaches for CRSS:** In CRSS scenarios, a key challenge is to distinguish between the target echoes and communication signals in the presence of noise and interference. Given the independent statistical characteristics of these two kinds of signals, it is advantageous to use ML techniques for signal classification, such as the independent component analysis (ICA). It is worth noting that these techniques could be applied to both coexistence and DFRC scenarios for designing the receivers.

**Physical layer security in CRSS:** Recent CRSS research has raised concerns of security and privacy issues. By sharing the spectrum with communication systems, military radar may unintentionally give away vital information to commercial users, or even worse to enemy eavesdroppers. To this end, physical layer security must be considered in CRSS scenarios, where a viable method is to let the radar actively transmit artificial noise to enemy targets to interfere with eavesdropping. Again, such schemes call for the design of novel beamforming/signaling approaches.

**Information theory aspects:** To gain further insight into the DFRC system, information theoretical analysis is indispensable for revealing the fundamental performance limits.

---

2 Note that CMC now becomes more and more important in 5G signaling. As the massive MIMO BS employs a huge number of RF chains for transmission, low-cost nonlinear PAs are deployed at the RF front-end and require CM transmission as well.

While some early contributions have been made towards the uplink systems, the downlink DFRC channel needs further investigation. Here the key point is to view radar targets as virtual energy receivers, and hence the DFRC transmission can be seen as the allocation of information and energy resources in the radar and communication channels. It is believed that such analysis could help us understand the intrinsic nature of the DFRC system and point us to essential system design criteria.

In the coming generations of wireless networks, where the frequency spectrum becomes one of the most valuable assets, communication and radar spectrum sharing will be an enabling solution that not only allows the efficient use of congested frequency bands, but also presents new designs of novel systems that are capable of accomplishing joint sensing and communication tasks. The concept of CRSS and the identified open problems provide scope for fruitful research for the years to come.

## References

1 H. Griffiths, L. Cohen, S. Watts, E. Mokole, C. Baker, M. Wicks, and S. Blunt. Radar spectrum engineering and management: Technical and regulatory issues. *Proc. IEEE*, 103(1):85–102, 2015.

2 H. Wang, J.T. Johnson, and C.J. Baker. Spectrum sharing between communications and ATC radar systems. *IET Radar Sonar Navig.*, 11(6): 994–1001, 2017.

3 F. Hessar and S. Roy. Spectrum sharing between a surveillance radar and secondary wi-fi networks. *IEEE Trans. Aerosp. Electron. Syst.*, 52(3): 1434–1448, 2016.

4 P. Kumari, J. Choi, N. González-Prelcic, and R.W. Heath. IEEE 802.11ad-based radar: An approach to joint vehicular communication-radar system. *IEEE Trans. Veh. Technol.*, 67(4):3012–3027, 2018.

5 C. Yang and H. Shao. Wifi-based indoor positioning. *IEEE Commun. Mag.*, 53(3):150–157, 2015.

6 N. Decarli, F. Guidi, and D. Dardari. A novel joint RFID and radar sensor network for passive localization: Design and performance bounds. *IEEE J. Sel. Topics Signal Process.*, 8(1):80–95, 2014.

7 G. Fortino, M. Pathan, and G. Di Fatta. BodyCloud: Integration of cloud computing and body sensor networks. In *4th IEEE International Conference on Cloud Computing Technology and Science Proceedings*, pages 851–856, 2012.

8 DARPA. Shared spectrum access for radar and communications (SSPARC), 2016.

9 F. Liu, L. Zhou, C. Masouros, A. Li, W. Luo, and A. Petropulu. Toward dual-functional radar-communication systems: Optimal waveform design. *IEEE Trans. Signal Process.*, 66(16):4264–4279, 2018.

10 D.W. Bliss. Cooperative radar and communications signaling: The estimation and information theory odd couple. In *Proc. IEEE Radar Conf.*, pages 0050–0055, 2014.

11 S.D. Blunt, J.G. Metcalf, C.R. Biggs, and E. Perrins. Performance characteristics and metrics for intra-pulse radar-embedded communication. *IEEE J. Sel. Areas Commun.*, 29(10):2057–2066, 2011.

**12** S. Kay and Z. Zhu. The complex parameter Rao test. *IEEE Trans. Signal Process.*, 64(24):6580–6588, 2016.

**13** R. Saruthirathanaworakun, J.M. Peha, and L.M. Correia. Opportunistic sharing between rotating radar and cellular. *IEEE J. Sel. Areas Commun.*, 30(10):1900–1910, 2012.

**14** F. Hessar and S. Roy. Spectrum sharing between a surveillance radar and secondary Wi-Fi networks. *IEEE Trans. Aerosp. Electron. Syst.*, 52(3): 1434–1448, 2016.

**15** H. Wang, J.T. Johnson, and C.J. Baker. Spectrum sharing between communications and ATC radar systems. *IET Radar Sonar Navig.*, 11(6): 994–1001, 2017.

**16** J.A. Mahal, A. Khawar, A. Abdelhadi, and T.C. Clancy. Spectral coexistence of MIMO radar and MIMO cellular system. *IEEE Trans. Aerosp. Electron. Syst.*, 53(2):655–668, 2017.

**17** F. Liu, A. Garcia-Rodriguez, C. Masouros and G. Geraci. Interfering channel estimation in radar-cellular coexistence: How much information do we need? *IEEE Trans. Wireless Commun.*, 18(9): 4238–4253, 2019.

**18** A. Babaei, W.H. Tranter, and T. Bose. A nullspace-based precoder with subspace expansion for radar/communications coexistence. In *Proc. IEEE Global Commun. Conf.*, pages 3487–3492, 2013.

**19** F. Liu, C. Masouros, A. Li, T. Ratnarajah, and J. Zhou. MIMO radar and cellular coexistence: A power-efficient approach enabled by interference exploitation. *IEEE Trans. Signal Process.*, 66(14):3681–3695, 2018.

**20** C. Masouros and G. Zheng. Exploiting known interference as green signal power for downlink beamforming optimization. *IEEE Trans. Signal Process.*, 63(14):3628–3640, 2015.

**21** R.M. Mealey. A method for calculating error probabilities in a radar communication system. *IEEE Trans. Space Electron. Telemetry*, 9(2): 37–42, 1963.

**22** G.N. Saddik, R.S. Singh, and E.R. Brown. Ultra-wideband multifunctional communications/radar system. *IEEE Trans. Microw. Theory Techn.*, 55(7):1431–1437, 2007.

**23** L. Han and K. Wu. Joint wireless communication and radar sensing systems-state of the art and future prospects. *IET Microw. Antennas Propag.*, 7(11):876–885, 2013.

**24** C. Sturm and W. Wiesbeck. Waveform design and signal processing aspects for fusion of wireless communications and radar sensing. *Proc. IEEE*, 99 (7):1236–1259, 2011.

**25** D. Gaglione, C. Clemente, C.V. Ilioudis, A.R. Persico, I.K. Proudler, and J.J. Soraghan. Fractional Fourier based waveform for a joint radar-communication system. In *Proc. IEEE Radar Conf.*, pages 1–6, 2016.

**26** L.B. Almeida. The fractional Fourier transform and time-frequency representations. *IEEE Trans. Signal Process.*, 42(11):3084–3091, 1994.

**27** D. Tse and P. Viswanath. *Fundamentals of Wireless Communication*. Cambridge University Press, 2005.

**28** A. Hassanien, M.G. Amin, Y.D. Zhang, and F. Ahmad. Dual-function radar-communications: Information embedding using sidelobe control and waveform diversity. *IEEE Trans. Signal Process.*, 64(8):2168–2181, 2016.

# 12

# The Role of Antenna Arrays in Spectrum Sharing

*Constantinos B. Papadias[1]\*, Konstantinos Ntougias[2], and Georgios K. Papageorgiou[3]*

[1] *The American College of Greece, Greece*
[2] *University of Cyprus, Cyprus*
[3] *Heriot Watt University, UK*

## 12.1 Introduction

The use of antenna arrays has been studied for almost a century now in a variety of wireless transmission and reception scenarios that range from military radar and target localization systems to TV and radio broadcasting systems and from a plethora of wireless networking setups that include all contemporary cellular, mobile, and local area broadband networks to a variety of automotive, fixed terrestrial, satellite, and space communication systems.

Key to the success of antenna arrays is their ability to shape radiation patterns (beams) that favor certain spatial directions (angles) over others. This feature enables them to handle co-channel interference (CCI) in the so-called spatial domain, thus giving rise to the spatial sharing of the wireless channel among different co-frequency signals. This is of major importance, since the traditional approach of separating the different transmissions in the time, frequency, or code domain to avoid the occurrence of CCI, which is attributed to the broadcast nature of radio transmissions and limits the throughput of wireless communication systems, results in poor utilization of the highly scarce and, as a consequence, expensive spectral resources.

In this chapter we are primarily interested in the role of antenna arrays in spectrum sharing, with emphasis on the spectrum sensing and shared spectrum access areas, for which specific representative examples are provided.

## 12.2 Spectrum Sharing

### 12.2.1 Spectrum Sharing from a Physical Viewpoint

Spectrum sharing is a term that has been broadly used in relation to spectrum access that goes beyond the conventional exclusive licensing paradigm, i.e. it refers to a scheme where two or more operators share a pool of spectral resources.

---

\* This work was performed when Dr. Papadias was with Athens Information Technology.

*Spectrum Sharing: The Next Frontier in Wireless Networks,* First Edition.
Edited by Constantinos B. Papadias, Tharmalingam Ratnarajah, and Dirk T.M. Slock.
© 2020 John Wiley & Sons Ltd. Published 2020 by John Wiley & Sons Ltd.

Before entering into the benefits that antenna array systems can offer to such a spectrum access method, we find it helpful to provide first a view of spectrum sharing that focuses on wireless transmission regardless of who is its originator, as defined below:

*Spectrum sharing, from a physical viewpoint, happens when different wireless transmitters access the same portion of the electromagnetic spectrum.*

This definition is agnostic to the nature of the wireless transmitters that share the spectrum, emphasizing instead the signal models that apply when spectrum sharing takes place. This should help better illuminate the role of antenna arrays in such contexts and alludes to their ability to assist other scenarios, including future use cases, that are beyond the scope of today's spectrum sharing systems or the focus of this chapter.

As per the aforementioned physical viewpoint definition, spectrum sharing may by applied in any of the following examples as a means to increase the spectral efficiency (SE):

- *When two nodes communicate with each other over a wireless link (duplexing)*: In contrast to the commonly utilized time-division/frequency-division duplex (TDD/FDD) schemes, where the downlink (DL) and uplink (UL) transmissions take place at different timeslots or make use of different frequencies, respectively, full-duplex (FD) communication represents a non-orthogonal duplexing paradigm wherein the same part of the spectrum is used simultaneously in both communication directions, thus resulting in higher SE. However, this aggressive duplexing method leads to severe self-interference that needs to be mitigated with advanced interference cancellation techniques [1].

- *When multiple co-located antennas send their signals into the same channel (MIMO transmission)*: By placing multiple antennas at the transmitting and receiving nodes of a wireless link, we can spatially multiplex different co-channel signals at the transmitter and separate them at the receiver via joint stream detection. When properly designed, such a multiple input multiple output (MIMO) communication system has the advantage of offering an important increase in the link's SE that grows linearly with the minimum number of antennas on each side of the link [2, 3]. The ability of the multi-antenna receiver to separate the different transmitted signals is due to the spatial degrees of freedom of its antenna array [4] and constitutes a key mechanism for spectrum sharing in broader contexts.

- *When multiple user terminals share the spectrum (multi-user MIMO transmission)*: The early efforts to multiplex the users of a cell in the space domain, as opposed to the spectrally less efficient orthogonal time-division/frequency-division/code-division multiple access (TDMA/FDMA/CDMA) approaches, were based on the shaping of beams to spatially direct the transmitted co-channel signals towards the intended terminals [5]. This space-division multiple access (SDMA) paradigm was met with skepticism, largely due to its inability to provide full orthogonality. Multi-user MIMO (MU-MIMO) addresses this issue by adding a precoding stage prior to transmission, thus enabling a multi-antenna base station (BS) to serve multiple terminals in the cellular DL on a single time-frequency resource over a so-called MIMO croadcast channel (BC) and at the same time alleviate the resulting intra-cell multi-user interference (MUI) or even prevent its occurrence [6]. A similar MIMO multiple access channel (MAC) setup applies to UL communication. In this case, multi-user detection takes place at the BS [7].

- *When multiple cells share the spectrum (multi-cell MU-MIMO transmission)*: According to the cellular principle, the same frequencies can be reused at cells that are sufficiently spaced to cope with the limited available spectrum and ensure at the same time that the inter-cell interference (ICI) that is caused by the concurrent transmission of co-channel signals at different cells is negligible [8]. The multi-cell extension of the MU-MIMO transmission concept facilitates more aggressive reuse of the available spectrum across the service area (even universal frequency reuse, which discards the need for using cellular planning as a means to control the ICI). Coordinated multi-point (CoMP) constitutes one variant of multi-cell MU-MIMO, where the BSs cooperate with each other to minimize or even eliminate both the intra-cell and the inter-cell CCI [9–12]. CoMP implementations in 4G long-term evolution advanced (LTE-A) networks mostly rely on centralized/cloud radio access network (RAN) setups to facilitate inter-BS communication [13, 14]. Massive MIMO represents the latest episode of multi-cell MU-MIMO, wherein the BSs are equipped with an excess (tens or even hundreds) of antennas to shape narrow (pencil-like) beams towards the intended receivers, thus reducing drastically both the intra-cell MUI and the ICI without the need for coordination among the BSs [15–17]. Massive MIMO implementations are commonly based on hybrid analog-digital transceivers to reduce the required hardware, power dissipation, and cost [18].

### 12.2.2  Spectrum Sharing from a Regulatory Viewpoint

The physical viewpoint of spectrum sharing that was introduced above departs from the conventional regulatory viewpoint, within which spectrum sharing is defined as follows:

*Spectrum sharing, from a regulatory viewpoint, happens when different users (i.e., operators) access the same portion of the electromagnetic spectrum.*

Based on the type of (or lack thereof) license possessed by these "players" for accessing the licensed part of the spectrum, we discern three types of users:

1. Operators with a license that provides them exclusive access rights.
2. Operators with a license that provides them limited spectrum access rights (e.g., subject to geographic restrictions or only valid during specific events).
3. Operators without any license to access the spectral segment of interest.

The first category refers to incumbents which lease an expensive multi-year license from their government/national regulatory authority (NRA) to operate in a given frequency spectrum, often at national level, on an exclusive use basis. These include both traditional cellular operators and various types of non-cellular operators, such as military, police, fire departments, and other civil agencies. The second category includes small-scale secondary cellular broadband access operators who provide service in a limited geographic area (e.g., within a municipality, an airport terminal, or an isolated island) and program making and special events (PMSE) operators who typically establish temporary networks for video and audio transmission during special short-term events (such as concerts, sports games, political rallies, etc.). Since these operators cannot afford to purchase an exclusive-use spectrum license, they typically sublet spectrum from incumbents under some spectrum usage rules that guarantee the protection of both "players" from harmful interference (e.g., as per the licensed shared access (LSA) model, see [19–23]). The third category consists of secondary operators that access the licensed spectrum of incumbents either orthogonally or in a non-orthogonal manner, i.e. either by detecting and subsequently exploiting vacant spectral slots or by using the same time-frequency resources with the incumbent but maintaining the power level of the resulting CCI below a predefined threshold. Of course the third category could also include "pirate" or other misbehaving users who deliberately or unwillingly, respectively, cause interference to the rightful owners of the spectrum.

We should add that unlicensed spectrum is another domain wherein spectrum sharing takes place. The regulation in this case consists of only some transmission power limitations and possibly some simple channel access directives, such as "listen before talk" (LBT). Characteristic examples include the various standards for the coexistence of LTE/LTE-A and Wi-Fi networks [24–27].

In this chapter we are mainly interested in the sharing of licensed spectrum among different "players". Defining as a primary user (PU) the operator who has an exclusive-use spectrum license and as a secondary user (SU) an operator without such license, there are three spectrum sharing paradigms within the regulatory framework [28, 29]:

1. *Interweave communication*: In this case, the SU is informed about the spectral activity at its vicinity by consulting a database, if such a registry has been made available by the NRA, or by using spectrum-sensing techniques, or via a combination of these methods to use channels that are not occupied at that time by any PU. The inability of any isolated receiver to sense reliably the spectral activity within the reach of its corresponding transmitter (such as in the well-known hidden node problem) has contributed significantly to the wide acceptance of the database approach, which was put forward by both Ofcom in the UK and the Federal Communications Commission (FCC) in the USA about a decade ago, over the spectrum-sensing approach, which was used in the early days of cognitive radio (CR) - see Chapters 1–4. Spectrum sensing is mainly used nowadays only as a means to further enhance/fine-tune the spectral activity data provided by the registry (see [30]).

2. *Underlay communication*: In this case, the SU transmits concurrently with the incumbent over the same frequency band, yet restricts itself to cause interference to incumbent receivers whose power is below a predefined threshold. This is also a paradigm that dates back to the first days of CR, giving rise to the "interference temperature" concept, i.e. a level of received interference power that can be accepted as being within tolerable noise levels. This implies usually that only low-power/short-range transmissions are allowed.

3. *Overlay communication*: This is a much more advanced concept for spectrum sharing between a secondary user and an incumbent user, so much so that it has yet to be adopted by any commercial entities: it prescribes that the secondary transmitter will coordinate its transmissions with those of the incumbent transmitter to provide sufficient spectral usage to both of them. From a mathematical point of view, such a configuration resembles the CoMP-based non-orthogonal multiple access case described in the previous subsection, which indeed provides higher SE than any uncoordinated approach. In practice, though, this would require close cooperation between the "players" (something that especially incumbents tend to avoid) and would also favor mostly similar types of service (air interface protocols). It is, however, a model that has been given a lot of attention in the research literature due to the hidden spectrum opportunities it offers.

## 12.3 Attributes of Antenna Arrays

Antenna arrays offer various gains in wireless communications, thanks to the spatial degrees of freedom that they provide. The key attributes of antenna arrays (which, as will be shown in the next section, can benefit spectrum sharing systems significantly) are summarized as follows [4]:

- *Signal power gain*: An antenna array has the ability to combine (coherently or not) the signals that impinge on or depart from its elements, thus shaping a beam which spatially focuses the radiated energy towards the direction of the intended receiver (main lobe) and reduces the radiated energy towards other directions (side lobes). Assuming an array with $N$ active elements, this results in a signal power gain of $N$. As the number of antenna elements grows, the beam gets narrower and the signal power gain increases accordingly. In the presence of no other artifact (e.g., fading or interference) apart from additive noise, which is assumed to be uncorrelated between the antenna elements, the signal power gain is equivalent to a signal-to-noise ratio (SNR) gain of $N$. This gain, in turn, is translated into a logarithmic gain of the considered link's capacity (equal to $\log N$ in the high-SNR regime or twice as much if a similar array of $N$ elements is used on the other side of the link as well).

- *Interference nulling ability*: An antenna array with $N$ active elements has also the ability to form up to $N - 1$ spatial nulls in certain directions. Our ability to choose these directions depends on the array topology itself (see [5]). Another way to quantify this gain is to compute the attained signal-to-interference-plus-noise ratio (SINR) when nulls are placed in the directions of the undesired interferers (or close to them).

- *Diversity gain*: In a fading environment, as is often the case in cellular communications, wherein the received signal strength may vary widely (e.g., by up to four orders of magnitude or 40 dB) depending on the receiver location, antenna arrays offer another unique opportunity: by combining signals received (or transmitted) from various antennas, they reduce the likelihood of a poor SNR due to the small correlation of the desired signal across neighboring antenna elements that are sufficiently spaced out (on the order of a few wavelengths or more). The gain that is attained in such cases over the average SNR of a corresponding single antenna link is called diversity gain and manifests itself as a reduction of the low tails of the SNR distribution. In the extreme asymptotic case of infinite receive antennas, the SNR distribution dully hardens, thus converting the random fading channel into a deterministic one.

- *Spatial multiplexing gain*: Alternatively, the $N$ degrees of freedom of an $N$-element antenna array can be used to reliably separate $K < N$ desired signals and null $N - K$ interfering ones. This property is behind the MIMO concept: by placing $M$ antennas at the transmitter and $N$ at the receiver, as correctly pointed out early on by Jack Winters in [31], the degrees of freedom under a spatial multiplexing operation can be as many as $\min(M, N)$ (for a full rank channel). That is, it is possible to transmit simultaneously $M \leq N$ data signals and separate them at the receiver, thus obtaining a linear SE gain equal to $\min(M, N)$, as shown by Telatar [2] and Foschini and Gans [3] (and earlier alluded to by Paulraj and Kailath in [32] with emphasis on co-located reception but distributed transmission), which goes way beyond the logarithmic gains achieved without spatial multiplexing.

## 12.4    Impact of Arrays on Spectrum Sharing

The antenna array attributes listed above are particularly useful in enabling and/or enhancing spectrum sharing in a variety of setups within the three standard paradigms mentioned in section 12.2.2. In the following we will briefly list the key mechanisms for this by considering separately the spectrum sensing and the shared spectrum access tasks. It should also be noted, even though this is beyond the scope of the chapter, that antenna arrays can facilitate in addition the detection of misbehaving nodes due to their higher spatial resolution and positioning capability (e.g., see [33]).

### 12.4.1    Spectrum Sensing

The key attributes offered by a set of collaborating (either co-located or distributed) antennas for spectrum sensing are listed below.

- *Better single-node sensing*: A node that is equipped with an antenna array can better sense the presence of activity in its intended spectrum in comparison to a single-antenna node. This is mainly due to the array gain, which enables the detection of transmitters whose power is too weak to be sensed against the background noise and interference by a single-antenna node (e.g., in hidden node scenarios), and the diversity gain, which allows sensing of signals that are severely faded in some of the antenna elements due to multipath propagation.
- *Two-dimensional (2D) sensing*: The directional sensing described above allows a 2D sensing in frequency/space (e.g., angular) domain to be performed. This is crucially important as it unveils a new dimension that is available for spectrum sharing. Hence, for example, interweave communication can be described under this context not in terms of orthogonal frequency segments but in the more general sense of disjointed frequency-angle segments ("slices"). Furthermore, such 2D sensing provides location information about the sensed transmitters, thus allowing us to acquire a more accurate mapping of the interfering nodes in space (whereas in the single antenna case, this task is more challenging).
- *Better collaborative sensing*: In the case of a collaborative sensing network, the nodes are distributed over a certain geographic area. This setup already facilitates the sensing task by providing (i) diversity gain (due to the disparate locations), (ii) better immunity to hidden-node problems (for the same reason), and (iii) better detection due to the increased dimensionality brought by the collaboration between the nodes. Adding antenna arrays to these nodes provides the further benefits of higher power gain per node, increased dimensionality of the received signal, and better localization capability.

### 12.4.2    Shared Spectrum Access

The key attributes offered by a set of collaborating antennas for shared spectrum access are listed below.

- *Interweave communication*: By forming nulls towards the directions of interferers, antenna arrays can, in principle, fully enable interweave communication solely in the spatial domain (or enhance it if orthogonality is partially achieved in the time or frequency domain).

- *Underlay communication*: By shaping radiation patterns that maximize the signal power in desired directions (e.g., secondary receivers) and minimize it in undesired directions (e.g., interfered primary receivers), antenna arrays can enable underlay communication in the spatial domain (or enhance it if other domains are used as well).
- *Overlay communication*: The cooperative transmission between primary and secondary transmitters in order to protect the PU from harmful interference and possibly meet some quality-of-service (QoS) target for the SU as well (e.g., under the CoMP or the interference alignment (IA) [34] paradigm) can be greatly enhanced when using antenna arrays by taking full advantage of their spatial degrees of freedom.

In the next two sections we provide representative examples of how antenna arrays can improve cooperative spectrum sharing and collaborative spectrum sensing.

## 12.5 Antenna-Array-Aided Spectrum Sharing

The use of antenna arrays as a means to allow the concurrent exploitation of a spectrum segment of interest by two or more operators has been studied manifold in the past, but mainly in limited setups where the SU has a single transmitter (e.g., see [35–39]).

The advanced resource allocation (RA) and interference management features of CoMP imply that this technology can act as an enabler of underlay spectrum sharing between multi-cell networks and incumbents that will provide substantial SE gains and meet the QoS requirements of both "players". Such a setup can also benefit from the cooperation between the two systems in terms of the sharing of control information such as the interference power threshold (IPT) level, channel state information (CSI), etc. (e.g., see the demo [40] that took place in the IEEE DySPAN 2015 5G Spectrum Sharing Challenge).

In this section, we present a simple RA scheme for coordinated beamforming (CBF) transmission in an underlay spectrum sharing setup, where the cooperating BSs exchange CSI and control information to serve disjoint groups of users in a coordinated manner and are informed by the incumbent about its IPT and the CSI of the corresponding cross channels, so that the sum-SE of the cellular network is maximized and the incumbent is protected from harmful interference. This work can form the basis for the derivation of low-complexity cooperative techniques that can be applied at the complex spectrum sharing setups of future networks, such as the ones that are considered in the European Commission's H2020 project PAINLESS (see http://painless-itn.com/) which involve portable access points [41].

### 12.5.1 System Setup

We consider an underlay spectrum sharing setup where the primary system (PS) is a single input single output (SISO) radio link and the secondary system (SS) is a network with $M$ cells. In each cell there is a BS with $N$ transmit antennas and $K \leq N$ single-antenna mobile stations (MS) that request service. Thus, there are $N_T = MN$ BS antennas and $K_T = MK$ single-antenna MSs (mobile users) in total. We denote the $m$th BS as $BS_m$ and the $k$th MS

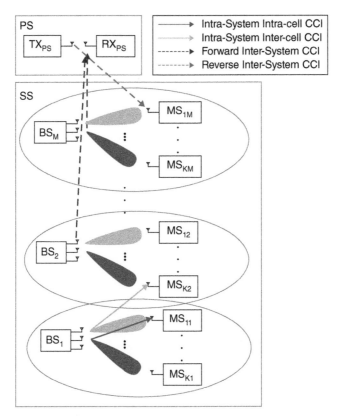

**Figure 12.1** System setup, notation, and types of interference.

in the $m$th cell as $MS_{km}$, $m \in \mathcal{M} = \{1, \ldots, M\}$, $k \in \mathcal{K} = \{1, \ldots, K\}$. The transmitter and the receiver of the PS are denoted simply as $TX_{PS}$ and $RX_{PS}$, respectively.

We identify four types of CCI: intra-system intra-cell CCI and intra-system inter-cell CCI, which are attributed to the transmissions of the BSs and are received by the MSs, forward inter-system (FIS) CCI, which is caused by the transmissions of the BSs and is received by $RX_{PS}$, and reverse inter-system (RIS) CCI, which occurs due to the transmission of $TX_{PS}$ and is received by the MSs.

Figure 12.1 illustrates the system setup, the notation, and the various types of interference.

### 12.5.2 Assumptions

For the above setup, we make the following assumptions:

- All transmissions are narrowband. Also, the SS utilizes universal frequency reuse (i.e., the same frequency is used across all cells).
- For convenience and without loss of generality, we focus on a single cooperation cluster, i.e. we assume that all BSs belong to the same cooperation cluster.
- Centralized CoMP is adopted, i.e. the RA decisions are taken by a master BS/central unit (CU) based on global CSI.

- The BSs have perfect knowledge of all the direct and cross channels within the SS as well as of the cross channels with the PS. Perfect CSI is also available at the nodes of the PS about their direct channel.
- Inter-BS communication is supported by an ideal transport network.
- Linear precoding is utilized at the SS due to its good balance between performance and computational demands.
- The frequency-flat quasi-static i.i.d. Rayleigh channel model is adopted.
- The samples of each signal or noise process are uncorrelated with each other. The zero-mean random processes are uncorrelated with each other as well.
- The MSs perform single-user detection and handle the RIS as additional noise. Moreover, they pass the composite received signal through a whitening filter.

### 12.5.3 System Model

The various direct and cross channels in the considered spectrum sharing setup are denoted as shown in Table 12.1. We note that when there are two lower indexes, the receiver is the corresponding MS. If there is also an upper index, it refers to the corresponding BS (type A direct channel), otherwise the transmitter is $TX_{PS}$ (type A cross channel). Notice that in these cases we use the symbol $h$ for the channel. On the other hand, when there is a single lower index or no index at all, the receiver is $RX_{PS}$. In the former case, the index refers to the corresponding BS (type B cross channel), whereas in the latter one the transmitter is $TX_{PS}$ (type B direct channel). Notice that in these cases we use the symbol $g$ for the channel.

The same indexing rules apply for the notation of the BF vectors, allocated powers, etc., but the order of the lower indexes is reversed, e.g. $\mathbf{w}_{mk}^{j}$ is the BF vector of $BS^{j}$ that is associated with $MS_{km}$.

#### 12.5.3.1 Secondary System
The SS is modeled as a MIMO interference broadcast channel (IBC) consisting of $M$ MIMO BCs formed between an $N$-antenna BS and $K$ single-antenna MSs each.

The complex baseband representation of the received signal at $MS_{km}$, $y_{km} \in \mathbb{C}$ ($k \in \mathcal{K}$; $m \in \mathcal{M}$), is [42]

$$y_{km} = \sum_{j=1}^{M} \sum_{l=1}^{K} (\mathbf{h}_{km}^{j})^{H} \mathbf{v}_{jl}^{j} + h_{km} \sqrt{P}d + n_{km}, \qquad (12.1)$$

**Table 12.1** Channel notation [42]

| Type of channel | Notation | Description |
| --- | --- | --- |
| Type A direct channel | $\mathbf{h}_{km}^{j}$ | Channel between $MS_{km}$ and $BS_{j}$ |
| Type A cross channel | $h_{km}$ | Channel between $MS_{km}$ and $TX_{PS}$ |
| Type B cross channel | $\mathbf{g}_{m}$ | Channel between $RX_{PS}$ and $BS_{m}$ |
| Type B direct channel | $g$ | Channel between $RX_{PS}$ and $TX_{PS}$ |

where the sample index has been omitted for convenience and $\mathbf{v}_{jl}^j \in \mathbb{C}^N$ is expressed as

$$\mathbf{v}_{jl}^j = \mathbf{w}_{jl}^j \sqrt{P_{jl}^j} s_{jl}^j. \tag{12.2}$$

The first term on the right-hand-side (RHS) of (12.1), which we will call $\breve{y}_{km}$ for convenience, can be decomposed into the sum of a data component, an intra-cell MUI component, and an ICI component caused by BS$_m$ serving MS$_{km}$, BS$_m$ serving the remaining $(K-1)$ users of the $m$th cell, and the remaining $(M-1)$ BSs serving each the $K$ users of its own cell, respectively, as follows [42]:

$$\breve{y}_{km} = (\mathbf{h}_{km}^m)^H \mathbf{v}_{mk}^m + \sum_{\substack{l=1 \\ i \neq k}}^{K} (\mathbf{h}_{km}^m)^H \mathbf{v}_{mi}^m + \sum_{\substack{j=1 \\ j \neq m}}^{M} \sum_{l=1}^{K} \left(\mathbf{h}_{km}^j\right)^H \mathbf{v}_{jl}^j. \tag{12.3}$$

The other two terms on the RHS of (12.1) represent, from left to right, the RIS CCI and the additive white Gaussian noise (AWGN) at MS$_{km}$, respectively.

Regarding notation, $\mathbf{h}_{km}^m \sim \mathcal{CN}(\mathbf{0}_{N \times N}, \mathbf{I}_N)$ is the channel between MS$_{km}$ and BS$_m$, $\mathbf{w}_{mk}^m \in \mathbb{C}^N$ denotes the BF vector used by BS$_m$ to serve user MS$_{km}$, $P_{mk}^m \in \mathbb{R}_+$ is the transmission power that is allocated to MS$_{km}$ by BS$_m$, and $s_{mk}^m \in \mathbb{C}$ is the symbol that is transmitted from BS$_m$ to MS$_{km}$. $h_{km} \in \mathcal{CN}(0, 1)$ is the channel between MS$_{km}$ and TX$_{PS}$. $P \in \mathbb{R}_+$ is the transmission power of TX$_{PS}$ and $d \in \mathbb{C}$ is the symbol that is transmitted by TX$_{PS}$ to RX$_{PS}$. Finally, $n_{km} \sim \mathcal{CN}(0, 1)$ is the AWGN at MS$_{km}$. Note that the BF vectors are normalized to unit power, i.e. $\|\mathbf{w}_{mk}^m\|^2 = 1$.

### 12.5.3.2  Primary System

The complex baseband representation of the received signal at RX$_{PS}$, $y \in \mathbb{C}$, is given by [42]

$$y = g\sqrt{P}d + \sum_{m=1}^{M} \sum_{k=1}^{K} (\mathbf{g}_m)^H \mathbf{v}_{mk}^m + z. \tag{12.4}$$

The first term on the RHS of (12.4) is the useful data signal component that is received at RX$_{PS}$, where $g \sim \mathcal{CN}(0, 1)$ is the channel between RX$_{PS}$ and TX$_{PS}$. The second term is the FIS CCI at RX$_{PS}$, where $\mathbf{g}_m \sim \mathcal{CN}(\mathbf{0}_{N \times N}, \mathbf{I}_N)$ is the channel between RX$_{PS}$ and BS$_m$. Finally, the third term $z \sim \mathcal{CN}(0, 1)$ is the AWGN at RX$_{PS}$.

### 12.5.4  Problem Formulation

#### 12.5.4.1  Sum-SE, SE, and SINR

The instantaneous sum-SE of the SS, i.e. the instantaneous sum-rate (SR) of the SS per unit of spectral bandwidth is given by [7]

$$R = \sum_{m=1}^{M} \sum_{k=1}^{K} R_{km}, \tag{12.5}$$

where $R_{km}$ is the instantaneous SE of $MS_{km}$:

$$R_{km} = \log_2(1 + \gamma_{km}), \quad k \in \mathcal{K}; \ m \in \mathcal{M}. \tag{12.6}$$

In (12.6), $\gamma_{km}$ refers to the instantaneous SINR at $MS_{km}$, which is given by [42]

$$\gamma_{km} = \frac{|\mathbf{v}^m_{mk}|^2}{\displaystyle\sum_{\substack{i=1 \\ i \neq k}}^{K} |\mathbf{v}^m_{mi}|^2 + \sum_{\substack{j=1 \\ j \neq m}}^{M} \sum_{l=1}^{K} |\mathbf{v}^j_{jl}|^2 + |h_{km}|^2 P + 1}, \tag{12.7}$$

where

$$|\mathbf{v}^m_{mk}|^2 = |(\mathbf{h}^m_{km})^H \mathbf{w}^m_{mk}|^2 P^m_{mk}. \tag{12.8}$$

The nominator in (12.7) corresponds to the power of the data signal component that is received at $MS_{km}$, while the terms at the denominator represent, from left to right, the power of the intra-cell MUI, ICI, RIS CCI, and AWGN components at $MS_{km}$.

### 12.5.4.2 Transmission Constraints

***Non-negative transmission power constraints:*** Each transmission from a BS to one of its $K$ users should have non-negative power. This transmission constraint is expressed as [42]

$$P^m_{mk} \geq 0, \quad k \in \mathcal{K}; \ m \in \mathcal{M}. \tag{12.9}$$

***Sum transmission power constraints:*** The transmissions within the SS are subject to a sum-power constraint (SPC) per BS, i.e. each BS is allowed to serve its $K$ users with a total power that does not exceed a maximum value $P_T$. Therefore, we have [42]

$$\sum_{k=1}^{K} P^m_{mk} \leq P_T, \quad m \in \mathcal{M}. \tag{12.10}$$

***Interference power constraint:*** The operation of the SS is subject to an interference-power constraint (IPC), i.e. the total power of the FIS CCI component that is received at $RX_{PS}$ should not exceed a threshold $P_I$, which represents the tolerable interference power level. By defining

$$a^m_{mk} = |(\mathbf{g}_m)^H \mathbf{w}^m_{mk}|^2, \quad m \in \mathcal{M}; \ k \in \mathcal{K}, \tag{12.11}$$

we can express the IPC constraint as [42]

$$\sum_{m=1}^{M} \sum_{k=1}^{K} a^m_{mk} P^m_{mk} \leq P_I. \tag{12.12}$$

### 12.5.4.3 Original Optimization Problem

We consider the joint determination of the BF vectors and allocated powers that maximize the sum-SE of the cellular network (SS) under the transmission power constraints (TPCs) and the IPC presented in section 12.5.4.2:

$$\min_{\substack{\mathbf{w}_{mk}^m, \, P_{mk}^m \\ m \in \mathcal{M}, \, k \in \mathcal{K}}} \quad -R = -\sum_{m=1}^{M}\sum_{k=1}^{K}\log_2(1+\gamma_{km}) \tag{12.13a}$$

s.t.

$$P_{mk}^m \geq 0, \quad m \in \mathcal{M}; \, k \in \mathcal{K}. \tag{12.13b}$$

$$\sum_{k=1}^{K} P_{mk}^m \leq P_T, \quad m \in \mathcal{M}. \tag{12.13c}$$

$$\sum_{m=1}^{M}\sum_{k=1}^{K} a_{mk}^m P_{mk}^m \leq P_I. \tag{12.13d}$$

This optimization problem is non-convex due to the coupled interference components in the objective function (i.e., the intra-system CCI terms at the denominator of the SINR of $MS_{km}$ in the expression of the sum-SE) [42].

#### 12.5.4.4 Relaxed Optimization Problem

***Block matrix representation of CBF:*** By stacking together the received signals at all MSs in a vector and ignoring, for convenience and without loss of generality, the RIS CCI, we can express the CBF transmission as follows:

$$\mathbf{y}_{SS} = \mathbf{HW}(\mathbf{P}_{SS})^{1/2}\mathbf{s} + \mathbf{n}. \tag{12.14}$$

In (12.14), $\mathbf{y}_{SS} \in \mathbb{C}^{K_T}$, $\mathbf{s} \in \mathbb{C}^{K_T}$, and $\mathbf{n} \sim \mathcal{CN}(\mathbf{0}_{K_T \times K_T}, \mathbf{I}_{K_T})$ are the vectors of received symbols at the $K_T$ MSs, transmitted symbols by the $M$ BSs, and AWGN samples at the $K_T$ MSs, respectively.

The composite channel matrix $\mathbf{H}$ is expressed as a block matrix [42]:

$$\mathbf{H} = \begin{bmatrix} \mathbf{H}_1 \\ \vdots \\ \mathbf{H}_M \end{bmatrix}, \tag{12.15}$$

where $\mathbf{H}_m$ is a block matrix with size $K \times M$ blocks that holds the channels of all users in the $m$th cell with $BS_m$ ($m \in \mathcal{M}$). The $(i,j)$th block of $\mathbf{H}_m$ is the channel between $MS_{im}$ and $BS_j$, $(\mathbf{h}_{im}^j)^H \sim \mathcal{CN}(\mathbf{0}_{N \times N}, \mathbf{I}_N)$ ($i \in \mathcal{K}$; $j \in \mathcal{M}$), and the $i$th row holds the channels between $MS_{im}$ and each one of the $M$ BSs:

$$\mathbf{H}_m = \begin{bmatrix} (\mathbf{h}_{1m}^1)^H & \cdots & (\mathbf{h}_{1m}^M)^H \\ \vdots & \ddots & \vdots \\ (\mathbf{h}_{Km}^1)^H & \cdots & (\mathbf{h}_{Km}^M)^H \end{bmatrix}. \tag{12.16}$$

It becomes apparent that $\mathbf{H}$ can be expressed as a $K_T \times M$ array of blocks $(\mathbf{h}_{im}^j)^H$, i.e. as a $K_T \times N_T$ matrix with entries of the form $(\mathbf{h}_{km}^j)_n$ that represent the scalar channel between $MS_{km}$ and the $n$th antenna of $BS_j$.

The precoding matrix $\mathbf{W}$ is expressed as a block matrix [42]:

$$\mathbf{W} = \begin{bmatrix} \mathbf{W}_1 & \cdots & \mathbf{W}_M \end{bmatrix}, \tag{12.17}$$

where $\mathbf{W}_m$ is a block matrix with size $M \times K$ blocks that holds the BF vectors of all BSs for all users in the $m$th cell ($m \in \mathcal{M}$). The $(i,j)$th block of $\mathbf{W}_m$ is the BF vector of $BS_i$ for serving $MS_{jm}$, $\mathbf{w}_{mj}^i \in \mathbb{C}^N$ ($i \in \mathcal{M}$; $j \in \mathcal{K}$), and the $j$th column holds the BF vectors of all BSs for $MS_{jm}$:

$$\mathbf{W}_m = \begin{bmatrix} \mathbf{w}_{m1}^1 & \cdots & \mathbf{w}_{mK}^1 \\ \vdots & \ddots & \vdots \\ \mathbf{w}_{m1}^M & \cdots & \mathbf{w}_{mK}^m \end{bmatrix}. \tag{12.18}$$

Note that since in CBF each BS serves only its own cell's users, $\mathbf{w}_{mj}^i = \mathbf{0}_N$ for $m \neq i$. Notice also that $\mathbf{W}$ can be expressed as a $M \times K_T$ array of blocks $\mathbf{w}_{mj}^i$, i.e. as a $N_T \times K_T$ matrix with entries of the form $(\mathbf{w}_{mk}^j)_n$ that represent the BF weight that is applied at the $n$th antenna of $BS_j$ for serving $MS_{km}$.

Finally, the power allocation matrix $\mathbf{P}_{SS}$ is defined as the block diagonal matrix [42]:

$$\mathbf{P}_{SS} = \begin{bmatrix} \mathbf{P}_1 & 0 & \cdots & 0 \\ 0 & \ddots & & 0 \\ 0 & \cdots & 0 & \mathbf{P}_M \end{bmatrix}, \tag{12.19}$$

where $\mathbf{P}_m \in \mathbb{R}^{K \times K}$ is the diagonal matrix that holds at its main diagonal the powers allocated by $BS_m$ to its users $MS_{km}$ ($m \in \mathcal{M}$; $k \in \mathcal{K}$), i.e. $\mathbf{P}_m = \mathrm{diag}(P_{m1}^m, \ldots, P_{mK}^m)$. It becomes apparent that $\mathbf{P}_{SS}$ is a $M \times M$ array of blocks $\mathbf{P}_m$, i.e. it is a matrix with dimensions $K_T \times K_T$.

***Centralized coordinated zero-forcing precoding:*** Let us consider the application of centralized coordinated zero-forcing (ZF) precoding in a spectrum-sharing-agnostic manner (i.e., by ignoring all types of inter-system CCI). The ZF precoding matrix $\mathbf{W}^{(ZF)}$ is obtained by calculating the Moore–Penrose pseudoinverse of $\mathbf{H}$ and normalizing it column-wise [11], that is,

$$\widetilde{\mathbf{W}}^{(ZF)} = \mathbf{H}^H (\mathbf{H} \mathbf{H}^H)^{-1}, \tag{12.20a}$$

$$\mathbf{W}^{(ZF)} = \frac{(\widetilde{\mathbf{W}}^{(ZF)})_{*j}}{\|(\widetilde{\mathbf{W}}^{(ZF)})_{*j}\|}, \quad j = 1, \ldots, K_T. \tag{12.20b}$$

***Relaxed optimization problem:*** By applying ZF precoding, we eliminate the intra-system CCI and thus convert the sum-SE maximization problem of (12.13) into a convex power allocation (PA) task which attains a unique solution [42]:

$$\min_{\substack{P_{mk}^m \\ m \in \mathcal{M}, \, k \in \mathcal{K}}} \quad -R = -\sum_{m=1}^M \sum_{k=1}^K \log_2(1 + \lambda_{mk}^m P_{mk}^m) \tag{12.21a}$$

s.t.

$$\text{TPCs: (12.13b), (12.13c)} \tag{12.21b}$$

$$\text{IPC: (12.13d),} \tag{12.21c}$$

where in (12.13d) the BF vectors are $(\mathbf{w}_{mk}^m)^{(\text{ZF})}$ and in (12.21a) they are $\lambda_{mk}^m P_{mk}^m = \gamma_{km}$, with $\lambda_{mk}^m$ given by

$$\lambda_{mk}^m = \frac{|(\mathbf{h}_{km}^m)^H (\mathbf{w}_{mk}^m)^{(\text{ZF})}|^2}{|h_{km}|^2 P + 1}. \tag{12.22}$$

### 12.5.5 Solution and Algorithm

The solution to the convex interference-constrained PA (ICPA) task of (12.21) is the multi-level water-filling (WF) scheme that is shown in Theorem 1 [42].

**Theorem 1** The solution of the problem presented in (12.21) is given by the ICPA scheme:

$$P_{mk}^m = \left( \frac{1}{\ln 2(v_m + \mu a_{mk}^m)} - \frac{1}{\lambda_{mk}^m} \right)^+, \quad m \in \mathcal{M}; \ k \in \mathcal{K}, \tag{12.23}$$

where $v_m$ and $\mu$ are the Lagrange multipliers.

*Proof:* See [42]. □

**Remark 12.1** It is interesting to note that this solution has a similar form to the solution derived in [35] for the scenario where the SS is a MIMO link and singular-value decomposition (SVD) precoding/combining is applied.

In order to calculate the Lagrange multipliers $v_1, \dots, v_M$ and $\mu$ which are related with the SPCs per BS and the IPC, respectively, we use an iterative algorithm that is based on the bisection method and is illustrated in Algorithm 1. Note that the positive constant $\delta_\mu$ controls the accuracy of the algorithm. The power levels $P_{mk}^m$ are computed according to Theorem 1.

#### 12.5.5.1 Solution for Other Linear Precoding Schemes

Let us consider the application of other heuristic linear precoding schemes. Maximum ratio transmission (MRT) is a common example. This is an selfish scheme which maximizes the receive SNR of the scheduled users but ignores the CCI. Hence, the achieved sum-SE floors in the interference-limited high-SNR regime due to the uncontrolled interference. The BF vectors in MRT are computed as follows [11]:

$$(\tilde{\mathbf{w}}_{mk}^m)^{(\text{MRT})} = \mathbf{h}_{km}^m, \quad m \in \mathcal{M}; \ k \in \mathcal{K}. \tag{12.24a}$$

$$(\mathbf{w}_{mk}^m)^{(\text{MRT})} = \frac{(\tilde{\mathbf{w}}_{mk}^m)^{(\text{MRT})}}{\|(\tilde{\mathbf{w}}_{mk}^m)^{(\text{MRT})}\|}, \quad m \in \mathcal{M}; \ k \in \mathcal{K}. \tag{12.24b}$$

Regularized ZF (RZF) precoding is another popular method. RZF is an extension of ZF precoding which leaves some amount of CCI unaffected in such a way that the SINR at the scheduled users is maximized in order to improve the performance of this "altruistic"

---

**Algorithm 1** ICPA algorithm for solving the optimization problem in (12.21).

---

1: **procedure** ICPA FOR CBF($\lambda_{mk}^m, \alpha_{mk}^m, P_T, P_I$)
2:     **Initialize:** $\mu_{\min}, \mu_{\max}$
3:     **while** $|\mu_{\max} - \mu_{\min}| > \delta_\mu$ **do**
4:         $\mu = \left(\mu_{\min} + \mu_{\max}\right)/2$
5:         **for** $m = 1$ to $M$ **do**
6:             Find min $\left(v_m\right), v_m \geq 0$ :
            $\sum_{k=1}^K \left(P_{mk}^m\right)^+ \leq P_T$
7:         **end for**
8:         Compute the power levels $P_{mk}^m$ according to Theorem 1
9:         **if** $\sum_{m=1}^M \sum_{k=1}^K a_{mk}^m P_{mk}^m \geq P_I$ **then**
10:             $\mu_{\min} = \mu$
11:         **else**
12:             $\mu_{\max} = \mu$
13:         **end if**
14:     **end while**
15:     **Output:** $P_{mk}^m$,   $m \in \mathcal{M}; k \in \mathcal{K}$
16: **end procedure**

---

scheme in the noise-limited low-/moderate-SNR regime. The RZF precoding matrix is computed as follows [11]:

$$\widetilde{\mathbf{W}}^{(\text{RZF})} = \mathbf{H}^H \left( \frac{1}{P_{mk}^m} \mathbf{I}_{K_T} + \mathbf{H}\mathbf{H}^H \right)^{-1}, \quad m \in \mathcal{M}; k \in \mathcal{K}. \tag{12.25a}$$

$$\mathbf{W}^{(\text{RZF})} = \frac{(\widetilde{\mathbf{W}}^{(\text{RZF})})_{*j}}{\|(\widetilde{\mathbf{W}}^{(\text{RZF})})_{*j}\|}, \quad j = 1, \dots, K_T. \tag{12.25b}$$

When these linear precoding strategies are utilized, the intra-system CCI is not eliminated. Hence, the considered PA tasks correspond to non-convex optimization problems. However, we can still apply the derived PA solutions from the case where ZF precoding is used as heuristic PA methods in these scenarios [11]. For RZF in particular we expect its performance to be almost identical to that of ZF in the high-SNR regime.

### 12.5.6 Performance Evaluation via Numerical Simulations

In this section we evaluate the performance of the proposed RA scheme for the underlay spectrum sharing setup under study in various use cases via numerical simulations performed in MATLAB. In each one of these use cases we consider a cellular network (cooperation cluster) comprising $M = 2$ cells (BSs) with $K = 2$ active single-antenna MSs in each cell, and we are interested in the average sum-SE of the cellular network (SS) $\overline{R}$ (in bits/s/Hz) versus the average SNR $\rho$ (which ranges from 0 to 30 dB with a 5-dB step) obtained after 100 simulation runs (i.e., channel realizations).

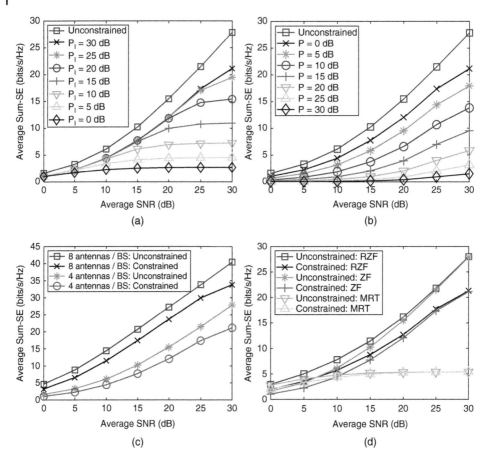

**Figure 12.2** Simulations results: standard WF-PA in stand-alone cellular network versus ICPA in the considered spectrum sharing setup. (a) Average sum-SE versus average SNR for various IPT values; (b) Average sum-SE versus average SNR for various PS transmission power values; (c) Average sum-SE versus average SNR for various numbers of antennas/BS; (d) Average sum-SE versus average SNR for coordinated MRT, ZF, and RZF.

The first use case is described by the following parameters: (i) each BS has $N = 4$ antennas, (ii) the IPT $P_I$ ranges from 0 to 30 dB with a 5-dB step, (iii) the SPC of each BS, $P_T$, is equal to the average SNR, (iv) the transmission power of $\text{TX}_{PS}$ is $P = 0$ dB, and (v) the linear precoding scheme that is applied at the SS is ZF. Figure 12.2a illustrates the performance of the SS for the various IPT values against the performance of an equivalent stand-alone cellular network (i.e., the performance of the SS in the absence of the PS), where the standard WF-PA scheme is utilized instead of the ICPA strategy (see [11]). We note that the performance loss caused by the requirement to meet the IPC is moderate for large IPT values (i.e., for relaxed IPCs). More specifically, for $P_I = 30$ dB we notice that in the spectrum sharing setup the cellular network reaches the same performance that the stand-alone cellular network reaches at about 3–4 dB higher average SNR value or, equivalently, as we see in Figure 12.2a, it achieves about 2.5–4 bits/s/Hz lower average sum-SE in comparison to the stand-alone cellular network for the same average SNR value in the high-SNR regime (i.e., from 15 to 25 dB average SNR). For smaller IPT values (i.e., for more stringent IPCs), the performance loss is larger. We also note that the average SNR that equals the IPT value signifies a point in the corresponding average sum-SE versus average SNR curve, after which the performance starts to decline, with a smaller or a larger rate depending on the IPT value, until the average sum-SE eventually floors (i.e., the slope of the curve gets smaller and smaller until the curve becomes pretty much parallel to the $x$ axis of the graph, which implies that the average sum-SE does not increase with the average SNR). Naturally, this flooring effect occurs for small average SNR values when the IPT value is small.

In the second use case, we fix the IPT at $P_I = 30$ dB and we vary instead the transmission power of TX$_{PS}$, $P$, from 0 to 30 dB with a 5-dB step. Recall that the RIS CCI affects the performance and that the MSs treat it as additional noise (i.e., the application of centralized ZF precoding in a spectrum-sharing-agnostic manner ignores this type of CCI, that is, the precoders do not null the RIS CCI). The power of this interference component, which is included in the denominator of the users' SINR, depends on $P$ (more specifically, the power of the RIS CCI received at MS$_{km}$ is $|h_{km}|^2 P$, $k \in \mathcal{K}$, $m \in \mathcal{M}$). As illustrated in Figure 12.2b, by raising $P$ from 0 to 5 dB, we lose about 4 dB or, equivalently, 2–3 bits/s/Hz in performance in the high-SNR regime. When $P$ becomes 10 dB, we lose another 4–5 dB or 2–4 bits/s/Hz in performance, approximately. It becomes apparent that the use of the proposed RA technique makes sense from a sum-SE point of view in setups/applications where the transmission power of the PS is relatively low.

In the third use case, we fix the transmission power of TX$_{PS}$ as well at $P = 0$ dB and we vary the number of BS antennas from $N = 4$ to $N = 8$. As depicted in Figure 12.2c, when we double the number of BS antennas, the performance loss in the spectrum sharing setup in comparison to the performance of the stand-alone cellular network is about 4 dB or 3–4 bits/s/Hz in the high-SNR regime. In that sense, the scenario that corresponds to $N = 8$ antennas/BS is not much different to the one that corresponds to $N = 4$ antennas/BS (i.e., similar performance trends are noticed). Of course, when more BS antennas are used, the achieved sum-SE is higher – this is true for both the stand-alone cellular network and the spectrum sharing setup under consideration. Also, the performance improvement in both cases is similar (i.e., about 10 dB or 10–12 bits/s/Hz).

In the fourth use case, we fix the number of antennas/BS at $N = 4$ and we study the performance of the considered systems (i.e., the cellular network in the underlay spectrum sharing setup versus the stand-alone cellular network) and corresponding PA strategies (i.e., ICPA versus the standard WF-PA) when MRT, ZF, and RZF precoding are applied. As can be seen in Figure 12.2d, in both setups ZF precoding performs worse than RZF for small to moderate average SNR values but its performance converges to that of RZF as the average SNR increases, whereas the sum-SE of MRT floors very quickly, as expected. Also, we notice that the performance of MRT is almost identical in these setups, since the sum-SE floors anyway (due to the intra-system CCI) even when there is no inter-system CCI.

## 12.6 Antenna-Array-Aided Spectrum Sensing

The community has been reluctant to adopt the conventional spectrum-sensing-based interweave model due to its inability to guarantee a certain QoS to both "players". LSA addresses this issue by forcing the licensee user (LU) to access the shared spectrum (or part of it) at predetermined locations and times where it is not utilized by the incumbent user (IU), in accordance with long-term spectral activity information that is made available by a database and commonly agreed between the involved "players" spectrum access rules. However, this conservative approach does not fully exploit the SE enhancement potential of spectrum sharing. The use of a number of sensing nodes that are equipped with directional antennas, in conjunction with the utilization of advanced collaborative spectrum sensing techniques, can further increase the SE of LSA systems by enabling the detection of additional spectrum access opportunities in a reliable manner.

In this section, we present examples of such novel spectrum sensing mechanisms and we evaluate their performance via over-the-air (OTA) experiments. In the considered setup, the sensing nodes are equipped with a so-called hex-antenna (HA). The prototype HA is composed of six single radio frequency (RF) parasitic antenna arrays, which collectively enable spectrum sensing over a 360° region in the azimuth. These antenna arrays offer a good tradeoff of directional gain versus cost and power consumption [43]. The cooperation between these directional sensing nodes enables us to reliably detect the spectral activity of the IU in the composite time-frequency-space domain.

### 12.6.1   Printed Yagi–Uda Arrays and Hex-Antenna Nodes

For the purposes of the presented demonstrations, we designed, simulated, and prototyped a number of planar printed parasitic antenna arrays that resonate in the 2.35 GHz LSA band. A 16 cm × 10 cm FR-4 dielectric board was used for the implementation of each such antenna array. As shown in Figure 12.3a, these arrays consist of: an active dipole that is placed on the front side of the FR-4 board along with its microstrip feeding lines that enable the mounting of an SMA connector for feeding the antenna array; six passive director dipoles that are placed in front of the active dipole on the same side of the FR-4 board to concentrate the emitted electromagnetic field and shape the radiation pattern in a desired way; and two passive reflector dipoles that are placed behind the active dipole on the back side of the FR-4 board to reflect the back-scattered radiation. Note that in the middle of each director or reflector dipole there is a small gap where analog loads (capacitors or inductors, respectively) are soldered to provide enhanced beam shaping control.

After establishing the initial design of the planar parasitic antenna array, the next step is the assembly of a 360°-azimuth spectrum sensing antenna. This is achieved by placing six such parasitic antenna arrays in a cyclic pattern to form a HA design, as illustrated in Figure 12.3b. Each such sector antenna is placed vertically pointing towards and two metal (copper) grounded plates are placed on either side of it to further confine the radiated electromagnetic field. This architecture results in six distinct 60° sectors which enable us to provide radio coverage over the whole 360° azimuth plane. Since only one of these sector antennas could be active at each given time, due to the hardware limitations of the transceiver platform, an RF switch is used to switch in a fast and sequential manner between the sector antennas (see Figure 12.3c).

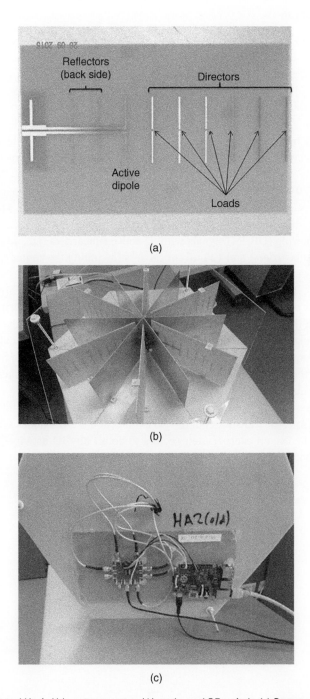

**Figure 12.3** Printed Yagi–Uda antenna array, HA node, and RF switch. (a) Prototype design of the parasitic Yagi–Uda antenna array; (b) Prototype HA node: note that each sector antenna is isolated by copper plates; (c) RF switch and controller on the back side of the HA node.

### 12.6.2 Test Setup

The testbed is based on the National Instrument's (NI) Universal Software Radio Peripheral (USRP) software-defined radio (SDR) platform and LabVIEW (Laboratory Virtual Instrument Engineering Workbench) development environment. The hardware is composed of two NI 2953R USRP radios, a host PC, and two HA nodes, as well as several controllers, RF switches, and monopole antennas.

As shown in Figure 12.4, the transmitter (Tx) and the receiver (Rx) of the IU are monopoles. Six monopole antennas, which are distributed over the test area, are also connected to the transmitter of the LU via an RF switch, while the LU receiver consists of two HA nodes.

A channel of 20 MHz bandwidth that is centered at the 2.35-GHz LSA band is considered. The transmitters and receivers were configured designed in LabVIEW. The applied communication standard, which is based on 802.11a, specifies the use of orthogonal frequency division multiplexing (OFDM) signals with 52 active subcarriers (i.e., 48 data subcarriers and four pilot subcarriers) and 12 null subcarriers (including the center subcarrier), for a total of $N_{sc} = 64$ subcarriers. The fast Fourier transform (FFT)/inverse FFT (IFFT) sample size is set to 64. The 52 active subcarriers are split into sets of 13 subcarriers which correspond to four distinct sub-bands with a bandwidth of 5 MHz each. The IU is allowed to hop between them and occupy any of them for one or more transmission frames (100 OFDM symbols). A preamble containing a known training sequence is always sent after every frame for synchronization and channel equalization purposes. The first OFDM symbol succeeding the preamble carries information about the type of modulation used for the duration of the next 99 symbols.

In phase 1 (sensing) of this experiment, which has a maximum duration of 10% of a transmission frame, the transmitter of the LU is silent and the HA nodes perform spectrum sensing consequently on each of their sectors to detect any IU activity. They then send either the individual decisions or the measurements of the $K = 12$ sectors to a fusion

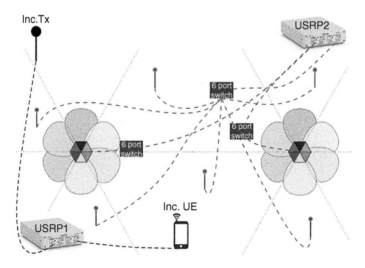

**Figure 12.4** Test setup.

center (host PC), which computes the sensing outcome based on this information. In phase 2 (transmission), the HA nodes play the role of the LU receiver and the best transmit antenna–receive sector pair is selected based on SINR measurements.

### 12.6.3 Collaborative Spectrum Sensing Techniques

Consider the hypotheses $\mathcal{H}_0$ and $\mathcal{H}_1$ which denote, respectively, the absence and the presence of IU activity in the channel of interest. The received signal at the $k$th sector antenna for the $i$th subchannel is given by [44–46]

$$
\begin{aligned}
\mathcal{H}_0 &: y_k^{(i)}(n) = z_k^{(i)}(n), \quad k = 1, \ldots, K; \ i = 1, \ldots, N_{sc}, \\
\mathcal{H}_1 &: y_k^{(i)}(n) = x_k^{(i)}(n) + z_k^{(i)}(n), \quad k = 1, \ldots, K; \ i = 1, \ldots, N_{sc},
\end{aligned}
\tag{12.26}
$$

where $x_k^{(i)}(n)$ is the transmitted signal component, which includes the effects of path loss, multipath fading, and time dispersion, $z_k^{(i)}(n)$ is the received AWGN component, which is assumed to be i.i.d. with zero mean and variance $\sigma^2$, and $n$ is the sample index. The goal of spectrum sensing is to correctly identify whether a signal is present or not in a given channel, based on the received samples (i.e., to decide between $\mathcal{H}_0$ and $\mathcal{H}_1$).

In this work, we consider two of the most popular spectrum sensing techniques, namely, maximum-eigenvalue-based detection (MED) and energy detection (ED) [44–46]. In MED, we collect the measurements from each sector over $L$ consecutive samples. The composite received signal can be expressed in compact form as

$$
\mathbf{y}_k^{(i)}(n) = \left[ y_k^{(i)}(n), \ldots, y_k^{(i)}(n - L + 1) \right]^T, \quad k = 1, \ldots, K; \ i = 1, \ldots, N_{sc}.
\tag{12.27}
$$

Then, we stack together the received signals from all sectors to obtain

$$
\mathbf{y}^{(i)}(n) = \left[ \mathbf{y}_1^{(i)}(n), \ldots, \mathbf{y}_K^{(i)}(n) \right]^T, \quad i = 1, \ldots, N_{sc}.
\tag{12.28}
$$

Next, we compute the sample covariance matrix of $\mathbf{y}^{(i)}$, $\mathbf{R}_y^{(i)}$, as follows:

$$
\mathbf{R}_y^{(i)} = \frac{1}{N} \sum_{m=0}^{N-1} \mathbf{y}^{(i)}(m) \left( \mathbf{y}^{(i)}(m) \right)^H, \quad i = 1, \ldots, N_{sc},
\tag{12.29}
$$

where $N$ is the number of samples. Finally, we compare the maximum eigenvalue of this matrix, $\lambda_{\max}^{(i)}$, to a given threshold $\lambda$ to obtain the sensing outcome:

$$
\lambda_{\max}^{(i)} \underset{\mathcal{H}_0}{\overset{\mathcal{H}_1}{\gtrless}} \lambda, \quad i = 1, \ldots, N_{sc}.
\tag{12.30}
$$

In ED, on the other hand, we compute the average received energy at each sector:

$$
E_k^{(i)} = \frac{1}{N} \sum_{n=0}^{N-1} \left| y_k^{(i)}(n) \right|^2, \quad k = 1, \ldots, K; \ i = 1, \ldots, N_{sc}
\tag{12.31}
$$

and we either use the OR rule to obtain the sensing outcome (i.e., the $i$th subchannel is declared busy if the measured average energy of at least one sector is above a given threshold and is declared idle otherwise):

$$
E_1^{(i)} \underset{\mathcal{H}_0}{\overset{\mathcal{H}_1}{\gtrless}} E_{\text{th}} \ || \ \cdots \ || \ E_K^{(i)} \underset{\mathcal{H}_0}{\overset{\mathcal{H}_1}{\gtrless}} E_{\text{th}}, \quad i = 1, \ldots, N_{sc},
\tag{12.32}
$$

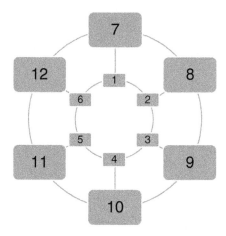

**Figure 12.5** Connectivity graph for the 12 sectors of the two HA nodes.

or we add together these values:

$$E_{sum}^{(i)} = \sum_{k=1}^{K} E_k^{(i)}, \quad i = 1, \dots, N_{sc} \tag{12.33}$$

and use the SUM rule to obtain the sensing outcome (i.e., the $i$th subchannel is declared busy if the sum of the measured average energies from all the sectors is above the given threshold and is declared idle otherwise):

$$E_{sum}^{(i)} \underset{\mathcal{H}_0}{\overset{\mathcal{H}_1}{\gtrless}} E_{th}, \quad i = 1, \dots, N_{sc}. \tag{12.34}$$

Alternatively, we can base the decision on sums of energy measurements from neighboring sectors instead of the total sum of energy measurements from all sectors. More specifically, we note that the sectors are connected according to the cyclic graph shown in Figure 12.5. Based on this graph, we form a $K \times K$ connectivity matrix $\mathbf{W}_c$ whose $(i,j)$th entry is 1 if the $i$th sector is connected to the $j$th sector and is 0 otherwise. Then, we compute the energy metric as $\mathbf{E}_{con} = \mathbf{W}_c \mathbf{E}$, where $\mathbf{E} = \begin{bmatrix} \mathbf{E}_1 & \cdots & \mathbf{E}_K \end{bmatrix}^T$ is a $K \times N_{sc}$ matrix and $\mathbf{E}_k$ is the $N_{sc}$-dimensional column vector that holds the energy measurements for the $N_{sc}$ subcarriers of the $k$th sector, $E_k^{(i)}, k = 1, \dots, K, i = 1, \dots, N_{sc}$. Referring to each set of connected sectors as a node, we notice that there are $K$ such nodes. The $(i,j)$th entry of the $K \times N_{sc}$ matrix $\mathbf{E}_{con}$ corresponds to the estimated energy of the $i$th node for the $j$th subchannel, $i = 1, \dots, K$, $j = 1, \dots, N_{sc}$. Finally, we obtain the decision according to the OR rule:

$$(\mathbf{E}_{con})_{1,i} \underset{\mathcal{H}_0}{\overset{\mathcal{H}_1}{\gtrless}} E_{th} \; || \; \cdots \; || (\mathbf{E}_{con})_{K,i} \underset{\mathcal{H}_0}{\overset{\mathcal{H}_1}{\gtrless}} E_{th}. \tag{12.35}$$

It can be readily seen that $(\mathbf{E}_{con})_{k,i} \geq (\mathbf{E}_k)_i = E_k^{(i)}$ for all $k = 1, \dots, K$ and $i = 1, \dots, N_{sc}$.

### 12.6.4 Experimental Results

This section presents the results of the OTA spectrum sensing experiment for two transmission scenarios in the high and low SNR regimes. In the corresponding figures, the test statistic is depicted in the left column (from top to bottom row: maximum eigenvalue,

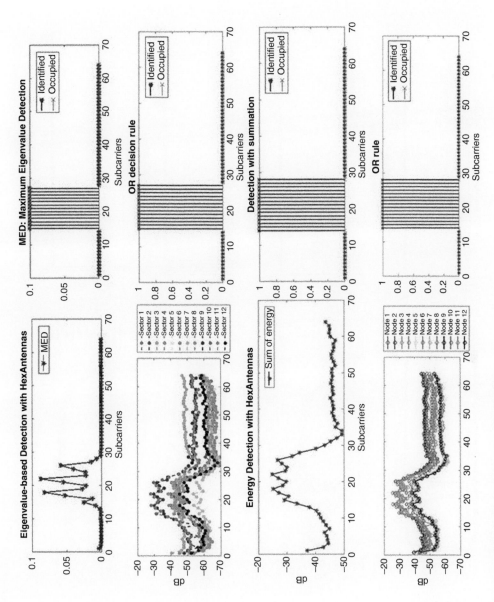

**Figure 12.6** Spectrum sensing results in the high SNR regime.

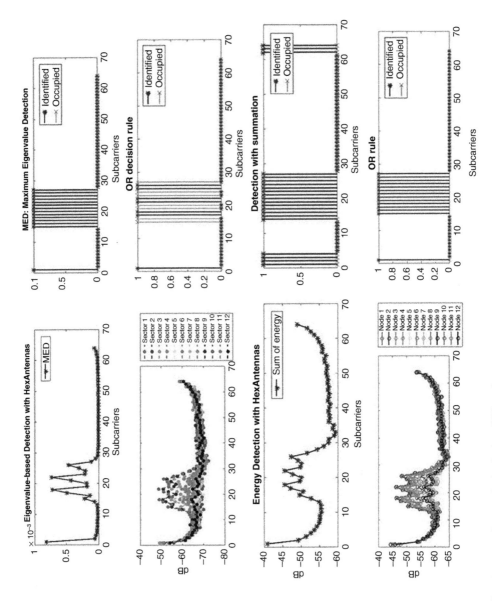

**Figure 12.7** Spectrum sensing results in the low SNR regime.

energy per sector, sum energy, and partial sum energy based on connectivity graph) and in the right column is illustrated the corresponding detection result.

### 12.6.4.1 Detection in High SNR

As illustrated in Figure 12.6, all considered spectrum sensing techniques correctly identified the occupied sub-band. However, ED with decision based on the SUM rule and ED with decision based on the connectivity graph resulted also in two false alarm errors.

### 12.6.4.2 Detection in Low SNR

As depicted in Figure 12.7, the MED method and the ED method with decision based on the connectivity graph correctly identified the occupied sub-band, although they resulted also in a false alarm error. The other ED variants, however, resulted in several misdetection errors. This is due to the dependency of these conventional ED methods' performance on the SNR. Thus, we note that the proposed connectivity-graph-based ED scheme substantially improves the robustness of ED.

## 12.7 Summary and Conclusions

In this chapter we presented the beneficial role that antenna arrays can offer in spectrum sharing, with emphasis on spectrum sensing and shared spectrum access problems, for which some representative examples were provided. More specifically, we presented experiments which show that cooperative spectrum sensing between directional nodes enables the detection of spectrum holes in a reliable manner. Such setups and techniques can be used to enhance the SE of LSA systems. Moreover, we showed that the use of simple centralized RA policies consisting of coordinated linear precoding schemes (e.g., ZF or RZF precoding) and WF-based PA strategies can be used as an enabler of underlay spectrum sharing, at least in cases where the IPT and the transmission power of the PS are relatively small.

Furthermore, by adopting, besides the conventional regulatory-oriented view of spectrum sharing, a more fundamental physical viewpoint and reviewing the key attributes of the antenna arrays, we also alluded to the potential of antenna arrays to impact potential future types of spectrum sharing approaches that have not yet been considered.

## Acknowledgments

We would like to acknowledge the contribution of Dr. Dimitrios Ntaikos, Mr Bobby Gizas and Ms Foteini Verdou, former employees of Athens Information Technology (AIT), in the prototyping, setup and testing of the experiments presented in section 12.6, which were conducted in the context of the EC FP7 project ADEL (https://cordis.europa.eu/project/rcn/189128/factsheet/en). We should also mention that this experimental work won the best demo award in EuCNC 2016.

# References

**1** Z. Zhang *et al.*, "Full-Duplex Wireless Communications: Challenges, Solutions, and Future Research Directions," *Proceedings of the IEEE*, vol. 104, no. 7, pp. 1369–1409, Jul. 2016.

**2** E. Telatar, "Capacity of Multi-Antenna Gaussian Channels," *European Transactions on Telecommunications*, vol. 10, no. 6, pp. 585–596, Nov. 1996.

**3** G. J. Foschini and M. J. Gans, "On Limits of Wireless Communication in a Fading Environment When Using Multiple Antennas," *Wireless Personal Communications*, vol. 6, no. 3, pp. 311–335, Mar. 1998.

**4** H. Boelcskei *et al.*, Eds., Space-Time Wireless Systems: From Array Processing to MIMO Communications. Cambridge University Press, 2006.

**5** A. Paulraj and C. B. Papadias, "Space-Time Processing for Wireless Communications," *IEEE Signal Processing Magazine*, vol. 14, no. 6, pp. 49–83, Nov. 1997.

**6** Viswanath, P. and Tse, D. N., "Sum Capacity of the Vector Gaussian Broadcast Channel and Downlink-Uplink Duality," *IEEE Transactions on Information Theory*, vol. 49, no. 8, pp. 1912–1921, Aug. 2003.

**7** H. Huang *et al.*, Eds., MIMO Communication for Cellular Networks. Springer-Verlag New York, 2012.

**8** V. H. MacDonald, "Advanced Mobile Phone Service: The Cellular Concept," *The Bell System Technical Journal*, vol. 58, no. 1, pp. 15–41, Jan. 1979.

**9** P. Marsch and G. Fettweis, Eds., Coordinated Multi-Point in Mobile Communications: From Theory to Practice. Cambridge University Press, 2011.

**10** B. Clerckx and C. Oestges, MIMO Wireless Networks, 2nd ed. Academic Press, Jan. 2013.

**11** E. Bjornson and E. Jorswieck, "Optimal Resource Allocation in Coordinated Multi-Cell Systems," *Foundations and Trends in Communications and Information Theory*, vol. 9, no. 2-3, pp. 113–381, Jan. 2013.

**12** E. Castaneda *et al.*, "An Overview on Resource Allocation Techniques for Multi-User MIMO Systems," *IEEE Communications Surveys & Tutorials*, vol. 19, no. 1, pp. 239–284, 2017.

**13** China Mobile, "C-RAN: The Road Towards Green RAN," White Paper, Dec. 2013.

**14** M. Artuso *et al.*, "Enhancing LTE with Cloud-RAN and Load-Controlled Parasitic Antenna Arrays," *IEEE Communications Magazine*, vol. 54, no. 12, pp. 183–191, Dec. 2016.

**15** T. L. Marzetta, "Noncooperative Cellular Wireless with Unlimited Numbers of Base Station Antennas," *IEEE Transactions on Wireless Communications*, vol. 9, no. 11, pp. 3590–3600, Nov. 2010.

**16** E. G. Larsson *et al.*, "Massive MIMO for next generation wireless systems," *IEEE Communications Magazine*, vol. 52, no. 2, pp. 186–195, Feb. 2014.

**17** E. Bjornson, J. Hoydis, and L. Sanguinetti, "Massive MIMO Networks: Spectral, Energy, and Hardware Efficiency," *Foundations and Trends in Signal Processing*, vol. 11, no. 3-4, pp. 154–655, Nov. 2017.

**18** A. F. Molisch *et al.*, "Hybrid Beamforming for Massive MIMO: A Survey," *IEEE Communications Magazine*, vol. 55, no. 9, pp. 134–141, Sep. 2017.

**19** WG FM53, "ECC Report 205: Licensed Shared Access," Tech. Rep., Feb. 2014.

**20** ETSI Technical Committee Reconfigurable Radio Systems (TC RRS), "ETSI TR 103 113 V1.1.1: Mobile Broadband Services in the 2300 MHz–2400 MHz Frequency Band under Licensed Shared Access Regime," System Reference Document, Jul. 2013.

**21** ——, "ETSI TS 103 154 V1.1.1: System Requirements for Operation of Mobile Broadband Systems in the 2300 MHz–2400 MHz Band under Licensed Shared Access (LSA)," Technical Specification, Oct. 2014.

**22** ——, "ETSI TS 103 235 V1.1.1: System Architecture and High Level Procedures for Operation of Licensed Shared Access (LSA) in the 2300 MHz–2400 MHz Band," Technical Specification, Oct. 2015.

**23** ——, "ETSI TS 103 379 V1.1.1: Information Elements and Protocols for the Interface Between LSA Controller (LC) and LSA Repository (LR) for Operation of Licensed Shared Access (LSA) in the 2300 MHz–2400 MHz Band," Technical Specification, Jan. 2017.

**24** R. Zhang *et al.*, "LTE-Unlicensed: The Future of Spectrum Aggregation for Cellular Networks," *IEEE Wireless Communications*, vol. 22, no. 3, pp. 150–159, Jun. 2015.

**25** B. Chen *et al.*, "Coexistence of LTE-LAA and Wi-Fi on 5 GHz with Corresponding Deployment Scenarios: A Survey," *IEEE Communications Surveys and Tutorials*, vol. 19, no. 1, pp. 7–32, Jul. 2016.

**26** 4G Americas, "LTE Aggregation and Unlicensed Spectrum," Tech. Rep., Nov. 2015.

**27** MulteFire Alliance, "MulteFire Release 1.0 Technical Paper: A New Way to Wireless," Tech. Rep., Jan. 2017.

**28** I. F. Akyildiz *et al.*, "A Survey on Spectrum Management in Cognitive Radio Networks," *IEEE Communications Magazine*, vol. 46, no. 4, pp. 40–48, Apr. 2008.

**29** S. Pandit and G. Singh, "An Overview of Spectrum Sharing Techniques in Cognitive Radio Communication System," *Wireless Networks*, vol. 23, no. 2, pp. 497–518, Feb. 2017.

**30** A. Morgado *et al.*, "Dynamic LSA for 5G Networks: The ADEL Perspective," in *European Conference on Networks and Communication (EuCNC)*, Paris, France, 2015, 29 June–2 July.

**31** J. H. Winters, "On the Capacity of Radio Communication Systems with Diversity in a Rayleigh Fading Environment," *IEEE Journal on Selected Areas in Communications*, vol. 5, no. 5, pp. 871–878, Jun. 1987.

**32** A. J. Paulraj and T. Kailath, "Increasing Capacity in Wireless Broadcast Systems Using Distributed Transmission/Directional Reception (DTDR)," Patent US Patent 5,345,599, Sep. 6, 1994.

**33** C. Galiotto *et al.*, "Unlocking the Deployment of Spectrum Sharing With a Policy Enforcement Framework," *IEEE Access*, vol. 6, pp. 11 793–11 803, Jan. 2018.

**34** V. Cadambe and S. Jafar, "Interference Alignment and Degrees of Freedom of the K-User Interference Channel," *IEEE Transactions on Information Theory*, vol. 54, no. 8, pp. 3425–3441, Aug. 2008.

**35** R. Zhang and Y. C. Liang, "Exploiting Multi-Antennas for Opportunistic Spectrum Sharing in Cognitive Radio Networks," *IEEE Journal of Selected Topics in Signal Processing*, vol. 2, no. 1, pp. 88–102, Feb. 2008.

**36** I. Turki, I. Kammoun, and M. Siala, "Beamforming Design and Sum Rate Maximization for the Downlink of Underlay Cognitive Radio Networks," in *International Wireless Communications and Mobile Computing Conference (IWCMC)*, Dubrovnik, Croatia, 2015, pp. 178–183, 24–28 Aug.

**37** V. D. Nguyen *et al.*, "An Efficient Precoder Design for Multiuser MIMO Cognitive Radio Networks With Interference Constraints," *IEEE Transactions on Vehicular Technology*, vol. 66, no. 5, pp. 3991–4004, May 2017.

**38** L. Claudino and T. Abrao, "Efficient ZF-WF Strategy for Sum-Rate Maximization of MU-MISO Cognitive Radio Networks," *AEU International Journal of Electronics and Communications*, vol. 84, pp. 366–374, Feb. 2018.

**39** L. Zhang, Y. Xin, and Y. C. Liang, "Weighted Sum Rate Optimization for Cognitive Radio MIMO Broadcast Channels," *IEEE Transactions on Wireless Communications*, vol. 8, no. 6, pp. 2950–2959, Jun. 2009.

**40** D. Ntaikos *et al.*, "Low-Complexity Air-Interface-Agnostic Cooperative Parasitic Multi-Antenna Spectrum Sharing System," in *IEEE International Symposium on Dynamic Spectrum Access Networks (DySPAN)*, Stockholm, Sweden, 2015, 2.

**41** Energy-autonomous portable access points for infrastructure-less networks (PAINLESS). [Online]. Available: http://painless-itn.com/

**42** K. Ntougias *et al.*, "Simple Cooperative Transmission Schemes for Underlay Spectrum Sharing Using Symbol-level Precoding and Load-controlled Arrays," *IEEE International Conference on Acoustics, Speech, and Signal Processing (ICASSP)*, Brighton, UK, 12–17 May, 2019.

**43** A. Kalis *et al.*, Eds., Parasitic Antenna Arrays for Wireless MIMO Systems. Springer-Verlag, New York, 2014.

**44** Y. Zeng and Y.-C. Liang, "Eigenvalue-Based Spectrum Sensing Algorithms for Cognitive Radio," *IEEE Transactions on Communications*, vol. 57, no. 6, pp. 1784–1793, Jun. 2009.

**45** E. Axell, G. Leus, and E. G. Larsson, "Overview of Spectrum Sensing for Cognitive Radio," in *2nd International Workshop on Cognitive Information Processing*, Elba, Italy, 2010, pp. 322–327, 14–16 June.

**46** B. Nadler, F. Penna, and R. Garello, "Performance of Eigenvalue-Based Signal Detectors with Known and Unknown Noise Level," in *IEEE International Conference on Communication (ICC)*, Kyoto, Japan, 2011, 5–9 June.

# 13

# Resource Allocation for Shared Spectrum Networks

*Eduard A. Jorswieck[1] and M. Majid Butt[2]*

[1] *TU Braunschweig, Braunschweig, Germany*
[2] *Nokia Bell Labs, Paris-Saclay, France*

## 13.1 Introduction

The coexistence of devices in dense wireless networks requires careful design of resource allocation algorithms for spectrum sharing for two reasons. First, the interference caused by sharing the same frequency bands at the same time on the same geographical location creates dependencies between the quality of service (QoS) of the different systems. Second, the conflicting interests of heterogeneous devices and their service level requirements in terms of data rate, reliability, latency, security, and energy efficiency lead to complicated resource assignment and allocation problems.

Spectrum licensing arose in the 1920s because of the constraints and special needs of the radio receivers used at that time. The radio transceivers were primitive in the sense that they could not distinguish between different transmissions on a single frequency; they were unable to learn and to adapt to their current situation, and they were limited in terms of their signal processing power and memory configuration. Therefore, the only way for multiple users to share the spectrum was to divide it into orthogonal chunks [42]. In the beginning, spectrum was divided into bands with large separation in-between (guard bands) to ensure that receivers could identify their signals. The regulators at different national or international levels harmonize the allocation of frequency spectrum radio services and are responsible for the assignment of particular bands to specific users in the form of licensing, an arrangement that is known as command and control. However, economists have long argued that market mechanisms should be applied to radio spectrum [10].

From a communications engineering point of view, different types of orthogonality in frequency, time, space or coding domain[1] were used for resource allocation depending on the type of interference: for users in one cell operated by one operator (intracell interference) time-division multiple access (TDMA) combined with frequency-division multiple access (FDMA) (used in global mobile communication systems) or code division multiple access (combined with TDMA/FDMA in third-generation (3G) systems) is applied to

---

1 Note that these dimensions are in general coupled and cannot be considered separately, in particular the spreading domain.

*Spectrum Sharing: The Next Frontier in Wireless Networks,* First Edition.
Edited by Constantinos B. Papadias, Tharmalingam Ratnarajah, and Dirk T.M. Slock.
© 2020 John Wiley & Sons Ltd. Published 2020 by John Wiley & Sons Ltd.

separate their signals at the receivers. For different sectors or cells, the intercell interference is controlled by applying different frequency reuse factors [33]. Fractional and adaptive frequency reuse is discussed in long-term evolution (LTE) and worldwide interoperability for microwave access (WiMAX) [41].

In the last 10–15 years the body of work on spectrum sharing in general and in particular on resource allocation for spectrum sharing has grown significantly. The following brief list contains only a small sample of references on this topic. The goal is to show the diversity and width of spectrum sharing approaches.

- In cellular communications, the demand for additional spectrum is greatly increasing, therefore the mobile cellular industry explores new approaches to utilize unlicensed and sparsely occupied spectrum. Coexistence and spectrum sharing shape 3GPP standards for 4G, 5G, and beyond [34].
- In cognitive radio scenarios (covered in Chapters 2–6), spectrum sharing and the interactions between primary and secondary systems are considered from different perspectives: spectrum sensing, interference temperature, resource allocation, incentives for primary links, contracts, and so on [25]. One important perspective is the business point of view in which the incentives for the primary link to share its spectrum with secondary systems are discussed [23].
- In unlicensed bands, (discussed in Chapter 14), e.g. in the 3.5-GHz spectrum, the coexistence between devices from Wi-Fi (IEEE 802.11), ZigBee (IEEE 802.15.4) or Bluetooth standards and the distributed channel assignment is studied (e.g., [44]).
- In heterogeneous small cell networks it is necessary to devise a spectrum sharing policy based on demands, fairness, and other key performance indicator (KPIs) which utilize a priority scheme in fulfilling operators' demands and envision a secure operator-specific information sharing policy where no critical information is exchanged between the operators [24].

Let us summarize the main observation from this short overview of the difference spectrum sharing scenarios:

> The limiting factor in coexisting wireless systems that share spectrum is interference, which is observed on the physical layer and requires careful and efficient interference-aware resource allocation.

In all of the above application scenarios the resource allocation becomes the important design tool to enable efficient coexistence between various wireless systems. In this chapter, some recent resource assignment and trading algorithms for spectrum sharing are reviewed as well as their properties in terms of computational and implementation complexity. Finally, current and future application scenarios are discussed in which spectrum and resource sharing lead to significant performance improvements compared to non-sharing approaches.

The chapter is organized as follows. First, the information theoretic background and basic observations and results regarding the achievable rate regions of the underlying interfere channel model are summarized. Then the types of spectrum sharing are discussed. The resource allocation problem is introduced afterwards. The three challenges regarding multi-objectives, conflicting utilities, and distributed implementation are explained and approached by a multi-objective programming (MOP) framework, game theoretic approaches, and stable matching-based resource allocation. The resource trading approach to spectrum sharing is reviewed before we conclude the chapter with a summary and outlook.

## 13.2 Information-theoretic Background

Spectrum sharing leads to interference at the air interface. On the physical layer, this problem is best analysed within the framework of the interference channel (IFC) [2]. Information-theoretic studies of the IFC have a long history [9, 11, 17, 38]. These references provide various achievable rate regions, which are generally larger in the more recent papers than in the earlier ones. For certain operating points the capacity is known [15], but the capacity region of the general IFC remains an open problem. For the special two-user IFC, a monograph summarizes all available results [43]. A deterministic approach to approximate the capacity region of the IFC [12] provides the capacity region within one bit. Recently, the characterizations of the capacity region or achievable rate regions have been used to compute efficient operating points of multi-antenna interference channels [8, 21]. The achievable rate region depends on the information available at the transmitters, and the cooperation at the transmitter and the receiver side [30]. Furthermore, the fading statistics influence the average and outage rate region.

In the following, we illustrate the usual transmission and coding strategies with an anecdotal example of a two-user two-carrier IFC with two antennas at the transmitters. The corresponding system model is shown in Figure 13.1. The signal model is described by

$$y_k(\ell) = \underbrace{\alpha_{k,k}(\ell)\sqrt{p_k(\ell)}}_{\text{useful signal}} + \underbrace{\alpha_{\bar{k},k}(\ell)\sqrt{p_{\bar{k}}(\ell)}}_{\text{interference signal}} + \underbrace{z_k(\ell)}_{\text{AWGN}}, \tag{13.1}$$

with link $k \in \{1, 2\}$, on carrier $\ell \in \{1, 2\}$ with effective channel $\alpha_{k,l}(\ell) = \mathbf{w}_k^H(\ell)\mathbf{h}_{kl}(\ell)$ from transmitter $k$ to receiver $l$ on carrier $\ell$ with beamforming vector $\mathbf{w}_k$ and MISO channel $\mathbf{h}_{kl}$, with the other link denoted by $\bar{k} = k + 1 \mod 2$. AWGN is modeled as zero-mean with variance $\sigma^2$.

From the signal model in (13.1), the dependencies between the two links are clearly visible. The positive effect of creating a large received signal power for the intended receiver automatically creates interference at the other receiver. These couplings are the reason for the conflict situation for resource allocation at the two transmitters on the two carriers.

One of the orthogonal transmission strategies is FDMA, where the two links use only one carrier exclusively. Standard point-to-point single-user channel coding schemes are used. The assignment of links $k$ to carriers $\ell$ is denoted by a matching $\mu$ [20]. With two carriers and two users, there are two possible matchings available, denoted by $\{1, 2\}$ and $\{2, 1\}$ which means link 1 mapped to carrier 1 and link 2 to carrier 2, i.e. $\mu(1) = 1, \mu(2) = 2$, and link 1 mapped to carrier 2 and link 2 to carrier 1, i.e. $\mu(1) = 2, \mu(2) = 1$.

An alternative to FDMA is to allow both links to share the spectrum and transmit on the same frequencies at the same time. The easiest coding and decoding strategy is to

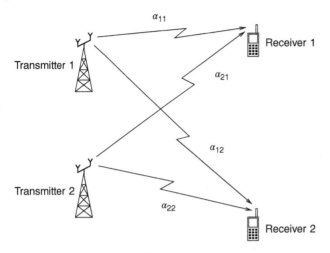

**Figure 13.1** System model for the two-user interference channel.

apply point-to-point single-user channel coding at the two links and simple single-user receivers where the interference from the other receiver is simply treated as additional noise. This strategy is referred to as treating interference as noise (TIN) [3]. The resulting signal-to-interference-plus-noise ratio (SINR) expression as a function of the power allocation for link $k$ on carrier $\ell$ is computed from (13.1) with $a_{kl}(\ell) = |\alpha_{kl}(\ell)|^2$ for $\ell \in \{1, 2\}$ as

$$\text{SINR}_k(\ell, \mathbf{p}(\ell)) = \frac{a_{kk}(\ell)p_k(\ell)}{\sigma^2 + a_{lk}(\ell)p_l(\ell)} \tag{13.2}$$

The corresponding achievable rates for the two links are a function of the power allocation $p_k(\ell)$ and read

$$R_k(\mathbf{p}) = \log\left(1 + \text{SINR}_k(1, \mathbf{p}(1))\right) + \log\left(1 + \text{SINR}_k(2, \mathbf{p}(2))\right). \tag{13.3}$$

Usually, the power allocation $\mathbf{p}$ has additional constraints, such as power constraints, therefore there is a feasible power set $\mathcal{P}$ for which $\mathbf{p} \in \mathcal{P}$. For general time sharing and alternative signal processing approaches to increase the achievable rate region of the interference channel, readers should refer to improper signaling and generalized time-sharing in [19].

The convex hull of the achievable rate region (if both carriers are used) is illustrated in Figure 13.2 in red and the corresponding FDMA operating points are shown as a green circle and a green square. The single-user operating points are shown as a blue plus and a blue cross on the axis for comparison.

The main observation from Figure 13.2 is that non-orthogonal resource allocation can achieve strictly higher rate tuples than orthogonal resource allocation. The red region is obtained from varying the power allocation for the two users over the two carriers and then taking the convex hull of all achievable rate tuples. In this case, FDMA achieves operating points inside the achievable rate region with TIN. It is important to stress that TIN is only one strategy which is complemented by successive interference cancellation (SIC) and simultaneous non-unique decoding (SND) [3]. This implies that the gain by

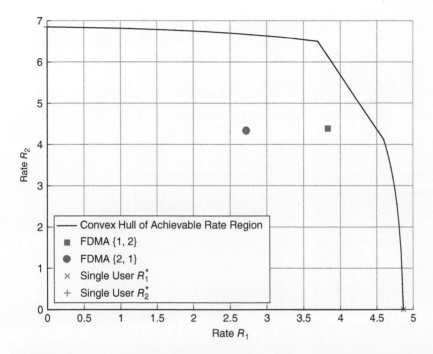

**Figure 13.2** Achievable rate region of the two-user two-carrier IFC with TIN in comparison with FDMA.

using a non-orthogonal resource allocation is even larger than in the example above. This suboptimality was observed for the interference channel among other channel models previously in [7].

Let us summarize the main lesson we have learned from this brief review of information theoretic background for spectrum sharing between wireless systems:

> Non-orthogonal resource allocation in general achieves strictly larger transmission rates than orthogonal resource allocation, in which a time-frequency resource is allocated exclusively to one link only.

Note that the non-orthogonal resource allocation and performance gains come with the cost of high processing requirements, increased complexity, possibly longer delays, increased signalling overhead, and other issues. It must be carefully verified that the overall system performance taking these costs into account is improved.

## 13.3 Types of Spectrum Sharing

There are different ways to classify the different spectrum sharing approaches. Let us start with a classification from the point of view of the lower technological layers, i.e. mainly from the physical and medium access control layer.

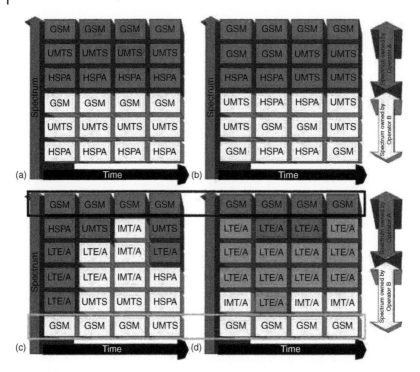

**Figure 13.3** Classification of spectrum sharing methods: (a) no spectrum sharing, (b) intra-operator spectrum sharing, (c) inter-operator orthogonal spectrum sharing, and (d) inter-operator non-orthogonal spectrum sharing (compare to Fig. 1 in [22]).

We follow the classification introduced in [22]. In the scenario without spectrum sharing (Figure 13.3a), the spectrum licenses are allocated statically for a rather long time period (auctions usually hold for 20 years or more) to one operator and to one radio access technology (RAT). The first step to flexible spectrum usage, which is already performed in many European countries, is intra-operator spectrum sharing (Figure 13.3b). This allows a dynamic allocation of RATs to portions of the owned licensed spectrum. The same frequency band might be used at different times of the day for different RATs. In orthogonal inter-operator spectrum sharing the spectrum bands can be relocated over time to different operators serving their users with different RATs (Figure 13.3c). However, each frequency band is allocated to at most one operator. This corresponds to the orthogonal resource allocation scenario for the two-user IFC from the last section. Finally, the most flexible way of spectrum sharing is non-orthogonal inter-operator spectrum sharing, where the shared bands can be assigned to multiple operators at the same time (Figure 13.3d). In addition, we can distinguish different types of spectrum sharing in time and frequency domain (dynamic combined vertical and horizontal sharing). In 3GPP study item Rel 15 is introduced to study 5G new radio (NR) operation in unlicensed spectrum and fair coexistence between NR, LTE, and Wi-Fi.

Another way to classify spectrum sharing approaches is to sort them by the underlying frequency bands. Licensed spectrum is exclusively assigned to one operator who then takes care of the conformal usage. For LTE there are currently over 40 bands globally assigned, more than 30 are FDD and more than 10 are TDD. Then there are frequency bands, which are unlicensed and traditionally used in a shared manner, e.g. 2.4 GHz, 5–7 GHz, and 57–71 GHz bands. Furthermore, there is shared spectrum allocated for new shared spectrum paradigms, e.g. 2.3 GHz in Europe or 3.5 GHz in the USA. For the regulatory bodies the main difference between these three types of spectrum types are the responsibilities. The licensed spectrum is assigned to one responsible operator, who has to care for its spectrum. Licensed shared access (LSA) falls into this category. In unlicensed spectrum, multiple different RATs interoperate and coexist in an uncoordinated way. Finally, the bands for new spectrum sharing should foster the coordinated sharing of these frequencies. The interested reader is referred to [40] for a survey on licensed spectrum sharing schemes for mobile operators.

## 13.4 Resource Allocation for Efficient Spectrum Sharing

In the scenarios of spectrum sharing between different wireless systems, there are the following challenges associated with resource allocation mechanisms:

1. The coexisting wireless links have different QoS requirements and this leads directly to a multi-objective problem formulation.
2. The resource allocation is a conflict problem between the wireless links, since their performance depends on the chosen resource allocations of other links in the network.
3. Usually, there is no central controller to perform the resource allocation, as in traditional single-operator networks or Cloud RAN (C-RAN) networks. If a central controller exists, as in LSA, it must rely on the truthful reporting and operation of the participating nodes.

All three challenges can be systematically approached and resolved by solid engineering tools: multi-objective programming, game theory, and stable matchings.

### 13.4.1 Multi-objective Programming

The first challenge can be illustrated by revisiting the two-user two-carrier IFC example above. The two links have two rates $R_1$ and $R_2$ which should be maximized simultaneously. This leads to the following multi-objective programming problem [31, 45]:

$$\max_{\mathbf{p} \in P}[R_1(\mathbf{p}), R_2(\mathbf{p})]. \tag{13.4}$$

Note that the constraint set $P$ can contain power constraints as well as more difficult minimum rate or energy efficiency or delay constraints, expressed as constraints of the general form $f_i(\mathbf{p}) \leq 0$.

Also note that the extension from the two-user to the general $K$ user case is straightforward [5, 31]. For ease of exposition and convenience of illustrations, we focus on the two-user case for the power allocation problem.

Collect the two rates in the vector $\mathbf{R}(\mathbf{p}) = [R_1(\mathbf{p}), R_2(\mathbf{p})]$. The operational meaning of the multiple objective in (13.4) is revealed with the following definition of *Pareto optimality*:

A point $\mathbf{p}^* \in P$ is Pareto optimal if and only if there does not exist another point $\mathbf{p} \in P$ such that $\mathbf{R}(\mathbf{p}) \geq \mathbf{R}(\mathbf{p}^*)$ and $R_k(\mathbf{p}) > R_k(\mathbf{p}^*)$ for at least one $k$. One approach to characterize the Pareto boundary of the achievable rate region is to solve the following single-objective problem instead:

$$\max_{\mathbf{p},\lambda} \lambda \quad \text{s.t.} \quad R_m(\mathbf{p}) \geq \lambda \rho_m, \quad m = 1, 2, \ \mathbf{p} \in P, \tag{13.5}$$

where $\rho_1$ and $\rho_2$ are the rate weights ($\rho_1, \rho_2 \geq 0, \rho_1 + \rho_2 = 1$). Programming problem (13.5) is called weighted Chebyshev problem. This problem can be solved by, for example, bisection [5]. Another systematic way to solve the MOP is scalarization, i.e. solving the weighted sum rate problem

$$\max_{\mathbf{p} \in P} w_1 R_1(\mathbf{p}) + w_2 R_2(\mathbf{p}) \tag{13.6}$$

for weights $w_1, w_2 \geq 0$ and $w_1 + w_2 = 1$. The descriptive intuition of (13.6) is that the weights determine a line from the origin with slope $\frac{w_2}{w_1}$. This line is shown in Figure 13.4 as dashed line. We then search for the intersection of this line with the Pareto boundary of the achievable rate region, therefore this approach is sometimes called the rate profile approach.

The same framework can also be applied to heterogeneous QoS requirements, such as energy efficiency, security, latency, and throughput in 5G and beyond networks [5].

In Figure 13.4, the same achievable rate region is shown as in Figure 1.2, including two operating points on the Pareto boundary obtained from the two explained solutions to the MOP, the rate profile point with equal weights (black cross) from (13.5), and the maximum sum rate point (blue circle) obtained from (13.6).

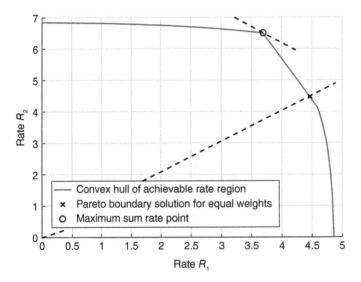

**Figure 13.4** Achievable rate region of the two-user two-carrier IFC and two operating points from the solutions to the MOP in (13.4): the maximum sum rate point shown as a blue circle is the solution to programming problem (13.6) with equal weights, and the Egalitarian operating point shown as a black cross is the solution to the programming problem (13.5) with equal weights.

## 13.4.2 Resource Allocation Games

The second challenge can be systematically modeled, studied, and solved by game theory [18]. Game theory is about optimization with multiple, conflicting objective functions. In conventional optimization, there is a single objective function that usually has a well-defined maximum or minimum. Finding this optimum point is then a matter of applying an appropriate numerical method. In game theory, the notion of optimality is not defined in terms of the maximum or minimum of a single cost function. Rather, the typical objective is to maximize two (or more) functions *jointly*, where the functions are coupled in such a way that increasing one of them necessarily means that the other must decrease. Game theory as a scientific discipline mostly evolved from work in economics during the 20th century. Economics continues to be an important application area of game theory, but more recently the theory has been successfully used in other fields as well, such as resource allocation in engineering problems (especially in communication systems [1, 28]).

Depending on availability and willingness to cooperate, the interaction of the links in the interference channel can be modeled as either a non-cooperative [26] or a cooperative [27] game.

A popular solution concept for non-cooperative games, which are described by the set of players (here the links), the strategies (here the power allocation), and the utility functions (here the rates), is the Nash equilibrium (NE). In order to compute the NE, the best responses of the players are helpful:[2]

$$\mathrm{br}_k(\mathbf{p}_{\overline{k}}) = \max_{\mathbf{p}_k \in \mathcal{P}_k} R_k(\mathbf{p}_k, \mathbf{p}_{\overline{k}}). \tag{13.7}$$

Under certain conditions on the channels [32], the best response dynamics starting from initial power allocation $\mathbf{p}_{\overline{k}}^0$ for link $\overline{k}$

$$\mathrm{br}_k\left(\mathrm{br}_{\overline{k}}(\dots(\mathrm{br}_k(\mathbf{p}_{\overline{k}}^0))\dots)\right) \tag{13.8}$$

will converge to the NE $(\mathbf{p}_1^N, \mathbf{p}_2^N)$, for which it holds for all $k \in \{1, 2\}$

$$R_k(\mathbf{p}_k^N, \mathbf{p}_{\overline{k}}^N) \geq R_k(\mathbf{p}_k, \mathbf{p}_{\overline{k}}^N) \tag{13.9}$$

for all $\mathbf{p}_k \in \mathcal{P}_k$. The best response dynamics in (13.8) is an iterative process that consists of measuring the received SINR values and adapting the transmit power according to (13.7). This iterative process can be executed in serial (by any order among the nodes) or in parallel (by all nodes simultaneously). The convergence of the best response dynamics to the NE, if it occurs, from any initial point, is called global stability. Whenever a non-cooperative game model is introduced, it is checked whether the NE exists, is unique, and fulfils global stability.

In cooperative games, players (here, systems) are allowed to bargain and strike deals with one another. The theory for cooperative games splits into *transferable utility* (the players can pay each other compensation) and *non-transferable utility* (no side payments are allowed). A fundamental point we must understand is that a player can be cooperative and rational at the same time, that is, cooperative is not the same as altruistic. The point is that even if

---

2 Here we assume that the power constraint set is separable as $\mathcal{P} = \mathcal{P}_1 \times \mathcal{P}_2$. Otherwise the correct solution concept is the generalized NE.

players are eventually interested in maximizing their own outcome, they may be willing to accept a bargaining solution that is found to be good enough for both.

The main result by Nash is the following theorem. Let $\mathcal{R}$ be a utility region. Suppose $\mathcal{R}$ is compact and convex and let $r_1^*, r_2^*$ be a so-called threat point. This point is the outcome that is achieved if the players cannot agree on any bargaining outcome. It may be taken, for example, as the NE of the game. Obviously, any meaningful threat point $r_1^*, r_2^*$ must lie inside $\mathcal{R}$. Next, consider a function which maps the set of possible utility regions and the set of possible threat points onto a bargaining solution $(\bar{r}_1, \bar{r}_2)$:

$$(\bar{r}_1, \bar{r}_2) = f(\mathcal{R}, r_1^*, r_2^*) \in \mathcal{R} \tag{13.10}$$

Nash's theorem states that the function $f(\cdot)$, and therefore the bargaining outcome, is uniquely defined under relatively general circumstances. Moreover, this outcome can be easily computed. Under the axioms of feasibility, Pareto-optimality, independence of irrelevant alternatives, symmetry, and independence to linear transform [13], Nash showed that $f(\cdot)$ is unique and given by

$$(\bar{r}_1, \bar{r}_2) = \max_{(r_1, r_2) \in \mathcal{R}} (r_1 - r_1^*)(r_2 - r_2^*). \tag{13.11}$$

The solution $(\bar{r}_1, \bar{r}_2)$ is called the Nash bargaining solution (NBS).

In Figure 13.5 we illustrate the operating points found by the non-cooperative solution (actually by playing the best response dynamics) NE and the cooperative solution NBS with two different conflict points: the origin and the found NE.

We can observe from the figure the impact of the SNR on the price of anarchy. At smaller SNR, the system is noise limited and the non-cooperative solution performs close to the Pareto boundary of the rate region. At high SNR, the system is interference limited and cooperation is required to achieve points close or on the Pareto boundary. The choice of the conflict point has an impact on the NBS as it is observed on the left-hand side of the figure.

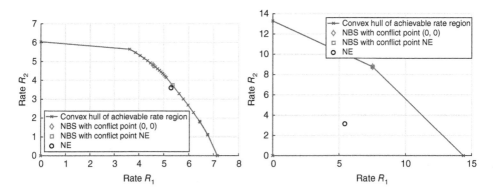

**Figure 13.5** Achievable rate region for the two-user two-carrier interference channel with three operating points: the NE and two NBS for [0, 0] and the NE as conflict point. The left-hand side is for an operating SNR of 10 dB and the right-hand side is for a different channel realization at a higher SNR of 20 dB.

### 13.4.3 Resource Matching for Spectrum Sharing

Finally, for the third challenge, the distributed assignment of links to resource blocks, we propose applying stable matchings [20]. We consider a general scenario with $K$ users and $N$ resources (undivisible) which can be exclusively allocated to any one user. The set of users is denoted by $\mathcal{K} = \{1, ..., K\}$ and the set of resources is denoted by $\mathcal{N} = \{1, ..., N\}$. User $k$ can have up to $q_k$ resources. The resource allocation problem is to match the users to the resources. This is a one-to-many matching problem. These types of problems have a long history since marriages (typically with quota $q_k = 1$) and college admissions ($q_k > 1$) are important and popular examples [35]. Regarding the college student terminology, we identify the students with resources and the colleges with users because one student can go only to one college and one resource is allocated to a single user. In [6] the two sides are spectrum providers (SPs) which correspond to users (colleges) and spectrum users (SUs) which correspond to resources (students).

The two-sided one-to-many matching market is illustrated in Figure 13.6. In the figure, there are four SPs (colleges, users) and five SUs (students, resources) with the following matching: $\{(1, A), (2, D), (3, B), (4, C), (4, E)\}$.

Each user has preferences on the resources based on local information. In wireless communication, the local information contains channel quality information and is given in terms of SINR values. The channel quality of user $k \in \mathcal{K}$ on resource $n \in \mathcal{N}$ is denoted as $\alpha_{k,n} \geq 0$. Thus each user has a preference relation $>_k$ over the subsets of resources. A resource $n \in \mathcal{N}$ is acceptable to user $k$ if the SINR leads to a user utility larger than zero, i.e. $\phi(\alpha_{k,n}) > 0$. The mapping $\phi : \mathbb{R}_0^+ \to \mathbb{R}_0^+$ maps the channel quality to a utility function taking the local context and information of the user into account. For Shannon capacity it is $\phi(x) = \log(1 + x)$ or for finite modulation and coding schemes it is usually a step function. This can lead to non-strict preferences, where neither matching is preferred to each other.

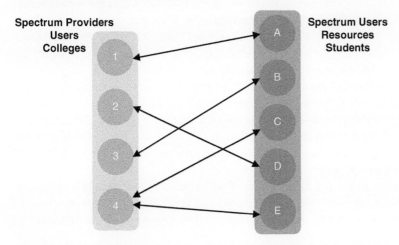

**Figure 13.6** Two-sided one-to-many matching market model with $K = 4$ users (colleges, SPs) and $N = 5$ resources (students, SUs).

Each resource also has preferences on the users based on local information. In wireless communication, the local information could contain channel quality, buffer state, or any context-related information available for resource allocation. Each resource $n \in \mathcal{N}$ has a preference relation $P_n$ over the set of users and being unused $(n)$. A user $k \in \mathcal{K}$ is acceptable to resource $n \in \mathcal{N}$ if $kP_n n$.

A resource allocation problem is specified by the tuple

$$(\mathcal{N}, \mathcal{K}, \mathbf{P}_{\mathcal{N}}, >_{\mathcal{K}}, \mathbf{q}) \tag{13.12}$$

consisting of the set of resources $\mathcal{N}$, the set of users $\mathcal{K}$, the set of preference relations of the resources $\mathbf{P}_{\mathcal{N}} = \{P_n\}_{n \in \mathcal{N}}$, the set of preference relations of the users $>_{\mathcal{K}} = \{>_k\}_{k \in \mathcal{K}}$, and the quota $q_k$, $1 \le k \le K$, describing how many resources a user $k$ can have at most.

A *matching* $\mu$ is a function from the set $\mathcal{N} \cup \mathcal{K}$ into the set of unordered families of elements of $\mathcal{N} \cup \mathcal{K}$ such that:

1. $|\mu(n)| = 1$ for every resource $n \in \mathcal{N}$ and $\mu(n) = n$ if $\mu(n) \notin \mathcal{K}$
2. $|\mu(k)| = q_k$ for every user $k \in \mathcal{K}$ and if the number of resources in $\mu(k)$, say $r$, is less than $q_k$, then $\mu(k)$ contains $q_k - r$ copies of $k$
3. $\mu(n) \in \mathcal{K}$ if and only if $n \in \mu(K)$.

The following definitions on the stability of a matching $\mu$ can be found in [35]. The matching $\mu$ is *blocked* by resource $n$ and user $k$ if resource $n$ strictly prefers $k$ to $\mu(n)$ and either (i) $k$ strictly prefers $n$ to some $n' \in \mu(k)$ or (ii) $|\mu(k)| < q_k$ and $n$ is acceptable to $k$. A matching is *individually rational* if for each resource $n \in \mathcal{N}$ it holds $\mu(n)P_n n$ or $\mu(n) = n$ and for each user $k \in \mathcal{K}$ it holds (i) $|\mu(k)| \le q_k$ and (ii) $n>_k n$ for every $n \in \mu(k)$. A matching is *stable* if it is individually rational and not blocked. A resource allocation mechanism is a systematic way of assigning resources to users. A *stable mechanism* is a mechanism that yields a stable matching for every resource allocation problem $(\mathcal{N}, \mathcal{K}, P_{\mathcal{N}}, >_{\mathcal{K}}, \mathbf{q})$.

The question whether there exists always a stable matchings was first answered positive and constructively in [14] by describing the algorithm which computes one stable matching. Every resource allocation problem has a stable matching. The so-called deferred acceptance (DA) algorithm finds one of these stable matchings. There are two variants available: the resource proposing and the user proposing algorithm.

The convergence of the DA algorithm for both proposing variants is guaranteed. As every proposing SU can at best propose to an SP once regardless of the decision (accept/reject) of the SP, the algorithm's convergence is guaranteed after a finite number of iterations. As there are $N$ proposing SUs and in each iteration $K$ SPs are available to be proposed, the computational complexity of the algorithm is $O(MN)$ provided that preference lists for all $n \in \mathcal{N}$ and $k \in \mathcal{K}$ are *a priori* available.

More information about the applications of stable matchings to wireless communications can be found in the review paper by [4]. An application of stable matching to multi-connectivity is illustrated in [39]. Here, we follow the spectrum sharing scenario from [6].

We evaluate the proposed stable matching approach in the context of the CBRS (see also Chapter 4). CBRS is based on a three-tiered sharing framework in which incumbent users – federal and non-federal – represent the highest tier and are protected from interference generated by the two lower tiers, priority access (PA) and general authorized access

(GAA). A priority access license (PAL) is defined as the authorization to use a 10-MHz channel in a single census tract for three years. In particular, PA users will be protected from interference generated by GAA use, while GAA users will receive no interference protection. PA users will be protected along the contour of the PAL protection area. Around each deployed citizens broadband service device (CBSD) a default protection contour will be determined based on a signal strength of −96 dBm in 10 MHz. PA licensees may opt to reduce their protection area. In fact, PA licensees may enter into spectrum leasing arrangements with approved entities for areas that are within their service area – the census tracts where they have a PAL – and outside of their protection areas.

Since different PA users have potentially different protection areas in the same census tract, each SU can rank PA licensees depending on, for example, the size of the available area, the match between the area of interest (orange and blue areas in Figure 13.7) and the available area, or the distance between the area of interest and the PA protection area. Although leasing agreements can be negotiated individually, in this paper we assume that they will be arranged in a common secondary market in which different PA licensees and SUs can express their preferences. On the other side, PA licensees could rank SUs based on different criteria.

To investigate the behavior of the proposed matching theory approach, we conduct extensive simulations in which the preferences of the SUs are randomly changed to simulate various combinations of SU preferences, resulting in different stable matchings. Since SU preferences depend on distance from the PA licensee protection area, they can change over time. In fact, while the PA licensee protection areas stay the same, the SUs areas of interest can vary due to traffic conditions. For example, SUs could be generally operating as GAAs, and temporarily be interested in acquiring spectrum resources and protection from interference in certain geographic areas in case of special events. We assume that the preferences of the SPs are fixed throughout the simulation period because the SU beamforming capabilities and the target market remain unchanged.

We can evaluate the performance of the many-to-one matching algorithm and show how it improves the matching statistics as compared to one-to-one matching, and minimizes

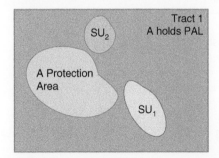 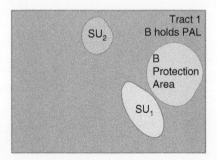

**Figure 13.7** PAL holders will be allowed to lease any bandwidth for any period of time and for any portion of their licensed geographic area within the scope of the PAL but outside of the PAL protection area. Green areas represent the protection areas of two PAL holders in the same census tract. The purple area is the region in which spectrum can be leased to spectrum users (SUs). The orange and blue areas represent the area of interests of two SUs. For example, $SU_2$ might prefer to lease spectrum from PA licensee B because of the larger distance between their areas of operation. For the same reason, for $SU_1$, the two PA licensees would be equivalent (cp. to Fig. 3 in [6]).

**Table 13.1** Many-to-one matching statistics for $K = 3$ and $N = 4$ in percentages: left-hand side, $q_k = 2$, right-hand side $q_A = 2, q_B = q_C = q_D = 1$.

| SU | 1 | 2 | 3 | 4 | SU | 1 | 2 | 3 | 4 |
|---|---|---|---|---|---|---|---|---|---|
| Pref 1 | 88 | 91 | 96 | 88 | Pref 1 | 58 | 76 | 85 | 62 |
| Pref 2 | 12 | 9 | 4 | 12 | Pref 2 | 23 | 14 | 9 | 20 |
|  |  |  |  |  | Pref 3 | 19 | 10 | 5 | 18 |

probability of unallocated spectrum for SUs when $K < N$. The preference list for the SP is given as follows:

$$P(A) = (1, 2, 3, 4) \quad P(B) = (2, 3, 4, 1) \quad P(C) = (3, 4, 1, 2). \tag{13.13}$$

Table 13.1 shows the matching statistics for $K = 3$ and $N = 4$ on the left-hand side when every SP $k$ can provide one spectrum slice each to at most two different SUs at every spectrum allocation instant, i.e. quota $q_k = 2$, while on the right-hand side, only SP $A$ offers two slices and the other SPs offer only one.

SU2 is the real beneficiary of this bias in spectrum availability from different SPs. SP $A$ has two spectrum slices available, which implies that the SUs who are the first and second preferences of SP $A$ will always get spectrum of their first choice if they request spectrum from SP $A$ as their first choice. The SUs who are the second preferences of SPs $B$ and $C$ do not get spectrum slice from SPs $B$ and $C$ with probability one even if they prefer SPs $B$ and $C$.

In summary, this section introduces three major challenges for resource allocation in spectrum sharing networks, namely, multiple objectives, conflicting interests, and distributed implementation. As solution approaches, we propose MOP, and non-cooperative and cooperative game theory solution concepts, namely NE and NBS, and stable matchings. The application of the three approaches is demonstrated by the two-user interference channel setup and in the context of CBRS. The main observations related to the challenges and solutions of spectrum sharing are collected in the following summary.

---

Spectrum sharing leads to major challenges for the resource allocation in terms of multiple objectives, conflicting interests, and distributed implementation. These can be resolved by multi-objective programming, game-theoretic approaches, and stable matchings, respectively.

---

## 13.5 Resource and Spectrum Trading

In this section we proceed and embed the interference channel setup from the last sections to an application scenario, in which two operators serve their users in the same geographical area. Adjacent base stations use a market in the vicinity to trade resources. The market has no authority over the base stations. Many local markets might exist, but we omit the question of which base stations trade through which markets. The base stations measure the load that is caused by the users. The core idea is to sell resources when the base station has or predicts low load and buy resources otherwise for a certain duration and a certain price. Thereby, we move forward from classical spectrum sharing [40] to micro-trading approaches as in [29] and [16].

We follow the system model from [37], and consider two operators $i \in \{A, B\}$ with each operating one evolved Node B (eNodeB) serving some users $UE_{ij}, j = 1, ..., K_i$. The number of available resource blocks (RBs) in the downlink (DL) is $N$. Each operator owns a part of the spectrum of $n_i$ RBs, $n_i < N, n_i \in \mathbb{N}_+$. A gap of $n_g$ RBs between both spectra is modeled as the guard band, i.e. the operators' operating bands are not necessarily adjacent. The whole spectrum satisfies

$$n_A + n_B + n_g = N. \tag{13.14}$$

In Figure 13.8 the spectrum sharing scenario is illustrated. Note that the input queues at the two evolved nodes B (eNBs) with the arrived bits for their respective UEs are also shown. The UEs are numbered according to their serving eNB. There are $K_A$ and $K_B$ UEs. Data arrives in the bit queues with rate $r_{ij}$ and fills a buffer with finite size $B_{ij}$ from eNB $i$ UE $j$.

In the following, we consider two types of spectrum trading: a practical spectrum trading implementation and a hypothetical upper bound achieved by simply merging the two operators (Figure 13.9).

The trading of the resources happens between the two eNBs directly. Selling of spectrum is done via an offer. If an offer is accepted by the other operator and acknowledged, it is called a contract. The contract has a temporal contract length of $t_D > 0$. For the formal definitions of these terms and the temporal flow, the interested reader is referred to [37].

There are many trading algorithms that one could think of. The decision whether to offer or to accept offered spectrum should be based on all the insights from the former sections

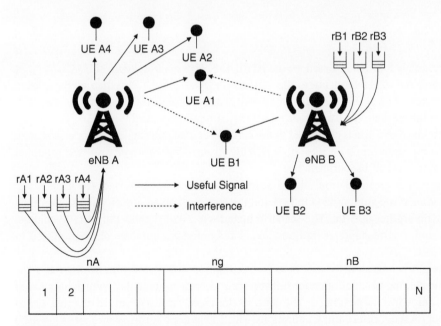

**Figure 13.8** Spectrum sharing scenario between two eNBs of two different operators and a number of UEs. Each operator owns part of the available spectrum $n_A$ and $n_B$ with guard bands $n_g$.

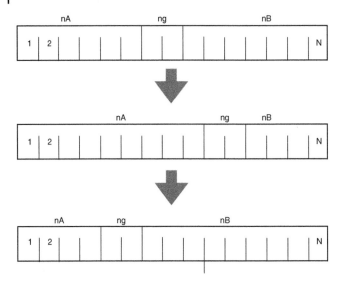

**Figure 13.9** A practical spectrum trading system: two operators exchange spectrum by increasing or decreasing their bandwidth. In the beginning, both operators have the same number of frequency blocks, $n_A = n_B$. Then operator $A$ increases its spectrum and $B$ decreases, so $n_A > n_B$. Finally, operator $B$ has more spectrum than operator $A$, $n_A < n_B$.

and on the current buffer status at the eNBs. In the heuristic example from [37], each eNB computes the so called estimated time to empty buffers (ETEB). It is the ratio of the sum of the current buffer states divided by the sum data rate on all frequency resources, i.e.

$$\text{ETEB}_A = \frac{\sum_{k=1}^{K_A} B_k}{\sum_{i=1}^{n_A} R_i},$$

(13.15)

and accordingly for eNB $B$. Based on the ETEB, the eNBs act as follows. If $ETEB_\ell \leq t_{sell} \leq t_{buy}$, then eNB $\ell$ offers spectrum to the market. If $t_{sell} < t_{buy} < ETEB_\ell$, then eNB $\ell$ tries to buy spectrum. If $t_{sell} < ETEB_\ell < t_{buy}$, then eNB $\ell$ does nothing and revokes possible offers sent from the market. The thresholds $t_{sell}$ and $t_{buy}$ are set heuristically.

The trading algorithm is compared to the extreme case of one single large operator owning the whole spectrum $n_A + n_B = n_S = N$. This virtual *super-operator* results in an upper bound on the achievable performance.

For the numerical simulations, the ns-3 simulation environment is used. The general system parameters are summarized in [37] and the software can be downloaded from http://code.nsnam.org/laa/ns-3-lbt. The scenario includes a mobility model, the 3GPP channel models, and a corresponding traffic model. The qualitative numbers are reproduced in Figure 13.10.

For the 95% percentile of the user perceived throughput (UPT), the gains from the extreme case of no sharing to super-operator full spectrum sharing is about 80%, which aligns well with former works. Depending on the contract duration, various gains are reported. Gains are higher for the shorter contract durations. This comes as no surprise since an eNB will be able to get its resources back quickly. However, short contracts have the drawback of a more

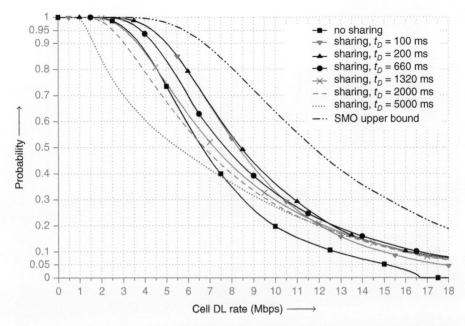

**Figure 13.10** The UPT of a user depending on the duration of a contract $t_D$ compared to the extreme cases of no sharing and the single super-operator with $n_S = n_A + n_B$ spectrum resources.

frequent reselling of resources, putting additional load on the backbone network through increased signaling, an effect that is not modeled here.

There are other performance metrics which are significantly improved by spectrum sharing, including the worst case delay. In addition to the quantitative gains, some qualitative gains are reported from [36]. An example of the user arrival rates and buffer states is shown in Figure 13.11. The *x* axis in all graphs refers to the time in seconds. The *y* axis corresponds to resource block usage and the buffer status of the users. There are 12 spectrum resources available, from which five are used by eNBs A and B, i.e. $n_A = n_B = 5$ in the baseline setup. For the spectrum sharing scenario, two spectrum blocks are shared additionally between the two operators, such that the maximum numbers of spectrum blocks for one eNB is 6. The buffer status of 12 UEs is shown.

From the graphical inspection in Figure 13.11 we can observe that the buffers of the 12 UEs in the baseline scenario are congested heavily around time 655–660 seconds and in particular the buffer of user 12 cannot be depleted for more than 15 seconds. However, for the spectrum sharing scenario, the buffer statistics visibly improve. In particular, user 12 can deplete its buffer within less than 10 seconds. Thereby, we confirm that the buffer statistics are improved by using this variant of simple queuing-aware spectrum trading. Let us summarize the main observation as follows:

> Spectrum sharing realized by heuristic spectrum trading on the network layer can achieve significant gains in terms of user throughput on system level. Depending on system load, the gain on average could be up to 120% and for the 95% percentile it could be up to 184% for the simplistic heuristic.

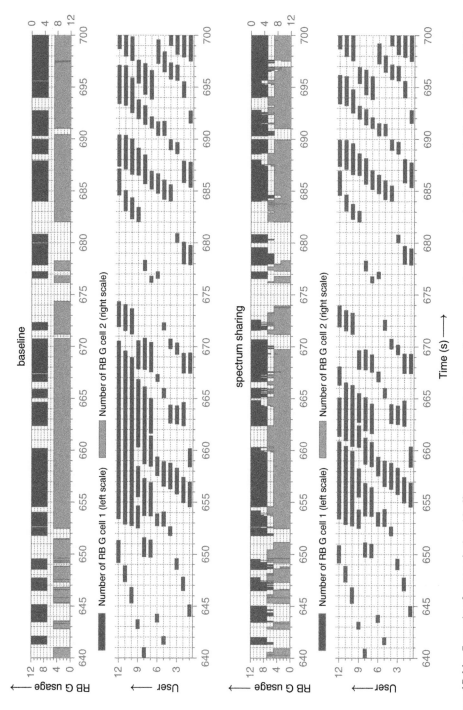

**Figure 13.11** Example of user arrival and buffer processes for the baseline scheme versus the spectrum sharing scenario (from Figure 6.1, [36]).

## 13.6  Conclusions and Future Work

Spectrum sharing is a technology that comprises physical layer and medium access control methods up to service and application layer techniques. It is a very good example of how fundamental results from information theory on achievable rate regions of interference channels lead to efficient and solid system design, including resource allocation and scheduling.

In this chapter, only the basic information theoretic results on the interference channel are reviewed and basic techniques as TIN and time sharing are mentioned. For future work, it is important to stress that more sophisticated coding schemes (e.g., rate splitting) and decoding schemes (successive interference cancellation and simultaneous non-unique decoding) lead to larger achievable rate regions and increase the gain of non-orthogonal compared to orthogonal resource allocation.

The basic types of spectrum sharing include a soft transition from intra-operator to non-orthogonal inter-operator spectrum sharing. On one hand, the more resources are shared, the more efficient is the resource allocation outcome. On the other hand, the more resources are shared, the higher is the complexity and signaling overhead to implement the resource allocation algorithm.

The approaches to model and solve the resource allocation problems in spectrum sharing scenarios include optimization and game theory as well as one matching variant for user to channel assignment. We stress that these techniques are not exhaustive and are only a fraction of the large body of work that has been reported.

The distributed implementation by a simple trading algorithm has shown that the spectrum sharing gains are available on a system level with heuristic sharing algorithms. The benefits include not only higher throughput but also smaller delays in packet delivery. For future work, practical implementations in test environments will show how the gains translate further under practical mobility and traffic scenarios.

In conclusion, with the sophisticated signal processing power available at modern base stations and in backhaul and fronthaul networks, we are able to implement flexible spectrum sharing algorithms. Spectrum sharing will lead to more efficient resource utilization, to lower operational cost of the network owner, and finally to more and better satisfied customers.

## References

1 Game theory in signal processing and communications, dedicated issue of IEEE Signal Processing Magazine, vol. 26, no.5, Sep. 2009.

2 R. Ahlswede. The capacity region of a channel with two senders and two receivers. *Annals of Probability*, 2(5):805–814, 10 1974. doi: 10.1214/aop/1176996549. URL https://doi.org/10.1214/aop/1176996549.

3 F. Baccelli, A. El Gamal, and D.N.C. Tse. Interference networks with point-to-point codes. *IEEE Transactions on Information Theory*, 57(5): 2582–2596, May 2011. ISSN 0018-9448. doi: 10.1109/TIT.2011.2119230.

**4** S. Bayat, Y. Li, L. Song, and Z. Han. Matching theory: Applications in wireless communications. *IEEE Signal Processing Magazine*, 33(6): 103–122, Nov. 2016. ISSN 1053-5888. doi: 10.1109/MSP.2016.2598848.

**5** E. Bjornson, E.A. Jorswieck, M. Debbah, and B. Ottersten. Multi-objective signal processing optimization: The way to balance conflicting metrics in 5G systems. *IEEE Signal Processing Magazine*, 31(6):14–23, Nov. 2014. ISSN 1053-5888. doi: 10.1109/MSP.2014.2330661.

**6** M.M Butt, I. Macaluso, E.A. Jorswieck, J. Bradford, N. Marchetti, and L. Doyle. Spectrum matching in licensed spectrum sharing. *Transactions on Emerging Telecommunications Technologies*, 29(10):e3476, 2018. doi: 10.1002/ett.3476. URL https://onlinelibrary.wiley.com/doi/abs/10.1002/ett.3476. e3476 ett.3476.

**7** G. Caire, D. Tuninetti, and S. Verdu. Suboptimality of TDMA in the low-power regime. *IEEE Transactions on Information Theory*, 50(4): 608–620, Apr. 2004. ISSN 0018-9448. doi: 10.1109/TIT.2004.825003.

**8** P. Cao, E.A. Jorswieck, and S. Shi. Pareto boundary of the rate region for single-stream MIMO interference channels: Linear transceiver design. *IEEE Transactions on Signal Processing*, 61(20):4907–4922, Oct. 2013. ISSN 1053-587X. doi: 10.1109/TSP.2013.2272922.

**9** A.B. Carleial. Interference channels. *IEEE Transactions on Information Theory*, 24:60–70, 1978.

**10** R H. Coase. The federal communications commission. *Journal of Law and Economics*, pages 1–40, 1959.

**11** M.H.M Costa. On the Gaussian interference channel. *IEEE Transactions on Information Theory*, 31:607–615, 1985.

**12** R.H. Etkin, D.N.C. Tse, and H. Wang. Gaussian interference channel capacity to within one bit. *IEEE Transactions on Information Theory*, 54(12):5534–5562, Dec. 2008. ISSN 0018-9448. doi: 10.1109/TIT.2008.2006447.

**13** D. Fudenberg and J. Tirole. Game Theory. MIT Press, 1993.

**14** D. Gale and L. Shapley. College admissions and the stability of marriage. *American Mathematical Monthly*, 69:9–15, 1962.

**15** A. El Gamal and Y.-H. Kim. Network Information Theory. Cambridge University Press, 2011.

**16** P. Grønsund, O. Grøndalen, K. Mahmood, and P.H. Lehne. Implementation of spectrum micro-trading for mobile operators in the spatial dimension. In *2014 European Conference on Networks and Communications (EuCNC)*, pages 1–5. Jun. 2014. doi: 10.1109/EuCNC.2014.6882654.

**17** T. Han and K. Kobayashi. A new achievable rate region for the interference channel. *IEEE Transactions on Information Theory*, 27:49–60, 1981.

**18** Z. Han, D. Niyato, W. Saad, T. Başar, and A. Hjørungnes. Game Theory in Wireless and Communication Networks: Theory, Models, and Applications. Cambridge University Press, 2011. doi: 10.1017/CBO9780511895043.

**19** C. Hellings and W. Utschick. Improper signaling versus time-sharing in the two-user Gaussian interference channel with TIN. *IEEE Transactions on Information Theory*, Aug. 2018. submitted.

**20** E.A. Jorswieck. Stable matchings for resource allocation in wireless networks. In *2011 17th International Conference on Digital Signal Processing (DSP)*, pages 1–8, Jul. 2011. doi: 10.1109/ICDSP.2011.6004983.

**21** E.A. Jorswieck, E.G. Larsson, and D. Danev. Complete characterization of the Pareto boundary for the MISO interference channel. *IEEE Transactions on Signal Processing*, 56(10):5292–5296, Oct. 2008.

**22** E.A. Jorswieck, L. Badia, T. Fahldieck, E. Karipidis, and J. Luo. Spectrum sharing improves the network efficiency for cellular operators. *IEEE Communications Magazine*, 52(3):129–136, Mar. 2014. ISSN 0163-6804. doi: 10.1109/MCOM.2014.6766097.

**23** D.M. Kalathil and R. Jain. Spectrum sharing through contracts for cognitive radios. *IEEE Transactions on Mobile Computing*, 12(10): 1999–2011, Oct. 2013. ISSN 1536-1233. doi: 10.1109/TMC.2012.171.

**24** M.G. Kibria, G.P. Villardi, K. Nguyen, K. Ishizu, and F. Kojima. Heterogeneous networks in shared spectrum access communications. *IEEE Journal on Selected Areas in Communications*, 35(1):145–158, Jan. 2017. ISSN 0733-8716. doi: 10.1109/JSAC.2016.2633014.

**25** S. Kusaladharma and C. Tellambura. An Overview of Cognitive Radio Networks, pages 1–17. John Wiley and Sons, 2017. ISBN 9780471346081. doi: 10.1002/047134608X.W8355. URL https://onlinelibrary.wiley.com/doi/abs/10.1002/047134608X.W8355.

**26** E.G. Larsson, E.A. Jorswieck, J. Lindblom, and R. Mochaourab. Game theory and the flat-fading Gaussian interference channel. *IEEE Signal Processing Magazine*, 26(5):18–27, Sep. 2009. ISSN 1053-5888. doi: 10.1109/MSP.2009.933370.

**27** A. Leshem and E. Zehavi. Cooperative game theory and the Gaussian interference channel. *IEEE Journal on Selected Areas in Communications*, 26(7):1078–1088, Sep. 2008. ISSN 0733-8716.

**28** A.B. MacKenzie and L.A. DaSilva. Game theory for wireless engineers. *Synthesis Lectures on Communications*, 1(1):1–86, 2006. doi: 10.2200/S00014ED1V01Y200508COM001. URL https://doi.org/10.2200/S00014ED1V01Y200508COM001.

**29** R. MacKenzie, K. Briggs, P. Gronsund, and P.H. Lehne. Spectrum micro-trading for mobile operators. *IEEE Wireless Communications*, 20(6):6–13, Dec. 2013. ISSN 1536-1284. doi: 10.1109/MWC.2013.6704468.

**30** I. Maric, R.D. Yates, and G. Kramer. Capacity of interference channels with partial transmitter cooperation. *IEEE Transactions on Information Theory*, 53:3536–3548, 2007.

**31** R.T. Marler and J.S. Arora. Survey of multi-objective optimization methods for engineering. *Structural and Multidisciplinary Optimization*, 26(6): 369–395, Apr 2004. ISSN 1615-1488. doi: 10.1007/s00158-003-0368-6. URL https://doi.org/10.1007/s00158-003-0368-6.

**32** J. Pang, G. Scutari, F. Facchinei, and C. Wang. Distributed power allocation with rate constraints in Gaussian parallel interference channels. *IEEE Transactions on Information Theory*, 54(8):3471–3489, Aug. 2008. ISSN 0018-9448. doi: 10.1109/TIT.2008.926399.

**33** T.S. Rappaport. Wireless Communications. Prentice Hall, 2 edition, 1996.

**34** A. Roessler. Impact of spectrum sharing on 4G and 5G standards a review of how coexistance and spectrum sharing is shaping 3GPP standards. In *2017 IEEE International Symposium on Electromagnetic Compatibility Signal/Power Integrity (EMCSI)*, pages 704–707, Aug. 2017. doi: 10.1109/ISEMC.2017.8077958.

**35** A.E. Roth and M.A.O. Sotomayor. Two-sided Matching: A study in game-theoretic modeling and analysis. Cambridge University Press, 1990.

**36** R. Schmidt. Gains of Spectrum Sharing in Future Cellular Wireless Networks: System Level Simulations using ns-3". Master's thesis, TU Dresden, School of Engineering, 01062 Dresden, Germany, 2017.

**37** R. Schmidt, A. Toyser, S. Naik, J. Nötzel, and E.A. Jorswieck. Network resource trading: Locating the contract sweet spot for the case of dynamic and decentralized non-broker spectrum sharing. In P. Marques, A. Radwan, S. Mumtaz, D. Noguet, J. Rodriguez, and M. Gundlach, editors, Cognitive Radio Oriented Wireless Networks, pages 3–14, Cham, 2018. Springer International Publishing. ISBN 978-3-319-76207-4.

**38** X. Shang, B. Chen, and M.J. Gans. On the achievable sum rate for MIMO interference channels. *IEEE Transactions on Information Theory*, 52: 4313–4320, 2006.

**39** M. Simsek, T. Hößler, E. Jorswieck, H. Klessig, and G. Fettweis. Multi-connectivity in multi-cellular, multi-user systems: A matching-based approach. *Proceedings of the IEEE*, 107(2): 394–413, 2018.

**40** R.H. Tehrani, S. Vahid, D. Triantafyllopoulou, H. Lee, and K. Moessner. Licensed spectrum sharing schemes for mobile operators: A survey and outlook. *IEEE Communications Surveys Tutorials*, 18(4):2591–2623, Fourthquarter 2016. ISSN 1553-877X. doi: 10.1109/COMST.2016.2583499.

**41** K.H. Teo, Tao Z., and Zhang J. The mobile broadband WiMAX standard [standards in a nutshell]. *Signal Processing Magazine, IEEE*, pages 144–148, Sep. 2007.

**42** K. Werbach. Open spectrum: The new wireless paradigm. *New America Foundation, Spectrum Policy Program*, Oct. 2002.

**43** S. Xiaohu and C. Biao. Two-user Gaussian interference channels: An information theoretic point of view. *Foundations and Trends® in Communications and Information Theory*, 10(3):247–378, 2013. ISSN 1567-2190. doi: 10.1561/0100000071. URL http://dx .doi.org/10.1561/0100000071.

**44** X. Ying, M.M. Buddhikot, and S. Roy. Coexistence-aware dynamic channel allocation for 3.5 GHz shared spectrum systems. In *2017 IEEE International Symposium on Dynamic Spectrum Access Networks (DySPAN)*, pages 1–2, Mar. 2017. doi: 10.1109/DyS-PAN.2017.7920771.

**45** L. Zadeh. Optimality and non-scalar-valued performance criteria. *IEEE Transactions on Automatic Control*, 8(1):59–60, Jan. 1963. ISSN 0018-9286. doi: 10.1109/TAC.1963.1105511.

# 14

# Unlicensed Spectrum Access in 3GPP

*Daniela Laselva[1], David López-Pérez[2], Mika Rinne[3], Tero Henttonen[4], Claudio Rosa[1], and Markku Kuusela[5]*

[1] *Nokia Bell Labs Aalborg, Denmark*
[2] *Nokia Bell Labs Dublin, Ireland*
[3] *Nokia Technologies, Finland*
[4] *Nokia Bell Labs CTO, Finland*
[5] *Nokia CSD Digital Automation, Finland*

## 14.1   Introduction

The use of license-exempt bands offers an opportunity for operator networks to exploit large amounts of additional spectrum without related cost. With this driver, the 3rd Generation Partnership Project (3GPP) has been studying alternative means for long-term evolution (LTE) and new radio (NR) operations to exploit unlicensed bands, provisioning standardized solutions for both LTE and NR at present [1]. For several years, the 3GPP approach has built on the use of complementary wireless local area network (WLAN) technology [2] by integrating, interworking or aggregating it under the control of the mobile operator network. These solutions involve control entities in the core network, in the radio access network or in both. An additional differentiator of these solutions is whether the WLAN itself is a known partner network of the operator or is an untrusted WLAN. In both cases, the motivation is to be able to use the large installed base of WLAN deployments and to quickly adapt to traffic changes in spotty areas. Recently, the 3GPP community got excited about standardizing a native LTE solution operable in the unlicensed bands, known as *licensed-assisted access (LAA)*. LAA operates as a carrier aggregation of component carriers present in the unlicensed bands, where the primary carrier is in the licensed band and additional component carriers can be operated in the unlicensed band in non-standalone fashion. For this reason, the 3GPP studied the coexistence of LTE carriers with known systems in the unlicensed bands, and specified mechanisms that at least meet or exceed the spectrum requirements. The studies showed that LTE carriers on an unlicensed band offer an efficient booster for high throughput, yet these make a fair and friendly neighbor both to the other LTE carriers and WLAN networks on the same spectrum. Standalone operations of 3GPP systems on unlicensed bands were first envisioned by industry alliances such as the Multe-Fire Alliance™ [3]. Recently, the required technical solutions for standalone operations of NR in the unlicensed spectrum are being addressed by 3GPP, as discussed in section 14.6.

*Spectrum Sharing: The Next Frontier in Wireless Networks,* First Edition.
Edited by Constantinos B. Papadias, Tharmalingam Ratnarajah, and Dirk T.M. Slock.
© 2020 John Wiley & Sons Ltd. Published 2020 by John Wiley & Sons Ltd.

This chapter will introduce the unlicensed spectrum access technologies from the 3GPP standard point of view according to the following structure. Section 14.2 describes LTE-WLAN aggregation (LWA), focusing on how WLAN can be aggregated to operate under the control of LTE and how the aggregation of traffic flows works for the radio bearers in the LTE convergence protocol. Special attention is given to describing the necessary network interfaces, how to manage the aggregation in the network, and how the existing networks can be updated to cope with the aggregation. For WLAN, we use the commodity of the Institute of Electrical and Electronics Engineers (IEEE) 802.11 standard (version 11ac) while yet recognizing that there is a new version (11ax) under development. That version is planned to include multi-user scheduling across multiple channels, but in each channel it still uses the same access mechanisms as the commodity WLAN here. Section 14.3 introduces the alternative of LTE-WLAN radio level integration over an Internet protocol (IP) secured tunnel (LWIP). Setting up the IP security (IPSec) tunnel is a burden, but it allows both the WLAN operations and achieving aggregated data rates to be hidden by means of traffic steering. LAA of LTE is presented in Section 14.4. This section describes what in LTE technology needed to change in order to enable its operation in an unlicensed band. Section 14.5 presents the performance metrics and technology analysis to summarize the characteristics of these unlicensed technologies and address their coexistence. Section 14.6 provides an outlook of the anticipated research and standardization directions in the context of NR operations in the unlicensed spectrum. Finally, section 14.7 concludes the chapter.

## 14.2 LTE-WLAN Aggregation at the PDCP Layer

LWA stands for LTE-WLAN aggregation, where the WLAN network operates under the control of an LTE evolved node B (eNB), which aggregates traffic flows for the radio bearer in the LTE packet data convergence protocol (PDCP) [4]. LWA operations are not visible to the core network since the standard user plane interface acts between the radio access network (RAN) and the core network without changes.

In LWA, WLAN access points can be co-located or non-co-located to an eNB, and they operate in any of the bands at around 2.4, 5, or 60 GHz. On one side, the non-co-located case allows higher density, independent placement, and, more importantly, leverage of existing WLAN deployments. However, it relies on the availability of high capacity and low latency backhaul to communicate with the WLAN access points. On the other side, the co-located case enables integration of WLAN to a small cell eNB in new deployments. Hence, this allows leveraging of the full potential of LWA, thanks to the timely availability of radio channel and load conditions of both accesses in the same network node. This allows the unlicensed component carriers to flexibly extend the capacity offering of the eNB. Given its deployment flexibility, LWA technology is applicable in a wide range of scenarios, including outdoor and indoor deployments such as public hotspots, enterprises, and shopping malls.

In the following, we present the LWA operations in detail, discussing how the aggregation is conveniently realized at a higher radio layer (i.e., PDCP) causing no impact on lower radio layers [i.e., the physical/media access control (MAC) of both LTE and WLAN], while maintaining the quality of service (QoS) and security level of LTE.

### 14.2.1 User Plane Radio Protocol Architecture

In this section, we review in detail the LWA radio protocol architecture as well as the baseline procedures. Although LWA bearers may be configured to include the delivery of both *downlink* and *uplink data* over WLAN, the following review will focus on the downlink communication due to its dominance in data rates. The presented operations are inverted in the uplink.

We start with the radio architecture, as illustrated in Figure 14.1, which shows the flexible support for the co-located scenario (left) and non-co-located scenario (right). For the latter, LWA adopts an open network interface, denoted Xw, between the eNB and a new WLAN termination (WT) function, which is located inside the WLAN network. The Xw interface procedures, terminated at the WT, are defined for interface setup, configuration, modification, and release in the control plane. Furthermore, these procedures support data delivery and delivery status in the user plane.

According to the figure, LWA integrates WLAN into the LTE radio protocol stack below the PDCP layer by reusing the LTE dual connectivity framework [1]. As for dual connectivity, the PDCP transmitter controls the handling of protocol data units (PDUs) across cellular and WLAN transmissions. In the downlink, the user equipment (UE) may receive PDCP PDUs over both LTE and WLAN interfaces, and the PDCP receiver at the UE performs their processing, including reassembling the PDUs from both interfaces, discarding the PDUs that are duplicated or late (exceeding a discard timer), and reordering them for in-sequence

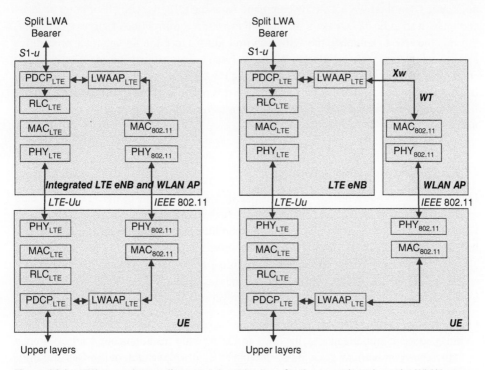

**Figure 14.1** LWA user-plane radio protocol architecture for the scenarios where the WLAN access point (AP) is co-located (left) and non-co-located (right) with the LTE eNB.

delivery to the upper layers. An LWA adaptation protocol (LWAAP) had to be supplemented to the legacy radio protocol stack to include the identifier of the LTE bearer, which carries the application data. Such an identifier is required to properly route, to the LTE protocol stack, the data traversing the WLAN [5]. Similarly, the EtherType field of the WLAN frame has to be populated with the value of 0x9E65 that has been specifically assigned by IEEE for the purpose of LWA operations, and whose presence ensures that the LWA PDUs can be identified, and thus can traverse transparently through the WLAN network. In addition, such a field allows the receiving UE to transfer the contents of the frames with the dedicated EtherType value from the WLAN chip directly to the LTE modem.

For the PDUs traversing the WLAN link, dedicated QoS and security design has been considered to ensure carrier grade service. First, LTE bearer level QoS is applied to any LWA bearer. This means that the QoS level in the WLAN delivery should reflect the LTE QoS level as closely as possible by mapping the bearer data to the most appropriate WLAN QoS level (access class) [2]. Second, the cellular security is applied to any PDCP PDU of an LWA bearer using the eNB created security keys, according to the PDCP ciphering mechanism [6]. Thus, the cellular security is also applied to those PDUs routed over WLAN to comply with the cellular security requirements over the entire data path of LWA.

LWA is applicable to UEs in the connected mode of LTE radio resource control (RRC), i.e., with data transmission and reception enabled with the network. For that, RRC has been extended to allow the UE to report WLAN measurements of the eNB-configured WLANs, which in turn permits the benefits of activating LWA and which WLAN to use [7] to be determined. On the LWA activation command from the eNB, the UE will attempt an association with an indicated access point and will provide a confirmation of success (or failure) towards the eNB to complete the activation procedure. Once the Xw interface procedures are also completed, with the selected WLAN WT, the transferring of data via WLAN for the configured radio bearer can be initiated. During LWA operations, the eNB remains in control of mobility inside the cellular network. For this purpose, the eNB can define a WLAN mobility set for each LWA UE, wherein the UE uses the mobility mechanisms of the WLAN network.

### 14.2.2 Bearer Type and Aggregation

LWA design offers a flexible aggregation capability, as it allows the bearer to schedule data towards LTE and WLAN at an *IP packet level* (carried as a PDCP PDU). When combined with its ability to consider instantaneous radio conditions and congestion situations in both links, it makes LWA-based aggregation rather promising, compared to other techniques, allowing simultaneous UE connectivity to multiple radio accesses [8, 9]. The framework enables any of the following operation modes and their dynamic re-evaluation (i.e., switching) to fully exploit the capacity gains from aggregation:

a) *Aggregation or split bearer mode*: Splitting the packets of a bearer to *both* LTE and WLAN radio access simultaneously. This aims to balance the instantaneous load or latency experienced on both accesses, thus allowing efficient aggregation of licensed and unlicensed spectrum. This mode enables the UE to reach a peak throughput theoretically equal to the sum of peak data throughputs that can be obtained via both links.

b) *Link switching mode (fast or slow)*: Switching the packets of a bearer to *either* LTE *or* WLAN link at a given time. This allows only the best performing radio access to be used at any one time. The selected access can change on a fast basis (e.g., hundreds of milliseconds) or slower (e.g., a few seconds), but it needs no RRC reconfiguration to indicate the change of the serving link because the receiver can process packets as they arrive from either link. This mode enables UE to reach a peak throughput equal to the momentary peak throughput obtained via either link.

c) *Offloading or switch bearer mode*: Switching packets of a bearer entirely to WLAN. This allows LTE resources to be released if congested unless the WLAN performance degrades too. This mode enables the UE to reach a peak throughput equal to the throughput obtained in WLAN. When needed, the switch back to LTE bearer is feasible and requires an RRC reconfiguration.

As a compromise between the user experienced performance and the device complexity, the 3GPP has introduced two new bearer types, namely *split LWA bearer* and *switched LWA bearer*. The split LWA bearer embraces all the above-mentioned modes by means of eNB scheduling, whereas the switched LWA bearer only supports the offloading mode to WLAN, thus removing the complexity required for aggregation [1]. The applicability of a bearer type depends on the UE capability. It is noted that data costs of aggregation are assumed to be associated with the LTE subscription for any of the LWA modes. If additional costs of WLAN are involved, they will be encountered when the UE selects the serving WLAN network and will depend on the selected aggregation mode. In the following, we will focus on the split LWA bearer.

### 14.2.3 Flow Control Schemes

In this section, the downlink flow control scheme that the eNB employs for LWA is reviewed [4, 11].

The flow control algorithm for LWA acts in the PDCP protocol layer and takes care of splitting the bearer data PDUs between the LTE and WLAN links. An optimal split requires a good estimation of the number of preceding PDCP bits successfully received at the UE through both links. The RLC status report available in RLC acknowledgment mode can be leveraged to determine the status of the LTE link. As no similar report was available for the WLAN link, a new feedback was introduced to be transferred over the Xw interface to indicate the WLAN status in term of WLAN data rate. However, to limit the impact on the WLAN infrastructure, an alternative feedback was provided as well in the form of an LWA status report. In this, a UE gives feedback to the eNB on PDCP PDUs traversing the WLAN network in a form that is more compact than the existing PDCP status report. The reporting periodicity can be set flexibly, where a more frequent report results in better performance due to more accurate statistics at the expense of higher overhead.

In the following, a flow control algorithm for LWA is presented, aiming at reducing the expected round-trip time of PDCP PDUs across the LTE and WLAN links. This metric is essential to achieve a minimized PDCP re-ordering delay and a faster in-sequence delivery to the upper layer. The latter plays a key role in improving the performance of any application running on transmission control protocol (TCP) or user datagram protocol (UDP).

It also minimizes the likelihood that the delivery to the upper layers includes gaps, which would trigger retransmissions by the TCP (as an end-to-end protocol). The delay estimates of the LTE path, ($d_{LTE}$) and WLAN path ($d_{WLAN}$) are obtained by mining RLC statistics and the LWA status reports, respectively. The algorithm decides whether a data unit arriving at the transmitting PDCP for a split bearer should be:

- forwarded to the LTE link ($d_{LTE} \leq d_{WLAN}$ and $d_{LTE} \leq d_{max}$)
- forwarded to the WLAN link via the Xw interface ($d_{LTE} > d_{WLAN}$ and $d_{WLAN} \leq d_{max}$) or
- discarded or held at the PDCP layer and its routing decision postponed, when ($d_{LTE} > d_{max}$ and $d_{WLAN} > d_{max}$), until congestion conditions are relieved in either link,

where $d_{max}$ is the maximum queuing delay limit, if any. On congestion detection, to avoid aggravating the situation, packets may be held or discarded at the PDCP layer, thus limiting the number of PDUs in flight between the splitting and reordering functions. This avoids both buffer underflow and overflow conditions due to rapid changes in the link capacity (which are known to occur in WLAN due to its way of sharing spectrum resources by contention). For more details on the flow control mechanism, please refer to [11].

## 14.3   LTE-WLAN Integration at IP Layer

LWIP stands for LTE WLAN radio level integration over an IPSec tunnel, and it was created to supplement LTE by exploiting the large installed base of WLAN access points that an operator may own or control [4]. Importantly, LWIP was designed with the intent of not imposing *any modifications* to those access points. This is achieved by letting LTE provide encapsulated data over the IPsec tunnel, which is transparent to the WLAN it traverses [1].

### 14.3.1   User Plane Radio Protocol Architecture

In WLAN, especially in large coverage multi-user environments, the uplink may become vulnerable because of congestion, which can be avoided by offloading uplink traffic to the LTE when critical [12]. To this end, the LWIP design aims to provide hotspot coverage with more consistent performance, typically for indoor scenarios, where WLAN has a significant footprint. It is therefore suitable for public venues and private enterprises, which have the need for improved coverage, reliability, and throughput compared to sole WLAN transmissions.

For setting up the IPSec tunnel, LWIP requires a security association to be negotiated between the UE and a RAN level security gateway (SeGW). Figure 14.2 depicts the radio protocol architecture and network elements for the scenario where the SeGW is co-located (left) or non-co-located (right) with the eNB [4]. In the co-located case, the interface between the eNB and the SeGW can be internal to the eNB, whereas in the latter case an extended Xw interface is utilized. The figure illustrates that the integration of flows between LTE and WLAN occurs above the LTE radio protocols, i.e., at the IP layer above PDCP. It can also be seen that the packet flows traversing the IPsec tunnel resemble IP packets and therefore both the LTE and IPsec flows utilize the same RAN-to-core network

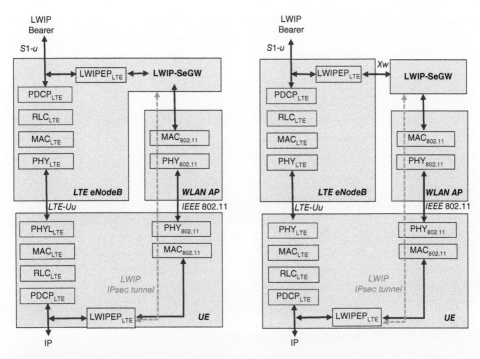

**Figure 14.2** LWIP user-plane radio protocol architecture for the scenarios where the LWIP-SeGW is co-located (left) and non-co-located (right) with the eNB.

interface, S1-u. The figure shows how each IP packet at the eNB is first encapsulated with the LWIP extension protocol (LWIPEP) [13] using the generic routing encapsulation (GRE) protocol [14]. Such encapsulation adds a dedicated header for bearer identification, similarly to the LWA adaptation protocol, described in section 14.2.1. Thereafter, the packets are delivered to the SeGW, which encapsulates them further into the IPsec tunnel and transparently passes them through the WLAN network to the UE. Finally, the UE extracts the IP packets from the IPsec tunnel for use by its IP stack.

Control plane functionality is configured by the RRC between the eNB and the UE [7], that is, LWIP uses the same control plane definitions of the WLAN mobility set and RRC measurement reporting for the eNB-controlled WLAN selection, as described for LWA in section 14.2.1.

After the UE completes regular WLAN association and authentication procedures to an WLAN access point within the RRC configured WLAN mobility set, the eNB can activate the LWIP functionality for a radio bearer. The activation can be done, for example, based on WLAN measurement reports from the UE. The activation is based on provisioning the necessary security information to both the UE and the SeGW [6]. This enables the UE to request the establishment of the IPsec tunnel from the SeGW and causes the IPSec tunnel to be set up according to the assigned security association [4]. Once the IPsec tunnel exists between the UE and the SeGW, the set of radio bearers configured for the LWIP operation can carry traffic via the IPsec tunnel over the WLAN. For the uplink flows delivered over WLAN, the SeGW terminates the IPSec tunnel and routes the flows to S1-u for both the co-located and non-co-located cases.

### 14.3.2  Flow Control Schemes

LWIP was initially designed to operate switched bearers only (without aggregation), which means that the WLAN and LTE links could not be simultaneously used for a bearer, contrary to LWA [4, 15]. However, different bearers could still make use of LTE and WLAN simultaneously.

Building on the LWA aggregation framework, LWIP was later enhanced to support aggregation, thus permitting the concurrent use of LTE and WLAN for a bearer. However, in contrast to LWA, fast switching and aggregation are still not feasible due to the absence of standardized operations such as fast feedback of the flow status, packet reordering, and duplicate discard functions at the IP layer [15].

Accounting for the current design limitations, the flow control algorithms for LWIP are described next based on *link-switching operations*, when assuming proprietary extensions therein.

Similarly to LWA, the UE can be configured to report the WLAN received signal strength indicator (RSSI) of the serving and neighboring WLAN access points to the eNB, and the eNB may perform link switching when the signal strength of the serving WLAN gets weaker or stronger relative to a threshold [15, 16]. As an example, the eNB may smoothly switch the bearer from WLAN to LTE when the UE moves away from the coverage footprint of the serving WLAN. However, just changing the serving WLAN access point (e.g., when moving into a different room) need not result in switching traffic to LTE. In addition to RSSI, other WLAN metrics of the hotspot 2.0 standard (such as the number of UEs associated with the serving access point or the backhaul data rate) acquired from the WLAN beacon or from the hotspot server can be considered for the switching decision. However, none of this information enables congestion to be detected in WLAN, and the potential gains of load balancing remain limited.

To enable dynamic switching between the LTE and WLAN links, proprietary extensions to LWIP are beneficial, namely, the packet reordering capability and flow control feedback at least are important for the IP packets, which are received through the two access technologies. Motivated by this objective, [15] provides an example of a flow control algorithm that introduces IP level packet reordering and duplicate discarding at the UE, which allow smoother interaction with UDP/TCP protocols, and channel probing to estimate the WLAN throughput (i.e., link quality) at the eNB.

In general, the eNB can obtain radio link statistics from the UE for the LTE link mining RLC statistics (e.g., MAC throughput), but these statistics are not available to the eNB for the WLAN link. Probing over the WLAN is therefore necessary as it allows feedback to be obtained from the UE to estimate its performance over WLAN. In [15], this estimate is obtained by sending IP probe packets from the eNB through the WLAN link to the UE and then observing their performance. During connection setup, large IP probe packets are sent over and the average number of probes lost, the average probe delay, and the average probe throughput are computed by the UE. During data transmission, small IP probe packets are periodically intercalated with the IP data packets, working as delimiters, to help the UE to calculate the average UE throughput ($U_u^{\text{avg}}$) between two consecutive active probes in a transmission opportunity. This information is then fed back from the UE to the eNB through a probe-acknowledgement using the LTE link, and then used for the selection and adjustment of the preferred access, as well as for detecting congestion in WLAN and subsequently switching the bearer data to LTE if necessary. In more detail, and taking the data transmission phase as an example, the eNB takes the average UE throughput samples ($U_u^{\text{avg}}$), puts them through a moving average filter ($\widehat{U}_u^{\text{avg}}$), and takes the following decisions:

- *Stall detection*: If ($x_u^{\text{stall}}$) consecutive probe-acknowledgements are missing, the eNB switches the traffic to the LTE link.
- *Inactivity detection*: If the average UE throughput in between a number of consecutive active probes is smaller than a threshold ($U_u^{\text{avg}} < U_u^{\text{min}}$), meaning that the UE generates a small amount of traffic, the eNB switches the traffic to the LTE link or enables WLAN-only transmission mode to save resources.
- *Congestion detection*: If the average UE throughput is smaller than a threshold ($\widehat{U}_u^{\text{avg}} < TPH_u^{\text{min}}$), the eNB switches the traffic to the LTE link.

Otherwise, by default, the eNB keeps the traffic on the WLAN link. For more details on the flow control mechanism, refer to [15].

## 14.4   LTE in Unlicensed Band

LAA leverages LTE radio access technology by extending its operations into the unlicensed 5-GHz band. This extension is made feasible by the LTE carrier aggregation framework, where LAA supports secondary cell operation in the unlicensed spectrum, yet always requires a primary cell in the licensed spectrum. The standardization of LAA was first introduced for downlink-only operation. Next, uplink transmissions were designed, initially consisting of sounding reference signals and scheduled uplink transmissions, and finally completed to carry autonomous uplink transmissions [1, 17, 18].

### 14.4.1   Spectrum and Regulations

LAA has so far been targeted at unlicensed spectrum deployments that utilize carrier extensions in the 5-GHz band. In many regions, depending on the country and the local area of use, large amounts of spectrum are available in the 5-GHz band (see Figure 14.3). For example, in Europe there are 455 MHz of spectrum available and in the USA there are even more. In China, Korea, and Japan large allocations are available as well. The LAA design complies at least with all regional regulatory requirements that affect the use of 5-GHz unlicensed spectrum. Yet it aims at efficient operation and fair coexistence between different devices, networks, and systems beyond the minimum requirements. In the following, the key requirements are reviewed and a comprehensive survey of the global regulatory requirements for the 5-GHz band can be found in [19].

The requirements mandate clear channel assessment (CCA) and listen-before-talk (LBT) mechanisms. LBT is a contention-based protocol that allows devices to share the same radio

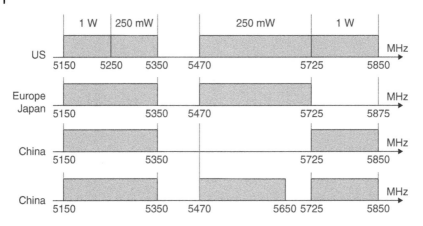

**Figure 14.3** 5-GHz unlicensed band availability by region.

channel without centralized coordination. Any transmission by a device on a radio channel is conditional and is provided only after having sensed the channel is idle. In addition, a device can occupy the channel only for a limited time before it is required to make a new assessment of the channel state by executing LBT again. By far, the most dominant unlicensed technology is WLAN defined by the IEEE 802.11 standard [2], and its devices and access points are typically certified by the Wi-Fi Alliance™. WLAN also uses LBT mechanisms to ensure fair coexistence between WLAN equipment. This also offers fairness and collision avoidance towards other radio technologies, possibly operating in the unlicensed band. The LAA LBT algorithm is further discussed in section 14.4.2.

Another regulatory requirement determines that all transmissions need to use the band efficiently, and therefore a device occupying a channel needs to occupy it with at least 80% of the nominal bandwidth. Furthermore, the maximum transmission power and the maximum power spectral density are subject to limitations. The maximum transmit power is limited to ranges between 200 mWatt and 1 Watt, depending on the frequency in use and depending on the region. In Europe, for example, the maximum power spectral density is limited to 10 dBm/MHz.

### 14.4.2 Channel Access

LAA introduces two types of channel access procedures, called Category 4 (Cat4) and Category 2 (Cat2) [20]. Cat4 LBT implements a channel access procedure with random back off and a variable size contention window. The transmitter starts its intention to transmit by sensing the channel during the slots of a defer duration $T_d$. The channel is sensed to be idle if the received power is below the energy detection threshold. In LAA, the energy detection threshold on a 20-MHz channel is typically set to –72 dBm. After sensing, the transmitter sets the value of the counter ($N$) to a random number generated between values zero and the contention window size. Next, the channel is sensed during the consecutive CCA slots, each having a duration of 9 μs. If the channel is sensed to be idle, $N$ is decreased. Every time the channel is sensed to be busy, the transmitter needs to sense the medium again as idle for an additional defer duration $T_d$ before it can start decreasing the counter

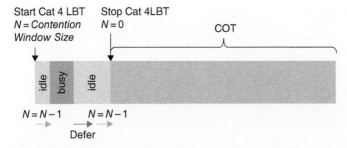

**Figure 14.4** Cat4 LBT procedure standardized by the 3GPP.

$N$ again. When $N$ reaches zero, the transmitter may initiate a transmission on the channel. The transmitter can subsequently use the channel up to a maximum time, denoted as the channel occupancy time (COT). The contention window size can vary between a minimum and a maximum value, as the transmitter increases and decreases the window size, based on the number of unsuccessful and successful transmission attempts on the medium. The values of the contention window size, the channel occupancy time, and the duration of the defer period $T_d$ depend on the channel access (priority) class, and they are set differently for uplink and downlink transmissions [20]. The Cat4 LBT algorithm and the channel access priority classes, standardized for LAA (see Figure 14.4), are very similar to those used for WLAN in IEEE 802.11. This ensures fair coexistence across devices using different technologies in the same band. Cat4 LBT is used by the eNB prior to any of its transmission on an unlicensed component carrier. Furthermore, it can be used by UEs for the uplink transmissions if they happen outside of the eNB-acquired COT.

In contrast, Cat2 LBT implements a scheme in which a device can start transmission after sensing the channel is idle once for a period of 25 µs. Cat2 LBT is used by the eNB prior to discovery signal transmissions (see section 14.4.4) and by the UE for uplink transmissions happening within the eNB-acquired channel occupancy time.

### 14.4.3 Frame Structure

Due to LBT, LAA does not follow either of the LTE frame structures (FSs) that are defined for frequency-division duplex (FDD; FS1) or time-division duplex (TDD; FS2). Instead, a new Frame Structure 3 (FS3) is introduced for the use on unlicensed carriers. FS3 is a dynamic frame structure that allows fast adaptation to traffic variations. In FS3, any subframe can be either uplink or downlink, and the eNB controls the downlink-to-uplink ratio by signaling (i) the type of the current and the next downlink subframes, as well as (ii) the offset to the start and duration of the next uplink burst within the eNB-acquired channel occupancy time, which follows the current downlink transmission. The operation of FS3 is illustrated in Figure 14.5. Three possible downlink subframe types are introduced: a normal subframe with 1 ms duration, a starting partial subframe, and a partial ending subframe. The partial starting and partial ending subframes are defined to increase the granularity of the channel access in LAA, thus better providing fair coexistence with WLAN. In particular, the partial ending subframe can be used to allow time to perform

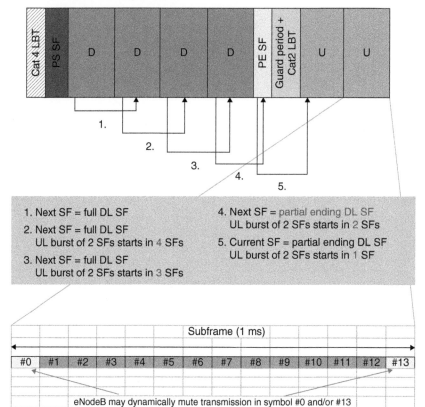

**Figure 14.5** LAA frame structure, including downlink burst and uplink burst within eNB-acquired COT.

LBT and time for the downlink–uplink switching, which is needed between the downlink and uplink transmission bursts.

During the uplink transmission bursts, the eNB may need to create gaps between the consecutive uplink subframes to enable LBT before the start of the next uplink transmission. This can happen, for example, in case of time multiplexing different UEs and when supporting transmission of sounding reference signals by multiple UEs. Therefore, LAA introduces dynamic control of muting of the first and last symbols of the uplink subframe. This muting information is dynamically signaled to the UE in the scheduling grant in case it is required for the scheduled uplink transmissions. For autonomous uplink transmissions, the starting position is signaled in the RRC configuration. More details can be found in section 14.4.5 on uplink enhancements.

### 14.4.4 Discovery Reference Signal and RRM

In unlicensed spectrum, continuous (periodic) transmission of synchronization signals, common reference signals or downlink control channels cannot be provided due to the channel access requirements. A new discovery reference signal (DRS) is therefore introduced for LAA, which is primarily used for cell search, time-frequency synchronization, and RRM measurements. The DRS is a discontinuous, periodic signal that consists of primary and secondary synchronization signals (PSS/SSS), common reference signals (CRS), and potentially channel state information reference signals (CSI-RS). DRS is transmitted periodically every 40, 80 or 160 ms and its subframe is similar to the LTE subframe #0 or #5 (of FS1 and FS2), except that it occupies only the first 12 (out of 14) orthogonal frequency division multiplexing (OFDM) symbols in a subframe.

In LAA, RRM measurements on unlicensed spectrum are required for the secondary cell configuration and cell (re)selection. The RRM measurements are based on the reception of the DRS, and the reporting consists of reference signal received power (RSRP) and reference signal received quality (RSRQ). To handle time uncertainty in the transmission of the DRS, the UE is additionally configured with a DRS measurement timing configuration (DMTC). DMTC consists of an offset and a periodicity of occurrence, and it has a fixed duration of 6 ms. DMTC enables the UE to set its measurement window properly and to capture the reference signals necessary for the measurements. The DRS structure and DMTC concept are visualized in Figure 14.6.

Due to the changing channel conditions, discontinuous transmissions, and hidden nodes and hence due to changing load conditions, the same RRM measurements that are used in the licensed band are not sufficient in the unlicensed band. In this regard, UEs can additionally be configured to measure and report the average RSSI and the channel occupancy within the configured RSSI measurement timing configuration (RMTC). The channel occupancy is defined as the percentage of time within the RMTC when the samples of the measured RSSI are above a defined threshold.

### 14.4.5 Uplink Enhancements

To fulfil the regulatory requirements on the occupied bandwidth and allow an efficient use of the UE transmission power, block-interleaved frequency division multiple access (B-IFDMA) has been selected as the baseline waveform for LAA uplink. At the same time, B-IFDMA allows the regulatory requirements to be met on the maximum allowed power spectral density. With B-IFDMA, a 20-MHz carrier is divided into ten interlaces, each interlace consisting of ten equally spaced physical resource blocks (PRB), as illustrated in Figure 14.7. The single carrier properties are maintained, although the cubic metric benefits are lost due to this PRB-level distributed and clustered allocation. Variable-sized physical uplink shared channel (PUSCH) allocations are obtained by allocating one or more interlaces to a single UE.

Due to the uncertainty of successful channel access, synchronous hybrid automatic-repeat-request (HARQ) operation with a fixed time relation between uplink (re)transmissions is not suitable in unlicensed spectrum, therefore asynchronous HARQ is used in LAA. This means that UEs need to rely on the uplink grants for uplink (re)transmissions.

Design targets for operation in unlicensed spectrum include flexible assignment of downlink and uplink resources supporting high uplink-to-downlink ratios, as well as uplink scheduling across the eNB transmission opportunities. The latter entails the eNB sending an uplink scheduling grant in the eNB-acquired channel occupancy time,

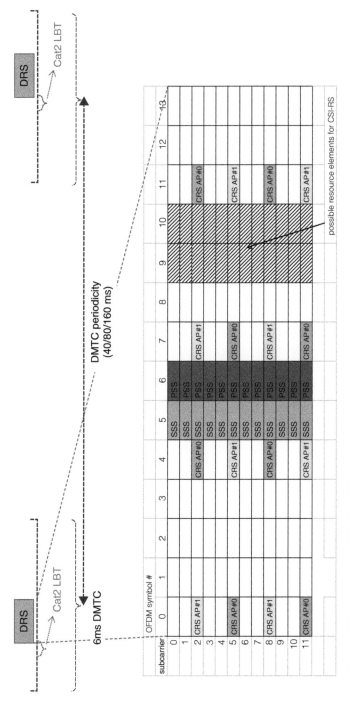

**Figure 14.6** LAA DRS structure and DMTC for RRM measurements.

**Figure 14.7** Schematic of B-IFDMA.

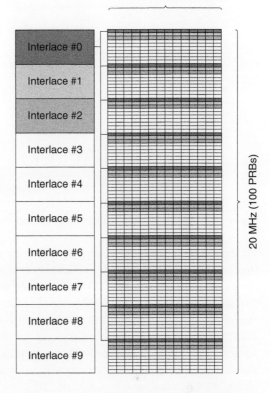

while the corresponding uplink transmission takes place in the following eNB-acquired channel occupancy time. To achieve these targets, the standard includes mechanisms for multi-subframe grants, flexible PUSCH timing, and two-stage scheduling in LAA uplink.

The eNB can schedule up to four consecutive uplink subframes using a single grant, and it can explicitly indicate the time of transmission relative to the transmission of that grant. If two-stage scheduling is enabled, the initial grant (Trigger A) provides all the information needed for transmission, except the exact timing of the transmission. The PUSCH transmission is triggered using a specific flag in the downlink common control channel (Trigger B), allowing for a significantly shorter time between such a trigger and the start of the PUSCH transmission. An example showing the high flexibility of LAA scheduling is illustrated in Figure 14.8.

The 3GPP standard includes support for autonomous uplink transmissions (AUL) in the unlicensed spectrum. AUL was primarily introduced to provide improved fairness for coexistence with the WLAN in case of many WLAN devices transmitting. However, due to the inherently faster and more autonomous nature of its channel access, AUL can also improve the uplink performance for both latency and throughput, especially in low load conditions. AUL is based on LTE semi-persistent scheduling, which avoids the steps of scheduling request and explicit grant, and where the UE is instead configured by higher layer signaling with pre-defined time occasions that can be used for autonomous transmissions. Both randomized and coordinated starting times within the first OFDM symbol of an uplink

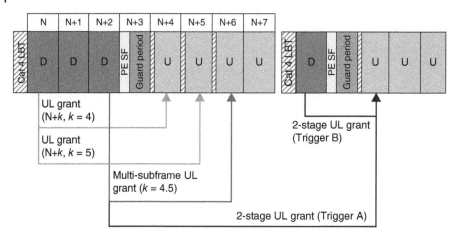

**Figure 14.8** LAA uplink (UL) scheduling framework.

subframe among the UEs are supported. Randomized starting positions enable time multiplexing of UEs allocated with the full transmission bandwidth (20 MHz), while the coordinated PUSCH starting time allows frequency multiplexing of AUL UEs having different frequency resources configured. Having a fixed time relation between the transmission and the retransmission process is not efficient due to LBT, and therefore a new uplink control information (UCI) field is introduced. UCI contains uplink HARQ information and information on the UE identity, channel access priority class, and channel occupancy time. The standard allows different operation conditions inside the eNB-acquired channel occupancy time (Cat2 LBT) and outside of it (Cat4 LBT).

## 14.5 Performance Evaluation

The aim of the performance evaluation presented in this section is twofold. First, it shows the performance merits of aggregating licensed and unlicensed spectrum, leveraging WLAN access points according to LWA and LWIP technologies. Second, it illustrates the advantages of LAA operating in the unlicensed spectrum compared to WLAN. Both performance analyses are evaluated by means of advanced quasi-static system simulations, following commonly accepted methodologies and assumptions according to the 3GPP technical reports [19, 21].

### 14.5.1 Aggregation Gains of LWA and LWIP

This section illustrates the gains when aggregating licensed and unlicensed spectrum, where the licensed spectrum is accessed by LTE and the unlicensed spectrum is accessed by WLAN, and the aggregating technology is LWA or LWIP, as defined by the 3GPP standard.

The evaluation is conducted in a simplified scenario for an enterprise (indoor) layout (see Figure 14.9). The scenario consists of one eNB operating in a 10-MHz channel in the 1.9-GHz band and connected via Ethernet backhaul to two WLAN access points, which

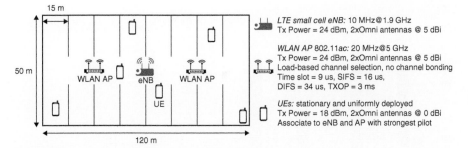

**Figure 14.9** Enterprise layout for LWA/LWIP evaluation.

use non-overlapping 20-MHz channels in the 5-GHz band. Further simulation assumptions follow the 3GPP recommendations in [19] and [21].

The network is under different load conditions with 1, 4, 20 or 32 stationary UEs that are uniformly present in the simulated layout. Since the benefits of aggregation are apparent for traffic types requesting high bit rates, such as video streaming, web-browsing, and file download, the traffic source for each UE is generalized for the analysis as a bidirectional transfer of discrete files carried over the TCP protocol. In the model, each UE independently requests a downlink file and delivers an uplink file, one file at a time. The interval between the end of a received file and the request for the next file follows an exponential distribution with the mean of 100 ms. The file size is set to 0.5 Mbyte in downlink and 0.25 Mbyte in uplink. The uplink traffic is assumed to be ideally carried over LTE, whereas the traffic over WLAN uplink is merely seen as network load [12].

The summary results focus on the traffic intense downlink, and the following features were evaluated for comparison:

- *LWA* operates with the configured split-bearer (aggregation) where the UE can receive packets (i.e., PDCP PDUs) of a bearer from LTE and WLAN simultaneously. The eNB determines the optimal split using the flow control algorithm described in section 14.2.3. LWA status reports are periodically sent every 10 ms [4, 11].
- *LWIP* operates with slow link-switching where the UE receives IP packets of a bearer from either LTE or WLAN. The eNB decides on the switching by the received signal strength-based flow control algorithm depicted in section 14.3.2 [15].
- *LWIP+* operates with link-switching that is decided based on the probing mechanism presented in section 14.3.2. At connection start, the WLAN link is probed with the probing packets (of 1500 bytes) at the rate of 5 Mbps for 100 ms, whereas during data transmission over WLAN the probing packets (of 160 bytes) are inserted in between data packets every 3 ms [15].
- *LTE-only* and *WLAN-only* operations show the baseline performance of sole LTE and sole WLAN, respectively.

The performance results of LWA, LWIP, and LWIP+ as well as the baselines are summarized in Figure 14.10 in absolute and relative terms. The results depict the session throughput of downlink files received per UE in terms of the mean (Figure 14.10) and cell-edge throughput, a.k.a. the fifth-percentile throughput (Figure 14.10).

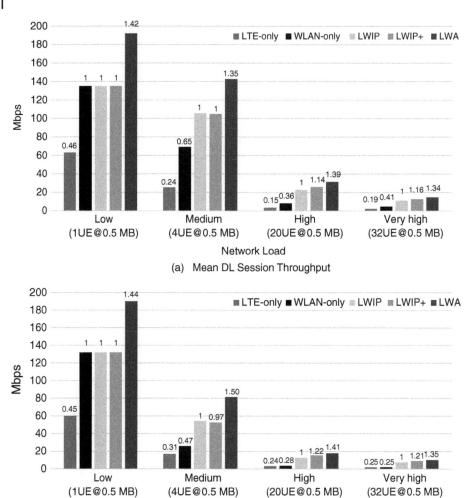

**Figure 14.10** Downlink session throughput of LTE-only and WLAN-only references, and the studied LWA, LWIP, and LWIP+ as a function of network load: (a) mean and (b) fifth percentile.

When the load is low (single UE in the scenario), LTE-only provides a mean throughput of 63 Mbps, while WLAN-only provides a mean throughput of 135 Mbps. In the low load case, the LWA ability to aggregate the licensed and unlicensed band can fully be exploited, which results in the average throughput of 192 Mbps. LWA gives 42% performance gain over the WLAN-only service. The performance of non-aggregating LWIP/LWIP+, which tends to use the WLAN link only, is comparable to the WLAN-only case. LWA gains over the other features already due to the larger bandwidth available, i.e., 30 MHz per UE in LWA contrary to 20 MHz per UE in LWIP/LWIP+ or 20 MHz per UE in WLAN-only, contrary to 10 MHz per UE in LTE-only.

For the medium load with four UEs, LWA and LWIP/LWIP+ provide substantial gains over LTE-only and WLAN-only. This is mostly due to the larger bandwidth available in the network, i.e., 50 MHz (for LWA and LWIP/LWIP+) instead of 40 MHz (for WLAN-only) and 10 MHz (for LTE-only). In this case, LWA outperforms LWIP/LWIP+ with 35% gains in the mean and 50% in the cell-edge due to the aggregation capability. LWIP+ performs on-par to LWIP since there is no severe congestion yet under the medium load.

When the load becomes high or very high, the gains of LWA and LWIP/LWIP+ tend to increase compared to the WLAN-only transmission, up to 3.7 times. This is because LWA and LWIP/LWIP+ can smartly use both networks, e.g., WLAN for downlink and LTE for uplink. Offloading uplink traffic from the unlicensed band to the LTE licensed band mitigates uplink contention and collisions that otherwise increase in a heavily loaded WLAN network [12]. In WLAN-only transmissions, all downlink and uplink traffic compete for the access medium by the well-known carrier sensing and collision avoidance algorithms, i.e., LBT [2], which degrade the performance of the WLAN network. LWA outperforms LWIP by up to 39% in average and 41% in cell-edge, respectively, due to its intelligent PDCP PDU splitting and bandwidth aggregation algorithms. The performance of LWIP+ is behind LWA, since it cannot aggregate bandwidth, whereas it outperforms LWIP by up to 14% in average and 22% in cell-edge due to its flow control algorithm, which based on its congestion awareness can switch smartly and provide load balancing.

Both LWA and LWIP+ significantly outperform LWIP, with the gain of LWA over LWIP, as said before, in the order of 40% in average and 40% in cell-edge across all studied load levels in the enterprise layout. Due to its aggregation capabilities, LWA also gains over LWIP+ across all studied cases. However, the relative gains of LWA over LWIP+ show a decreasing trend with the increasing load, i.e., the average gains decrease from 40% to 20%, as there is less opportunity for aggregation in LWA, and the gains from load balancing in LWIP+ start to dominate.

Finally, Figure 14.11 shows how the system throughput substantially differs across the studied features for the same settings due to the closed-loop properties of the traffic model.

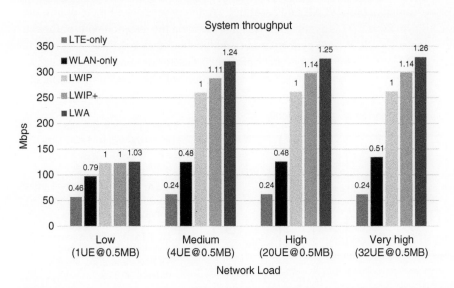

**Figure 14.11** System throughput aggregated across all the UEs in the network for the LTE-only and WLAN-only references, and for the studied LWA, LWIP, and LWIP+ shown as a function of network load.

The results indicate that LWA outperforms other solutions in this comparison. LWA offers about 25% gains compared to LWIP and about 10% gains compared to LWIP+ in the mean throughput, for all load values, except in the low load case, where the throughput stays constrained by the small amount of offered traffic.

This analysis is shown for the traffic intense downlink communication. However, aggregation can provide similar benefits also when aggregating the uplink traffic, since the same mechanisms apply.

### 14.5.2 Performance Advantages of LAA

The focus of this section is to illustrate the advantages of LAA operating in the unlicensed spectrum as compared to WLAN. The performance evaluation covers both capacity and coverage analyses, and considers the coexistence of LAA with WLAN.

The capacity evaluation is conducted in the 3GPP defined indoor scenario [19], assuming International Telecommunications Union (ITU) indoor hotspot (InH) propagation and consisting of a single-floor building with four equally spaced small cells per operator, centered along the shorter dimension of the building (see Figure 14.12). Evaluations for both one- and two-operator cases are shown, where the two-operator evaluation covers both a single technology deployment (two LAA operators or two WLAN operators) and a mixed technology deployment (one LAA operator and one WLAN operator).

The capacity evaluation is performed under the traffic of file transfer protocol (FTP) Model 3, where the number of users is kept fixed and data calls of size 0.5 Mbyte per user are dynamically generated according to a Poisson process, independently for downlink and uplink. The evaluation is performed under various load conditions, the load being adjusted by varying the session arrival rate per user, whilst keeping the average uplink/downlink traffic ratio constant at 20:80. To simplify the comparison, a single-channel deployment of 20 MHz is assumed in both WLAN and LAA. In LAA, the ratio of downlink to uplink subframes within an 8-ms COT is dynamically adapted based on the instantaneous traffic conditions within each cell. WLAN performance evaluation assumes 802.11ac standard with request to send/clear to send disabled, cyclic prefix of 0.8 μs, maximum channel occupancy time of 4 ms, Minstrel rate control, and frame aggregation with block acknowledgement.

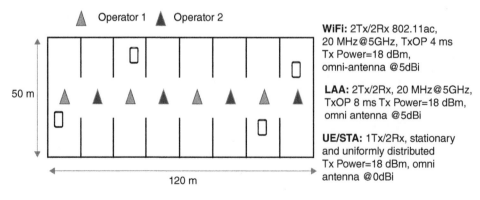

**Figure 14.12** Indoor scenario for capacity evaluation.

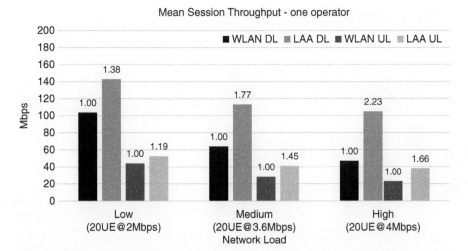

**Figure 14.13** Mean UE session throughput comparison between LAA and WLAN, one-operator case, split for the downlink (DL) and uplink (UL).

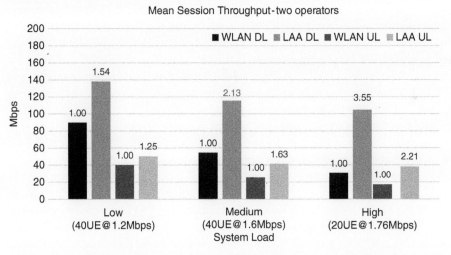

**Figure 14.14** Mean UE session throughput comparison between LAA and WLAN, two-operator case with single technology deployment, split for the downlink (DL) and uplink (UL).

A capacity comparison between LAA and WLAN is illustrated in Figure 14.13 for one operator, and in Figures 14.14 and 14.15 for two operators, assuming single technology deployment or mixed technology deployment, respectively. The performance is measured as mean UE session throughput under various load conditions. The traffic load is broadly characterized as low, medium, and high corresponding to average buffer occupancy (see [19]) of less than 15%, 15–50% and greater than 50%, respectively.

In the one-operator case, LAA provides gains of 1.4–2.2× in downlink and 1.2–1.7× in uplink over WLAN. Relative gains over WLAN are higher at higher load, suggesting LAA can handle high system load better compared to the WLAN. One way to interpret the results

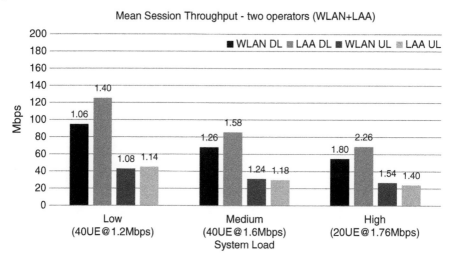

**Figure 14.15** Mean UE session throughput comparison between LAA and WLAN, mixed two-operator case (LAA + WLAN). Performance gains are relative to WLAN performance in two WLAN operator case.

is that LAA can achieve targeted user bit-rate in clearly higher system load than WLAN. For example, for a target data rate of 100 Mbps in downlink and 40 Mbps in uplink, LAA can handle approximately twice as much traffic as WLAN. These significant performance gains of LAA over WLAN result from several gain mechanisms. One important gain mechanism is the scheduled uplink approach, which leads to smaller contention overhead compared to the WLAN and improved performance at high load. In addition, HARQ and fast link adaptation based on instantaneous channel conditions make LAA robust against interference and noise, and further contribute to the high spectral efficiency.

With two operators assuming a single technology deployment, the relative gains of LAA over WLAN are increased further compared to the one operator case: In the downlink, the gains of LAA over WLAN are 1.5–3.6× and in the uplink they are 1.3–2.2×, with again the gains being higher for higher traffic loads. When more competing nodes (i.e., access points and/or UEs) are added to the network, the higher uplink contention overhead for WLAN implies that WLAN nodes start to block each other when the load increases, which leads to a performance reduction compared to LAA. The dependency of WLAN congestion on the number of nodes trying simultaneously to access the medium is a well-known bottleneck of the WLAN performance (see [22]).

When comparing the performance in the LAA operator network in the mixed technology deployment relative to the scenario consisting of two WLAN operators, it is observed that LAA provides gains of 1.4–2.3× in downlink and 1.1–1.4× in uplink over WLAN. The gain reduction compared to the single technology deployment is due to more aggressive channel access of WLAN, which makes the WLAN network a less friendly neighbor than the LAA network. The results also show that the LAA network can coexist fairly with the WLAN network: the performance in the WLAN network is improved if WLAN is replaced with LAA, with gains ranging from 1.1× to 1.8× in downlink and 1.1× to 1.5× in uplink. When comparing the performance in the LAA network against the performance in the WLAN network

**Figure 14.16** Coverage comparison between LAA and WLAN.

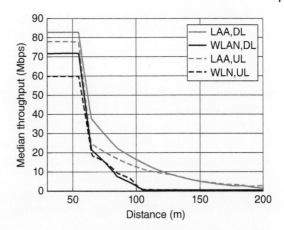

in the mixed technology deployment, the LAA network has 25–30% better performance than the WLAN network in the downlink over the entire range of simulated loads, whereas in the uplink performance in the LAA and WLAN networks is very similar.

As coverage is an important performance measure for outdoor deployments, a coverage evaluation of LAA and WLAN is presented next. This is conducted in the 3GPP outdoor scenario, assuming ITU urban micro (UMi) propagation [19]. This coverage evaluation assumes a single-cell single-UE deployment, and only considers the median data rate [without multiple input multiple output (MIMO)] as a function of the distance between the UE and the access point or eNB. The result of the coverage comparison is illustrated in Figure 14.16 for both link directions. It can be observed that in downlink LAA provides data rate of 10 Mbps at approximately 120 m from the transmitter, while WLAN reaches the same throughput at approximately 80 m. Hence, in the downlink the achieved coverage area of LAA at 10 Mbps is more than double compared to WLAN. In the uplink, the corresponding coverage area of LAA is approximately 70% larger than WLAN. The main enablers for the extended coverage with LAA are the adoption of HARQ and fast link adaptation, which make LAA more robust against interference and noise.

## 14.6 Future Technologies

### 14.6.1 5G New Radio in Unlicensed Band

The 3GPP has studied new radio access on the unlicensed spectrum (NR-U) and has concluded on its feasibility. The subsequent normative work has focused on the 5-GHz band like LAA. Future applicability to higher frequency bands is foreseen as well (e.g., unlicensed 60 GHz). To guarantee a fair coexistence between different radio access technologies operating in the unlicensed spectrum, channel access in NR-U is based on LBT principles to ensure coexistence in IEEE 802.11 for WLAN and in 3GPP for LAA. Besides the licensed-assisted operation in the secondary cells on the unlicensed spectrum, NR-U will support dual connectivity thereof and standalone deployments. The standalone deployments of NR-U will offer further business opportunities for the telecom industry, allowing leverage of the assets of NR to verticals, and serving industry players who do not have a spectrum license.

As compared to other radio access technologies operating in the 5-GHz unlicensed bands, NR-U is anticipated to improve spectral efficiency, latency, and reliability, thus enabling 5G use cases in the shared and unlicensed spectrum by inheriting the technology components of the NR, namely:

- scalable OFDM-based air interface, supporting higher subcarrier spacing and shorter symbol duration
- flexible TDD frame structure with fast signaling
- lean carrier design with reduced overhead from reference symbols and control channels
- native support for massive MIMO techniques providing higher capacity and data rates, and better reliability
- advanced channel coding, reducing processing time in both the UE and eNB
- efficient and cost-effective wideband (>20 MHz) operation.

### 14.6.2  The Role of WLAN in the 5G System

While addressing the relevance of integrating non-3GPP access networks (i.e., dominantly WLAN) to the 5G system (5GS), 3GPP has studied different approaches, and has taken the learnings from the LTE standard. So far, it has been decided that WLAN will integrate outside of the next generation (5G) RAN (NG-RAN) via the NG interface directly to the 5G core network [23, 24]. Only untrusted WLAN is considered relevant because it exploits a large variety of any kind of WLAN deployments without assumptions of operator partnering. Currently, the aim is to standardize the management object (to appear in [25]) for the assistance of WLAN discovery and for WLAN selection policies and priorities. While these integration solutions are mainly an architectural issue and the matter of defining interfaces, from the radio access point of view and from the spectrum sharing point of view they pose similar challenges to integrating uncoordinated WLAN to 3GPP operation (LAA or NR-U) on the same unlicensed band.

## 14.7  Conclusions

In this chapter, several 3GPP concepts and standard technologies to operate in the unlicensed spectrum were reviewed. We classified the technologies according to their capability to supplement the licensed spectrum operations either by leveraging the existing WLAN footprint (LWA, LWIP/LWIP+) or by adapting the 3GPP standard (LAA).

For each 3GPP technology – with the design choices of protocol architectures, procedures, mobility, and security – we identified how to enable a flexible use of both the licensed and unlicensed spectrum. Special attention was given to the architectural network enhancements needed to make the deployments cope with the functionality. The numerical analysis unveiled that LWA and LWIP achieve significant gains of aggregating spectrum. LAA was observed to be both high performing and friendly thanks to its mechanisms of coexistence towards any system on the unlicensed band. For this reason, the objective has been to extend the LAA design as a building block, and 3GPP is defining the operation of the coming 5G networks in the unlicensed spectrum in the form of NR-U. first commercial deployments and deployments in the testing phase have been reported (e.g., [26]), and LAA seems to be dominant over the other technologies. This furthermore provides a solid basis for the foreseen upgrades to 5G NR-U.

# References

**1** 3GPP (2018) TS 36.300, *Evolved Universal Terrestrial Radio Access (E-UTRA) and Evolved Universal Terrestrial Radio Access Network (E-UTRAN), Overall Description, Stage 2,* V15.3.0, Oct. 2018.

**2** IEEE 802.11-2016 (2016) *Part 11: Wireless LAN Medium Access Control (MAC) and Physical Layer (PHY) Specifications.*

**3** MulteFire Alliance (2017) White paper: MulteFire release 1.0 technical paper: A new way to wireless. Available at https://www.multefire.org/white-papers/.

**4** D. Laselva, D. López-Pérez, M. Rinne and T. Henttonen (2018) 3GPP LTE-WLAN Aggregation Technologies: Functionalities and Performance Comparison. *IEEE Communications Magazine,* 56 (3), 195–203.

**5** 3GPP (2018) TS 36.360, *Evolved Universal Terrestrial Radio Access (E-UTRA); LTE-WLAN Aggregation Adaptation Protocol (LWAAP) specification,* V15.3.0.

**6** 3GPP (2018) TS 33.401, *3GPP System Architecture Evolution (SAE); Security architecture,* V15.5.0, Sep. 2018.

**7** 3GPP (2018) TS 36.331, *Evolved Universal Terrestrial Radio Access (E-UTRA); Radio Resource Control (RRC): Protocol specification,* V15.3.0, Sep. 2018.

**8** S. Singh, S. Yeh, N. Himayat and S. Talwar (2016) Optimal traffic aggregation in multi-RAT heterogeneous wireless networks. *Proceedings of the IEEE ICC,* 626–631, May 2016.

**9** S. Borst, A. Ö. Kaya, D. Calin and H. Viswanathan (2016) Optimal path selection in multi-RAT wireless networks. *Proceedings of the IEEE INFOCOM Workshop,* 592–597, April 11, 2016, San Francisco, CA, USA.

**10** I. Balan, E. Perez, B. Wegmann and D. Laselva (2016) Self-optimizing adaptive transmission mode selection for LTE-WLAN aggregation. *Proceedings of the IEEE PIMRC,* September 4–7, 2016, Valencia, Spain.

**11** D. López-Pérez, D. Laselva, E. Wallmeier et al. (2016) Long term evolution-wireless local area network aggregation flow control. *IEEE Access,* 4, 9860–9869.

**12** D. López-Pérez J. Ling, B.H. Kim et al. (2016) Boosted Wi-Fi through LTE Small Cells: The Solution for an All-Wireless Enterprise. *Proceedings of the IEEE PIMRC,* September 4–7, 2016, Valencia, Spain.

**13** 3GPP (2016) TS 36.361, *Evolved Universal Terrestrial Radio Access (E-UTRA); LTE-WLAN Radio Level Integration Using IPsec Tunnel (LWIP) encapsulation: Protocol specification,* V14.0.0, Dec. 2016.

**14** G. Dommety (2000) Key and Sequence Number Extensions to GRE. *IETF RFC 2890,* Sep. 2000.

**15** D. López-Pérez, J. Ling, B.H. Kim et al. (2017) LWIP and Wi-Fi Boost Flow Control. *Proceedings of the IEEE WCNC,* March 19–22, 2017, San Francisco, CA, USA.

**16** S.T.V. Pasca, S. Patro, B.R. Tamma and A.A. Franklin (2017) Tightly coupled LTE Wi-Fi radio access networks: A demo of LWIP. *Proceedings of the COMSNETS*, January 4–8, 2017, Bengaluru, India.

**17** H.-J. Kwon, J. Jeon, A. Bhorkar et al. (2017) Licensed-assisted access to unlicensed spectrum in LTE release 13. *IEEE Communication Magazine*, 55 (2), 201–207.

**18** A. Mukherjee, J. Cheng, S. Falahati et al. (2016) Licensed-assisted access LTE: Coexistence with IEEE 802.11 and the evolution toward 5G. *IEEE Communications Magazine*, 54 (6), 50–57.

**19** 3GPP (2015) TR 36.889, *Feasibility Study on Licensed-Assisted Access to Unlicensed Spectrum*, V13.0.0, Mar. 2015.

**20** 3GPP (2018) TS 37.213, *Physical layer procedures for shared spectrum channel access*, V15.1.0, Sep. 2018.

**21** 3GPP (2013) TR 36.872, *Small cell enhancements for E-UTRA and E-UTRAN: Physical layer aspects*, V12.1.0, Dec. 2013.

**22** L.G.U. Garcia, I. Rodriguez, D. Catania and P. Mogensen (2013) IEEE 802.11 Networks: A Simple Model Geared Towards Offloading Studies and Considerations on Future Small Cells. *Proceedings of the IEEE VTC*, September 2–5, 2013, Las Vegas, USA.

**23** 3GPP (2018) TS 23.501, *System Architecture for the 5G System*, V15.3.0, Sep. 2018.

**24** 3GPP (2018) TS 24.502, *Access to the 3GPP 5G Core Network (5GCN) via non-3GPP access networks*, V15.1.1, Sep. 2018.

**25** 3GPP (2018) TS 23.402, *Aspects, Architecture Enhancements for non-3GPP accesses*, V15.3.0, Mar. 2018.

**26** GSA (2018) *Report Evolution from LTE to 5G. Update*, Nov. 2018.

# 15

# Performance Analysis of Spatial Spectrum Reuse in Ultradense Networks

*Youjia Chen[1], Ming Ding[2], and David López-Pérez[3]*

[1] *Fuzhou University, Fuzhou, PR China*
[2] *Commonwealth Scientific and Industrial Research Organisation (CSIRO), Eveleigh, Australia*
[3] *Nokia Bell Labs, Dublin, Ireland*

## 15.1 Introduction

From 1950 to 2000, the wireless network capacity has increased around 1million fold, in which an astounding 2700× gain was achieved through aggressive spatial spectrum reuse (SSR) via network densification using smaller cells [8, 10, 13]. Generally speaking, SSR indicates that multiple cells within a given area of interest simultaneously reuse a given chunk of spectrum. If the SSR grows linearly, the wireless network capacity has the potential to grow linearly too, as each cell can make an independent and equal contribution to it, provided that the inter-cell interference remains constant. The aforementioned 2700× gain stands as a glorious testimony of the feasibility and fulfillment of such potential.

In the first decade of 2000, network densification continued to fuel the 3rd Generation Partnership Project (3GPP) 4G long-term evolution (LTE) networks, and is expected to remain as one of the main forces to drive the 5G new radio (NR) beyond 2020 [10]. In particular, the orthogonal deployment of dense small cell networks (SCNs), in which small cells and macrocells operate in different frequency bands [10], has gained much momentum in the past years. This is because such deployment can provide a much larger SSR, while incurring a simplified network management, due to the avoidance of inter-tier interference between small cells and macrocells.

New concepts such as *spectrum sharing*, where multiple independent entities (e.g., governments, operators, individuals) simultaneously use a specific radio frequency band in a specific geographical area, possibly using different radio access network (RAN) technologies [e.g., Citizens Broadband Radio Services (CBRSs)], are pushing the concept of SSR further. The principal objective of spectrum sharing is a more efficient usage of scarce spectrum resources, at less cost and latency than required to clear spectrum the

*Spectrum Sharing: The Next Frontier in Wireless Networks*, First Edition.
Edited by Constantinos B. Papadias, Tharmalingam Ratnarajah, and Dirk T.M. Slock.
© 2020 John Wiley & Sons Ltd. Published 2020 by John Wiley & Sons Ltd.

old-fashioned way, and it can take many forms, e.g. coordinated and uncoordinated. From a deployment perspective, spectrum sharing may also lead to a significant network densification since it may imply an increased number of base stations (BSs) and user equipment (UE) reusing a frequency band in an unit area.

Bearing in mind this implication of spectrum sharing, it is important to note that as we walk down the path of network densification and gradually enter the realm of ultra-dense networks (UDNs), things start to deviate from the traditional understanding. The capacity scaling law observed in sparse networks may not apply to UDNs. In particular, a fundamental question arises: *Is there a limit to the SSR?* In other words, when we deploy thousands or millions of small cell BSs per square kilometer, is activating all BSs on the same time/frequency resource the best strategy, as we have practiced in the last half century? In this chapter, we provide an answer to this question, providing a mathematical framework to show the existence of an optimum SSR operation point. This answer also provides important insights to the spectrum sharing community, indicating that there is an optimum reuse point, and that more is not necessarily better.

## 15.2 Network Scenario and System Model

In this section we present the network scenario and the system model considered in this book chapter.

### 15.2.1 Network Scenario

We consider the downlink (DL) of a cellular network with BSs deployed on a plane according to a homogeneous Poisson point process (PPP) $\Phi$ with a density of $\lambda$ BSs/km$^2$. Active DL UEs are also Poisson distributed in the considered network with a density of $\rho$ UEs/km$^2$. Here, we only consider active UEs in the network because non-active UEs do not trigger any data transmission and thus they can be safely ignored in the analysis. Note that the total number of UEs in cellular networks is usually much higher than the number of active UEs. In fact, the number of active UEs with data to transmit at a given time slot and on a given frequency band may not be very many. A typical active UE density in populated scenarios is around $\rho = 300$ UEs/km$^2$ [10].

In practice, a BS will enter into idle mode if there is no UE connected to it, which reduces the interference to UEs in neighboring BSs as well as the energy consumption of the network. Since UEs are randomly and uniformly distributed in the network, a widely accepted assumption is that the active BSs should follow another homogeneous PPP distribution $\tilde{\Phi}$ [11], the density of which is $\tilde{\lambda}$ BSs/km$^2$. Note that $\tilde{\lambda} \leq \lambda$ and $\tilde{\lambda} \leq \rho$, since one UE is served by at most one BS. Also note that a larger $\rho$ requires more active BSs to serve the more active UEs, thus leading to a larger $\tilde{\lambda}$. The authors in [7, 11] showed that the expression

$$\tilde{\lambda} = \lambda \left[ 1 - \frac{1}{\left( 1 + \frac{\rho}{q\lambda} \right)^q} \right] \tag{15.1}$$

used to calculate $\tilde{\lambda}$ is accurate for UDNs, where an empirical value of 3.5 was suggested for $q$.

### 15.2.2 Wireless System Model

The horizontal distance between a BS and a UE is denoted by $r$. Moreover, the absolute antenna height difference between a BS and a UE is denoted by $L$. Thus, the three-dimensional (3D) distance between a BS and a UE can be expressed as

$$w = \sqrt{r^2 + L^2}. \tag{15.2}$$

Note that the value of $L$ is in the order of several meters [1].

Following [6], we adopt a general path-loss model consisting of multiple pieces of functions. Note that the realistic LoS probability functions usually take complicated mathematical forms, e.g. in the 3GPP standards [1]. Therefore, to achieve both analytical tractability and result accuracy it is desirable to approximate such complicated LoS probability functions as a few pieces of elementary functions, e.g. linear functions. Such piecewise LoS probability function is well captured by the path-loss model used in this chapter. In more detail, the path loss $\zeta(w)$ is a multi-piece function of $w$ written as

$$\zeta(w) = \begin{cases} \zeta_1(w), & \text{when } L \leq w \leq d_1 \\ \zeta_2(w), & \text{when } d_1 < w \leq d_2 \\ \vdots & \vdots \\ \zeta_N(w), & \text{when } w > d_{N-1} \end{cases}, \tag{15.3}$$

where each piece $\zeta_n(w), n \in \{1, 2, \ldots, N\}$ is modeled as

$$\zeta_n(w) = \begin{cases} \zeta_n^{\mathrm{L}}(w) = A_n^{\mathrm{L}} w^{-\alpha_n^{\mathrm{L}}}, & \text{if LoS, with a probability } \mathrm{Pr}_n^{\mathrm{L}}(w) \\ \zeta_n^{\mathrm{NL}}(w) = A_n^{\mathrm{NL}} w^{-\alpha_n^{\mathrm{NL}}}, & \text{otherwise.} \end{cases}, \tag{15.4}$$

where

- $\zeta_n^{\mathrm{L}}(w)$ and $\zeta_n^{\mathrm{NL}}(w), n \in \{1, 2, \ldots, N\}$, are the $n$th piece path-loss functions for the LoS and the NLoS cases, respectively
- $A_n^{\mathrm{L}}$ and $A_n^{\mathrm{NL}}$ are the path-losses at a reference 3D distance $w = 1$ for the LoS and the NLoS cases, respectively
- $\alpha_n^{\mathrm{L}}$ and $\alpha_n^{\mathrm{NL}}$ are the path-loss exponents for the LoS and the NLoS cases, respectively.

In essence, in the $n$th piece, $\zeta_n(w)$ equals to $\zeta_n^{\mathrm{L}}(w)$ if there is a line-of-sight (LoS) transmission or to $\zeta_n^{\mathrm{NL}}(w)$ if there is a non-line-of-sight (NLoS) one, where the corresponding probabilities of having or not a LoS in such $n$th piece is given by $\mathrm{Pr}_n^{\mathrm{L}}(w)$, which is the $n$th piece LoS probability function that a transmitter and a receiver separated by a 3D distance $w$ have a LoS path. $\mathrm{Pr}_n^{\mathrm{L}}(w)$ is assumed to be a *monotonically decreasing function* with respect to $w$. Existing measurement studies have confirmed this assumption [1].

In addition, we assume a practical user association strategy in which each UE is connected to the BS providing the maximum average received signal strength (i.e., with the largest $\zeta(w)$) [3, 6]). It is very important to note that in our previous work [5] and some other work, e.g. [2, 14], it was assumed that each UE is associated with its closest BS. Such an assumption is not appropriate for the considered path-loss model in (15.3) because in practice a UE connects to the BS offering the largest received signal strength [1]. This BS does not necessarily have to be the nearest one to the UE. It could be a farther one with a strong LoS path. Moreover, with this association criterion, i.e., the maximum average received signal strength, the formulation of active BS density in (15.1) still works according to [7].

Finally, we assume that the BS transmission power is a constant value $P$, each BS/UE is equipped with an isotropic antenna, and the multi-path fading between a BS and a UE is modeled as independently identical distributed (i.i.d.) Rayleigh fading [3, 6, 14]. Note that a more accurate/general multi-path modeling is Rician fading. However, it is important to note that the simulation results in [4] and the analytical results in [9] show that Rician fading does not qualitatively change the conclusions of this UDN performance analysis, only quantitatively. Thus, we embrace Rayleigh fading for tractability reasons in this chapter.

## 15.3  Performance Analysis of Full Spectrum Reuse Network

In this section we present the coverage probability and the area spectral efficiency for a UDN with a full spectrum reuse strategy in this chapter.

### 15.3.1  The Coverage Probability

First, we investigate the coverage probability, which is defined as the signal-to-interference-plus-noise ratio (SINR) of a typical UE at the origin $o$ above a threshold $\gamma$:

$$p^{\mathrm{cov}}(\lambda, \rho, \gamma) \triangleq \Pr[\mathrm{SINR} > \gamma], \tag{15.5}$$

where the SINR is computed by

$$\mathrm{SINR} = \frac{P\zeta(w)h}{I_{\mathrm{agg}} + P_{\mathrm{N}}}, \tag{15.6}$$

where $h$ is the channel power gain, which is modeled as an exponentially distributed random variable (RV) with a mean of 1 due to our consideration of Rayleigh fading. As a result, the channel coefficient is a complex Gaussian RV, and thus, the channel power gain, i.e. the squared magnitude of this complex Gaussian RV, follows an exponential distribution. Moreover, $P$ and $P_{\mathrm{N}}$ are the BS transmission power and the additive white Gaussian noise (AWGN) power at each UE, respectively, and $I_{\mathrm{agg}}$ is the cumulative interference given by

$$I_{\mathrm{agg}} = \sum_{i:b_i \in \tilde{\Phi} \setminus b_o} P\beta_i g_i, \tag{15.7}$$

where $b_o$ is the BS serving the typical UE, and $b_i$, $\beta_i$, and $g_i$ are the $i$th interfering BS, the path loss from $b_i$ to the typical UE, and the multi-path fading channel gain associated with this link (also exponentially distributed RVs), respectively. Here, $\tilde{\Phi}$ denotes the set of active BSs, since only they inject interference into the network.

Considering the adopted multi-piece path-loss model in (15.3), which includes LoS and NLoS transmissions, the coverage probability $p^{\text{cov}}(\lambda, \rho, \gamma)$ can be formulated as

$$
p^{\text{cov}}(\lambda, \rho, \gamma) = \sum_{n=1}^{N} (T_n^{\text{L}} + T_n^{\text{NL}})
$$

$$
= \sum_{n=1}^{N} \int_{\sqrt{d_{n-1}^2 - L^2}}^{\sqrt{d_n^2 - L^2}} \Big( \Pr[\text{SINR}_n^{\text{L}}(r) > \gamma] \cdot f_{R,n}^{\text{L}}(r) \tag{15.8}
$$

$$
+ \Pr[\text{SINR}_n^{\text{NL}}(r) > \gamma] \cdot f_{R,n}^{\text{NL}}(r) \Big) \, dr,
$$

where $T_n^{\text{L}}$ and $T_n^{\text{NL}}$ denote the coverage probabilities contributed by the LoS and NLoS transmissions in the $n$th piece distance range, respectively. The derivations of the items in (15.8) are summarized as follows.

**Lemma 1** The probability density function (PDF) $f_{R,n}^{\text{L}}(r)$ and $f_{R,n}^{\text{NL}}(r)$ for $\sqrt{d_{n-1}^2 - L^2} < r \leq \sqrt{d_n^2 - L^2}$, which denotes the probabilities that the signal comes from a LoS path and a NLoS path with horizontal distance $r$, respectively. can be expressed as

$$
f_{R,n}^{\text{L}}(r) = \exp\left( -\int_0^{r_1} \left(1 - \Pr^{\text{L}}(\sqrt{u^2 + L^2})\right) 2\pi u \lambda du \right)
$$

$$
\times \exp\left( -\int_0^r \Pr^{\text{L}}(\sqrt{u^2 + L^2}) 2\pi u \lambda du \right) \Pr_n^{\text{L}}(\sqrt{r^2 + L^2}) 2\pi r \lambda, \tag{15.9}
$$

$$
f_{R,n}^{\text{NL}}(r) = \exp\left( -\int_0^{r_2} \Pr^{\text{L}}(\sqrt{u^2 + L^2}) 2\pi u \lambda du \right) \times
$$

$$
\exp\left( -\int_0^r \left(1 - \Pr^{\text{L}}(\sqrt{u^2 + L^2})\right) 2\pi u \lambda du \right) \left(1 - \Pr_n^{\text{L}}(\sqrt{r^2 + L^2})\right) 2\pi r \lambda, \tag{15.10}
$$

where

$$
r_1 = \arg_{r_1} \left\{ \zeta^{\text{NL}}(\sqrt{r_1^2 + L^2}) = \zeta_n^{\text{L}}(\sqrt{r^2 + L^2}) \right\}, \tag{15.11}
$$

$$
r_2 = \arg_{r_2} \left\{ \zeta^{\text{L}}(\sqrt{r_2^2 + L^2}) = \zeta_n^{\text{NL}}(\sqrt{r^2 + L^2}) \right\}. \tag{15.12}
$$

*Proof:* See Appendix 15.1.

**Lemma 2** The probability that the typical UE is covered by an LoS signal with a horizontal distance $r$, i.e. $\Pr[\text{SINR}_n^{\text{L}}(r) > \gamma]$, is given by

$$
\Pr[\text{SINR}_n^{\text{L}}(r) > \gamma] = \exp\left( -\frac{\gamma P_{\text{N}}}{P \zeta_n^{\text{L}}(\sqrt{r^2 + L^2})} \right) \mathcal{L}_{I_{\text{agg}}}^{\text{L}} \left( \frac{\gamma}{P \zeta_n^{\text{L}}(\sqrt{r^2 + L^2})} \right), \tag{15.13}
$$

where

$$
\mathcal{L}_{I_{\text{agg}}}^{\text{L}}(s) = \exp\left( -2\pi \tilde{\lambda} \int_r^{+\infty} \frac{\Pr^{\text{L}}(\sqrt{u^2 + L^2})u}{1 + (sP\zeta^{\text{L}}(\sqrt{u^2 + L^2}))^{-1}} du \right)
$$

$$
\times \exp\left( -2\pi \tilde{\lambda} \int_{r_1}^{+\infty} \frac{[1 - \Pr^{\text{L}}(\sqrt{u^2 + L^2})]u}{1 + (sP\zeta^{\text{NL}}(\sqrt{u^2 + L^2}))^{-1}} du \right). \tag{15.14}
$$

The probability that the typical UE is covered by an NLoS signal with a horizontal distance $r$, i.e. $\Pr[\text{SINR}_n^{\text{NL}}(r) > \gamma]$, is given by

$$\Pr[\text{SINR}_n^{\text{NL}}(r) > \gamma] = \exp\left(-\frac{\gamma P_N}{P\zeta_n^{\text{NL}}(\sqrt{r^2 + L^2})}\right) \mathcal{L}_{I_{\text{agg}}}^{\text{NL}}\left(\frac{\gamma}{P\zeta_n^{\text{NL}}(\sqrt{r^2 + L^2})}\right), \quad (15.15)$$

where

$$
\begin{aligned}
\mathcal{L}_{I_{\text{agg}}}^{\text{NL}}(s) &= \exp\left(-2\pi\tilde{\lambda}\int_{r_2}^{+\infty} \frac{\Pr{}^{\text{L}}(\sqrt{u^2 + L^2})u}{1 + (sP\zeta^{\text{L}}(\sqrt{u^2 + L^2}))^{-1}}\,du\right)\\
&\quad \times \exp\left(-2\pi\tilde{\lambda}\int_{r}^{+\infty} \frac{[1 - \Pr{}^{\text{L}}(\sqrt{u^2 + L^2})]u}{1 + (sP\zeta^{\text{NL}}(\sqrt{u^2 + L^2}))^{-1}}\,du\right).
\end{aligned} \quad (15.16)
$$

*Proof*: See Appendix 15.2.

Note that in (15.13) and (15.15),

- $\exp\left(-\frac{\gamma P_N}{P\zeta_n^{\text{L}}(\sqrt{r^2+L^2})}\right)$ and $\exp\left(-\frac{\gamma P_N}{P\zeta_n^{\text{NL}}(\sqrt{r^2+L^2})}\right)$ are the probabilities that the signal power beats the noise power by a factor of at least $\gamma$.
- $\mathcal{L}_{I_{\text{agg}}}^{\text{L}}\left(\frac{\gamma}{P\zeta_n^{\text{L}}(\sqrt{r^2+L^2})}\right)$ and $\mathcal{L}_{I_{\text{agg}}}^{\text{NL}}\left(\frac{\gamma}{P\zeta_n^{\text{NL}}(\sqrt{r^2+L^2})}\right)$ are the probabilities that the signal power beats the aggregate interference power by a factor of at least $\gamma$. Specifically, in the expression of $\mathcal{L}_{I_{\text{agg}}}^{\text{L}}(s)$ given by (15.14), the first (second) term of the product calculates the probability that the signal power beats the aggregate interference power from all LoS (NLoS) BSs by a factor of at least $\gamma$.
- Considering the assumption that $h$ follows an exponential distribution, we can invoke $\Pr[h > \gamma(a + b)] = \Pr[h > \gamma a]\,\Pr[h > \gamma b], (a, b \in \Re^+)$, and thus the product of the two probabilities in the above bulletins yields the probability that *the signal power beats the sum power of the noise and the aggregate interference* by a factor of at least $\gamma$.

**Theorem 1** Considering the general path-loss model in (15.3) and the adopted user associated strategy, we derive the limit of the coverage probability when the BS density increases towards infinite, i.e. $\lim_{\lambda \to +\infty} p^{\text{cov}}(\lambda, \rho, \gamma)$, as

$$
\begin{aligned}
\lim_{\lambda \to +\infty} p^{\text{cov}}(\lambda, \rho, \gamma) &= \lim_{\lambda \to +\infty} \Pr\left[\frac{P\zeta_1^{\text{L}}(L)h}{I_{\text{agg}} + P_N} > \gamma\right]\\
&= \exp\left(-\frac{P_N\gamma}{P\zeta_1^{\text{L}}(L)}\right) \lim_{\lambda \to +\infty} \mathcal{L}_{I_{\text{agg}}}^{\text{L}}\left(\frac{\gamma}{P\zeta_1^{\text{L}}(L)}\right),
\end{aligned} \quad (15.17)
$$

where $\lim_{\lambda \to +\infty} \mathcal{L}_{I_{\text{agg}}}^{\text{L}}(s)$ with $s = \frac{\gamma}{P\zeta_1^{\text{L}}(L)}$ is given by

$$
\begin{aligned}
\lim_{\lambda \to +\infty} \mathcal{L}_{I_{\text{agg}}}^{\text{L}}(s) &= \exp\left(-2\pi\rho\int_0^{+\infty} \frac{\Pr{}^{\text{L}}(\sqrt{u^2 + L^2})u}{1 + (sP\zeta^{\text{L}}(\sqrt{u^2 + L^2}))^{-1}}\,du\right)\\
&\quad \times \exp\left(-2\pi\rho\int_0^{+\infty} \frac{[1 - \Pr{}^{\text{L}}(\sqrt{u^2 + L^2})]u}{1 + (sP\zeta^{\text{NL}}(\sqrt{u^2 + L^2}))^{-1}}\,du\right).
\end{aligned} \quad (15.18)
$$

*Proof*: See Appendix 15.3.

From the above result, a new SINR invariance law is revealed, summarized in the following theorem.

**Theorem 2**  If $L > 0$ and $\rho < +\infty$, then $\lim\limits_{\lambda \to +\infty} p^{\mathrm{cov}}(\lambda, \rho, \gamma)$ approaches a constant that is independent of $\lambda$ in UDNs.

*Proof:* The two terms on the right-hand side of (15.17) are both independent of $\lambda$.  □

### 15.3.2  The Area Spectral Efficiency

Next, we investigate the network capacity performance in terms of the area spectral efficiency (ASE) in bps/Hz/km$^2$, which is defined as [6]

$$A^{\mathrm{ASE}}(\lambda, \rho, \gamma_0) = \tilde{\lambda} \int_{\gamma_0}^{+\infty} \log_2(1 + \gamma) f_\Gamma(\lambda, \rho, \gamma) d\gamma, \tag{15.19}$$

where $\tilde{\lambda}$ from (15.1) represents the density of active BSs that make an effective contribution to the ASE, $\gamma_0$ is the minimum working SINR of a practical SCN, and $f_\Gamma(\lambda, \rho, \gamma)$ is the PDF of the SINR $\gamma$ observed at the typical UE for a particular pair of values $\{\lambda, \rho\}$.

Based on the definition of $p^{\mathrm{cov}}(\lambda, \rho, \gamma)$ in (15.5) and the partial integration theorem, (15.19) can be reformulated as

$$A^{\mathrm{ASE}}(\lambda, \rho, \gamma_0) = \frac{\tilde{\lambda}}{\ln 2} \int_{\gamma_0}^{+\infty} \frac{p^{\mathrm{cov}}(\lambda, \rho, \gamma)}{1 + \gamma} d\gamma + \tilde{\lambda} \log_2(1 + \gamma_0) p^{\mathrm{cov}}(\lambda, \rho, \gamma_0). \tag{15.20}$$

From Theorem 15.2 and the expression of the ASE in (15.20), a wireless capacity scaling law can be concluded in the following.

**Theorem 3**  If $L > 0$ and $\rho < +\infty$, then $\lim\limits_{\lambda \to +\infty} A^{\mathrm{ASE}}(\lambda, \rho, \gamma_0)$ approaches a constant that is independent of $\lambda$ in UDNs, which is given by

$$\lim_{\lambda \to +\infty} A^{\mathrm{ASE}}(\lambda, \rho, \gamma_0) = \frac{\rho}{\ln 2} \int_{\gamma_0}^{+\infty} \frac{\lim\limits_{\lambda \to +\infty} p^{\mathrm{cov}}(\lambda, \rho, \gamma)}{1 + \gamma} d\gamma$$
$$+ \rho \log_2(1 + \gamma_0) \lim_{\lambda \to +\infty} p^{\mathrm{cov}}(\lambda, \rho, \gamma_0), \tag{15.21}$$

where $\lim\limits_{\lambda \to +\infty} p^{\mathrm{cov}}(\lambda, \rho, \gamma)$ is given in (15.17) and proved to be independent of $\lambda$ in Theorem 15.2.

From the above constant wireless capacity scaling law, it can be seen that *the network densification should be stopped at a certain level for a given UE density $\rho$.* Furthermore, for a given UE density, there exists an optimal BS density $\lambda^*$ that can maximize the ASE, i.e.

$$\begin{aligned} \underset{\lambda}{\text{maximize}} \quad & A^{\mathrm{ASE}}(\lambda, \rho, \gamma_0) \\ \text{s.t.} \quad & p^{\mathrm{cov}}(\lambda, \rho, \gamma) \geq p_0, \end{aligned} \tag{15.22}$$

where the constraint means that the BS density should satisfy a minimum requirement of coverage probability to ensure an acceptable user experience.

The solution $\lambda^*$ would answer the fundamental question, *For a given UE density $\rho$, how dense should a UDN be?* Moreover, it raises another fundamental question, i.e. *Should we activate all BSs in a given area, or is there an optimal frequency reuse strategy?*

## 15.4 Performance with Multi-channel Spectrum Reuse

Compared with the aggressive SSR approach, in which all the BSs are activated in the same time/frequency resource, let us consider a spectrum sharing network, where BSs are uniformly allocated to $M$ channels, with a resulting BS density of $\frac{\lambda}{M}$ per channel, and discuss what ensues.

First, let us focus on the formulation of coverage probability, denoted by $\hat{p}^{\text{cov}}(\lambda, \rho, \gamma, M)$. The density of active BSs in the network, i.e. $\tilde{\lambda}$ in (15.1), remains the same as the BS and UE densities remain unchanged. Second, since all BSs can be activated to serve UEs, every UE can still be served by the BS providing the strongest signal strength. As such, the signal part does not change. In more detail, the PDFs of $f_{R,n}^{L}(r)$ in (15.9) and $f_{R,n}^{\text{NL}}(r)$ in (15.10) do not change.

However, with the proposed approach, the density of interfering BSs in the same channel is reduced by a factor of $M$, since the density of BSs deployed in one channel is $\frac{\lambda}{M}$. That is, $\frac{\tilde{\lambda}}{M}$ should replace $\tilde{\lambda}$ in the formulations of $\mathcal{L}_{I_{\text{agg}}^{L}}(s)$ in (15.14) and $\mathcal{L}_{I_{\text{agg}}^{\text{NL}}}(s)$ in (15.16). Therefore, compared with the full SSR strategy, the SINR and thus the coverage probability for the typical UE will improve when the $M$-channel spectrum sharing strategy is adopted, i.e. with a frequency reuse factor of $\frac{1}{M}$.

Denoting by $\hat{A}^{\text{ASE}}$ the ASE of the network with the $M$-channel spectrum reuse strategy, this key performance indicator can be formulated as

$$
\begin{aligned}
\hat{A}^{\text{ASE}}(\lambda, \rho, \gamma_0, M) &= \frac{\tilde{\lambda}}{M} \int_{\gamma_0}^{+\infty} \log_2(1+\gamma) \hat{f}_{\Gamma}(\lambda, \rho, \gamma, M) d\gamma \\
&= \frac{\tilde{\lambda}}{M \ln 2} \int_{\gamma_0}^{+\infty} \frac{\hat{p}^{\text{cov}}(\lambda, \rho, \gamma, M)}{1+\gamma} d\gamma + 1/M \, \tilde{\lambda} \log_2(1+\gamma_0) \hat{p}^{\text{cov}}(\lambda, \rho, \gamma_0, M).
\end{aligned}
\tag{15.23}
$$

In the above formulation, $M$ means that each BS only uses $\frac{1}{M}$ of the frequency resource, and according to the above discussion the density of active BSs $\tilde{\lambda}$ remains the same.

As a result of the above trade-off, the increase of UEs' SINR and the decrease of spectrum resource for each BS. In a nutshell, there exists an optimal spectrum reuse factor $M$ to maximize the network performance, which is similar to that in (15.22), and can be formulated as

$$
\begin{aligned}
\underset{M}{\text{maximize}} \quad & \hat{A}^{\text{ASE}}(\lambda, \rho, \gamma_0, M), \\
\text{s.t.} \quad & \hat{p}^{\text{cov}}(\lambda, \rho, \gamma_0, M) \geq p_0, \\
& 1 \leq M \leq \lambda,
\end{aligned}
\tag{15.24}
$$

where $p_0$ is the minimum coverage probability that the network requires.

## 15.5 Simulation and Discussion

According to Tables A.1-3–A.1-7 of [1] and [12], we adopt the following parameters to deal with a 3GPP case: $\alpha^{L} = 2.09$, $\alpha^{\text{NL}} = 3.75$, $A^{L} = 10^{-10.38}$, $A^{\text{NL}} = 10^{-14.54}$, $P = 24$ dBm, and $P_{N} = -95$ dBm (with a noise figure of 9 dB).

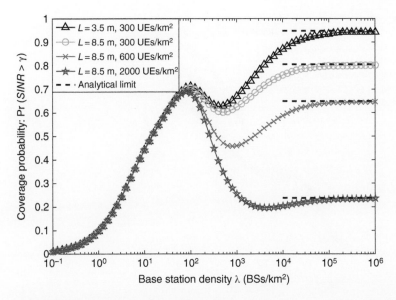

**Figure 15.1** The coverage probability $p^{cov}(\lambda, \rho, \gamma)$ versus $\lambda$ for the 3GPP case with $\gamma = 0$ dB and various values of $\rho$ and $L$.

### 15.5.1 Performance with Full Spectrum Reuse Strategy

In Figure 15.1, we display the coverage probability for this 3GPP case with $\gamma = 0$ dB and various values of $\rho$ and $L$. Here, solid lines, markers, and dash lines represent analytical results derived by (15.8), simulation results, and $\lim_{\lambda \to +\infty} p^{cov}(\lambda, \rho, \gamma)$ derived in (15.17), respectively. From this figure, we can observe the following:

- When the BS density is around $\lambda \in [10^{-1}, 10^2]$ BSs/km², the network is noise-limited, and thus $p^{cov}(\lambda, \rho, \gamma)$ increases with $\lambda$ as the network is lightened up with more BSs and the signal power benefits from LoS transmissions.
- When the BS density is at around $\lambda \in [10^2, 10^3]$ BSs/km², $p^{cov}(\lambda, \rho, \gamma)$ decreases as $\lambda$ increases. This is because of the transition of a large number of interfering paths from NLoS to LoS, which accelerates the growth of the aggregate inter-cell interference.
- When $\lambda \in [10^3, 10^5]$ BSs/km², $p^{cov}(\lambda, \rho, \gamma)$ continuously increases as $\lambda$ increases. This is because the signal power continues to increase with the network densification, while the interference power becomes bounded, as only BSs with active UEs are turned on and thus the number of interfering BSs is bounded by the number of active UEs.
- When $\lambda > 10^5$ BSs/km², $p^{cov}(\lambda, \rho, \gamma)$ gradually reaches its limit, characterized by (15.17), which verifies the SINR invariance law in Theorem 15.2. This is because the signal power also becomes bounded as the UEs cannot get infinitely close to their serving BSs due to the antenna height, $L$.

In Figure 15.2, we plot the ASE results for the 3GPP case with $\gamma_0 = 0$ dB, $L = 8.5$ m, and various values of $\rho$.

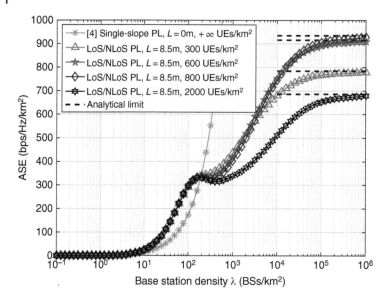

**Figure 15.2** The ASE $A^{\mathrm{ASE}}(\lambda, \rho, \gamma_0)$ versus $\lambda$ for the 3GPP case with $\gamma_0 = 0$ dB, $L = 8.5$ m, and various values of $\rho$.

In this figure, $A^{\mathrm{ASE}}(\lambda, \rho, \gamma_0)$ is calculated from the results of $p^{\mathrm{cov}}(\lambda, \rho, \gamma)$ using (15.20). Because the analysis on $p^{\mathrm{cov}}(\lambda, \rho, \gamma)$ has been validated in Figure 15.1, we only show the analytical results of $A^{\mathrm{ASE}}(\lambda, \rho, \gamma_0)$ here. From the figure, we can observe the following:

- The *linear capacity scaling law* in [2], based on simplistic assumptions on channel modeling and UE density, shows an optimistic but unrealistic future for 5G UDNs.
- The ASE crawls (does not increase quickly) when $\lambda \in [10^2, 10^3]$ BSs/km$^2$ due to the degradation of the coverage probability (see Figure 15.1) caused by the transition of a large number of interfering paths from NLoS to LoS.
- For a given $\rho$, the value of $A^{\mathrm{ASE}}(\lambda, \rho, \gamma_0)$ approaches the limit in (15.21) when $\lambda \to +\infty$, and thus the *constant wireless capacity scaling law* in Theorem 15.3 is validated for UDNs with a non-zero $L$ and a finite $\rho$.

### 15.5.2 Performance with Multi-channel Spectrum Reuse Strategy

To show the performance impact of different spectrum sharing strategies, in the following we plot the coverage probability and ASE performance for a different number of channels $M$.

In Figure 15.3, it can be seen that the coverage probability performance is improved with an increased $M$ due to the increased SINR. The improvement is prominent compared with the co-channel deployment ($M = 1$), i.e. the aggressive SSR strategy. Hence, a proper spectrum reuse factor $M$ can be leveraged to improve signal quality and network coverage.

In contrast to the coverage probability, Figure 15.4 shows that *there is an optimum channelization that maximizes the ASE*. For example, the optimum number of channels is

- $M = 2$ for the case with 600 UEs/km$^2$ and 1000 BSs/km$^2$

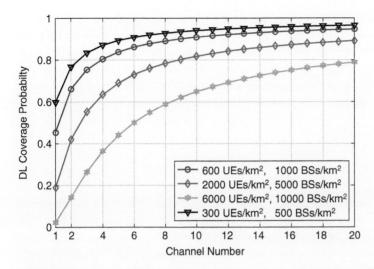

**Figure 15.3** The coverage probability $\hat{p}^{cov}(\lambda, \rho, \gamma, M)$ versus $M$ with $\gamma = 0$ dB and various values of $\lambda$ and $\rho$.

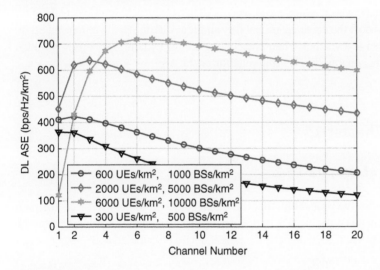

**Figure 15.4** The ASE $\hat{A}^{ASE}(\lambda, \rho, \gamma_0, M)$ versus $M$ with $\gamma_0 = 0$ dB and various values of $\lambda$ and $\rho$.

- $M = 3$ for the case with 2000 UEs/km$^2$ and 5000 BSs/km$^2$
- $M = 6$ for the case with 6000 UEs/km$^2$ and 10000 BSs/km$^2$.

Analyzing these results one can see that there are cases in which deploying BSs in multiple channels is not beneficial in terms of ASE. For the case with 300 UEs/km$^2$ and 500 BSs/km$^2$, the ASE monotonously decreases with $M$. These results show that the tradeoff between the increased SINR and the decreased available bandwidth in each channel determines the ASE performance. When the network deployment results in medium or low UE SINRs, the $M$-channel spectrum reuse strategy can greatly boost the

ASE due to the enhanced SINR. In this case, it is important to note that a large $M$ is not necessarily a good choice, since it leads to a very limited bandwidth available for each cell and therefore a limited network throughput. In contrast, when the network deployment already results in large UE SINRs, a co-channel deployment ($M = 1$) becomes the best policy to maximize the system throughput, as we have practiced in the past decades. In the future UDNs, an $M$ value larger than one might become a default configuration for cellular networks.

## 15.6 Conclusion

In this chapter, the stochastic geometry theory was used to analyze the network performance in terms of coverage probability and ASE, especially in dense and ultra-dense networks. A sophisticated system model considering LoS and NLoS transmissions, antenna heights, finite UE densities, and active/sleep BSs was adopted, which captures essential network characteristics. Furthermore, by comparing the performance of a network with the conventional full spectrum reuse strategy and the multi-channel spectrum reuse strategy, we investigated the optimal spectrum reuse strategy. In 3G/4G, the full spectrum reuse has been widely adopted. However, when the network is UDN, A multi-channel spectrum reuse strategy can greatly boost the coverage probability and the ASE due to the enhanced SINR. This chapter has also shown that when considering a multi-channel spectrum strategy, there exists an optimal channelization that maximizes the ASE.

## Appendix for Chapter 15

## 15.A.1 Proof of Lemma 15.1

The PDF $f_{R,n}^L(r)$ denotes the probability that the signal comes from the $n$th piece LoS path with a distance $r$, which is the joint PDF of the serving distance $r$ and the event, $B_n^L$, that the signal comes from the $n$th piece LoS path. Therefore, it can be calculated as

$$f_{R,n}^L(r) = f_{R,n|B_n^L}(r|B_n^L) \; \Pr[B_n^L], \tag{15.A.1}$$

where $\Pr[B_n^L] = \Pr_n^L(\sqrt{r^2 + L^2})$ according to (15.A.4) and $f_{R,n|B_n^L}(r|B_n^L)$ should characterize the joint event of the following three independent subevents:

1. Since the BSs follow an HPPP with density $\lambda$, for the typical UE, its serving BS $b_o$ exists at a horizontal distance $r$ from it and the corresponding unconditional PDF of $r$ is $2\pi r\lambda$.
2. There should be no LoS BS that can provide a better link to the typical UE than the LoS BS $b_o$, the probability of which is

$$p_n^L(r) = \exp\left(-\int_0^r \Pr^L(\sqrt{u^2 + L^2})2\pi u\lambda du\right). \tag{15.A.2}$$

3. There should be no NLoS BS that can provide a better link to the typical UE than the LoS BS $b_o$, the probability of which is

$$p_n^{\mathrm{NL}}(r) = \exp\left(-\int_0^{r_1}(1 - \mathrm{Pr}^{\mathrm{L}}(\sqrt{u^2 + L^2}))2\pi u\lambda du\right),\qquad(15.\mathrm{A}.3)$$

where $r_1$ is the horizontal distance at which an NLoS BS has the same signal reception level as $b_o$. Hence, $r_1$ can be computed by (15.A.11).

Consider the joint probability of the three independent subevents above, we have

$$f_{R,n|B_n^{\mathrm{L}}}(r|B_n^{\mathrm{L}}) = p_n^{\mathrm{NL}}(r)p_n^{\mathrm{L}}(r)2\pi r\lambda.\qquad(15.\mathrm{A}.4)$$

Then, substituting $f_{R,n|B_n^{\mathrm{L}}}(r|B_n^{\mathrm{L}})$ into (15.A.1), the proof completes. In a similar way, we can obtain $f_{R,n}^{\mathrm{NL}}(r)$.

## 15.A.2 Proof of Lemma 15.2

Adopting the definition of SINR in (15.A.6) and substituting the formulation of the cumulative interference in (15.A.7) and the path-loss model in (15.A.4), we have

$$\mathrm{Pr}[\mathrm{SINR}_n^{\mathrm{L}}(r) > \gamma] = \mathrm{Pr}\left[\frac{P\zeta_n^{\mathrm{L}}(\sqrt{r^2 + L^2})h}{I_{\mathrm{agg}} + P_{\mathrm{N}}} > \gamma\right]$$

$$= \mathbb{E}_{[I_{\mathrm{agg}}]}\left\{\mathrm{Pr}\left[h > \frac{\gamma(I_{\mathrm{agg}} + P_{\mathrm{N}})}{P\zeta_n^{\mathrm{L}}(\sqrt{r^2 + L^2})}\right]\right\}\qquad(15.\mathrm{A}.5)$$

$$= \mathbb{E}_{[I_{\mathrm{agg}}]}\left\{\overline{F}_H\left(\frac{\gamma(I_{\mathrm{agg}} + P_{\mathrm{N}})}{P\zeta_n^{\mathrm{L}}(\sqrt{r^2 + L^2})}\right)\right\}$$

where $\mathbb{E}_{[X]}\{\cdot\}$ denotes the expectation operation taking the expectation over the variable $X$ and $\overline{F}_H(h)$ denotes the complementary cumulative distribution function of RV $h$. Since we assume $h$ to be an exponential RV, we have $\overline{F}_H(h) = \exp(-h)$ and thus (15.A.5) can be further derived as

$$\mathbb{E}_{[I_{\mathrm{agg}}]}\left\{\overline{F}_H\left(\frac{\gamma(I_{\mathrm{agg}} + P_{\mathrm{N}})}{P\zeta_n^{\mathrm{L}}(\sqrt{r^2 + L^2})}\right)\right\}$$

$$= \exp\left(-\frac{\gamma P_{\mathrm{N}}}{P\zeta_n^{\mathrm{L}}(\sqrt{r^2 + L^2})}\right)\mathbb{E}_{[I_{\mathrm{agg}}]}\left\{\exp\left(-\frac{\gamma I_{\mathrm{agg}}}{P\zeta_n^{\mathrm{L}}(\sqrt{r^2 + L^2})}\right)\right\}\qquad(15.\mathrm{A}.6)$$

$$= \exp\left(-\frac{\gamma P_{\mathrm{N}}}{P\zeta_n^{\mathrm{L}}(\sqrt{r^2 + L^2})}\right)\mathcal{L}_{I_{\mathrm{agg}}}^{\mathrm{L}}\left(\frac{\gamma}{P\zeta_n^{\mathrm{L}}(\sqrt{r^2 + L^2})}\right),$$

where $\mathcal{L}_{I_{\mathrm{agg}}}^{\mathrm{L}}(s)$ is the Laplace transform of RV $I_{\mathrm{agg}}$ evaluated at $s$ on the condition of event $B^{\mathrm{L}}$ that the typical UE is associated with a BS with a LoS path. Based on the presented UAS, we can derive $\mathcal{L}^{L}_{I_{\text{agg}}}(s)$ as

(a) $\mathcal{L}^{L}_{I_{\text{agg}}}(s) = \mathbb{E}_{[I_{\text{agg}}]}\{\exp(-sI_{\text{agg}})|B^{L}\}$

$$= \mathbb{E}_{[\Phi,\{\beta_i\},\{g_i\}]}\left\{\exp\left(-s\sum_{i\in\Phi/b_o}P\beta_i(w)g_i\right)\bigg|B^{L}\right\}$$

$$= \exp\left(-2\pi\lambda\int_{r}^{+\infty}\Pr^{L}(\sqrt{u^2+L^2})[1-\mathbb{E}_{[g]}\{\exp(-sP\zeta^{L}(\sqrt{u^2+L^2})g)\}]u\,du\right.$$

$$\left.-2\pi\lambda\int_{r_1}^{+\infty}[1-\Pr^{L}(\sqrt{u^2+L^2})][1-\mathbb{E}_{[g]}\{\exp(-sP\zeta^{NL}(\sqrt{u^2+L^2})g)\}]u\,du\right)$$

$$= \exp\left(-2\pi\lambda\int_{r}^{+\infty}\frac{\Pr^{L}(\sqrt{u^2+L^2})u}{1+(sP\zeta^{L}(\sqrt{u^2+L^2}))^{-1}}du\right)\exp\left(-2\pi\lambda\int_{r_1}^{+\infty}\frac{[1-\Pr^{L}(\sqrt{u^2+L^2})]u}{1+(sP\zeta^{NL}(\sqrt{u^2+L^2}))^{-1}}du\right).$$

$$(15.A.7)$$

Note that (a) considers the interference coming from LoS and NLoS paths. Plugging $s = \frac{\gamma}{P\zeta_n^{L}(\sqrt{r^2+L^2})}$ into (15.A.7) and further plugging (15.A.7) into (15.A.6), we can obtain the general expression of $\Pr\left[\frac{P\zeta_n^{L}(\sqrt{r^2+L^2})h}{I_{\text{agg}}+P_N}>\gamma\right]$ shown in (15.A.13).

## 15.A.3 Proof of Theorem 15.1

As $\lambda\to+\infty$, we have that the horizontal distance $r$ from the typical UE to its serving BS $b_o$ approaches zero, i.e. $r\to0$ and $w\to L$ in (15.A.2). Consequently, the path loss of this link should be dominantly characterized by the first-piece LoS path loss function (i.e., $\zeta_1^{L}(w)$). For instance, $L$ is smaller than $d_1=67.75$ m of the 3GPP case [1], which supports the use of $\zeta_1^{L}(w)$ in this case. Thus, $\lim_{\lambda\to+\infty}p^{\text{cov}}(\lambda,\rho,\gamma)$ can be derived as

$$\lim_{\lambda\to+\infty}p^{\text{cov}}(\lambda,\rho,\gamma) = \lim_{\lambda\to+\infty}\Pr[\text{SINR}>\gamma|\zeta(w)=\zeta_1^{L}(L)] \qquad (15.A.8)$$

$$= \lim_{\lambda\to+\infty}\Pr\left[\frac{P\zeta_1^{L}(L)h}{I_{\text{agg}}+P_N}>\gamma\right]. \qquad (15.A.9)$$

Moreover, from (15.A.1), we have that $\lim_{\lambda\to+\infty}\tilde{\lambda}=\rho$. Substituting $s=\frac{\gamma}{P\zeta_1^{L}(L)}$ and $\lim_{\lambda\to+\infty}\tilde{\lambda}=\rho$ into the calculation of $\mathcal{L}^{L}_{I_{\text{agg}}}(s)$ in (15.A.14), we can obtain the result of $\lim_{\lambda\to+\infty}\mathcal{L}^{L}_{I_{\text{agg}}}\left(\frac{\gamma}{P\zeta_1^{L}(L)}\right)$ in (15.18) and this concludes our proof.

## References

**1** 3GPP. TR 36.828: Further enhancements to LTE Time Division Duplex for Downlink-Uplink interference management and traffic adaptation, Jun. 2012.

**2** J. G. Andrews, F. Baccelli, and R. K. Ganti. A tractable approach to coverage and rate in cellular networks. *IEEE Transactions on Communications*, 59(11):3122–3134, 2011.

**3** T. Bai and R. W. Heath. Coverage and rate analysis for millimeter-wave cellular networks. *IEEE Transactions on Wireless Communications*, 14(2): 1100–1114, 2015.

**4** M. Ding and D. López-Pérez. On the performance of practical ultra-dense networks: The major and minor factors. In *2017 15th International Symposium on Modeling and Optimization in Mobile, Ad Hoc, and Wireless Networks (WiOpt)*, pages 1–8, 2017.

**5** M. Ding, D. López-Pérez, G. Mao, P. Wang, and Z. Lin. Will the area spectral efficiency monotonically grow as small cells go dense? In *2015 IEEE Global Communications Conference (GLOBECOM)*, pages 1–7, Dec. 2015.

**6** M. Ding, P. Wang, D. López-Pérez, G. Mao, and Z. Lin. Performance impact of LoS and NLoS transmissions in dense cellular networks. *IEEE Transactions on Wireless Communications*, 15(3):2365–2380, 2016.

**7** M. Ding, D. López-Pérez, G. Mao, and Z. Lin. Performance impact of idle mode capability on dense small cell networks. *IEEE Transactions on Vehicular Technology*, 66(11): 10446–10460, 2017.

**8** M. Dohler, R. W. Heath, A. Lozano, C. B. Papadias, and R. A. Valenzuela. Is the PHY layer dead? *IEEE Communications Magazine*, 49(4):159–165, April 2011.

**9** A. H. Jafari, D. López-Pérez, M. Ding, and J. Zhang. Performance analysis of dense small cell networks with practical antenna heights under Rician fading. *IEEE Access*, 6: 9960–9974, 2018.

**10** D. López-Pérez, M. Ding, H. Claussen, and A. H. Jafari. Towards 1 Gbps/UE in cellular systems: Understanding ultra-dense small cell deployments. *IEEE Communications Surveys Tutorials*, 17(4):2078–2101, 2015.

**11** S. Lee and K. Huang. Coverage and economy of cellular networks with many base stations. *IEEE Communications Letters*, 16(7):1038–1040, 2012.

**12** Spatial Channel Model AHG. Subsection 3.5.3, Spatial Channel Model Text Description V6.0, Apr. 2003.

**13** W. Webb. Wireless Communications: The Future. Wiley, New York, 2007.

**14** X. Zhang and J. G. Andrews. Downlink cellular network analysis with multi-slope path loss models. *IEEE Transactions on Communications*, 63 (5):1881–1894, 2015.

# 16

# Large-scale Wireless Spectrum Monitoring: Challenges and Solutions based on Machine Learning

*Sreeraj Rajendran and Sofie Pollin*

KU Leuven, Belgium

The new generation of wireless technologies, making use of both licensed and unlicensed frequency bands, promises improved throughput, latency, and reliability, enabling the creation of novel applications. The fifth-generation (5G) wireless technologies are envisioned to be more diverse and heterogeneous. Densified small cells are becoming fundamental for these high throughput and low latency networks to support the new high bandwidth applications such as live data streaming, video calls, and vehicle-to-vehicle communications. Manual wireless spectrum management and analysis will be inefficient and suboptimal in such dense and heterogeneous wireless environments. In addition, the number of unauthorized wireless spectrum usages and anomalies is increasing every year in the form of uncertified wireless devices, fake base stations, unintentional transmitter leakages, and easily available spectrum jammers. The key question is then: *Who is going to monitor this complex and dense electromagnetic space and how will we do it cost-efficiently?* Even if we know how to sample the RF spectrum cost-effectively, how can we analyze it, and translate samples to spectrum knowledge?

## 16.1 Challenges

Radio spectrum is one of our most precious and widely used natural resources. With the advent of new wireless communication technologies, spectrum usage has become very complex, resulting in airwave congestion and other interference issues [5]. Diverse spectrum regulations across countries have also contributed to this chaotic spectrum usage when non-standardized wireless devices cross country borders. In addition, easily available illegal wireless jammers or low-cost software-defined radio (SDR) devices which are capable of generating custom wireless signals are making the problem worse. Furthermore, illegal repeaters, which are used to boost mobile coverage, can adversely affect the cell planning of mobile operators, resulting in poor coverage and dropouts [34]. Unintentional and intentional jamming of localization services such as a global navigation satellite system (GNSS), which is used for a wide range of applications such as automated vehicle navigation, airplane landing procedures, and maritime vessel tracking, is increasing with the availability

*Spectrum Sharing: The Next Frontier in Wireless Networks,* First Edition.
Edited by Constantinos B. Papadias, Tharmalingam Ratnarajah, and Dirk T.M. Slock.
© 2020 John Wiley & Sons Ltd. Published 2020 by John Wiley & Sons Ltd.

of easy jamming devices [32, 35]. On the other hand, densified small cells are becoming fundamental for the new high throughput and low latency requirements [15]. Such dense and heterogeneous deployment makes the enforcement and management of the wireless spectrum usage difficult. Automated monitoring, real-time analysis, and detection of anomalous behaviors in the spectrum is becoming more crucial than ever before.

Spectrum resource monitoring is important for end users, operators, spectrum regulatory bodies, and military applications. Each use case has its own specific needs and challenges. The grand challenge is how to design a cost-effective solution that meets the requirements of all potential end users. Users might be interested in electrosmog or optimization of their indoor Wi-Fi network. Regulatory bodies might be keen on enforcing spectrum regulation. Operators might be concerned about coverage maps over time for optimizing their cell networks or refarming of their frequency bands. Military applications might be the most challenging, requiring the detection and positioning of any signal hidden on purpose. Novel operators might be interested in Internet of Things (IoT) cases, such as cooperative detection of signals transmitted by low-cost/low-power transceivers.

Wireless spectrum monitoring is an interesting field of research which has attracted both academia and industry, and there have been many initiatives in the past. A few initiatives are funded by National Science Foundation (NSF) projects in the USA [1, 4, 11]. A metro scale spectrum observatory [29], an initiative from the University of Washington, is also sponsored by NSF. Radiohound [12] is another spectrum monitoring initiative concentrating on mobile environment monitoring using low-cost SDR. Hawkeye360[1] is a recent organization planning to build low earth orbit satellites to monitor global radio frequency activity across air, land, and sea, and assist with emergencies.

A few other spectrum monitoring solutions are proposed in the literature. Some examples include Microsoft Spectrum Observatory,[2] which allows users to sense the spectrum using expensive sensors, Google Spectrum[3] for measurements on TV white spaces, and the IBM Horizon[4] project that proposed a generic decentralized architecture to share IoT data. While Google Spectrum and IBM Horizon fail to cover a large part of the spectrum as they are application specific deployments, Microsoft Spectrum Observatory fails to enable large-scale sensor deployment mainly due to the cost of the sensing stations.

Even though a majority of wireless researchers, industries and spectral regulators are keen to develop a worldwide spectrum monitoring infrastructure, the research community has not succeeded in deploying one. The multidisciplinary nature of the spectrum monitoring solution is one of the main challenges that prevents the realization of such a system, which in turn requires proper integration of new disruptive technologies. The infrastructure should flexibly address the variability and cost of the used sensors, large spectrum data management, sensor reliability, security and privacy concerns, which can also target a wide variety of the use cases as mentioned before. In section 16.2 we will propose a spectrum monitoring system that relies on a large-scale data center with low cost sensors deployed by the crowd as a promising method to enable spectrum monitoring.

Spectrum monitoring is more than sampling the spectrum. Once the infrastructure is there to obtain spectrum samples, the main challenge remains the interpretation and analysis of the spectrum data. Early spectrum analysis approaches focused on the

---

1 http://www.he360.com.
2 http://observatory.microsoftspectrum.com/.
3 https://www.google.com/get/spectrumdatabase/.
4 https://bluehorizon.network/documentation/sdr-radio-spectrum-analysis.

well-defined use case of primary user presence detection or modulation recognition. Recently, more complex RF signal analysis is being considered going from radar to modulation recognition or spectrum anomaly detection. In section 16.3 we will focus on two very concrete spectrum analysis examples where the use of machine learning is promising.

## 16.2 Crowdsourcing

Crowdsourcing is one way to achieve a large-scale sensor deployment which is effectively used in one of the successful spectrum monitoring framework, Electrosense [25]. Electrosense is a non-profit organization based in Switzerland which aims at improving the way the radio frequency spectrum is used. The initiative's goal is to sense the entire wireless spectrum in populated regions of the world and to make the data available in real time for different stakeholders. Electrosense is an open initiative to which everyone can contribute with spectrum measurements and access the collected data. Worldwide deployments are plausible when the crowdsourcing paradigm is combined with low cost sensors, as shown in Figure 16.1. Big data solutions are combined with the power of crowdsourcing to solve many

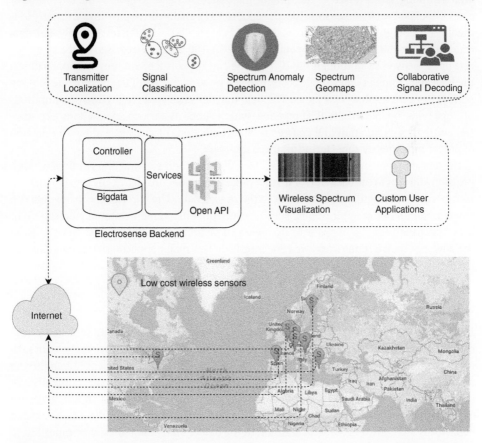

**Figure 16.1** High-level overview of the Electrosense network. Low-cost sensors collect spectrum information which is sent to the Electrosense backend. Different algorithms are run on the collected information in the backend and the results of these algorithms are provided to the users as a service through an open application programming interface (API).

of the spectrum monitoring issues in Electrosense. The sensing devices could be low-cost SDR dongles connected to embedded devices like a Raspberry Pi or high-end SDR devices connected through a more powerful machine. The software that runs on the Electrosense nodes is released as open source.[5] The Electrosense network is currently in its beta stage where volunteers have deployed sensors across seven countries in Europe. Through Electrosense, an Open Spectrum Data as a Service (OSDaaS) model was introduced to address the usability of the spectrum data for a wide range of stakeholders, including wireless operators, spectrum enforcement agencies, and military and generic users.

## 16.3 Wireless Spectrum Analysis

The wireless spectrum data in general refers to the wide range of wireless information, including radio signal power levels, frequency occupancy, specific signal features, wireless channel/environment information, and the device details connected to a particular network. In the spectrum sensing context we refer to, the spectrum data in its raw sensed form, after sampling, is expressed in the time or frequency domain. To control the data transfer costs associated with the sensing, Electrosense sensors enable three pipelines with very low, medium, and high data transfer costs, namely, Feature, PSD, and IQ, respectively. The data transfer rate required for the IQ pipeline is in the 30–100 Mbps range based on the sampling rate of the sensor. Even though the IQ pipeline can support a broad range of applications, the huge data transfer and storage costs involved prevent the use of the IQ pipeline for long periods even for a subset of sensors. The data transfer rates are brought down to hundreds of Kbps with the help of the PSD pipeline, which supports various applications like spectral occupancy calculation and coarse anomaly detection. Electrosense allows a selection of proper pipelines by considering all these tradeoffs.

The sensed spectrum data is processed locally for decoding, extracting features or compression and sent to the backend for further processing. The core idea of any sensing method is to extract meaningful information from the sensed data. The wireless spectrum data can aid a large number of applications, such as spectrum enforcement, anomaly detection, or even spectrum coverage maps over frequency, time and space dimensions. To emphasize the power of such a framework, we concentrate on two major applications throughout this chapter: anomaly detection and wireless signal classification.

### 16.3.1 Anomaly Detection

Detecting anomalous behavior in the wireless spectrum is a demanding task due to the sheer complexity of electromagnetic spectrum use. Wireless spectrum anomalies can take a wide range of forms, from the presence of an unwanted signal in a licensed band to the absence of an expected signal, which makes manual labeling of anomalies difficult and suboptimal. To achieve the vision of automated spectrum awareness, a wireless spectrum anomaly detector which can continuously monitor the spectrum and detect unexpected behavior is vital. Furthermore, in addition to the detection of anomalies, it is important to

---

5 https://github.com/electrosense/es-sensor.

understand the cause of an anomaly. This ranges from an unexpected transmission in the analyzed band that can be classified [26], to the absence of an expected signal.

Wireless anomaly detection to some extent has been addressed in wireless sensor networks in the past [22, 24, 37]. These techniques make use of derived expert features from very low rate sensor data such as temperature and pressure instead of high volume radio physical layer data, as is our interest. An anomaly detector for dynamic spectrum access (DSA) is presented in [14], where distributed power measurements via cooperative sensing are used for anomaly detection. The proposed detector is limited to authorized user anomaly detection only, for the specific case of DSA. Similarly, [9] makes use of hidden Markov model (HMM) on spectral amplitude probabilities that can detect interference on the channel of interest, again in the DSA domain.

In [20], the authors present a recurrent anomaly detector based on predictive modeling of raw IQ data. The authors used a LSTM model for predicting the next four IQ samples from the past 32 samples and an anomaly was detected based on the prediction error. Even though this model works on raw physical layer data which require no expert feature extraction, it is still not sufficiently automated and generic for practical anomaly detection. First, different copies of the same model need to be trained for different wireless bands such that the model is able to predict anomalies specific to the band of interest. For instance, an LTE signal in the FM broadcast band is definitely an anomaly, thus preventing a single model to be trained on both bands. Second, the model does not extract any interpretable features to understand the cause of the anomaly. In [33], the authors extend this prediction idea to spectrograms and test the model on some synthetic anomalies. A reconstruction-based anomaly detector on vanilla deep autoencoders is presented in [6]. This model lacks interpretable feature extraction properties like class labels, which implies the need for training multiple copies of the same model on different bands.

An adversarial autoencoder (AAE)-based model which fills the shortcomings of these state-of-the-art models is proposed in [27]. First, it is shown that a *single model can be trained over multiple bands* in an unsupervised fashion, avoiding the need for multiple copies of models on various bands. Second, the same model can be *trained in a semi-supervised fashion for extracting interpretable features* such as signal bandwidth and position. Third, the reconstructed signal from the proposed model can be used for *localizing anomalies* in the wireless spectrum. Furthermore, the model enables various other advantages such as *wireless data compression* and *signal classification*, which are significant contributions in contrast to the state-of-the-art models [6, 20, 33].

A deep learning model based on AAE is used to achieve the aforementioned properties, as shown in Figure 16.2. A long short-term memory (LSTM) layer with 512 cells is used as the encoder for extracting interpretable features while a convolutional neural network (CNN)-based decoder is employed for reconstructing the input data from the extracted features. The AAE architecture is trained in a semi-supervised fashion to make the features more interpretable while the reconstruction is fully unsupervised. Two layer feed forward networks with 256 cells and relu activations are employed in both discriminators. The LSTM output is fed through a softmax layer for signal classification and a linear layer for extracting the latent features.

The discriminators $(D_s)$ are neural networks that evaluate the probability that the latent code $\mathbf{z}$ is from the prior distribution $p(\mathbf{z})$ that we are trying to impose rather than a sample

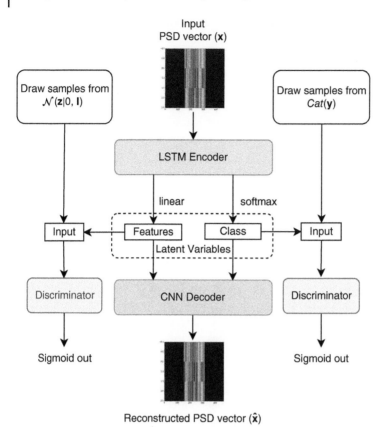

**Figure 16.2** Model architecture for anomaly detection.

from the output of the encoder ($E$) model. The discriminator receives **z** from both the encoder and the prior distribution, and is trained to distinguish between them. The encoder is trained to confuse the discriminators into believing that the samples it generates are from the prior distribution. Thus the encoder is trained to reach the solution by optimizing both networks by playing a min-max adversarial game that is expressed in [7] as

$$\min_{E} \max_{D_s} \mathbb{E}_{\mathbf{z} \sim p(\mathbf{z})}[log(D_s(\mathbf{z}))] + \mathbb{E}_{\mathbf{x} \sim p_{data}}[log\,(1 - D_s(E(\mathbf{x})))] \tag{16.1}$$

Generative models try to model the underlying distributions of the input data, the latent variables, which are further used for data reconstruction. In a spectrum anomaly detector with interpretable features (SAIFE), the input power spectral density (PSD) data is assumed to be generated by the latent *class* variable which comes from a categorical distribution with number of categories $k = $ *number of frequency bands* and the continuous latent *features* from a Gaussian distribution of zero mean and unit variance; $p(\mathbf{y}) = Cat(\mathbf{y})$ and $p(\mathbf{z}) = \mathcal{N}(\mathbf{z}|0, \mathbf{I})$.

Three scores are used to detect whether the input data frame is anomalous or not:

1. *Reconstruction loss*: This error measures the similarity between the input data and the reconstructed data defined as $R_l = \sum_{i=0}^{N} |\mathbf{x} - \hat{\mathbf{x}}|$ where $\mathbf{x}$ is the frame input, $\hat{\mathbf{x}} = D(\mathbf{z})$ is the decoder frame output, and $N$ is the number of data points in the frame.
2. *Discriminator loss*: The discriminator in the AAE model is trained to distinguish between the samples from the prior distribution and the samples generated by the encoder. We use the same discrimination loss used during the training process, which is defined as $D_l = \sigma(\mathbf{z}, 1)$, where $\sigma$ is the sigmoid cross entropy. The loss from both continuous ($D_{lcont}$) and categorical ($D_{lcat}$) discriminators is used to compute the final anomaly score.
3. *Classification error*: The class labels predicted by the encoder are cross-checked with the original band of interest to detect the presence of other known but unexpected signals in a selected frequency band.

A simple n-sigma threshold is employed on the reconstruction and discriminator loss based on the mean and standard deviation values from the training data. The symbol $\vee$ in the following equation represents *logical or* operation. An input data frame is classified as anomalous if $A_{score}$ is *True*:

$$A_{score} = (R_l > (\mu_{R_{lt}} + n * \sigma_{R_{lt}}))$$
$$\vee ((\mu_{D_{lcont}} - n * \sigma_{D_{lcont}}) > D_{lcont} > (\mu_{D_{lcont}} + n * \sigma_{D_{lcont}}))$$
$$\vee ((\mu_{D_{lcat}} - n * \sigma_{D_{lcat}}) > D_{lcat} > (\mu_{D_{lcat}} + n * \sigma_{D_{lcat}}))$$
$$\vee (Class_{Encoder} ! = Class_{input}) \tag{16.2}$$

The threshold value $n$ is selected empirically based on the expected true positive rate and false detection rate.

We use two spectrum data sets along with one synthetic anomaly set to evaluate the performance of the used model. A synthetic spectrum dataset is necessary to understand the performance of the model in a controlled environment. The synthetic data set consists of four different signal types with signal parameters as reported in Table 16.1. The signals are (i) *single-cont*: single continuous signal with random bandwidth, signal-to-noise ratio (SNR) and center frequency, (ii) *single-rshort*: pulsed signals in time with similar parameters

**Table 16.1** Synthetic signal dataset parameters

| Type | Single-cont, single-rshort, mult-cont, dethop |
| --- | --- |
| Input frame size | $6 \times 64$ |
| SNR range | 5 dB to +20 dB |
| Number of training samples | 6000 vectors |
| Number of test samples | 6000 vectors |

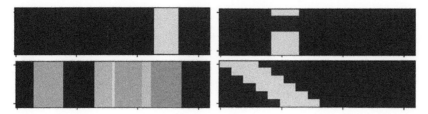

**Figure 16.3** Sample signals *single-cont*, *single-rshort*, *mult-cont*, and *dethop* from synthetic signal dataset (time on *y* axis and frequency on *x* axis).

**Table 16.2** Synthetic anomaly dataset parameters

| Type | Scont, randpulses, wpulse, oclass |
|---|---|
| Input frame size | $6 \times 64$ |
| SNR range | $-20\,dB$ to $+20\,dB$ |
| Number of training samples | 6000 vectors |
| Number of test samples | 6000 vectors |

as *single-cont*, (iii) *mult-cont*: multiple continuous signals with possible overlap, and (iv) *dethop*: random bandwidth and SNR signals with deterministic shifts/hops in frequency as depicted in Figure 16.3. Similarly, four synthetic signals, (i) *scont*: same as single-cont, (ii) *randpulses*: random pulsed transmissions on the given band, (iii) *wpulse*: pulsed wideband signals covering the entire frequency, and (iv) *oclass*: signals from other classes in synthetic dataset, are used as anomalies as reported in Table 16.2.

### 16.3.2 Performance Comparisons

To evaluate the performance of SAIFE, the anomaly detection performance is compared against various state-of-the-art algorithms such as one-class support vector machines (OSVMs), isolation forest (IFO) [13], lightweight on-line detector of anomalies (LODA) [23], and robust covariance (RCOV) [28]. The average anomaly detection accuracy of these algorithms over different frequency bands are plotted in Figure 16.4. On average SAIFE performs better than all other algorithms for all anomalies on all synthetic frequency bands. *Oclass* anomaly performance is quite good when compared to other algorithms as SAIFE performs explicit frequency band classification as one of the features. Figure 16.5 shows the receiver operating characteristic (ROC) curves for all anomalies on the *det-hop* channel for all algorithms. Anomaly signals similar to the original signals are intentionally selected to thoroughly analyze the detection capabilities of the model. Improving the number of features of SAIFE from 20 to 100 can also help to improve the detection accuracy to some extent, as shown in Figure 16.4.

To understand the performance on detecting real anomalies, the model is tested on the real-world Electrosense dataset. The data is collected through the Electrosense API[6] with a

---

6 https://electrosense.org/open-api-spec.html.

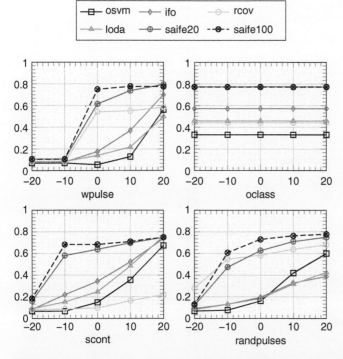

**Figure 16.4** Anomaly detection accuracies for different anomalies with a constant false alarm rate of 10% averaged over four different frequency bands. For *oclass* anomaly, anomaly vectors are randomly selected from other classes without specific SNR-based evaluation, resulting in one detection accuracy value (plotted as a line for uniformity). Anomaly SNR on the *x* axis and detection accuracy on the *y* axis.

spectral resolution of 10 kHz and time resolution of 60 seconds. These sensors are low-cost RTLSDRs configured at a sampling rate of 2.4 MS/s with omni-directional antennas which are deployed indoors. The sensors follow sequential scanning of the spectrum with an fast Fourier transform (FFT) size set to 256, giving a frequency resolution close to 10 kHz. With a FFT size of 256 and sensor ADC bit-width of 8, we get an effective bitwidth of 12, resulting in a theoretical dynamic range of 74 dB. Practical dynamic range depends on the ADC front-end stages and the noise level, which may vary between 60 and 65 dB. Five FFT vectors are averaged for reducing the thermal noise of the receiver. The model is trained on 7 days of data from one of the Electrosense sensors and tested on the next 500 hours for anomalies with a detection threshold of $3\sigma$ ($n = 3$).

The number of detected anomalies, based on $A_{score}$, along with a few sample anomalies for seven frequency bands are shown in Figure 16.6. For instance, the model detects unexpected missing transmissions (top right) and some out-of-band transmissions (top left). It can be noticed that after 230 hours the 192–197 MHz bands started giving more anomalous detections. Visual inspection of the anomalous PSD patches in this band revealed transmission pattern variations. These detected variations could be either because of the transmitter behavior changes or from the position/antenna changes of the sensor. The model

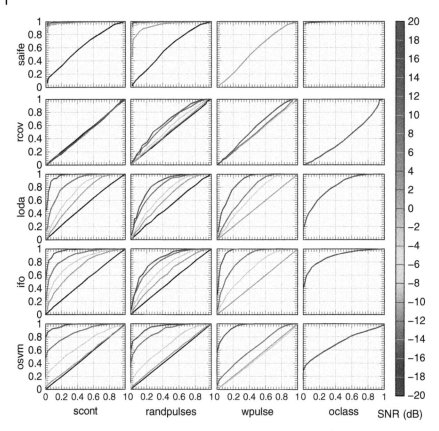

**Figure 16.5** ROC curves for different detection algorithms on *det-hop* synthetic band for various anomalies.

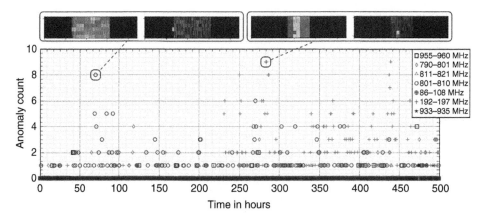

**Figure 16.6** Detected anomalies for a duration of 500 hours from one of the Electrosense sensors. Sample input data (left) and the localized anomaly (right) for some sample anomalies are also plotted for a subset of the frequency bands.

also provides the flexibility to add these anomalous detections to the training set, enabling incremental learning, if the user believes that the behavior is normal. Incorporating this user feedback and enabling automated retraining of models on these kind of anomalous behaviors will be addressed in future work.

In this section we have formulated wireless spectrum anomaly detection as an unsupervised learning problem where the model tries to learn the normal spectrum data distributions and detect the uncommon patterns as anomalous. As long as new anomaly patterns are not present in the training data in abundance, the model can detect these anomalies independent of the number of new patterns. SAIFE makes the basic assumption that non-frequent behavior is anomalous, which is not always true. For example, transmissions in industrial, scientific, and medical (ISM) bands can be very sparse which are not anomalies. Similarly, a pirate transmitter or transmission duty cycle limit violations in a licensed band is anomalous even when they are very frequent. This prevents modeling wireless anomaly detection as a fully unsupervised learning problem. We will solve wireless spectrum anomaly detection, in a semi-supervised setup, by formulating it as a crowdsourced active learning problem in the near future. Once we detect an anomalous transmission, it is important to understand more about the anomalous signal. This is where wireless signal classification comes into the picture.

### 16.3.3 Wireless Signal Classification

Wireless signal classification is an area of huge interest mainly due to its value in a wide set of applications such as anomaly detection, interference detection, and wireless spectrum enforcement. Classification models try to map the input signal to the corresponding modulation or technology type for further decoding and analysis. These models should be robust against channel, time, SNR, and symbol rate variations. In addition, the speed of signal classification is also an important factor. Classification models in supervised and semi-supervised settings are detailed in the following subsections.

#### 16.3.3.1 Fully Supervised Models
Recently, a few fully supervised deep learning models have been proposed for wireless signal classification [21, 26, 36] that perform better than the expert feature based classifiers, even in low SNR conditions. While the CNN models [21] try to learn a wide range of matched filters for different SNRs, the LSTM models try to capture the temporal amplitude and phase variations of the input signal. Detailed explanations about these models can be found in the original papers.

A brief performance comparison of these deep learning models with other classification techniques on a standard radio machine learning (RadioML) modulation classification dataset [19] is presented in Figure 16.7. As the used sample length is only 128 (standard RadioML dataset), expert features cannot be efficiently generated before feeding it to traditional machine learning methods like support vector machines (SVMs) or random forest. All the models are fed with the same raw training and test data for this comparison. Random

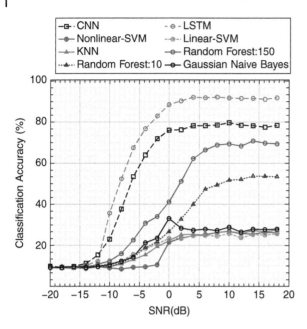

**Figure 16.7** Wireless classification accuracy of CNN and LSTM deep learning models compared with traditional machine learning methods on the RadioML dataset.

forest with 150 decision trees is able to provide close to 70% of accuracy at very high SNR conditions while others could reach only around 26%. It could be clearly noticed that the deep learning models perform better than the other standard techniques when fed with the raw sensed data from the in-phase and quadrature phase (IQ) pipeline. First, these models can classify signals very efficiently with a very low number of symbols, usually with hundreds of samples (tens of modulated symbols) when compared to the classical cyclostationary-based expert feature models which require samples in the thousands range (hundreds of modulated symbols) for averaging. Second, the deep learning models perform better even in low SNR conditions. Finally, in typical modulation or technology classification problems very deep models are not required to achieve good performance. Two layer CNN or LSTM models can easily abstract features for reaching close to 90% accuracy.

The parameter space of typical wireless modulations is huge as there are multiple modulation parameters or wireless specifications that can be varied, such as the symbol rate, sample rate, pulse shaping filter parameters, and so on. Even though the fully supervised models can give very good classification accuracy on wireless signals with parameters it has seen in the training phase, labeling the data for a wide range of parameters is very tedious. This points to a pressing need for semi-supervised models.

### 16.3.3.2 Semi-supervised Models

A few papers in the literature have looked into wireless signal classification in a semi-supervised manner. In [17], the authors perform a supervised bootstrapping of sparse representation, that is training a deep learning model in a supervised fashion with known labels, and then proceed with the idea that by discarding the last softmax-layer of the fully supervised model the features generated have the ability to separate additional classes. The model is shown to be able to extract abstract features which can be used to separate unseen modulations with simple clustering techniques. However, the model makes use of non-labeled data only for clustering and not for training. This prevents the model from providing guaranteed clustering for all unseen modulations with drastic parameter variations.

Another work [8] makes use of sparse coding and dictionary learning to perform blind signal classification. Sparse coding is an unsupervised method to learn a dictionary of over-complete basis vectors that can represent data efficiently. Each basis vector in the dictionary is also known as an atom. During the training stage a few dictionary atoms are generated in a semi-supervised fashion by using well-known pulse shaping filters and similarity matching [2] is used for finding the best suited modulation. In a realistic scenario selecting parameters for generating supervised dictionary is difficult as filter shape is not the only parameter that can vary. Also the model is tested only on an AWGN channel and proper analysis is not done with fading and other hardware effects on the dictionary generation. To address the shortcomings, generic models that can use both labeled and unlabeled samples and can learn relevant features are required for proper classification. In addition, similar to the fully supervised deep learning models, they should do automated feature extraction avoiding the manual expert feature extraction requirements.

To enable the specified requirements, a mutual information maximizing generative adversarial network with layers specific to the wireless domain is selected [31]. This model is an advanced version of the Information Maximizing Generative Adversarial Networks (InfoGAN) [3] model with a semi-supervised setting. Used layers in the generator and discriminator are detailed in Figure 16.8, where convolutional layers are used in the generator and LSTM in the discriminator. Being a generative model, the generator makes signals which resemble the actual input data while the discriminator tries to segregate them. In addition, a softmax layer is added to the discriminator to enable signal classification along with segregation. A discriminator which is trained on such generated and real inputs is seen to be more robust as the amount of signal variation the model sees is much larger. In-depth mathematical details about the model can be found in [31] and are omitted here in order to avoid repetition. As shown in Figure 16.9, the trained discriminator on instantaneous amplitude and phase of the received signal is able to achieve an average accuracy of 80% only by using 10% labeled samples, which is only 10% lower than the accuracy of the fully supervised discriminator. On the other hand, a fully supervised model with the same LSTM layers could achieve only 50% accuracy using 10% of labeled samples. The discriminator also reaches accuracy close to the fully supervised model by only using 20% labeled samples. Currently the model is tested only on pure modulation schemes without any error correction or encoding schemes. Detailed classification tests for different encoding schemes will be included in future work.

### 16.3.3.3 Performance-friendly Models

Deep layers and inherent non-linearities make deep learning models processor-intensive for low-end sensor deployment. The typical average depth of the models currently used in the wireless sensing is two, which prevents us from doing much network pruning without sacrificing the performance. Quantizing the networks weights and activations is one other method to reduce the required memory footprint for holding intermediate results. For instance, binarized networks can exceptionally reduce the intermediate memory footprints and replace most of the arithmetic operations with bitwise operations [10].

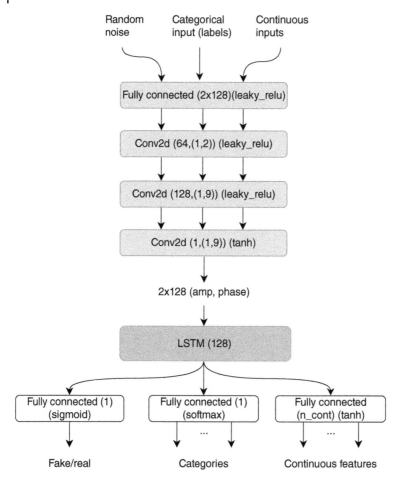

**Figure 16.8** Generator and discriminator details.

Even though quantized models can provide computational gains, the performance degradation of these models should be analyzed in detail. To validate this, the accuracy comparison plots of various quantized models for wireless signal classification are summarized in Figure 16.10. In [10] the authors had already noticed that binarizing LSTMs results in very poor accuracy. However, models with binarized CNNs have been reported to provide accuracy close to their full precision variants. Also by allowing more quantization levels on the LSTM models, a higher accuracy can be achieved while still reducing the computational cost. Two quantized LSTM model variants were tested, one with ternary weights (−1, 0, +1) and full precision activation (TW_FA) and the other with ternary weights and four bits activation (TW_4BA). The accuracy results of these models are summarized in Figure 16.10. Results show that LSTM models with ternary weights and 4 bit activation can provide close to 80% accuracy, reducing the very high computational power required for full precision models. Binary CNN models also provided an accuracy level only 10% below the full precision variants. We believe the classification accuracy can be further improved by proper hyper-parameter tuning and longer training.

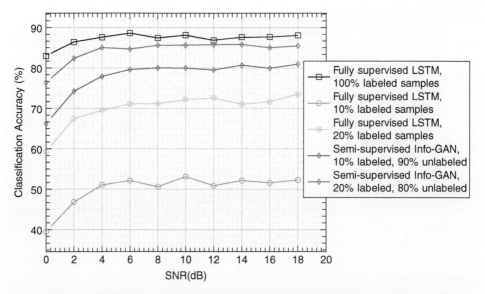

**Figure 16.9** Classification results for a fading channel with receiver effects for ['BPSK', 'QPSK', 'QAM16', 'QAM64', 'PAM4', '8PSK', 'CPFSK', 'GFSK'] from the RadioML dataset.

**Figure 16.10** Wireless signal classification accuracy of two layer quantized models on the RadioML dataset.

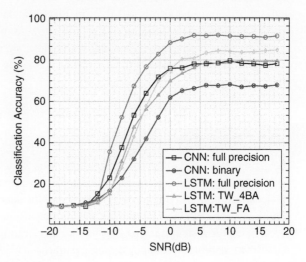

## 16.4 Future Research Directions

Low-cost SDR hardware and advances in big data have helped in developing large-scale spectrum monitoring frameworks. New machine learning models are being developed actively for understanding the signals better, for instance very deep models for wireless signal classification [18]. Gradually, deep learning models are becoming an ingrained part of future wireless technologies, especially for sensing and end-to-end communication systems [16]. The following are some open problems related to wireless spectrum monitoring where new solutions can create a huge impact.

### 16.4.1 Machine Learning

*Interpretable feature extraction*: In section 16.3 we only looked at one discrete interpretable feature, namely, category of the signal. For a wireless signal there is a wide range of other continuous features such the bandwidth, symbol rate, channel occupancy, duty cycle and so on. The semi-supervised InfoGAN model presented in this paper also extracts some continuous features that can also be trained in a semi-supervised manner. The expressive power of alternative models based on variational autoencoders in terms of extracting disentangled features from the wireless data is also an interesting research direction that can help automated spectrum interpretation.

*Expert feature extraction models*: Deep learning models for wireless signal classification presently use layers which basically apply non-linearity after simple multiply-accumulate-add operations while it is well established in the research community that cyclic cumulants, which are generated by time-shifted multiplication and averaging of the input itself, perform well in the expert feature space. Models that can take advantage of these features in the internal layers might be useful in improving the performance of current models. These models might also be useful for multi-label classification setup, which can further aid interference analysis.

*Models for sub-sampled wireless signals*: Recently there has been a surge in interest in subsampled wireless classification models that can enable low-energy spectrum sensing [30]. It should be also noted that, in Electrosense, the deployed low-cost spectrum sensors' SDR has many limitations, including low sampling bandwidth (max. 3.2 MS/s), which prevents it from receiving high bandwidth signals like Wi-Fi or LTE. Efforts to develop deep learning models that can work efficiently on subsampled signals are worth pursuing.

### 16.4.2 Anomaly Geo-localization

Localizing detected anomalies geographically is very important to mitigate imminent threats in high priority wireless communication scenarios such as airports. Anomaly geo-localization is also important to understand the source of intentional or unintentional jamming events. When compared to conventional localization, one of the major challenges in localizing these signals is that they are not cooperating. Localizing unknown interfering signals with complex frequency hopping patterns is a difficult and an interesting problem to tackle in the future.

### 16.4.3 Crowd Engagement and Sustainability

An important challenge that we see for large-scale deployments is to improve the engagement of new users. What are the best incentives that can motivate new users to join the network and share spectrum data? How can the crowd-sourced model be kept cost-effective in terms of the backend processing and storage costs? These are worthwhile problems to be tackled for a self-sustaining spectrum monitoring framework.

## 16.5 Conclusion

Automated monitoring of wireless spectrum over frequency, time, and space is still a difficult research problem. In this chapter we addressed some of the main challenges of large-scale wireless spectrum monitoring. A crowdsourced spectrum monitoring framework, which leverages state-of-the-art SDR and big data architectures, can help to address these challenges and democratize spectrum awareness. In addition, new machine learning models were presented which can be used to interpret sensed spectrum data effectively, as shown by the anomaly detection and signal classification examples. A semi-supervised deep learning setup was also presented based on the latest deep learning research which achieves performance close to fully supervised models with only 20% of the labeled samples. Furthermore, the performance of quantized models was analyzed and it was shown that, in principle, considerable computational performance can be achieved at a cost of 10% classification accuracy loss. The rise of deep learning models for wireless applications is inevitable and the future looks bright for the entire industrial and research wireless sectors.

## References

1 Ayon Chakraborty and Samir R Das. Measurement-augmented spectrum databases for white space spectrum. In *Proceedings of the 10th ACM International on Conference on emerging Networking Experiments and Technologies*, pages 67–74. ACM, 2014.

2 Sheng Chen, Stephen A Billings, and Wan Luo. Orthogonal least squares methods and their application to non-linear system identification. *International Journal of control*, 50(5):1873–1896, 1989.

3 Xi Chen, Yan Duan, Rein Houthooft, John Schulman, Ilya Sutskever, and Pieter Abbeel. Infogan: Interpretable representation learning by information maximizing generative adversarial nets. In *Advances in neural information processing systems*, pages 2172–2180, 2016.

4 Samir Das, Himanshu Gupta, Peter Milder, and Petar Djuric. NSF project #1642965: Specsense: Bringing spectrum sensing to the masses, October 2016.

5 Paul Denisowski. Recognizing and resolving lte/catv interference issues. *White Paper, Rohde and Schwarz*, 2011.

6 Qingsong Feng, Yabin Zhang, Chao Li, Zheng Dou, and Jin Wang. Anomaly detection of spectrum in wireless communication via deep auto-encoders. *The Journal of Supercomputing*, 73(7):3161–3178, 2017.

7 Ian Goodfellow, Jean Pouget-Abadie, Mehdi Mirza, Bing Xu, David Warde-Farley, Sherjil Ozair, Aaron Courville, and Yoshua Bengio. Generative adversarial nets. In *Advances in neural information processing systems*, pages 2672–2680, 2014.

8 Y. Gwon, S. Dastangoo, H. T. Kung, and C. Fossa. Blind signal classification via sparse coding. In *2016 IEEE Global Communications Conference (GLOBECOM)*, pages 1–6, Dec 2016. doi: 10.1109/GLOCOM.2016.7841634.

9 W. Honghao, J. Yunfeng, and W. Lei. Spectrum anomalies autonomous detection in cognitive radio using hidden markov models. In *2015 IEEE Advanced Information*

*Technology, Electronic and Automation Control Conference (IAEAC)*, pages 388–392, Dec 2015. doi: 10.1109/IAEAC.2015.7428581.

**10** I. Hubara, M. Courbariaux, D. Soudry, R. El-Yaniv, and Y. Bengio. Quantized Neural Networks: Training Neural Networks with Low Precision Weights and Activations. *ArXiv e-prints*, September 2016.

**11** Sneha Kasera, Neal Patwari, and Jeff Phillips. NSF project #1564287: Detecting and localizing spectrum offenders using crowdsourcing, August 2016.

**12** N. Kleber, A. Termos, G. Martinez, J. Merritt, B. Hochwald, J. Chisum, A. Striegel, and J. N. Laneman. Radiohound: A pervasive sensing platform for sub-6 ghz dynamic spectrum monitoring. In *2017 IEEE International Symposium on Dynamic Spectrum Access Networks (DySPAN)*, pages 1–2, March 2017. doi: 10.1109/DyS-PAN.2017.7920764.

**13** Fei Tony Liu, Kai Ming Ting, and Zhi-Hua Zhou. Isolation-based anomaly detection. *ACM Transactions on Knowledge Discovery from Data (TKDD)*, 6(1):3, 2012.

**14** S. Liu, Y. Chen, W. Trappe, and L. J. Greenstein. Aldo: An anomaly detection framework for dynamic spectrum access networks. In *IEEE INFOCOM 2009*, pages 675–683, April 2009. doi: 10.1109/INFCOM.2009.5061975.

**15** Marja Matinmikko, Matti Latva-Aho, Petri Ahokangas, Seppo Yrjölä, and Timo Koivumäki. Micro operators to boost local service delivery in 5g. *Wireless Personal Communications*, 95(1):69–82, 2017.

**16** T. O'Shea and J. Hoydis. An introduction to deep learning for the physical layer. *IEEE Transactions on Cognitive Communications and Networking*, 3(4):563–575, Dec 2017. ISSN 2332-7731. doi: 10.1109/TCCN.2017.2758370.

**17** T. J. O'Shea, N. West, M. Vondal, and T. C. Clancy. Semi-supervised radio signal identification. In *2017 19th International Conference on Advanced Communication Technology (ICACT)*, pages 33–38, Feb 2017. doi: 10.23919/ICACT.2017.7890052.

**18** T. J. O'Shea, T. Roy, and T. C. Clancy. Over-the-air deep learning based radio signal classification. *IEEE Journal of Selected Topics in Signal Processing*, 12(1):168–179, Feb 2018. ISSN 1932-4553. doi: 10.1109/JSTSP.2018.2797022.

**19** Timothy J O'Shea and Nathan West. Radio machine learning dataset generation with gnu radio. In *Proceedings of the GNU Radio Conference*, volume 1, 2016.

**20** Timothy J O'Shea, T Charles Clancy, and Robert W McGwier. Recurrent neural radio anomaly detection. *arXiv preprint arXiv:1611.00301*, 2016.

**21** Timothy J O'Shea, Johnathan Corgan, and T Charles Clancy. Convolutional radio modulation recognition networks. In *International Conference on Engineering Applications of Neural Networks*, pages 213–226. Springer, 2016.

**22** Animesh Patcha and Jung-Min Park. An overview of anomaly detection techniques: Existing solutions and latest technological trends. *Computer Networks*, 51(12):3448–3470, 2007. ISSN 1389-1286. doi: https://doi.org/10.1016/j.comnet.2007.02.001. URL http://www.sciencedirect.com/science/article/pii/S138912860700062X.

**23** Tomáš Pevný. Loda: Lightweight on-line detector of anomalies. *Machine Learning*, 102(2):275–304, 2016.

**24** S. Rajasegarar, C. Leckie, and M. Palaniswami. Anomaly detection in wireless sensor networks. *IEEE Wireless Communications*, 15(4):34–40, Aug 2008. ISSN 1536-1284. doi: 10.1109/MWC.2008.4599219.

**25** S. Rajendran, R. Calvo-Palomino, M. Fuchs, B. Van den Bergh, H. Cordobes, D. Giustiniano, S. Pollin, and V. Lenders. Electrosense: Open and big spectrum data. *IEEE Communications Magazine*, 56(1): 210–217, Jan 2018. ISSN 0163-6804. doi: 10.1109/MCOM.2017.1700200.

**26** S. Rajendran, W. Meert, D. Giustiniano, V. Lenders, and S. Pollin. Deep learning models for wireless signal classification with distributed low-cost spectrum sensors. *IEEE Transactions on Cognitive Communications and Networking*, pages 1–1, 2018. doi: 10.1109/TCCN.2018.2835460.

**27** S. Rajendran, W. Meert, V. Lenders, and S. Pollin. Saife: Unsupervised wireless spectrum anomaly detection with interpretable features. *arXiv preprint arXiv:1807.08316*, 2018.

**28** Peter J Rousseeuw and Katrien Van Driessen. A fast algorithm for the minimum covariance determinant estimator. *Technometrics*, 41(3): 212–223, 1999.

**29** S. Roy, K. Shin, A. Ashok, M. McHenry, G. Vigil, S. Kannam, and D. Aragon. Cityscape: A metro-area spectrum observatory. In *2017 26th International Conference on Computer Communication and Networks (ICCCN)*, pages 1–9, July 2017. doi: 10.1109/ICCCN.2017.8038427.

**30** C. M. Spooner, A. N. Mody, J. Chuang, and M. P. Anthony. Tunnelized cyclostationary signal processing: A novel approach to low-energy spectrum sensing. In *MILCOM 2013 - 2013 IEEE Military Communications Conference*, pages 811–816, Nov 2013. doi: 10.1109/MILCOM.2013.143.

**31** A. Spurr, E. Aksan, and O. Hilliges. Guiding InfoGAN with Semi-Supervision. *ArXiv e-prints*, July 2017.

**32** STRIKE3. STRIKE3, Initial Findings from the STRIKE3 GNSS Interference Monitoring Network. https://www.gps.gov/governance/advisory/meetings/2018-05/dumville.pdf. Accessed: 2018-07-01.

**33** Nistha Tandiya, Ahmad Jauhar, Vuk Marojevic, and Jeffrey H Reed. Deep predictive coding neural network for rf anomaly detection in wireless networks. *arXiv preprint arXiv:1803.06054*, 2018.

**34** Telstra. Telstra. https://exchange.telstra.com.au/illegal-mobile-repeaters/. Accessed: 2018-07-01.

**35** Sarang Thombre, M Zahidul H Bhuiyan, Patrik Eliardsson, Björn Gabrielsson, Michael Pattinson, Mark Dumville, Dimitrios Fryganiotis, Steve Hill, Venkatesh Manikundalam, Martin Pölöskey, et al. Gnss threat monitoring and reporting: Past, present, and a proposed future. *The Journal of Navigation*, 71(3):513–529, 2018.

**36** Nathan E. West and Timothy J. O'Shea. Deep architectures for modulation recognition. *CoRR*, abs/1703.09197, 2017. URL http://arxiv.org/abs/1703.09197.

**37** Miao Xie, Song Han, Biming Tian, and Sazia Parvin. Anomaly detection in wireless sensor networks: A survey. *Journal of Network and Computer Applications*, 34(4):1302–1325, 2011. ISSN 1084-8045. doi: https://doi.org/10.1016/j.jnca.2011.03.004. URL http://www.sciencedirect.com/science/article/pii/S1084804511000580. Advanced Topics in Cloud Computing.

# 17

# Policy Enforcement in Dynamic Spectrum Sharing

*Jung-Min (Jerry) Park, Vireshwar Kumar, and Taiwo Oyedare*

*Virginia Tech, USA*

## 17.1 Introduction

Fully realizing the vision of *dynamic spectrum sharing* (DSS) requires the adoption of fundamentally new spectrum access paradigms. For instance, in DSS, a heterogeneous mix of wireless systems of differing access priorities, quality of service requirements, and transmission characteristics needs to coexist without causing harmful interference to each other. In these novel paradigms, when different stakeholders share a common resource (such as in spectrum sharing), security and enforcement become critical considerations that are essential to the welfare of all stakeholders. Policy enforcement is especially a paramount consideration when sharing government (including military) spectrum with non-government (commercial) systems. Hence, to securely and efficiently employ innovative spectrum access technologies, the spectrum regulatory authorities throughout the world have emphasized the need to adopt new regulatory policies which could be enforced with the help of frameworks, such as a spectrum monitoring system, which was discussed in Chapter 16.

In this chapter, we review the critical security and privacy threats to the harmonious and efficient functioning of DSS ecosystems and their countermeasures. First, a taxonomy for classifying the threats is discussed. The taxonomy considers fundamental mechanisms for enabling coexistence (i.e., spectrum sensing-driven mechanism or database-driven mechanism) as well as the points of attack with respect to the five-layer protocol stack. For each threat category, representative security and privacy threats, and their relation to other types of threats are described. Furthermore, the existing proposals for threat countermeasures and spectrum policy enforcement are discussed. The enforcement mechanisms are discussed in the context of two distinct approaches: *ex ante* and *ex post* enforcement. The former represents actions that are designed to "prevent" or reduce the likelihood of a potentially harmful interference event, while the latter denotes "punitive" measures designed to punish malicious behavior after a potentially harmful interference event has occurred. The chapter concludes by discussing the research and regulatory challenges that need to be addressed to ensure policy enforcement in DSS.

*Spectrum Sharing: The Next Frontier in Wireless Networks,* First Edition.
Edited by Constantinos B. Papadias, Tharmalingam Ratnarajah, and Dirk T.M. Slock.
© 2020 John Wiley & Sons Ltd. Published 2020 by John Wiley & Sons Ltd.

## 17.2 Technical Background

In the perspective of policy enforcement, there are three major attributes associated with the DSS model: user, coexistence, and security attributes. These attributes are briefly discussed below. They will be utilized later to present a classification of threats and their countermeasures.

**User attributes** In spectrum sharing, users of different access priorities share a common resource, namely, spectrum, within a clearly defined hierarchy. On one hand, the licensed shared access (LSA) model adopted in Europe employs a two-tier sharing structure (incumbent tier 1 users and licensee tier 2 users); on the other hand, the spectrum access system (SAS) adopted in the USA employs a three-tier structure (incumbent tier 1 users, priority access license tier 2 users, and general authorized access tier 3 users) [51]. For the following discussions in this chapter, we follow the two-tier spectrum sharing model in which users are broadly classified into two categories: *incumbent/primary users* (PUs) and *secondary users* (SUs). The PUs have access priority over the SUs, and may consist of government users and licensed users. The SUs have secondary (i.e., subordinate) rights to spectrum, and typically consist of unlicensed opportunistic users.

**Coexistence attributes** There are two different mechanisms for enabling the harmonious coexistence of heterogeneous wireless systems in a shared spectrum ecosystem: spectrum sensing and geolocation databases. In a sensing-driven spectrum sharing scenario, SUs become cognizant of the surrounding *radio frequency* (RF) environment through either standalone or cooperative spectrum sensing [70], and their transmission behavior is dictated by spectrum sensing results. Note that SU radios need to have sufficient intelligence to use transmission parameters that are compliant with regulatory spectrum policies. Radios with such capabilities are often referred to as *cognitive radios* (CRs). In a database-driven spectrum sharing application, SUs are required to obtain spectrum availability information from a geolocation database which may also prescribe policies to access the shared spectrum (e.g., maximum allowed transmission power) [33, 52]. In some DSS ecosystems, both geolocation databases and sensing-based mechanisms are utilized in tandem to enable harmonious spectrum sharing between users of different access priorities [23].

**Security attributes** To ensure the viability of spectrum sharing, the following security and enforcement requirements must be met [7, 57]:

- *Confidentiality*: The data communicated between users and the database should not be disclosed to unauthorized users.
- *Integrity*: The data stored in the database and communicated among users should be protected from malicious alteration, insertion, deletion or replay.
- *Availability*: The users should have access to the database and/or the spectrum when it is required.
- *Authentication*: The network components, including the database and the mobile terminals, should be able to establish and verify their identity.

- *Non-repudiation*: The users should not be able to deny either having received or sent a message. Also, they should not be able to deny having accessed the spectrum at a specified location and time.
- *Compliance*: The network should be able to detect non-compliant behavior causing harmful interference.
- *Access control*: No user should be able to access either the database or the spectrum without proper credentials.
- *Data privacy*: Along with the data stored in the geolocation databases, the users' sensitive data, should be properly protected.
- *Operational privacy*: Sensitive operational attributes (e.g., location) of the users should be preserved.

## 17.3 Security and Privacy Threats

In the DSS paradigm, SUs may need to employ *software-defined radios* (SDRs) to harmoniously coexist with PUs as well as other SUs. Unlike a legacy radio, which is hardware or firmware-based, a SDR enables a user to readily re-configure its transmission parameters, allowing for greater flexibility. However, this "programmability" of SDRs also significantly increases the possibility of "rogue" or malfunctioning SUs. In this section, the security and privacy issues that pose the greatest threats to spectrum sharing are presented.

The threats to spectrum sharing can be classified into two broad categories based on the spectrum sharing approach that the attacks target: threats to sensing-driven spectrum sharing and threats to database-driven spectrum sharing. Based on this classification, a taxonomy of threats is presented in Figure 17.1 to provide a systematic discussion of the topic and to offer a clear picture of the known security and privacy issues.

### 17.3.1 Sensing-driven Spectrum Sharing

The following threats exploit the vulnerabilities in the spectrum sensing-based mechanism which is utilized for enabling a spectrum sharing ecosystem.

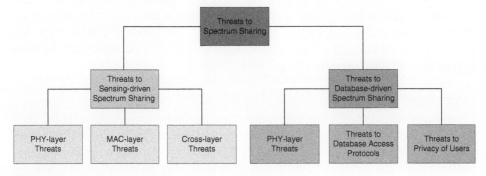

**Figure 17.1** Taxonomy of threats to spectrum sharing.

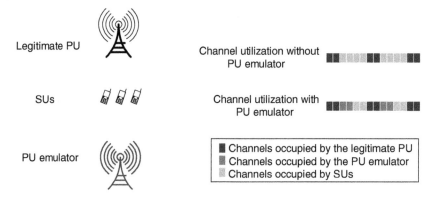

**Figure 17.2** Primary user emulation attack.

### 17.3.1.1 PHY-layer Threats

Threats in this subcategory directly impact the physical (PHY)-layer mechanisms, most notably spectrum sensing. Spectrum sensing by SUs can be manipulated by a rogue transmitter to either hijack their spectrum or affect their spectrum sharing decisions, e.g. through *PU emulation* (PUE) attacks [15, 17]. In a PUE attack (shown in Figure 17.2), a malicious SU emulates a PU's transmission characteristics in order to gain illegitimate access to the spectrum and/or prevent other SUs from accessing the spectrum. The PUE attack can also be used as a tool to launch more sophisticated attacks [54].

An approach for enhancing the accuracy of spectrum sensing is to employ cooperative spectrum sensing and centralized decision making [40]. In this approach, multiple users sense and send their observations about the RF environment to a fusion center. The fusion center ingeniously combines the reported information to make the final decision regarding the presence/absence of the PU's transmissions. Another approach for collaborative sensing is to employ cooperative spectrum sensing and distributed decision making. In this approach, no fusion center is used, instead each SU makes the decision about the presence of a PU based on its own observations and those shared by other SUs.

Both sensing approaches mentioned above are prone to *spectrum sensing data falsification* (SSDF) attack in which rogue SUs send false observations about the RF environment [14, 20, 60]. An illustration of the SSDF attack is presented in Figure 17.3. Due to the SSDF attack, legitimate SUs in the network may acquire an inaccurate perception of the RF environment and make decisions that may cause interference to PUs. Also, rogue SUs may violate the spectrum sharing policies and transmit selfishly on convenient channels, causing harmful interference to PUs and other SUs [42].

### 17.3.1.2 MAC-layer Threats

There are numerous attacks that may compromise the media access control (MAC)-layer mechanisms of spectrum sharing. In a multi-hop CR network, a pre-defined frequency channel – called the cognitive control channel – is used by SUs to exchange control information, e.g. channel negotiation and spectrum hand-off. A rogue transmitter may jam this channel with little effort and cause *denial of service* (DoS) to SUs [72]. This is referred to as *control channel corruption* (CCC) attack. An alternative technique for enabling coexistence

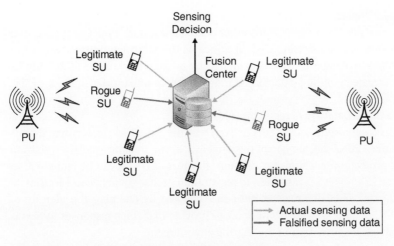

**Figure 17.3** Spectrum sensing data falsification attack.

of SUs and coordinating the use of channels among SUs is to utilize beacons. Again, in this mechanism a malicious transmitter can launch a *beacon falsification* (BF) attack which may compromise critical functionalities, such as inter-cell spectrum contention and inter-cell synchronization [12]. Some SUs may also implement a *carrier sense multiple access with collision avoidance* (CSMA/CA) protocol in which a SU backs off by a random time after sensing the transmission from another SU. If there is a collision of packets transmitted by any two SUs, the SUs double the back-off window before retransmission. In this protocol, a malicious user may utilize a small back-off window and gain priority over other users [61]. This is called the *small back-off window* (SBW) attack.

### 17.3.1.3 Cross-layer Threats

In some scenarios, multiple attacks can be conducted in tandem to exploit vulnerabilities in two or more layers of the protocol stack. These attacks are called cross-layer attacks. For instance, in DSS utilizing the CSMA/CA protocol, a malicious user can launch an SSDF attack (at the PHY-layer) and an SBW attack (at the MAC-layer) in a coordinated fashion [64]. This coordination makes it difficult to detect either of the two attacks. Also, this cross-layer attack is more successful than a single-layer attack in diminishing the overall SU channel utilization. Another example of a cross-layer attack is known as a *Lion* attack which targets the PHY and transport layers [34]. In a Lion attack, a malicious user launches a PUE attack and forces the target SUs to carry out frequency hand-offs. The transmission interruptions caused by the frequency hand-offs lead to very poor throughput at the transport layer since the *transmission control protocol* (TCP) is quite sensitive to variations in delay and bandwidth.

### 17.3.2 Database-driven Spectrum Sharing

The threats described in this subsection impinge the security and privacy of users in spectrum sharing enabled by geolocation databases.

#### 17.3.2.1 PHY-layer Threats

Undesirable interference from rogue transmitters can significantly impact SU spectrum utilization in database-driven sharing [33]. Specifically, the information about spectrum availability provided by databases can be exploited by a rogue transmitter to amplify its ability to launch targeted jamming attacks and hide its non-compliant transmissions [71].

#### 17.3.2.2 Threats to the Database Access Protocol

The *database access protocol attacks* (DAPAs) refer to the varied set of security threats related to the access control mechanism of the database [8, 58]. In database-driven sharing without suitable protection mechanisms, different flavours of DAPA can be launched. In a masquerade attack, a rogue SU can listen to registration exchanges between a legitimate SU and the database, and later register with the database by claiming the identity of the legitimate SU. Spoofing a database in order to provide malicious responses to SUs is another type of attack that can disrupt the sharing environment.

Furthermore, the attacker may compromise the integrity of a SU's query and/or database's response. If an attacker is able to change some of the information in the SU's query (e.g., the location of the SU or its capabilities), the database will respond with incorrect information about available spectrum or maximum allowed transmission power. The attacker may also directly modify the available spectrum or power level information carried in the database response. Additionally, selectively jamming database queries/responses may cause a DoS to the SUs. Further, if a database includes a mechanism by which spectrum allocated to a SU can be revoked by sending a revoke message, malicious users can pretend to be the database and send a revoke message to that SU terminating or unfairly limiting spectrum access of the SU.

#### 17.3.2.3 Threats to the Privacy of Users

Although using geolocation databases for spectrum sharing has many advantages over the sensing-based approach, it poses a potentially serious privacy problem. There is the possibility that through sophisticated inference techniques, SUs can obtain knowledge beyond what is revealed directly by the database's responses. This type of attack is referred as a *database inference attack* (DIA) [6], which compromises the operational privacy of the PUs.

For instance, SUs, through seemingly innocuous queries to the database, can infer various attributes of PUs. Some of these attributes include PU identity (e.g., the call sign of the transmitter in the Federal Communications Commission (FCC) Consolidated Database System), geolocation (i.e., latitude and longitude), antenna parameters, transmission power, transmit protection contours (co-channel and adjacent channel), and times of operation. When the incumbent systems are commercial systems, such as TV spectrum, the inference of these attributes is not an issue. However, when the incumbents are government, possibly military, systems, then the information revealed by the databases may result in a serious breach of operational privacy. For instance, the operational privacy of PUs is an especially critical concern for sharing of federal government spectrum in the 3.5-GHz band with non-government systems in the USA.

Another issue that may arise as a result of using geolocation databases for spectrum sharing is the compromise of operational privacy of SUs. For instance, since SUs need

**Table 17.1** Security features compromised by threats.

| Security feature | Threats | | | | | | |
|---|---|---|---|---|---|---|---|
| | PUE | SSDF | CCC | BF | SBW | DAPA | DIA |
| Confidentiality | | | × | | | × | |
| Integrity | | × | × | × | × | × | |
| Availability | × | × | × | × | × | × | |
| Authentication | × | | × | × | | × | |
| Non-repudiation | × | | | × | | | |
| Compliance | × | | | × | | | |
| Access control | × | | | × | | × | |
| Data privacy | | | | | | × | |
| Operational privacy | | | | | | | × |

to send their location information to the database to receive information on the set of available channels in their region, their location privacy may be threatened by an untrustworthy database. An advanced attack on the SU's location privacy is called *spectrum utilization-based location inference* (SULI) attack which allows an attacker to infer the location of the SU from the channels utilized by it [26].

Table 17.1 summarizes the threats to spectrum sharing discussed in this section by highlighting the security features that they compromise (denoted by ×).

## 17.4 Enforcement Approaches

Enforcing spectrum access control in legacy radios (e.g., cellular phones) is relatively straightforward since the spectrum access policies are an inseparable part of the radio's firmware and platform. Also, making controlled changes to a legacy radio's behavior would require an adversary to have very specialized technical expertise in the radio's firmware and hardware. Unfortunately, the reconfigurability of SDRs/CRs not only makes them vulnerable to unauthorized modifications, but also makes it difficult to enforce spectrum policies.

Considering the wide landscape of the threats, we discuss a battery of countermeasures for spectrum policy enforcement by classifying them into two broad categories: *ex ante* (preventive) and *ex post* (punitive) enforcement. The taxonomy of enforcement approaches is illustrated in Figure 17.4. The objective of *ex ante* enforcement is to prevent or reduce the probability of a policy violation causing harmful interference or loss of user privacy. On the other hand, the objective of *ex post* enforcement is to identify and/or punish malicious or selfish users after a policy violation has occurred. A real-world policy enforcement framework may need to employ a combination of specific *ex ante* and *ex post* enforcement approaches.

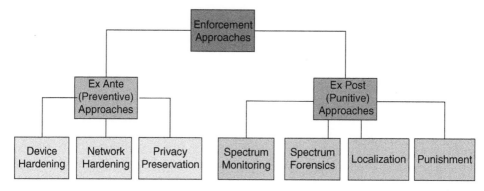

**Figure 17.4** Taxonomy of enforcement approaches for spectrum sharing.

### 17.4.1 *Ex Ante* (Preventive) Approaches

The *ex ante* approaches can be divided into three classes: device hardening, network hardening, and privacy preservation.

#### 17.4.1.1 Device Hardening

Device hardening is an important step in ensuring policy enforcement in DSS. It follows the concept of target hardening, i.e. strengthening the security of SDRs/CRs to deter or delay the threats. This technique is discussed below by differentiating between software and/or hardware-based approaches.

*Software-based approach* The most prominent software-based approach for enforcing policy control is to employ *policy-based* CRs. Policy-based CRs adapt with evolving spectrum access policies and constantly changing application requirements by decoupling the policies from device-specific implementation and optimization. These radios can invoke situation-appropriate adaptive actions based on policy specifications and the current spectrum environment [28]. In order to regulate and enforce proper transmission behavior, policy-based CRs need mechanisms to interpret and enforce spectrum access policies. Each transmission from the policy-based CRs needs to be evaluated against those policies to determine the legality of the transmission parameters. Within a policy-based CR, the aforementioned tasks are carried out in real time by a software module called the *policy reasoner*. There are two major types of policy reasoners: rule-based and ontology-based policy reasoners.

Rule-based policy reasoners utilize logic programming techniques to encode the axioms and rules, and enforce policy conformance [4, 58, 59, 62]. Using the rule-based approach simplifies the design of the policy reasoner because the reasoning complexity is sufficiently low in most applications to meet the real-time processing requirements of the radio. However, they do not support the sharing of the policy structure among different policy authors (i.e., regulatory authorities), limiting the interoperability of the policy-based CRs across different regulatory policy domains. Also, complex spectrum policies are difficult to specify and manage with rule-based policies.

To overcome these limitations of rule-based policy reasoners, the IEEE 1900.5 Standard for Policy Language Requirements and System Architectures for Dynamic Spectrum Access

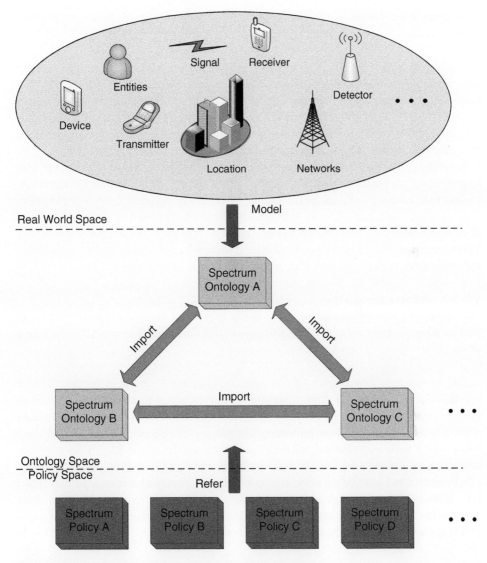

**Figure 17.5** Components of an ontology-based policy reasoner.

Systems prescribes the use of an ontology-based policy language for managing the functionality and behavior of DSS networks [1, 5, 39]. Managing ontologies to support the formal representation of spectrum policies in the ever-changing DSS ecosystem is significantly easier than managing rule-based policy enforcement. Figure 17.5 illustrates the components of an ontology-based policy reasoner utilized for spectrum access policies.

***Middleware-based approach*** A *secure radio middleware* (SRM) layer can be implemented between the operating system and the hardware [46]. The SRM layer checks all software transmission requests that are sent to the hardware layer to make sure that configurations such as transmission power, frequency, and type of modulation conform with policies in a policy database. Unlike a software-based policy reasoner that provides feedback to the radio's software, the SRM layer simply discards non-conforming requests.

**Hardware-based approach** An effective hardware-based approach is to use tamper-resistant techniques to protect a radio's software against unauthorized modifications [67]. The tamper-resistant module is designed to thwart static attacks (i.e., static information extracted by examining the software code) and to protect partially against dynamic attacks (i.e., dynamic information extracted while the software code executes). This approach is also effective in enforcing countermeasures against rogue transmission [42, 43].

In addition to tamper-resistant techniques, the integrity assessment of an SDR can also be performed by a hardware dedicated to power fingerprinting [30]. This mechanism is able to detect the execution of a tampered routine by closely monitoring the power consumption of the radio platform. Also, an independent power-check module can be implemented at the hardware of the SDR transceiver to control its maximum transmission power [47]. These hardware-based approaches are designed to prevent transmissions that cause harmful interference to PUs/SUs even if the radio's software is compromised.

### 17.4.1.2 Network Hardening

The concept of network hardening refers to the preventive measures required to protect PUs from interference, SUs from DoS, and geolocation databases from threats to the access protocol.

**Protecting PUs from interference** As discussed in Chapters 2 and 4, a popular *ex ante* approach to protect the PU from undesirable interference is to employ the concept of *exclusion zones* [66]. An exclusion zone is a spatial region in which no in-band emissions from SUs are permitted. This protection boundary can also be dynamically adjusted based on the radio environment, network conditions, and corresponding PU interference protection requirement [9]. The dynamic exclusion zones can be realized with the help of a dedicated/crowd-sourced network with spectrum *environmental sensing capability* (ESC) [55]. Dynamically adjusting the PU's protection boundary allows more SUs to operate closer to the PU, resulting in an improvement in spectrum utilization efficiency while also ensuring that the PU is adequately protected from interference.

**Protecting SUs from jamming** The traditional well-known anti-jamming techniques, such as *direct-sequence spread spectrum* (DSSS) and *frequency hopping spread spectrum* (FHSS), are insufficient for preventing jamming attacks in the spectrum sharing ecosystem. This is because the spectrum information disseminated in either the sensing-based or the database-based sharing mechanism is available to all SUs, including rogues SUs/jammers. Hence, novel countermeasures must complement the traditional techniques by considering dynamic channel allocation mechanisms and jammer inference mechanisms [71].

**Protecting geolocation databases** The access control mechanism of the geolocation databases must be protected using state-of-the-art cryptographic primitives for encryption and authentication [8, 35, 37]. Additionally, a distributed architecture for storing and disseminating information is an essential aspect for securing the database-driven DSS [2].

### 17.4.1.3 Privacy Preservation

The application of the DSS paradigm in many scenarios is limited by the privacy concerns of PUs and SUs. These concerns can be mitigated by employing the following measures which thwart an adversary from directly gaining or inferring such information that could compromise the privacy of users.

**Protecting PU privacy** In the database-driven sharing mechanism, the operational privacy of PUs cannot be addressed by tightly controlling access to the database, since all SUs need access to it to enable spectrum sharing. A more viable approach is to "obfuscate" the information revealed by the database in its responses to SU queries [6, 11, 18]. For instance, to infer the location of the PU, a rogue SU may exploit the *allowed transmission power* values which are inherently provided by the database in response to the SU's queries. The true power values can be masked using two obfuscation strategies: *perturbation with additive noise* and *perturbation with transfiguration* [6]. In perturbation with additive noise, the database adds random noise values to the actual power values and responds with these modified power values. In perturbation with transfiguration, the database modifies the structure of the exclusion zone by employing randomly shaped contours in place of the actual circular contour, and then responds with the power values corresponding to these randomly shaped contours.

Figure 17.6 presents an illustration of the impact of perturbation with additive noise and transfiguration on the normalized values of the location privacy of the PU and the spectrum utilization of SUs. The location privacy is measured by the metric called *inaccuracy*, which is defined as the expected distance between the estimate of the PU location inferred through queries and the PU's true location. The spectrum utilization is measured by the metric called *area sum capacity*, which is defined as the sum of channel capacity values of the SUs in the region served by the database. Figure 17.6 illustrates that the obfuscation strategies need to be performed in an intelligent manner such that a certain level of privacy is assured to the PU while supporting an efficient use of the spectrum. For instance, the

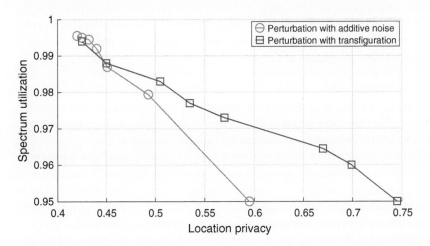

**Figure 17.6** Trade-off between location privacy of the PU and spectrum utilization of SUs [6].

database may keep track of the information revealed through its past responses so that it can leverage such a history to compute an optimal response to the current query.

Another approach is to utilize the attribute-based encryption for the responses [48]. In this approach, the PU obtains attribute credentials based on its operational specifications and then utilizes these credentials for encrypting the responses. The encrypted responses can only be decrypted by the SU with qualified attribute credentials. This approach is computationally expensive, but it does not adversely affect the spectrum utilization.

**Protecting SU privacy**  The location-based services, such as DSS, rely on accurate, continuous, and real-time streams of the users' location data. However, with the potential of mishandling of such information by an untrusted database, these services pose a significant privacy risk to SUs. Techniques for mitigating such a risk include sending a space- or time-obfuscated version of the users' actual locations [26, 32], hiding some of the users' locations by using mix zones [24], sending fake queries which are indistinguishable from real queries and issued from fake locations to the database [16], applying $k$-anonymity to location privacy [27], and utilizing private information retrieval techniques [31]. Again, these privacy-preserving techniques for SUs bring forth the delicate trade-off between privacy and efficient spectrum utilization.

### 17.4.2  *Ex Post* (Punitive) Approaches

To counter threats which may bypass *ex ante* enforcement approaches, it is crucial to deploy a multi-pronged *ex post* enforcement approach for detection and remedy [25]. The *ex post* enforcement procedure can be divided into four stages: spectrum monitoring, spectrum forensics, localization, and punishment.

#### 17.4.2.1  Spectrum Monitoring
The logical first step in *ex post* enforcement for an enforcement entity (e.g., the FCC Enforcement Bureau) is to perform data collection by spectrum monitoring, which refers to the procedure of recording spectral RF emissions. Spectrum monitoring helps in verifying policy compliance in DSS by detecting interference events. It can be practically realized by combining interference detection results from dedicated sensors [25, 55] and crowd-sourced sensors [21, 45]. A detailed discussion on spectrum monitoring is presented in Chapter 16.

#### 17.4.2.2  Spectrum Forensics
Spectrum forensics refers to the procedure of leveraging the data obtained from spectrum monitoring to gather actionable evidence (which may be tenable in a court of law) of rogue transmission by a SU. This can be performed by uniquely identifying or authenticating rogue transmitters. Ideally, the enforcement entity would want to carry out the identification using a PHY-layer scheme because it is not the intended recipient of the transmitted signals, and it has little or no knowledge of the higher-layer parameters of the SU. Also, a PHY-layer scheme enables the enforcement entity to quickly distinguish between compliant and rogue transmitters without having to complete higher-layer processing.

Spectrum forensic schemes can be broadly divided into two categories: identification and authentication approaches. Schemes in the first category utilize the *intrinsic* characteristics of the waveform or communication medium (e.g., transmitter-unique

RF signal characteristics) as unique signatures to identify transmitters. They include RF fingerprinting and electromagnetic signature identification [13, 36]. Although these identification approaches have been shown to work in controlled lab environments, their sensitivity to environmental factors, such as temperature changes, channel conditions, and interference, limit their efficacy in real-world scenarios. Moreover, they have been shown to be vulnerable to impersonation attacks [19]. Hence, more mature and refined detection/identification techniques, such as those based on machine leaning [41, 50], are needed for robust spectrum forensics.

Schemes in the second category enable a transmitter to *extrinsically* embed an authentication signal (e.g., message authentication code or digital signature) in the message signal and enable a receiver to extract it. Such schemes include PHY-layer watermarking [22, 29, 38] and transmitter authentication [43, 44, 68, 69]. For this approach to be viable, all SU radios must incorporate a mechanism for authenticating their waveforms and employ tamper-resistant mechanisms to prevent hackers from circumventing the mechanism.

### 17.4.2.3 Localization

After the identification of the malfunctioning or rogue transmitter (by analyzing its signal), the logical next step in *ex post* enforcement is to localize the non-compliant transmitter. The location of an authorized user who may be required to report its location can be verified by the regulatory framework once its identity is established. On the other hand, it is unlikely that the rogue transmitter would provide any cooperation for its location estimation. Thus, the localization in DSS has to be achieved via a non-interactive technique, e.g. by measuring the *received signal strength* (RSS) and *time difference of arrival* (TDOA) [15, 21, 49, 63]. The information about the distances measured between the rogue transmitter and the receivers in the spectrum monitoring system can be merged at the regulator to localize the rogue transmitter.

### 17.4.2.4 Punishment

The aim of punishment is to impose a *penalty* for the non-compliant behavior [3, 20, 25, 53, 65]. Therefore, the efficacy of deterrence against rogue transmissions not only depends on the probability of a bad actor getting caught, but also on the severity of punishment when the perpetrator is caught. To be effective, the penalty has to be sufficiently large to offset the benefits from non-compliance [25]. It should also be proportional to the harm caused due to non-compliance. Additionally, the implications of imperfect enforcement mechanisms need to be taken into account as the risk of punishing compliant users may deter the prospects of spectrum sharing.

In a spectrum sharing ecosystem, there are two methods for penalizing non-compliant transmitters: restricting access to spectrum and charging economic penalties. In the first method, the resource allocation strategy takes into account the compliance behavior of the SU. A rogue SU may not be allowed to access the spectrum for an amount of time that is commensurate with the severity of the infraction. This can be achieved by revoking the license/permit of the rogue transmitter and curtailing its operating rights [25]. The second method is to handle the punishment financially. Those causing the harm are charged commensurately with the severity of the harm. The collected amount can be paid to those who suffered due to the rogue transmitter. In this way, it can act as one of the benefits to legitimate SUs for their compliance.

## 17.5   Open Problems

To motivate future work, we briefly discuss important open research and regulatory challenges in realizing a secure and efficient DSS ecosystem.

### 17.5.1   Research Challenges

*Challenges in* **ex ante** *approaches*   The development of a flexible and descriptive policy language, which can be used to specify spectrum access policies for DSS, is a challenge that needs the attention of the research community. Another important challenge related to spectrum policies includes the development of advanced algorithms for executing policy inference and reasoning tasks carried out by policy-based cognitive radios. Another open problem is the use of ontology-based policies for enforcement while meeting the real-time processing requirements of the radio.

*Challenges in* **ex post** *approaches*   Traditional *ex post* enforcement techniques for wireless systems relied on transmitter specifications (transmission power, antenna parameters, bandwidth, and sensitivity) to detect and prevent harmful interference. However, these traditional approaches are less effective in DSS since the dynamic spectrum access enables the flexibility/mobility of radios in time, space, and spectral domains, which exacerbates the problem of security and enforcement [10, 56, 65]. Hence, novel mechanisms need to be developed for monitoring, forensics, and localization of transmitters in DSS.

### 17.5.2   Regulatory Challenges

*Enforcement and privacy*   There is an interesting tradeoff between enforcement and privacy that exists in the context of shared spectrum access. The collaboration of wireless nodes to monitor the neighboring nodes can help detect, locate, and punish policy-violating transmitters. However, privacy considerations need to be addressed before such solutions can be adopted.

*Adjudication*   In *ex post* enforcement, the locus of adjudication is another critical problem that remains unaddressed [66]. The adjudicating entity must have jurisdiction to adjudicate interference events. At present, there is no clearly defined process for resolving certain types of interference events. For example, for an event that occurs in the 1695–1710 MHz band in the USA, a civil court may refer the matter to the FCC for resolution, but the FCC has no jurisdiction over federal bands and the National Telecommunications and Information Administration is ill-equipped to deal with civil disputes.

*Regulation and enforcement*   There is a fundamental tradeoff between spectrum regulations and enforcement. Tighter regulations can reduce the need for enforcement, but such an approach incurs a significant cost: tighter regulations can create a regulatory ecosystem that discourages investment in research and deployment of wireless innovation. Finding an optimal tradeoff between regulations and enforcement is a challenge that the regulatory community will need to address in the coming years.

## 17.6   Summary

In this chapter we focused on the engineering aspects of spectrum enforcement and security. Going forward, we believe that building an optimal policy enforcement framework will require a skillful combination of *ex ante* and *ex post*, centralized and decentralized, and general and application-specific enforcement components that co-evolve with markets and regulatory policy frameworks within a complex ecosystem. Building such a complex enforcement framework will require a greater understanding of not only the engineering challenges, but also of the ramifications of the enforcement solutions in terms of legal, economic, and regulatory policy aspects.

## References

**1** IEEE standard for policy language requirements and system architectures for dynamic spectrum access systems. IEEE Std 1900.5, 2012.

**2** G. Aggarwal, M. Bawa, P. Ganesan, et al. Two can keep a secret: A distributed architecture for secure database services. In *The Second Biennial Conference on Innovative Data Systems Research (CIDR)*, 2005.

**3** G. Atia, A. Sahai, V. Saligrama, et al. Spectrum enforcement and liability assignment in cognitive radio systems. In *IEEE International Symposium on Dynamic Spectrum Access Networks (DySPAN)*, pages 1–12, 2008.

**4** B. Bahrak, A. Deshpande, M. Whitaker, and J. Park. BRESAP: A policy reasoner for processing spectrum access policies represented by binary decision diagrams. In *IEEE International Symposium on Dynamic Spectrum Access Networks (DySPAN)*, pages 1–12, 2010.

**5** B. Bahrak, J. Park, and H. Wu. Ontology-based spectrum access policies for policy-based cognitive radios. In *IEEE International Symposium on Dynamic Spectrum Access Networks (DySPAN)*, pages 489–500, 2012.

**6** B. Bahrak, S. Bhattarai, A. Ullah, et al. Protecting the primary users' operational privacy in spectrum sharing. In *IEEE International Symposium on Dynamic Spectrum Access Networks (DySPAN)*, pages 236–247, 2014.

**7** G. Baldini, T. Sturman, A. R. Biswas, et al. Security aspects in software defined radio and cognitive radio networks: A survey and a way ahead. *IEEE Communications Surveys & Tutorials*, 14(2):355–379, 2012.

**8** E. Bertino and R. Sandhu. Database security-concepts, approaches, and challenges. *IEEE Transactions on Dependable and Secure Computing*, 2(1):2–19, 2005.

**9** S. Bhattarai, A. Ullah, J. Park, et al. Defining incumbent protection zones on the fly: Dynamic boundaries for spectrum sharing. In *IEEE International Symposium on Dynamic Spectrum Access Networks (DySPAN)*, pages 251–262, 2015.

**10** S. Bhattarai, J. Park, B. Gao, et al. An overview of dynamic spectrum sharing: Ongoing initiatives, challenges, and a roadmap for future research. *IEEE Transactions on Cognitive Communications and Networking*, 2(2): 110–128, 2016.

**11** S. Bhattarai, P. R. Vaka, and J. Park. Thwarting location inference attacks in database-driven spectrum sharing. *IEEE Transactions on Cognitive Communications and Networking*, 4(2):314–327, 2018.

**12** K. Bian and J. Park. Security vulnerabilities in IEEE 802.22. In *Proceedings of the 4th Annual International Conference on Wireless Internet*, pages 1–9, 2008.

**13** V. Brik, S. Banerjee, M. Gruteser, and S. Oh. Wireless device identification with radio-metric signatures. In *Proceedings of the 14th ACM International Conference on Mobile Computing and Networking (MobiCom)*, pages 116–127, 2008.

**14** R. Chen, J. Park, and K. Bian. Robust distributed spectrum sensing in cognitive radio networks. In *IEEE International Conference on Computer Communications (INFOCOM)*, pages 1876–1884, 2008.

**15** R. Chen, J. Park, and J. H. Reed. Defense against primary user emulation attacks in cognitive radio networks. *IEEE Journal on Selected Areas in Communications*, 26(1):25–37, 2008.

**16** R. Chow and P. Golle. Faking contextual data for fun, profit, and privacy. In *Proceedings of the 8th ACM Workshop on Privacy in the Electronic Society (WPES)*, pages 105–108, 2009.

**17** T. C. Clancy and N. Goergen. Security in cognitive radio networks: Threats and mitigation. In *International Conference on Cognitive Radio Oriented Wireless Networks and Communications (CrownCom)*, pages 1–8, 2008.

**18** M. A. Clark and K. Psounis. Trading utility for privacy in shared spectrum access systems. *IEEE/ACM Transactions on Networking (TON)*, 26(1): 259–273, 2018.

**19** B. Danev, H. Luecken, S. Čapkun, and K. Defrawy. Attacks on physical-layer identification. In *Proceedings of the Third ACM Conference on Wireless Network Security (WiSec)*, pages 89–98, 2010.

**20** L. Duan, A. W. Min, J. Huang, and K. G. Shin. Attack prevention for collaborative spectrum sensing in cognitive radio networks. *IEEE Journal on Selected Areas in Communications*, 30(9):1658–1665, 2012.

**21** A. Dutta and M. Chiang. "See something, say something" crowdsourced enforcement of spectrum policies. *IEEE Transactions on Wireless Communications*, 15(1):67–80, 2016.

**22** C. Fei, D. Kundur, and R. H. Kwong. Analysis and design of secure watermark-based authentication systems. *IEEE Transactions on Information Forensics and Security*, 1(1):43–55, 2006.

**23** V. Frascolla, A. J. Morgado, A. Gomes, et al. Dynamic licensed shared access-a new architecture and spectrum allocation techniques. In *IEEE 84th Vehicular Technology Conference (VTC-Fall)*, pages 1–5, 2016.

**24** J. Freudiger, R. Shokri, and J. P. Hubaux. On the optimal placement of mix zones. In *International Symposium on Privacy Enhancing Technologies*, pages 216–234, 2009.

**25** C. Galiotto, G. K. Papageorgiou, K. Voulgaris, et al. Unlocking the deployment of spectrum sharing with a policy enforcement framework. *IEEE Access*, 6:11793–11803, 2018.

**26** Z. Gao, H. Zhu, Y. Liu, M. Li, and Z. Cao. Location privacy in database-driven cognitive radio networks: Attacks and countermeasures. In *IEEE International Conference on Computer Communications (INFOCOM)*, pages 2751–2759, 2013.

**27** B. Gedik and L. Liu. Protecting location privacy with personalized k-anonymity: Architecture and algorithms. *IEEE Transactions on Mobile Computing*, 7(1):1–18, 2008.

**28** A. Ginsberg, W. D. Horne, and J. D. Poston. Community-based cognitive radio architecture: Policy-compliant innovation via the semantic web. In *IEEE Symposium on New Frontiers in Dynamic Spectrum Access Networks (DySPAN)*, pages 191–201, 2007.

**29** N. Goergen, T. C. Clancy, and T. R. Newman. Physical layer authentication watermarks through synthetic channel emulation. In *IEEE International Symposium on Dynamic Spectrum Access Networks (DySPAN)*, pages 1–7, 2010.

**30** C. R. A. González and J. H. Reed. Power fingerprinting in SDR integrity assessment for security and regulatory compliance. *Analog Integrated Circuits and Signal Processing*, 69(2-3):307–327, 2011.

**31** M. Grissa, B. Hamdaoui, and A. A. Yavuz. Unleashing the power of multi-server PIR for enabling private access to spectrum databases. *IEEE Communications Magazine*, 56(12):171–177, 2018.

**32** M. Gruteser and D. Grunwald. Anonymous usage of location-based services through spatial and temporal cloaking. In *Proceedings of the 1st International Conference on Mobile Systems, Applications and Services (MobiSys)*, pages 31–42, 2003.

**33** D. Gurney, G. Buchwald, L. Ecklund, et al. Geo-location database techniques for incumbent protection in the TV white space. In *IEEE Symposium on New Frontiers in Dynamic Spectrum Access Networks (DySPAN)*, pages 1–9, 2008.

**34** J. Hernandez-Serrano, O. León, and M. Soriano. Modeling the lion attack in cognitive radio networks. *EURASIP Journal on Wireless Communications and Networking*, 2011:1–10, 2011.

**35** M. Jakobi, C. Simon, N. Gisin, et al. Practical private database queries based on a quantum-key-distribution protocol. *Physical Review A*, 83(2): 022301, 2011.

**36** K. Kim, C. M. Spooner, I. Akbar, and J. H. Reed. Specific emitter identification for cognitive radio with application to IEEE 802.11. In *IEEE Global Telecommunications Conference (GLOBECOM)*, pages 1–5, 2008.

**37** S. Kirrane, A. Mileo, and S. Decker. Access control and the resource description framework: A survey. *Semantic Web*, 8(2):311–352, 2017.

**38** J. E. Kleider, S. Gifford, S. Chuprun, and B. Fette. Radio frequency watermarking for OFDM wireless networks. In *IEEE International Conference on Acoustics, Speech, and Signal Processing*, volume 5, pages 397–400, 2004.

**39** M. M. Kokar and L. Lechowicz. Language issues for cognitive radio. *Proceedings of the IEEE*, 97(4):689–707, 2009.

**40** A. Kortun, T. Ratnarajah, M. Sellathurai, et al. On the performance of eigenvalue-based cooperative spectrum sensing for cognitive radio. *IEEE Journal of Selected Topics in Signal Processing*, 5(1):49–55, 2011.

**41** M. Kulin, T. Kazaz, I. Moerman, and E. De Poorter. End-to-end learning from spectrum data: A deep learning approach for wireless signal identification in spectrum monitoring applications. *IEEE Access*, 6: 18484–18501, 2018.

**42** V. Kumar, J. Park, and K. Bian. Blind transmitter authentication for spectrum security and enforcement. In *Proceedings of the ACM Conference on Computer and Communications Security (CCS)*, pages 787–798, 2014.

**43** V. Kumar, J. Park, and K. Bian. PHY-layer authentication using duobinary signaling for spectrum enforcement. *IEEE Transactions on Information Forensics and Security*, 11(5):1027–1038, 2016.

**44** V. Kumar, J. Park, and K. Bian. Transmitter authentication using hierarchical modulation in dynamic spectrum sharing. *Journal of Network and Computer Applications*, 91:52–60, 2017.

**45** Vireshwar Kumar, He Li, Jung-Min (Jerry) Park, and Kaigui Bian. Enforcement in spectrum sharing: Crowd-sourced blind authentication of co-channel transmitters. In *IEEE International Symposium on Dynamic Spectrum Access Networks (DySPAN)*, 2018.

**46** C. Li, A. Raghunathan, and N. K. Jha. An architecture for secure software defined radio. In *Design, Automation and Test in Europe (DATE)*, pages 448–453, 2009.

**47** X. Li, J. Chen, and F. Ng. Secure transmission power of cognitive radios for dynamic spectrum access applications. In *Conference on Information Sciences and Systems (CISS)*, pages 213–218, 2008.

**48** J. Liu, C. Zhang, H. Ding, et al. Policy-based privacy-preserving scheme for primary users in database-driven cognitive radio networks. In *IEEE Global Communications Conference (GLOBECOM)*, pages 1–6, 2016.

**49** S. Liu, Y. Chen, W. Trappe, and L. J. Greenstein. Non-interactive localization of cognitive radios based on dynamic signal strength mapping. In *Sixth International Conference on Wireless On-Demand Network Systems and Services*, pages 85–92, 2009.

**50** K. Merchant, S. Revay, G. Stantchev, and B. Nousain. Deep learning for RF device fingerprinting in cognitive communication networks. *IEEE Journal of Selected Topics in Signal Processing*, 12(1):160–167, 2018.

**51** M. D. Mueck, S. Srikanteswara, and B. Badic. Spectrum sharing: Licensed shared access (LSA) and spectrum access system (SAS). *Intel White Paper*, 2015.

**52** R. Murty, R. Chandra, T. Moscibroda, and P. Bahl. Senseless: A database-driven white spaces network. *IEEE Transactions on Mobile Computing*, 11(2):189–203, 2012.

**53** V. Muthukumar and A. Sahai. Fundamental limits on ex-post enforcement and implications for spectrum rights. *IEEE Transactions on Cognitive Communications and Networking*, 3(3):491–504, 2017.

**54** T. R. Newman, T. C. Clancy, M. McHenry, and J. H. Reed. Case study: Security analysis of a dynamic spectrum access radio system. In *IEEE Global Telecommunications Conference (GLOBECOM)*, pages 1–6, 2010.

**55** T. T. Nguyen, A. Sahoo, M. R. Souryal, and T. A. Hall. 3.5 GHz environmental sensing capability sensitivity requirements and deployment. In *IEEE International Symposium on Dynamic Spectrum Access Networks (DySPAN)*, pages 1–10, 2017.

**56** J. Park, J. H. Reed, A. A. Beex, et al. Security and enforcement in spectrum sharing. *Proceedings of the IEEE*, 102(3):270–281, 2014.

**57** S. Parvin, F. K. Hussain, O. K. Hussain, et al. Cognitive radio network security: A survey. *Journal of Network and Computer Applications*, 35(6):1691–1708, 2012.

**58** B. Patil, A. Mancuso, and S. Probasco. Protocol to access white space (PAWS) databases: Use cases and requirements, 2013. URL https://tools.ietf.org/html/rfc6953. Accessed: November 1, 2018.

**59** F. Perich and M. McHenry. Policy-based spectrum access control for dynamic spectrum access network radios. *Web Semantics: Science, Services and Agents on the World Wide Web*, 7(1):21–27, 2009.

**60** A. S. Rawat, P. Anand, H. Chen, and P. K. Varshney. Collaborative spectrum sensing in the presence of byzantine attacks in cognitive radio networks. *IEEE Transactions on Signal Processing*, 59(2):774–786, 2011.

**61** M. Raya, I. Aad, J-P. Hubaux, and A. E. Fawal. DOMINO: Detecting MAC layer greedy behavior in IEEE 802.11 hotspots. *IEEE Transactions on Mobile Computing*, 5(12):1691–1705, 2006.

**62** A. Toninelli, J. Bradshaw, L. Kagal, and R. Montanari. Rule-based and ontology-based policies: Toward a hybrid approach to control agents in pervasive environments. In *Semantic Web and Policy Workshop*, pages 42–54, Sept. 2005.

**63** H. Wang, Z. Gao, Y. Guo, and Y. Huang. A survey of range-based localization algorithms for cognitive radio networks. In *2012 2nd International Conference on Consumer Electronics, Communications and Networks (CECNet)*, pages 844–847, 2012.

**64** W. Wang, Y. Sun, H. Li, and Z. Han. Cross-layer attack and defense in cognitive radio networks. In *IEEE Global Telecommunications Conference (GLOBECOM)*, pages 1–6, 2010.

**65** M. B. H. Weiss, W. H. Lehr, L. Cui, and M. Altamaimi. Enforcement in dynamic spectrum access systems. In *Telecommunications Policy Research Conference*, 2012.

**66** M. B. H. Weiss, M. Altamimi, and M. McHenry. Enforcement and spectrum sharing: A case study of the 1695–1710 MHz band. In *8th International Conference on Cognitive Radio Oriented Wireless Networks (CROWNCOM)*, pages 7–12, 2013.

**67** S. Xiao, J. Park, and Y. Ye. Tamper resistance for software defined radio software. In *33rd Annual IEEE International Computer Software and Applications Conference (COMPSAC)*, pages 383–391, 2009.

**68** L. Yang, Z. Zhang, B. Y. Zhao, et al. Enforcing dynamic spectrum access with spectrum permits. In *Proceedings of the Thirteenth ACM International Symposium on Mobile Ad Hoc Networking and Computing (MobiHoc)*, pages 195–204, 2012.

**69** P. L. Yu, J. S. Baras, and B. M. Sadler. Physical-layer authentication. *IEEE Transactions on Information Forensics and Security*, 3(1):38–51, 2008.

**70** T. Yucek and H. Arslan. A survey of spectrum sensing algorithms for cognitive radio applications. *IEEE Communications Surveys & Tutorials*, 11(1):116–130, 2009.

**71** H. Zhu, C. Fang, Y. Liu, et al. You can jam but you cannot hide: Defending against jamming attacks for geo-location database driven spectrum sharing. *IEEE Journal on Selected Areas in Communications*, 34(10): 2723–2737, 2016.

**72** L. Zhu and H. Zhou. Two types of attacks against cognitive radio network MAC protocols. In *International Conference on Computer Science and Software Engineering*, volume 4, pages 1110–1113, 2008.

# 18

# Economics of Spectrum Sharing, Valuation, and Secondary Markets

*William Lehr*

*Massachussetts Institute of Technology, USA*

## 18.1 Introduction

Advances in wireless technology that increase spectrum agility provide much of the technical foundation needed for spectrum sharing among heterogeneous networks. The emergence of robust markets for spectrum sharing will change the economics of wireless with both predictable and, as yet, uncertain implications for how spectrum is valued. This chapter will explore how the emergence of secondary spectrum sharing markets will impact wireless industry economics and the valuation of spectrum resources.

Earlier chapters in this volume have addressed many of the emerging trends in wireless technologies that are enabling more flexible and dynamic ways to use and share scarce radio spectrum. Ensuring that this valuable resource is used efficiently and directed toward its highest value uses for society requires the co-evolution of technology, wireless markets, and regulatory policies. New technology gives rise to new market opportunities and business models, which in the case of wireless requires new spectrum management frameworks. The focus of this chapter is on understanding what the advances in information, computing and telecommunications (ICT) technologies, leading us toward 5G, imply for the economics and future of spectrum management.

The transition to a global digital economy is driving transformative and disruptive changes across all sectors of the economy and society. The rise of big data, artificial intelligence, cloud computing, autonomous vehicles, the Internet of Things (IoT), the sharing economy, and a host of other transformations are pushing ICT capabilities ever deeper into the fabric of our economic and daily lives. This is driving exponential growth in demand for improved network connectivity and sensing capabilities and capacity. Increasingly, that means growing demand for wireless, which in turn implies continued growth in the demand for spectrum resources. It is no longer feasible to meet this growing demand with new allocations of fresh radio spectrum, although extending access to ever higher

*Spectrum Sharing: The Next Frontier in Wireless Networks,* First Edition.
Edited by Constantinos B. Papadias, Tharmalingam Ratnarajah, and Dirk T.M. Slock.
© 2020 John Wiley & Sons Ltd. Published 2020 by John Wiley & Sons Ltd.

bandwidths in the millimeter-waves and beyond is certainly part of the solution. Meeting the growing demand for spectrum will require sharing spectrum more intensively.[1]

This chapter reviews the basic economics of spectrum as a resource and explains how regulatory and technical trends have increased both the need and opportunities for sharing spectrum more intensively. It provides a review of the different ways in which spectrum may be valued in dollar terms, the challenges to using such estimates, and the factors that contribute to making some spectrum usage rights more valuable than others. When viewed as an economic asset, spectrum is best understood as a bundle of property or usage rights that establish the terms under which potential users of the spectrum may use it. These usage rights are given form by the technologies, regulatory policies, and markets that comprise the wireless ecosystem. They may be altered and transferred by changes in technology, regulatory policies, or markets. Altering these rights and managing how they are used is central to understanding how spectrum may be shared.

A central theme in this chapter is that 5G will require spectrum to be shared among many types of heterogeneous users, uses, and networks along multiple dimensions to accommodate the needs of the diverse users, uses, usage contexts, and market/technical environments that are expected to co-exist in the 5G future. The chapter will discuss the need for and some of the challenges associated with the rise of more robust markets for secondary spectrum trading. Such markets will be increasingly important to accommodate the changing business models associated with 5G and will be rendered increasingly feasible by the sorts of advances in technology and sharing frameworks discussed in greater detail in earlier chapters. Collectively the trends toward softwarization, smaller cells, and more flexible, dynamic, and granular capabilities facilitating agile spectrum management will help fuel the demand for and the capabilities needed for the growth of more robust secondary markets.

The biggest challenges in transitioning to a world of more dynamic spectrum sharing are not principally technical, but are regulatory and economic. The goal of regulatory policies should be to eliminate barriers to spectrum being efficiently valued by market processes to a feasible extent. So long as demand outstrips supply, spectrum will be scarce and the opportunity cost of using spectrum will be determined by the next best use not currently being served. Today, much of the scarcity is artificial and due to legacy regulatory frameworks and the evolving nature of wireless markets. Some users (e.g., government) have access to spectrum that is significantly under-utilized, while others are forced to spend billions of dollars to acquire spectrum rights. The emergence of many promising 5G business opportunities and innovations may be foreclosed without more affordable spectrum access options. In the former case, spectrum is effectively priced too low, while in the latter, potentially too high. With the transition to 5G, much of the technical capabilities and demand cases for sharing spectrum more intensively should exist. This will greatly expand the supply of usable spectrum. Ceteris paribus, that should cause spectrum scarcity rents to decrease on average; however, whether that happens, will depend on the relative speeds with which demand and

---

1 Even if extending access to millimeter-wave and beyond spectrum were to provide an ample excess supply of spectrum, that would alleviate but not eliminate the need to share spectrum more intensively. Expanded sharing of spectrum goes hand-in-hand with the transition to market-based spectrum management, and is needed to facilitate smoother transitions when legacy uses for particular bands need to be transitioned to newer and more socially valuable uses.

supply grow. Managing this process will require the co-evolution of technology, regulatory policies, and markets. It is too complex and uncertain for the centralized administrative management, or command and control (C&C) spectrum management that characterized legacy approaches. In the future, markets will need to play a bigger role in ensuring that spectrum is efficiently allocated to users and usage cases so as to maximize the benefits from using spectrum for society and the economy.[2]

## 18.2 Spectrum Scarcity, Regulation, and Market Trends

The value of radio spectrum derives from how it is used. It is an essential input for wireless sensing and communication services. Spectrum is regarded as a national resource that is managed by the government on behalf of the public interest. If allocated in terms of frequencies, the quantity of spectrum is a finite resource. However, it is also a renewable resource that can be shared,[3] with its capacity limited only by the extent to which the wireless signals intended for different users interfere with each other.

Interference occurs at the receivers, and a perfect receiver should be able to disambiguate its intended signal if that signal differs along any dimension of the electrospace (in frequency, time, space, or direction) from other transmissions arriving at the receiver, at least in principle.[4] In reality, this is often not the case due to various impairments in the propagation environment, the lack of perfect channel knowledge or of the interference characteristics, limitations in the receiver sensitivity, or a host of other technical issues. The limits of sharing, and hence the capacity of spectrum, are determined by the state of technology.

Advances in radio technologies, many of which have been documented in other chapters in this volume, have significantly expanded technical options for sharing spectrum along multiple dimensions. Advances such as software and cognitive radios, smart antennas, and new architectures for radio networks (e.g., small cells) have allowed wireless networks to be much more agile in adapting to complex wireless environments to facilitate sharing the spectrum more intensively. In addition to expanding the capacity of the spectrum to support ever more users and usage within occupied spectrum, further advances in technology promise to extend commercial uses into the millimeter, and eventually sub-millimeter (terahertz), bands. Thus, technology has been continuously expanding the potential supply of available spectrum capacity across all bands.

Although we are far from approaching the technical capacity limits of our spectrum resources, usable spectrum remains an economically scarce – and hence valuable – resource

---

2 It is worth noting that the transition from C&C to markets does not eliminate the need for spectrum regulation. There will still be a need to enforce property rights and ensure that markets operate efficiently (e.g. are not subject to excess market power or suffer from other market failures). Relative to traditional C&C regulation, market-based regulation is more light-handed, allowing more scope for market participants to work things out among themselves in the marketplace and subject to the Darwinian forces of competition.

3 Like bandwidth or water, but unlike oil or food, spectrum that is used (consumed) by one user/use can be reused.

4 Matheson and Morris [1] characterized a seven-dimensional electrospace, consisting of frequency, time, spatial location $(x, y, z)$, and direction of travel (azimuth, elevation angle).

across most frequency bands.[5] The state of commercially available wireless technologies, business models, and regulatory policies limits the extent to which spectrum is currently able to be shared, thereby contributing to spectrum scarcity.

A significant contributor to spectrum scarcity has been legacy spectrum management regimes that have imposed significant impediments to sharing spectrum more intensively, thereby resulting in significant *artificial scarcity*. Historically, limitations in the capabilities of available technologies and the range of users seeking access to spectrum resulted in spectrum being assigned in specific frequency bands for specific uses and technologies. For example, analog broadcast licenses were granted to radio and television stations to operate high-powered, omnidirectional broadcast antenna stations in the spectrum below 1 GHz. License areas were sized to permit reception by low-quality receivers at the edges of the license territories. Although this architecture made sense at the time, and the relative abundance of spectrum made it feasible to exclusively assign spectrum, it has proved extremely challenging to transition legacy broadcast spectrum for use by higher-valued mobile broadband services today.

The prevailing regulatory model was C&C, by which regulators specified the usage, technology, assignment of licenses, and other important terms that limited opportunities to innovate and expand sharing options [2, 3]. In a now famous article from 1959, economist Ronald Coase pointed out that it would be more efficient to shift from administrative assignment to market-based assignment for scarce spectrum resources [4, 5].[6] Nevertheless, it has taken decades to make significant progress in reforming regulatory models to enable more market-based spectrum management. Regulatory reforms have increased opportunities for more flexible spectrum usage models to emerge across many bands. Two of the most significant steps toward market-based spectrum management are the introduction of auctions for flexible, exclusive-use licenses and the expansion of unlicensed spectrum.

In the USA, the first auction was for 2G cellular licenses in the 900-MHz MHz personal communication service (PCS) band in 1994,[7] and unlicensed usage expanded following the allocation of spectrum in the industrial, scientific, and medical (ISM) bands after 1985.[8] These two licensing frameworks represent two distinct models for enabling market forces to play a bigger role in spectrum management. The result is that each has helped sustain

---

5 In certain bands, spectrum efficiency (for the given available bandwidth and antennas) is actually reaching its limits (e.g. with the use of advanced cooperative multiple input multiple output (MIMO), turbo coding, etc.).

6 Although it is generally desirable to transition toward more market-based spectrum management, it is likely both desirable and, in any case, unavoidable that C&C management will continue to be used in certain contexts and for certain bands. Addressing the needs of government for national security and public safety may be inconsistent with market-based allocation in all contexts. For example, first responders can pre-empt normal traffic patterns and assert priority access with "lights and sirens," and similar arrangements are needed for spectrum management.

7 In a move to allow greater scope for service providers to make market-based choices, PCS licensees in the USA were allowed to choose their cellular technology, whereas in Europe and many other markets, regulators mandated that cellular providers adopt global system for mobile (GSM) technology.

8 ISM spectrum was allocated in the 900-, 2400-, and 5000-MHz bands in 1985 and was used for cordless phones, garage door openers, wireless local area networks (WLANs), and other wireless devices. The 802.11a and 802.11b Wi-Fi standards were finalized in 1999. Unlicensed uses for low-power devices were first permitted by the FCC in 1938 [6].

exponential growth in wireless, spectrum sharing, and investment in enhancing spectrum efficiency.

Under the exclusive, flexible use model, the licensee is granted property rights to determine how the licensed spectrum should be utilized, protection from interference, and the right to exclude unaffiliated users. Under the unlicensed model, users are granted a property right for non-exclusive use to the spectrum, and must tolerate potential interference from other authorized users of the spectrum.[9] In both cases, spectrum users are subject to regulatory controls over such things as their transmission power to limit interference to other authorized users and uses. In both cases, the users may be viewed as possessing a bundle of spectrum usage rights that include both entitlements (e.g., right to use the spectrum as authorized) and obligations (e.g., limitations on causing interference to other users).[10]

The exclusive-use licensed model is well-suited to the traditional business models employed by cellular providers offering wireless services to large numbers of subscribers over a large geographic coverage area. The legacy network architectures (macro-cells) and technologies used by cellular providers depended on having predictable quality spectrum access over their serving area. This was crucial both to allow cellular providers to support the base station hand-offs required for highway-speed mobile communications and for their business models that relied on the promise of (near) ubiquitous service availability anywhere in their geographic markets. Moreover, with older technologies, the radios, networks, and services were co-specialized investments (e.g., hardware radios required access to specific frequencies).[11] Long-term exclusive licenses provided cellular operators with a predictable supply of an essential resource, thereby protecting the cellular operator's investment in network facilities and in developing its cellular business model. Moreover, exclusive licenses that are transferrable provide strong incentives to the licensees to use the spectrum efficiently.[12]

The unlicensed-use model is better suited to equipment-based, end-user-deployed wireless business models. This is best exemplified by the success of wireless LANs based on Wi-Fi. The property right to access the spectrum that is essential for the device to be usable

---

9 It is important to realize that all spectrum usage rights may be thought of as property rights. Property rights describe a set of rights and obligations that pertain to the use of "property," which in the case of spectrum is not tangible in the way land is. The ability to transfer spectrum rights, to exclude other users (or not), the right to enforcement of interference protection rights, the obligation to avoid emitting radiation beyond what is allowed, etc. are all part of the property rights bundles. In the case of unlicensed, users do not have the right to exclude other users (a right that does adhere to holders of exclusive-use, licensed spectrum), but they have enforceable property rights to use the spectrum.

10 For various takes on viewing spectrum as a bundle of rights, see Faulhaber and Farber [7], Doyle et al. [8], Hatfield and Weiser [9], and Weiss et al. [10].

11 Investments are co-specialized when they are more valuable when used together. In extreme cases, they may only be valuable if used together (e.g., a right and left shoe or the railroad spur to the mouth of a mine). When assets are co-specialized, joint-ownership is often desirable to avoid hold-up or coordination problems and to realize the optimal value of the assets. When radio hardware was purpose-built to only operate at a specific frequency, the hardware and the networks that made use of it were tightly bound to the specific spectrum licenses. Frequency agile radios reduce this co-specialization which makes it feasible to separate the investment in radios and the investment in spectrum licenses in particular bands. That allows network operators to mix-and-match radios with spectrum and to allow innovations in the radios and in how the spectrum is accessed to be partially decoupled.

12 When licenses are not readily transferrable to be used with other technologies or for other uses (as was the case with broadcast spectrum), the incentives to use the spectrum efficiently are greatly reduced.

is granted on a non-exclusive basis to the device owner. In this case, the incentive to use the spectrum efficiently is indirect since individual users have non-exclusive rights and do not internalize the interference externality that they may impose on other, unaffiliated users of the spectrum. This provides a weaker economic incentive to adopt sharing solutions that minimize interference than confronts the licensee of exclusive-use spectrum.[13] Although unlicensed users do not have a right to interference protection from other authorized users, they also do not have to pay for access to the spectrum. That trade-off has proved very attractive for many wireless users and uses.

The risk that other users would interfere with each other proved not to be a hindrance to significant Wi-Fi growth. Indeed, today most of the data traffic associated with cellular smartphones is carried via Wi-Fi, thereby benefiting from the availability of wired broadband back-haul and the lower opportunity cost associated with using unlicensed spectrum.[14] This off-loading of cellular traffic to Wi-Fi helps free up licensed spectrum resources, enhancing the user's experience when using cellular services, and thereby contributing to the growth of demand and usage for both licensed cellular and unlicensed spectrum. Indeed, it is worth remembering that the first iPhone that launched the smartphone revolution in 2007 was not even a 3G telephone. Apple's iconic design made use of Wi-Fi to provide mass market consumers with a compelling (nomadic) mobile broadband data experience that helped propel smartphone adoption from a niche "road warrior" market to the mass market. Once launched, consumer demand for mobile data services and the stimulus that provides for demand for increased wireless data networking capacity – and hence spectrum resources – has continued at exponential rates.

Both the exclusive licensed and unlicensed spectrum management models have helped sustain significant spectrum sharing. In the case of cellular networks, the operators manage the sharing on behalf of their many subscribers, facilitating shared access for subscribers moving around their coverage areas using a diverse array of licensed spectrum assets. Since acquiring the spectrum resources is expensive (typically via auction but sometimes through mergers and acquisitions (M&A) transactions or spectrum leases), cellular operators have an incentive to efficiently balance investments in networking infrastructure (smaller cells to facilitate spatial reuse or more flexible radios to enable integration of less expensive spectrum resources) and additional spectrum resources. Artificial scarcity that makes spectrum too expensive may induce inefficient investment in network infrastructure. Partial estimates of the value of such usage to society are feasible by looking at the total revenues of cellular operators.

---

13 Although users of unlicensed spectrum have less strong incentives to use the spectrum efficiently, they still benefit from and have incentives to expand the capacity of unlicensed spectrum, even if only to benefit themselves (and affiliated users) by adopting more advanced and spectrally efficient unlicensed technologies. In many situations (campuses, suburban homes), the only users that unlicensed may share with are affiliated users.

14 In many locations, Wi-Fi provides faster data rate performance than 3G/4G cellular services, and since many mobile data plans are subject to data caps that make them more expensive to use than Wi-Fi broadband (which is typically not subject to data caps), end-users have an incentive to rely on Wi-Fi rather than cellular service when available. Most smartphones on the market today can support multiple radio connections, including Wi-Fi and various 3G/4G options that may be user-selected or controlled by the operator or application, depending on the circumstances.

In the case of unlicensed networking, end-users typically deploy base stations for coverage in local areas such as offices, homes, or hot-spots (e.g., coffee shops). Although the users do not pay directly for the spectrum (and hence have little incentive to economize on their use of spectrum to avoid interfering with other unlicensed users), unlicensed users pay indirectly when interference from other users (congestion costs) either reduces the value of their usage or induces them to invest in more interference-tolerant (and typically more expensive) network equipment. Furthermore, since Wi-Fi networks are often composed of a single or few access points within a small coverage area, it is not unreasonable to anticipate that unaffiliated users might negotiate directly to manage interference among themselves.[15] For example, Wi-Fi 802.11b networks operating at 2.4 GHz share spectrum with microwave ovens. In the event that a microwave oven is causing interference, then it is likely that both the Wi-Fi base station and the microwave belong to the same home-owner, who can easily address the interference problem by either not using the oven at the same time as the Wi-Fi or by relocating the Wi-Fi base station. The point of this example is to highlight that the interference challenges emerging in different wireless contexts (associated with different usage or business models, different frequencies, or different technologies) may call for different spectrum management strategies.

Because no one pays directly for the spectrum used to support unlicensed use, estimating the value to society of unlicensed uses is more difficult than for exclusively licensed spectrum. The social value of unlicensed use must be inferred indirectly from estimates of the value of the uses enabled by unlicensed spectrum. These estimates are important to policymakers engaged in spectrum reform.

Arguments over whether it makes more economic sense for society to allocate spectrum as unlicensed or licensed continue to rage. Cellular operators have been strong advocates of the licensed model, whereas other industry participants like Google, Microsoft, Cisco, and Intel have argued in favor of expanded access for unlicensed and other models such as licensed shared access (LSA).[16] Comparing cellular revenues to equipment sales for unlicensed, Hazlett [11] concluded that the licensed-spectrum awarded to cellular operators had generated an order-of-magnitude higher surplus than unlicensed, whereas Thanki [12], using an alternative approach that focused on the role of unlicensed in enabling several important applications, produced surplus several times larger than Hazlett's estimates.[17]

While these debates are expected to continue, they all point to the conclusion that spectrum is a valuable resource that remains economically – if not technologically – scarce and that addressing the heterogeneous demand of wireless users (commercial and government, communications and sensing) and uses (legacy and new, long and short range, variable data

---

15 Originally, most Wi-Fi access points were quasi-fixed and their range was limited, hence the number of Wi-Fi access points likely to be located in the same area was relatively small. That limited the number of unaffiliated Wi-Fi transmitters that would be expected to share the spectrum, potentially causing the capacity of the spectrum to be interference limited. With the widespread proliferation of mobile Wi-Fi devices (e.g., cellular providers offering tethering services) and Wi-Fi base stations with wider-area coverage capabilities, Wi-Fi access has become increasingly congested in many high-usage locations, raising calls for better models for coordinating shared access in "unlicensed-like" settings.

16 See Chapter 6 for a discussion of LSA.

17 Other examples of dueling estimates of the economic value created by the use of spectrum for licensed or unlicensed spectrum are provided by Cooper [13], Deloitte [14], Katz [15], Lewin et al. [16], Milgrom et al. [17], Thanki [18], and Ofcom [19].

rates, asymmetric and symmetric connectivity, planned and ad hoc) of all kinds will require us to expand shared access to spectrum.

Moreover, it is reasonable to expect that expanding shared access will require enabling diverse business and regulatory frameworks that will accommodate shared access among heterogenous users in the same spectrum. For example, enabling ultrawide band (UWB) unlicensed operations below the noise floor for licensed bands, TV white space access in broadcast bands, or unlicensed access in the 5-GHz bands used by satellites provide examples of how regulators have expanded sharing options for heterogeneous radio networks.

Perhaps the boldest experiment is the one associated with the launch of the three-tiered sharing model for the Citizens Broadband Radio Service (CBRS) at 3.5 GHz, underway in the USA.[18] This is especially interesting because it provides a framework for sharing and management among multiple tiers of spectrum users with heterogeneous protection rights and operating under diverse business models in the same spectrum band. This includes government incumbent users, as well as two new classes of commercial users: priority access license (PAL) and general authorized access (GAA) users. The PAL users are granted exclusive usage rights when the incumbent users are not using the spectrum, while the GAA users have unlicensed access rights to use spectrum not in use by the incumbents or PAL users. A spectrum access system (SAS) that will rely on a mix of database and sensing capabilities to manage real-time access to the shared spectrum is under development to support sharing among these heterogeneous wireless users.[19]

This effort is important because it provides a regulatory model for managing the sharing among a potentially arbitrary number of users with heterogeneous property rights over spectrum access. Enabling this capability is important for promoting market-based spectrum management, in which spectrum resources may be traded as bundles of property rights. As already explained, both the exclusively licensed and unlicensed regulatory frameworks represent property rights models that facilitate spectrum sharing. Intermediate models are also possible, and some of these may include elements of C&C management (e.g., to facilitate sharing between government and commercial users, as may be desirable to support public safety uses). The CBRS framework was also noteworthy in mixing sharing among commercial and government users, each of which operate under distinctly different business models [22]. This step is important for the transition to market-based spectrum management to enable spectrum to be appropriately valued at its true opportunity cost (rather than based on the artificial scarcity that arises when spectrum resources are arbitrarily assigned in ways that preclude government and commercial users from sharing the spectrum).

When successfully deployed, the SAS should facilitate easier tracking and transitioning of property rights to accommodate future changes in spectrum usage associated with changing technologies and market conditions.[20] Even if the ultimate regulatory model used to

---

18 See Chapter 4 and https://www.fcc.gov/wireless/bureau-divisions/broadband-division/35-ghz-band/ 35-ghz-band-citizens-broadband-radio for further discussion of CBRS.
19 See Lehr [20, 21] or Weiss et al. [10] for a discussion of the CBRS and the role of more flexible management capabilities for spectrum rights bundles in promoting spectrum sharing.
20 Once we have figured out how to support sharing among three tiers of users with different property rights with respect to access, excludability, and interference protection, we can consider adding additional tiers or allowing market players to negotiate their own exchanges of property rights. This capability would facilitate the transitioning from legacy to newer technologies.

manage 3.5-GHz spectrum in the USA ends up different from the multi-tiered framework originally proposed for the CBRS, the progress made in developing the new regulatory framework for dynamic spectrum sharing among heterogeneous wireless networks should prove applicable to other bands in the USA and elsewhere.[21]

What these regulatory developments point to are the recognition of the growing demand for access to spectrum resources from all types of users and for all types of uses (by application, technology, frequency band, and business model) and the increasing commercial availability of cost-effective technologies to allow spectrum to be shared more intensively on a more granular basis in time and space. On-board and network-based spectrum sensing technologies, software and cognitive multi-protocol radio technologies, and MIMO and beam-forming antennas are facilitating dynamic spectrum access (DSA), bringing the promise of real-time spectrum management ever closer. The latest long-term evolution (LTE) standards for cellular give operators the flexibility in the radio access network (RAN) to make use of diverse spectrum resources to seamlessly support applications on a general-purpose Internet protocol (IP) network. Wireless applications are increasingly capable of being unbundled from the underlying network infrastructure such that an end-to-end communication may traverse a mix of wired and wireless hops. Increasingly, network radios and end-devices are capable of operating in multiple frequencies, including actively making use of both licensed and unlicensed spectrum to flexibly support on-demand resource provisioning. For example, licensed spectrum can be used to support more quality-of-service-sensitive applications like real-time voice or control services, while unlicensed spectrum may be utilized for downloading resource-intensive video.

The net effect of these trends is *to make spectrum resources more fungible.* In the days of C&C regulation, broadcast spectrum could not be used for voice telephony, and in the days of heterogeneous 2G networks, customers of one cellular provider in the USA had to fall back to a 1G advanced mobile phone system (AMPS) to roam on other cellular providers' networks. Today, mobile broadband users can make telephone calls using over-the-Internet voice over IP (VoIP) services like Skype, without caring what the underlying wireless connectivity is (what frequency, what protocol, etc.). In that sense, the spectrum usage value associated with supporting wireless voice telephony is more fungible.

Additionally, as operators move toward smaller cell architectures to facilitate spectrum spatial reuse, economize on scarce device battery power, and take advantage of available wired backhaul, the relative difference in propagation performance for sub-1 GHz and higher mid-band spectrum at 3–5 GHz becomes less important, rendering these frequencies closer substitutes.[22] While the trends in wireless technology and architectures and markets are moving spectrum toward becoming closer substitutes, there are limits to how

---

21 The progress made with the CBRS band relied heavily on the earlier work done to enable TV white space sharing in the broadcast bands below 1 GHz. Similarly, the multi-tiered sharing framework advanced for CBRS may prove useful for managing spectrum in the 5 GHz and higher bands that will be reformed in coming years.

22 Reed and Tripathi [23], commenting on the rationale for weighting megahertz of spectrum held in higher frequency bands lower when computing the spectrum market shares for implementing spectrum cap rules, noted that with the transition to smaller cell architectures, the benefits of lower frequency (below 1 GHz) spectrum relative to mid-band spectrum (3–5 GHz) were less important. With small cells, the key is to provide capacity rather than coverage (since, by definition, small cells offer less coverage). Thus, there is greater scope for substituting among bands.

far spectrum fungibility may progress. The underlying physics imply that the spectrum in different frequency bands and locations will remain, at most, imperfect substitutes [24].

Terrestrial spectrum is a local resource. That is, access to spectrum in New York is not a substitute for spectrum in California.[23] Even more important, sub-3-GHz spectrum, mid-band spectrum (3–10 GHz) and high-band spectrum (above 10 GHz) have significantly different propagation characteristics. There is much more spectrum available at higher frequencies and the shorter wavelengths enable smaller antennas, but the non-line-of-sight (NLOS) performance of higher frequency spectrum is greatly reduced. In some cases, technologies like MIMO can help turn that deficiency into an asset by exploiting the additional information that may be extracted from multiple signals that follow different paths from transmitter to receiver, whereas massive MIMO type beamforming could make up for the steeper propagation loss and smaller antenna apertures in high-band frequencies.

Lower frequency spectrum, especially below 1 GHz, is particularly valuable to cost-effectively provide coverage over large areas where the reduced carrying capacity of the spectrum is not a significant constraint. That includes rural areas with less dense demand. Higher mid-band frequencies are useful to provide a mix of coverage and capacity to meet the requirements of today's 3G and 4G mobile broadband services. To meet the order of magnitude improvements in performance and capacity called for by 5G, expanded access to high-band spectrum above 10 GHz and eventually up to and including terahertz ranges above 300 GHz will be needed to ensure that over-crowding of the spectrum does not block the continued growth of wireless services. Spectrum in these higher frequencies typically requires narrow beam widths and line-of-sight (LOS) transmission paths, which limits its range. Such spectrum is likely to have its greatest application in providing wide-channel bandwidth capacity for short-range data transmissions from small cells.

## 18.3 Estimating Spectrum Values

As noted earlier, spectrum is valuable because it is an essential input in the production of valuable products and services. Observing that value is challenging because much of the usage is not priced (e.g., unlicensed) or is bundled with other inputs (networks), the value of which is difficult to separate from the spectrum resources it is used in combination with. Moreover, even when the private value of the spectrum is observable, that is less than the value to society since much of the consumer surplus is not directly observable. Furthermore, if exclusive control of the spectrum helps protect market power, then part of the private value of spectrum may derive from the monopoly profits that such control enables. When applicable, that can provide a motivation for hoarding spectrum to raise rivals' costs and denying access to socially valuable users or uses that might otherwise share the excess spectrum.

---

23 An advantage of space-based wireless is that its extremely wide coverage area makes it usable in many more geographic locations, but at the cost of relying on a space-based platform that has meant higher latency communications and reduced flexibility in upgrading technologies. New satellite designs suggest some of these constraints may be relaxing, which would render satellite wireless usage interesting.

As with any asset, there are multiple approaches that may be used to estimate its dollar value: (i) analyze market transaction data to determine the value for different spectrum bundles, (ii) build an engineering cost model to estimate a business model based on using a specific bundle of spectrum assets, (iii) develop a general equilibrium model of supply and demand to estimate a market equilibrium price for spectrum, (iv) infer the value of spectrum indirectly from changes in the value of another asset (e.g., analyze changes in firm equity values attributable to fluctuations in spectrum value), or a hybrid approach of the above. All of these approaches suffer from the challenge of estimating the benefits of a future that does not yet exist (but might with more innovative 5G capabilities enabled) and of estimating benefits that are not readily measured in dollar terms (e.g., improved quality of life, civil engagement, or safety). As a consequence, all of the approaches suffer from the problem of "the drunk looking for his keys under the streetlight." He looks there because that is where the light is, not necessarily where the keys are. That problem notwithstanding, of these approaches, the first two are the most common approaches.

Transaction data includes data on M&A involving spectrum assets, data on spectrum leases, and, most importantly, data from the bids and prices paid for spectrum licenses acquired via spectrum auctions.

A reason for using auctions to allocate spectrum is because they are believed to offer an efficient mechanism for assigning spectrum to the user who will generate the highest value use, assuming that the market for auction bids is competitive [25–27]. Of course, if the auction is poorly designed or if the markets for wireless services are insufficiently competitive, then using auctions to assign spectrum may not be efficient. Furthermore, in the face of rapidly changing technology and markets, even if the initial assignment of the spectrum is efficient at the time of the auction, it may no longer be efficient later in the life of a long-term spectrum license.

As we will discuss further below, in the absence of robust secondary markets to allow for post-auction spectrum transfers, the hope that spectrum auctions will induce efficient spectrum utilization decisions by the licensee is greatly reduced. The auction payments represent a sunk cost for the licensee that is only (partially) reversible if the licensee can sell the spectrum after the auction. Without an ability to transfer the spectrum, the licensee will be partially immunized from bearing the opportunity cost of the spectrum. Moreover, if good secondary markets exist, then efficient assignment can be realized even without an auction by relying on an initial lottery and then an efficient secondary market to determine who should actually make use of the licensed spectrum. A benefit of the auction is that it allows any scarcity rents that the auction yields to be captured by the government on behalf of the public, rather than by an arbitrary lottery winner.[24] Auctions have also proved useful in effecting the regulatory transition from one framework and class of usage to another (e.g., from government to commercial, from broadcast television to mobile broadband). Finally, whether auctions are used or not, robust secondary markets can help address problems that may arise, such as licenses that are associated with either too large or too small coverage

---

24 Scarcity rents accrue to the owner of an asset when the available supply of a resource is less than the prevailing demand. The fact that the resource is worth more may have nothing to do with anything the resource owner may have done, and so arguably the benefits of the higher prices may be something that should be shared with society. In such cases, windfall taxes could be used to capture any excess profits for the public that might otherwise accrue to the lottery winners.

areas. A licensee who acquires a license with too much (too little) spectrum may off-load excess (acquire more) spectrum via the secondary market [28].

Historically, different types of licenses traded on the basis of different price metrics. For example, TV broadcast licenses were sold with local TV stations. TV stations in desirable television markets (typically, dense metropolitan markets) were much more valuable than in rural markets. The value of those stations was not directly affected by changes in the demand for spectrum for other services like mobile broadband since those licenses could not be used for other services. Finally, as with data about M&A involving spectrum assets, the reported data is rarely standardized (often available only from trade press reports that may not be adequately detailed) and usually includes other non-spectrum assets (e.g., customer accounts and network infrastructure).

The spectrum transaction data that comes closest to reflecting direct market-based valuations of the private value for spectrum is associated with the exclusive, flexible-use licenses used by cellular operators. To compare spectrum valuations across different license territories and frequency bands, the most common metric used is $/MHz-POP, derived from dividing the value of a spectrum transaction ($) by the total population (POP) in the licensed coverage area times the bandwidth (measured in megahertz). These normalizations are sensible for rough approximations: POP provides a rough measure of the addressable subscriber market associated with the licensed spectrum, while MHz provides a rough measure of the capacity of the spectrum.

At best the $/MHz-POP metric for standardizing spectrum values is an imprecise measure since it does not account for inherent differences associated with the propagation characteristics of different spectrum, other license features (duration, rights to transferability, current state of commercial availability of technology, etc.). The POP metric does not directly account for obvious market and demographic factors that impact both revenue potential and costs associated with using the spectrum to provide wireless services to end-users. For example, rural licenses with less dense populations may be associated with higher-networking costs, but also with a premium for lower-frequency spectrum (below 1 GHz) because of the usefulness of such spectrum for providing wider-area coverage with fewer cell sites. License territories with richer populations or where existing network (and spectrum) capacity is constraining further growth are more valuable. Furthermore, in a world of increased spectrum sharing among heterogeneous users/uses, dynamic spectrum access, and with the commercialization of higher frequencies for wireless services, the $/MHz-POP metric is becoming an increasingly noisy indicator of spectrum value.

Nevertheless, several studies of auction data using the $/MHz-POP metric prove useful in illustrating several characteristics of spectrum values. Connolly et al. [29] and Wallsten [30] used data from US spectrum auctions since 1996 to analyze the econometric effect of different factors on spectrum values. Wallsten used data for all 69 000 licenses that were auctioned in the 80 auctions that took place from 1996 through 2011, whereas Connolly et al. focused on a smaller subset of the licenses (about 7000) that were awarded from 1996 through 2015 for use by mobile applications. Wallsten found that licenses with more megahertz are more valuable, but not on a $/MHz basis, which seems surprising and may be due to unobserved factors (e.g., lower frequency spectrum which is also typically associated with smaller megahertz licenses earning a significant premium relative to higher frequency spectrum licenses). Both Wallsten and Connolly et al. found that licenses with higher POP are more valuable, which is logical since licenses covering more POP signal larger addressable market demand potential. Connolly et al. also found that license values increase with median income and POP density, which further accentuates the demand potential. Both studies found that paired spectrum is more valuable than unpaired, which reflects the legacy bias of mobile network technologies to use separate channels for upstream and downstream traffic between the handsets and the cellular base stations. Wallsten also found that policy uncertainty lowers license values, whereas increased flexibility increased license values. The fact that CMRS licenses, which are flexible, are more valuable is found by Connolly et al. Both studies found that spectrum value increased over time as markets for wireless services and applications expanded and demand soared.

An interesting point noted by Connolly et al. is that the discount associated with higher frequency spectrum has decreased as a consequence of technological advances that make using such spectrum less costly[25] and demand for additional, less-crowded spectrum has increased. Finally, Wallsten found that license prices (measured as $/MHz-POP) are higher for smaller territory licenses, which challenged prior presumptions that larger area licenses are more valuable. Both of these results are encouraging for the 5G future because of the expected transition toward smaller cells and expanded use of higher-frequency spectrum.

The other common approach for valuing spectrum assets is to use engineering cost or business models to estimate the tradeoffs from using different spectrum assets. Examples of studies that have estimated the costs of building wireless networks with different spectrum resources include Oughton and Frias [31], who provide a detailed cost-model for building out 5G small cell infrastructure across the UK, Frias et al. [32], who consider the total cost of ownership of different portfolios of spectrum assets, Johansson et al. [33], who model the costs of supporting heterogeneous wireless networks, and Bouras et al. [34], who model the costs of dense cell deployments. Gomez and Weiss [35] offer one of the few papers to examine how different technical features may limit spectrum substitutability, thereby reducing the fungibility of different spectrum rights. These studies highlight how spectrum and non-spectrum assets may partially substitute for each other. For example, relying on lower-frequency spectrum allows cell sites to be spaced further apart to provide equivalent area coverage or more advanced base stations with smart antennas can expand the capacity of existing spectrum resources.

## 18.4 Growing Demand for Spectrum

We are in the midst of transitioning to the digital economy of Smart(er)-X, where X refers to any aspect of our social or economic lives that can be enhanced by taking advantage of embedded ICT technology. Smart highways (with autonomous vehicles), smart healthcare (with non-invasive real-time monitoring), smart power grids (with dynamic load management of renewable energy), and smart supply chains (with just-in-time inventory control) are some of the examples of smart X. There are Smart-X opportunities to be explored across

---

25 For example, widespread commercialization of MIMO techniques and more intelligent antenna designs facilitates using higher-frequency spectrum.

all sectors of the economy, from manufacturing to finance, from education to agriculture. The horizon vision of the ICT infrastructure needed to fuel this transformation is one of *everything, always (24/7) connect(able) to networked digital communication, computing, and storage resources wherever and whenever wanted.*

Realizing this vision requires expanded access to radio frequency spectrum for remote sensing and communications from a growing universe of wireless users and uses, using heterogeneous networks and technologies. These range from narrowband to wideband, short to long range, planned to ad hoc, legacy to new technologies, all of which need to coexist and share the radio frequency spectrum. The uses and users have diverse usage requirements and business models.

Different applications may require different types of wireless support. For example, video conferencing requires high-speed, two-way bandwidth and the service has to meet tight end-to-end latency bounds, whereas most entertainment video is more latency tolerant and most of the traffic is one-way, downstream to the user. The type of data may also impact the types of service guarantees and security needed (e.g., healthcare versus entertainment data).

Different wireless usage models have different spectrum requirements. For example, some users may have a high tolerance for interference because either their radios are sufficiently sophisticated to be able to sustain communications in very noisy environments or they have applications that are very disruption tolerant.[26] Other users may be very intolerant of noise. This may be users with poor quality receivers [e.g., legacy global positioning system (GPS) or television receivers] or certain sensing applications that are trying to extract low power signals (e.g., radio telescopes).

Different users may operate under different business models, impacting their demand for spectrum resources. For example, commercial service providers are typically investor financed and profit driven, whereas government users are typically budget financed and mission driven. Business and consumer end-users do not acquire spectrum assets directly, but provide infrastructure and equipment that supports wireless usage in both unlicensed and licensed spectrum (e.g., handsets, small cells, and backhaul).

The exponential growth in ICT services of all types is driving heterogeneous growth in demand for spectrum resources. Since supply has not been able to keep pace with demand growth, spectrum prices have tended to rise over time across most bands, but as with many other assets, spectrum prices have been subject to bubbles and prices vary widely across countries, bands, and time. For example, Marsden et al. [36] summarized spectrum auction pricing across markets since 2000 (see Figure 18.1). This data shows how the tech-bubble around 2000 on the cusp of operators preparing for 3G fueled a bubble in spectrum demand. With the crash in dot.com stock prices that began in March 2000, spectrum prices also fell. In addition, poor auction design in the UK and Germany resulted in excessively high prices being paid for 3G spectrum at the peak of the bubble.

As the data in the figure illustrates, prices have been trending upward since around 2008. Whether this upward trend continues or it levels off or declines will depend in part on our collective success in moving toward more efficient spectrum usage models that

---

26 Cognitive radios and other types of waveform agile radios may adjust their power, frequency or other radio parameters to adapt to radio conditions in real time. Low value or very delay tolerant applications may be disruption tolerant and able to function adequately in noisy environments.

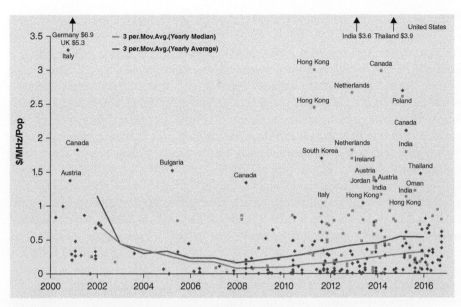

Notes: Green = Prices for coverage bands below 1 GHz) (700 MHz, 800 MHz, 850 MHz and 900 MHz); Blue = Prices for capacity bands above 1 GHz (PCS, AWS,1800 MHz, 2.1 GHz and 2.6 GHz).

Prices per MHz pop are adjusted for inflation and were converted to USD using IMF purchasing power parity (PPP) rates. Prices are also adjusted for licence duration, based on a standard 15 years, using a 5% discount rate.

Source: NERA Economic Consulting Global Spectrum Auction Database.

**Figure 18.1**   Global trends in spectrum prices, by band and auction, 2000–2016. Source: Figure 4, reproduced from [36].

more aggressively exploit sharing opportunities and expand access to additional (higher frequency) spectrum. Thus, the future trend in average spectrum prices (across bands, locations) is uncertain. Perhaps more interesting, however, are predictions about whether the variance in spectrum value across bands and locations will narrow or increase in the future. A case might be made for either outcome.

## 18.5   5G Future and Spectrum Economics

5G is the vision of the next generation of wireless infrastructure that is needed to support the transition to a digital economy [37]. It calls for an order-of-magnitude improvement in wireless performance along multiple dimensions (data rates, connected devices, latency, reliability, etc.). In addition to requiring expanded access to spectrum resources, realizing the 5G vision will have important implications for the wireless business ecosystem, and consequently spectrum economics [38].

First, there is the trend toward smaller cell wireless network architectures that is driven by the need to meet the performance requirements of 5G as well as to facilitate spatial reuse of the spectrum [39, 40]. For example, meeting the latency goals of 5G will require pushing computing resources closer to the edge of the wireless network. Furthermore, as noted already, as cell sizes get smaller, some of the differences in propagation characteristics between lower and higher frequency spectrum become less important. Taking advantage of higher frequency spectrum will in many cases require using smaller cells.

Second, the wireless ecosystem is becoming increasingly agile in being able to support applications seamlessly using diverse spectrum assets. This trend is driven, in part, by the need to support the diverse ICT usage cases associated with Smart-X, which requires a wireless ecosystem capable of supporting heterogeneous networking technologies (wired and wireless, mobile and fixed, legacy and new) that may vary by location and evolve over time. To reduce spectrum costs and support these heterogeneous demands for wireless connectivity, network operators are adopting capabilities to support agile management of diverse spectrum assets. All of the big four national mobile network providers in the USA (Verizon, AT&T, Sprint, and T-Mobile) have portfolios of spectrum assets across multiple bands that vary in coverage by location. Each of them has indicated that they need additional spectrum. 4G and 5G technologies are designed to provide the capability to flexibly and dynamically manage spectrum on a more granular basis in time and space. For example, the CBRS PAL licenses are expected to be allocated in 10-MHz chunks within the 3.5-GHz band, but not to specific frequencies, which means that PAL licensees will have to have frequency tunable radios.[27] Mobile broadband services offered by new entrants (in the market for mobile broadband service) like Google and Comcast make use of Wi-Fi connectivity when available, and roll-over to cellular services (purchased at wholesale) when that is not available. At the same time, most end-user radio equipment is capable of supporting wireless communications using multiple wireless options, including multiple unlicensed bands as well as cellular bands.[28]

Third, there has been an ongoing general trend toward *softwarization* across ICT systems, including wireless networking. Softwarization is the process by which ICT functionality is moved out of hardware into software. This has been enabled by ever-faster, lower-cost hardware for both general purpose computation (computer processing units), specialized processing (video processors), and programmable chips, and by the development of the necessary software tools and platforms. Replacing hardware solutions with software facilitates customization and faster innovation. It enables *virtualization*, the capability of creating a virtual machine platform that can simulate the operations of different hardware and software environments and isolate those simulations from the underlying hardware and software on which it is deployed and from other virtual machines that may share those resources. Virtualization supports network slicing where the resources of a network can be allocated to different users/uses in a flexible, customizable way.

Softwarization and virtualization also support *delocalization*, where ICT functionality can be remotely controlled or delivered from where the resources supporting the ICT functionality are actually located. This allows network operators to realize scale and scope economies by consolidating computing and storage resources in data centers, thereby lowering costs and enhancing efficiency. This raises the potential to manage access to spectrum resources remotely.

---

27 This has been characterized as buying tickets to a movie theater that guarantees you a seat, but not a specific assigned seat.

28 For example, many user devices like eReaders, tablets, and smartphones support unlicensed broadband connections via BlueTooth, multiple flavors of Wi-Fi, and have options for cellular connectivity as well.

Increased softwarization, virtualization, and delocalization are enabling faster ICT innovation and facilitating the deployment of more capable and flexible networking infrastructures and expanding the usage/application domains (Smart-X opportunities) for using ICTs and connecting them wirelessly. This is driving increased demand for heterogeneous wireless applications and enabling new types of business models. The rise of the sharing economy, cloud services, and IoT applications are helping to drive increased demand for wireless connectivity of all types.

For example, cloud services from providers such as Amazon, IBM, Microsoft, and Google are contributing to lowering the cost to accessing computation capabilities for all types of businesses. Their services allow customers the flexibility to scale their demand for computing and communication services. Virtualization allows cloud providers to share scale and scope economies not available to individual enterprises. These arise from limiting data center costs for equipment, power, ensuring reliability, and the backbone networking services needed to tie the data centers together and make them accessible to end-users. By outsourcing their ICT service needs to cloud service providers, businesses can reduce their maintenance costs and turn what otherwise might be lumpy fixed cost investments into variable costs that can scale more easily with their user needs. Moreover, as cloud services have evolved to become more user-friendly, the specialized ICT expertise required to make use of them is becoming less important and an ecosystem of intermediary service providers (market research firms, consultants, business process providers) is emerging to expand direct and indirect access to cloud services to businesses of all sizes. At this point, cloud services have evolved sufficiently that analysts are talking about "Everything-as-a-Service," or "XaaS," which highlights the rich portfolio of specialized and general-purpose cloud services available to allow businesses to outsource virtually all ICT functions to the cloud, turning them from capital investments to service purchases.

Although the cloud is increasingly attractive to more users, some users may still prefer to self-provision their ICT. Wireless makes that increasingly feasible in ways not achievable in a wired-only world. Unlicensed spectrum and a thriving ecosystem of equipment and applications makes it feasible for users to deploy all sorts of wireless ICT applications and solutions from video surveillance to WLANs. Moreover, the near ubiquitous availability of wired broadband access services in most developed markets means that small cell, end-user access points can readily be connected to wider-range networks and computing resources when desired. Furthermore, mesh networking and software to manage multiple access points makes it increasingly viable for end-users to deploy wireless edge-networks should those be preferred to the services offered by cellular or other wide-area wireless network service providers.

The trends toward smaller cells, more agile spectrum management capabilities, and softwarization have important economic implications for the wireless ecosystem. For example, the trend toward smaller cells is causing the worlds of fixed and mobile broadband, unlicensed and licensed spectrum use, government and commercial users, and end-user-deployed and service-provider-deployed networking to converge.

Fixed and mobile broadband are converging because the improved quality of mobile broadband services in a 5G future will allow mobile services to compete as closer substitutes to (historically much faster) fixed broadband services. Additionally, smaller cells require deeper penetration of wired (typically fiber) infrastructure closer to the edge to provide backhaul. Denser neighborhood fiber reduces the cost of deploying both faster mobile and fixed services to more locations. Moreover, to better compete with mobile services and address the growing challenges of over-the-top services, fixed broadband providers are offering mobile application support to end-users.[29] Finally, new fixed-wireless broadband solutions (potentially using millimeter-wave spectrum) are further blurring the boundaries between the network architectures and service offerings of mobile and fixed broadband providers.[30]

The trend toward smaller cells and increased spectrum agility in equipment and provider networks is also driving the convergence of unlicensed and licensed spectrum usage models. As the cell sizes shrink, Wi-Fi and LTE become closer substitutes. As noted earlier, it is already the case that most of the data traffic associated with smartphones is carried on Wi-Fi because of the benefits to both cellular subscribers and service providers when cellular traffic can be off-loaded to Wi-Fi. Moreover, the typical usage models for resource intensive wireless is nomadic. That is, leaning-in applications or applications requiring displays that may not be readily portable, like video conferencing or watching television (on big displays), are typically experienced in relatively fixed locations that do not require support for high-speed, continuous user mobility across wide coverage areas. Legacy cellular networks were designed to provide such support for relatively low-speed voice and mobile broadband services at highway speeds, which requires seamless base station handoffs, a capability that is more difficult for Wi-Fi networks operating in unlicensed spectrum. Options for supporting mobile broadband will blend both types of network architectures and the competition between legacy and new architectures and operators should provide stronger market incentives toward efficiency and toward selecting the best network and spectrum depending on the local context.

The convergence of government and commercial users and usage is driven by the need and high capital costs of supporting ubiquitous connectivity for all types of users. Ensuring widespread availability of 5G infrastructure will be capital intensive and require substantial new investment in new wireless infrastructure and spectrum resources. Deploying dense small cell networks to provide support for ubiquitous coverage will present a significant challenge. One category of government users with an especially pressing need for 5G connectivity are public safety providers and first responders. The challenge of meeting their needs are amplified because they need the capabilities wherever a problem arises, which may be anywhere and not just where commercial demand for capacity is most likely to occur. Moreover, the performance requirements for first responders engaged in life-critical activities are typically more stringent. It will be challenging enough to deploy one 5G infrastructure in many locations, and even more challenging to deploy two such

---

29 For example, Comcast's Xfinity residential service, which provides fixed broadband, paid television programming, and telephone service bundles, now allows its subscribers to access their media content, telephone services, and broadband connectivity remotely. Subscribers can use an application running on a smartphone, tablet or personal computer to access their content or telephone service using any available broadband Internet access service, and allows subscribers to roam on Xfinity Wi-Fi access points distributed throughout Comcast's serving areas when away from home.

30 For example, two industry-led alliances, the Telecom Infra Project (TIP) and the Open Radio Access Network (ORAN), are both focused on developing open access radio access network solutions to promote a more vigorous and open ecosystem [41].

infrastructures (one for commercial and another for government users). Furthermore, in light of the advances in softwarization, virtualization, network slicing, and spectrum agility, it will not be technically necessary to deploy separate infrastructures. In many situations, sharing between government and commercial users will be facilitated by the fact that their heaviest demands for capacity may be uncorrelated (i.e., commercial uses are interrupted during emergency situations).

Finally, the transition to 5G should help drive the convergence of end-user deployed and service-provider deployed wireless networking infrastructure. As network-supported/connected ICT capabilities are pushed closer to the edge and in light of the transition to smaller cell architectures, end-user provided resources will represent a larger share of the total investment in wireless infrastructure. The traffic and users served by a small cell will be less than for a larger macrocell. A small cell uses less spectrum resources (which cost less). At the same time, access to the antenna site, power for the small cell, and even the hardware supporting the small cell may be provided by the end-user or be more directly under the end-user's control. The ability to delocalize functionality means that service providers can manage end-user provided equipment remotely, if so desired. At the same time, growing availability of X-as-a-Service cloud services means that end-users who so wish can self-provision more of the functionality that they might previously had to rely on service providers for.

These 5G architectural and convergence trends will also facilitate the emergence of new classes of business models for spectrum users. These will include new types of wireless virtual network providers and new local infrastructure providers, both of which should increase demand for support for shared spectrum access.

Realizing Smart-X will require a lot of support for wireless services, but in addition it will require significant domain-specific knowledge and capabilities. The challenges and investment required to put in place the sort of Smart-X solutions that will be enabled by 5G and the transition to a digital economy will involve much more than just improving the quality and expanding options for wireless access. It is reasonable to expect that many of the firms that will compete to provide Smart-X solutions to end-users and their suppliers with the necessary domain expertise may choose to specialize in the vertical sectors where they have relevant and specialized domain capabilities. This implies that many of the new business opportunities that 5G will enable will arise in vertical sectors and be pursued by firms with vertical niche experience, whether the X in Smart-X refers to intermodal transport services, building and environmental control systems, healthcare, finance, education, agriculture, etc.

These vertical service 5G providers are likely to have specialized requirements for wireless access support, depending on the focus of their niche markets. For example, smart healthcare solution providers are likely to have different cybersecurity requirements, supply-chain relationships, and wireless connectivity requirements than entertainment media providers. Addressing these vertical niches is likely to create opportunities for new types of specialized mobile virtual network operators (MVNOs) that cannot justify incurring the full costs of managing facilities-based wireless networks. For these firms, supporting heterogeneous wireless solutions is not a core competency but a necessary capability whose availability they may need to ensure for their customers. Moreover, being too committed to any particular wireless solution may introduce "channel" conflicts[31] and lock them in with sunk costs that will limit their strategic flexibility. The rise of such vertical niche MVNOs will increase demand for shared spectrum since these MVNOs are likely to seek to meet their wireless needs via a portfolio of wholesale solutions. The ability to address this demand will be facilitated by the significant expansion in wireless networking capacity, the trend toward more agile spectrum management capabilities, smaller cells, and virtualization that will characterize the transition to 5G.

Another important change that may be expected with the transition to 5G is the emergence of new types of shared infrastructure providers of local facilities. Enabling 5G will require significant new capital investment in the deployment of small cell infrastructure. For example, the siting of small cells requires a lot of complementary infrastructure (site access, power, management of interconnection to wider-area networks) that may make more sense to manage locally where the spectrum is actually being used and the complementary networking assets are physically located. For many of the Smart-X applications (such as smart cities), the natural manager and deployer of much of the 5G infrastructure may be the city or municipality. The city may be able to justify the investment costs on the basis of specialized applications such as IoT for public safety (e.g., using sensors to detect gunshots and enable faster responses), for traffic management (e.g., to reduce congestion, improve public transportation, and manage parking), or for monitoring critical infrastructure (e.g., repair statuses of roads and bridges). In stadiums, factories, malls or other shared-use venues, the owner of the venue may be well-suited for deploying the infrastructure.

In addition to the above business models, there is growing interest in so-called "neutral hosts" business models. Such providers may focus on providing the natural monopoly elements of local 5G infrastructure, offering shared access to wider-area mobile network providers or end-users. Softwarization and virtualization techniques that are used at the core of the networks can also be used in edge components to support the sharing of these elements. For example, the resources of software-enabled base stations may be dynamically reconfigured to support multiple wireless networks or sliced to provide on-demand access to local 5G capabilities. Mobile network operators (MNOs) have demonstrated their willingness to outsource components of their networks and share these with other MNOs already in the case of large coverage area cell sites. Historically, MNOs built out their mobile telephone networks by building their own cell towers. Today, most macrocell towers are owned by third parties who lease space on the towers to multiple MNOs. The towers support multiple base station radios. In a world of software radios, the towers can be smaller and the radios themselves can be shared. Opportunities for such outsourcing and shared use of facilities are likely to expand in the transition to 5G.

---

31 For a Smart-X solution provider in a niche market without an investment stake in wireless infrastructure, facilitating flexible options for their customers (e.g., to choose among wireless network service providers or self-provisioning) may avoid conflicts with customer interests. A channel conflict arises when a vertically integrated provider is induced to sell the integrated bundle when it would be better for the customer to mix-and-match the products from different suppliers. Such problems arise often in business. For example, the telephone switch manufacturer Alcatel-Lucent sold switches both to carriers and to large corporate customers. The sales to the latter cannibalized the sales to the former, resulting in a sales channel conflict. Alcatel-Lucent responded by splitting into two companies.

The economic implication of these 5G transition effects for spectrum is that there will be a growing demand for access to heterogeneous spectrum resources from all sorts of participants with all sorts of wireless networking needs that are likely to change over time and differ by location. This implies increased demand for more granular dynamic spectrum management capabilities in an increasingly crowded wireless ecosystem. The implication of this is that locking-in spectrum assignments to specific technologies, uses, or users is likely to be increasingly costly for society. At the same time, and partially in response to these changing demand demographics, the technical capabilities to support more dynamic and flexible spectrum management are being deployed. These complementary developments on both the demand and supply sides of the evolving wireless ecosystem are pushing us toward a world of shared spectrum, where the sharing may occur flexibly across multiple dimensions. This includes more options for real-time spectrum provisioning, options for frequency hopping to take advantage of white space and support seamless and transparent connectivity for higher-level applications, and options for customized access for different classes of users or uses (e.g., commercial and public safety, planned and ad hoc networking, etc.).

## 18.6 Secondary Markets and Sharing

An important and needed development for supporting the more dynamic 5G future of spectrum sharing is the emergence of more robust secondary markets for trading bundles of spectrum rights. Secondary market trading is an essential mechanism to enable spectrum resources to be shared across heterogeneous networks, business models, and uses. If efficient secondary markets did exist for spectrum, such markets would serve a number of important functions. First, such markets would provide a continuously available mechanism for balancing aggregate spectrum supply and demand over time, and thereby help ensure that spectrum is continuously assigned to its most valuable use. The transaction data provided by such markets would be useful for estimating the value of spectrum and a signal to all market participants of mismatches between supply and demand. The availability of such market price signals would help promote competition and enable market participants to better target their wireless investments and business planning. Spectrum users would be forced to confront the opportunity cost of using the spectrum and the price signals would impose stronger incentives to use spectrum efficiently.

The emergence of efficient secondary markets for spectrum would expand opportunities to further unbundle spectrum and networking infrastructure, facilitating more options to efficiently mix-and-match users, technologies, and networks with spectrum resources. This would enable emerging 5G businesses scale their spectrum assets as their businesses grow. In effect, the rise of efficient secondary markets will help make spectrum more commodity-like. Finally, such markets could also provide the basis for derivative financial securities that could allow better mechanisms to insure against market, technical, or regulatory uncertainty regarding future fluctuations in spectrum supply and demand.[32]

---

32 Derivative securities such as futures, options, and forward contracts in financial security and commodity markets provide an important business tool for managing risk. The emergence of secondary markets for spectrum could give rise to spectrum financial derivatives.

For there to be efficient spectrum markets, there needs to be a liquid supply of spectrum and willing competitive population of both buyers and sellers able to conclude spectrum trades at low transaction costs. This is a tall order that is far from today's reality and likely never to be fully realized. First, although trends like softwarization, small cells, and spectrum agility may make spectrum more commodity-like, the fundamental inherent differences in the propagation characteristics of spectrum, the heterogeneous requirements of wireless users, and the location-specificity of spectrum resources ensure that spectrum assets will never be perfectly substitutable. This points to what might be reasonably expected from future secondary markets, that is, we should not expect one homogeneous market, but rather many secondary markets, potentially different ones for different types of transactions (long-term leases versus real-time access), by location (perhaps local band managers), by frequency (separate markets in below 1 GHz and higher frequencies), etc. Although secondary spectrum markets do already exist, they are in nascent form.[33]

It is also premature to believe that there is a competitive population of potential buyers and sellers. The fact that more trading does not occur today could be because spectrum is already efficiently allocated and the efficient matching of spectrum assets with users is sufficiently long term that trades should be relatively infrequent and transaction volume low. Alternatively, it could be because transaction costs are too high and that might be because the secondary markets are not efficient. Transaction costs may be high because of regulatory, technical or other market impediments to trading that are precluding the markets from being efficient. For example, those with spectrum assets might feel that the relative scarcity of spectrum protects their market power from competition or that the risks of future spectrum scarcity (that cannot be addressed later through secondary market transactions) induces them to hoard spectrum. Or perhaps the lack of sufficiently liquid supply forecloses the emergence of new wireless ventures that otherwise might provide the population of potential buyers. While spectrum is a necessary input, it is far from the only input and some assurance of access to a reasonable supply of spectrum is a necessary condition to justify the investment in the other complementary assets.

Also, much of the spectrum that is arguably under-utilized is currently allocated to users/uses (broadcast television licenses or government users) who are precluded by spectrum regulations or other rules from selling their excess spectrum. For example, US laws that are designed to protect government agencies from engaging in financial deals outside the standard budget process preclude agencies from negotiating directly with commercial spectrum users who might otherwise be interested in buying or leasing government spectrum.[34] These problems illustrate how the emergence of secondary markets is a chicken/egg story and why we should expect secondary markets for spectrum to take time to grow and evolve with the evolution of the 5G ecosystem.

At this stage in the evolution of 5G, it remains unclear precisely what form the needed secondary markets should take. For example, unlicensed spectrum is not exclusive and is currently free to use. Any authorized compliant radio may utilize unlicensed spectrum.

---

33 Mayo and Wallsten [42] found evidence of significant trading of spectrum assets and the prevalence of MVNOs suggests that there are wholesale options for accessing spectrum resources (although MVNO access is typically bundled with wireless network services).

34 See, for example, IDA [43], which documents some of the challenges that would need to be overcome to allow government agencies to trade their spectrum.

This raises the potential problem of the Tragedy of the Commons.[35] However, that could be addressed if access were priced and that might be accomplished with a tax or fee for unlicensed devices. The fee could be established by a secondary market in unlicensed device medallions, like the medallions used to manage the supply of taxicabs. While this would require changing how unlicensed spectrum is regulated, it highlights how regulatory reforms could be used to create a market in access to unlicensed spectrum that would internalize the aggregate congestion externality that arises with open and unlimited access.

Alternatively, a framework like the multi-tiered rights model for the CBRS and its SAS framework might be further extended to allow markets to emerge offering variable classes of interference protection and sharing opportunities. Lehr [20] outlined how the right to exclude other users could be offered as an additional right that could be separately priced from different levels of technical interference protection. Users who want stronger interference protections would pay more, and if additionally they wanted to exclude other users either as added insurance of interference protection or for other reasons, they would pay even more.[36]

The secondary markets for spectrum resources that may emerge may be more akin to the markets for real estate. Consider, for example, the market for temporary housing which ranges from owner-occupied houses to apartments to hotel rooms. There are intermediaries like Hotels.com and AirBnB that create platforms for commoditizing/transacting rooms. There are many models for how these may work, including various ala carte rental or purchase options, bulk purchases of rooms, or forward contracts for reservations that allow the diverse accommodation needs of users to be flexibly addressed. In the case of spectrum, analogously diverse business model approaches may be needed to address 5G wireless spectrum demand. Whereas in the market for rooms there are realters of houses, in spectrum there might be license brokers; where there are hotel chains that provide rooms as a service, in wireless there are large full-service MNOs. Also, in the housing markets there are lots of specialty providers and in 5G there may be neutral hosts. Finally, as there are market-makers like AirBnB facilitating end-users to directly match rooms with those seeking them, so *ad hoc* sharing arrangements for dynamic spectrum access of Wi-Fi access points may emerge. Obviously, not all of these business models for sharing spectrum will prove successful. Figuring out what models for secondary market trading make the most sense ought to be left to the market to sort out.

---

35 Hazlett [11] discusses both the Tragedy of the Commons and the Tragedy of the Anticommons. The former arises when open access of a shared resource results in over-consumption because no one takes into account the adverse impact on others of one's own consumption. The Tragedy of the Anticommons exists when resource rights are so distributed that no-one can organize an effective plan for administering access.
36 For example, Lehr [20] proposed adding an option to exclude the licensing framework for PAL licenses under the CBRS three-tiered framework. A PAL licensee would have a guarantee of interference protection from GAA users, while also having an obligation to avoid interfering with incumbents. This highlights the multi-tiered interference protections enabled by the CBRS framework. Should a PAL licensee want to exclude GAA users, potentially separately from any interference protection concerns, the licensee would pay to exercise the option. This separation of rights would provide a mechanism for market-based differential pricing for shared spectrum (a PAL licensee is protected against interference from GAA and other PAL licensees, but must avoid interfering with incumbent users). Precisely how this might work in practice for spectrum was not worked out, but the sorts of thinking that may be appropriate in considering how spectrum might be more efficiently shared in the future and the role that secondary markets may play in enabling such sharing was suggested.

## 18.7 Conclusion

We are in the midst of a (decades long) process of transitioning from a rigid spectrum management framework characterized as C&C in which spectrum resources were tightly co-specialized with wireless network infrastructures, technology and usage models toward a more market-based world. This transition is needed because of the increasing complexity and need for continuous re-optimization of the allocation and assignment of spectrum resources in light of the continuing exponential growth in wireless services of all types and the rapid pace of market and technical change. The transition to 5G is illustrative of these trends.

In this environment, the inefficiency of relying on centralized, administrative management of spectrum resources becomes even more evident. It is too slow, confronts too many asymmetric information problems, and is too prone to capture. These economic features accentuate the benefits of decentralizing resource management to competitive markets. This was a point recognized by Coase in 1959 that is even more applicable to the coming world of 5G. However, then as now, the ideal of perfectly efficient competitive markets is unlikely to ever be fully realizable. Active spectrum regulation is likely to be necessary indefinitely, but that should not block the general movement toward increased reliance on market forces.

Facilitating this transition requires the emergence and evolution of robust secondary markets. As these markets grow, the market price signals they will provide will help serve as signals for market participants of the opportunity cost of using spectrum for alternative uses and will help sustain the further growth of such markets. Indeed, the lack of good data on the relative value of spectrum in different uses is itself an impediment to the growth of secondary markets, since would-be buyers and sellers need such pricing data to decide whether to participate in the market.

In the past, the tight binding of spectrum resources to specific network architectures, technologies, services, and usage models was re-enforced by the state of the available technology (e.g., limited capabilities to share spectrum without interference and high costs of trying to share), business models (e.g., purpose-built networks tailored to specific usage cases), and regulatory regimes (that limited the ability of users to alter their business model, which specified specific technologies, coverage, build-out, service requirements, and other aspects of the business[37]). In this framework, it was very difficult to separate (unbundle) spectrum from the networks or their associated usage cases.

Over time, networks have become more capable, allowing unbundling of spectrum from applications to a significant extent (e.g., voice telephony can be supported over very heterogeneous spectrum – in frequency, channel, cell-size, etc.). The transition to LTE with 4G created an IP data platform that facilitates unbundling applications on the network side, while at the same time providing more flexibility to mix-and-match spectrum resources on the RAN side. Concurrent regulatory reforms have expanded access to flexible-use/exclusive licenses for mobile broadband and to additional unlicensed spectrum. Both models facilitate different types of complementary spectrum sharing models.

---

37 For example, grant of broadcast licenses came with public service obligations for children's programming and political advertising, as well as coverage requirements.

An important next step in the evolution of spectrum secondary markets, which has already begun, is to enable secondary markets with more robust capabilities to endogenize and reformulate spectrum usage rights in response to changing market and technological requirements. Increasingly, the technology is becoming available to enable this to take place on a more granular basis (in time, geospacer, and context). Sorting out which sharing models make the most sense is best accomplished by leaving it to markets, if markets can be sustained that are sufficiently liquid, competitive, and low-transaction cost. Ensuring these last properties will require continued attention and the co-evolution of wireless technologies, business models, markets, and regulatory policies.

# References

1 R. Matheson and A. Morris (2012) The technical basis for spectrum rights: Policies to enhance market efficiency. *Telecommunications Policy*, 36(9), 783–792.

2 FCC (2002a) Report of the Spectrum Efficiency Working Group. Federal Communications Commission, Washington, DC, Rep. ET Docket 02-135, November 2002. Available at https://www.fcc.gov/document/spectrum-policy-task-force (accessed 12 February 2019).

3 Ofcom (2004) Spectrum Framework Review: a Consultation on Ofcom's views as to how radio spectrum should be managed. *UK Office of Communications (Ofcom)*, 23 November 2004. Available at http://stakeholders.ofcom.org.uk/binaries/consultations/sfr/summary/sfr.pdf (accessed 12 February 2019).

4 R. Coase (1959) The Federal Communications Commission. *Journal of Law and Economics*, 2 (October), 1–40.

5 T. Hazlett (2001) The Wireless Craze, the Unlimited Bandwidth Myth, The Spectrum Auction Faux Pas, and the Punchline to Ronald Coase's 'Big Joke'. AEI-Brookings Joint Center for Regulatory Studies, *Working Paper* 01-01, January 2001.

6 FCC (2002b) Report of the Unlicensed Devices and Experimental Licenses Working Group, Spectrum Policy Task Force. Federal Communications Commission (FCC), 15 November 2002. Available at https://www.fcc.gov/sptf/files/EUWGFinalReport.doc (accessed 12 February 2019).

7 G.R. Faulhaber and D. Farber (2002) Spectrum management: property rights, markets, and the commons. *AEI-Brookings Joint Center for Regulatory Studies Working Paper* 02-12, 6.

8 L. Doyle, J. Kibiłda, T. Forde, and L. DaSilva (2014) Spectrum Without Bounds, Networks Without Borders, *Proceedings of the IEEE*, 102(3): 351–365.

9 D. Hatfield and P. Weiser (2005) Property rights in spectrum: Taking the next step. First IEEE International Symposium on New Frontiers in Dynamic Spectrum Access Networks (DySPAN 2005), *Baltimore Maryland*, 43–55.

10 M. Weiss, W. Lehr, A. Acker, and M. Gomez (2015) Socio-technical considerations for Spectrum Access System (SAS) design. *In IEEE Dynamic Spectrum Access Networks (DySPAN) IEEE International Symposium* 2015, 35–46.

11 T. Hazlett (2005) Spectrum tragedies. *Yale Journal on Regulation*, 22: 242–274.

**12** R. Thanki (2012) The Economic Significance of Licence-Exempt Spectrum to the Future of the Internet. White Paper, June 2012. Available at http://bit.ly/2tkcesj (accessed 12 February 2019).

**13** M. Cooper (2012) Efficiency Gains and Consumer Benefits of Unlicensed Access To The Public Airwaves: The Dramatic Success of Combining Market Principles and Shared Access. Silicon Flatirons White Paper, January 2012. Available at https://ecfsapi.fcc.gov/file/7521479487.pdf (accessed 12 February 2019).

**14** Deloitte (2014) The Impact of Licensed Shared Use of Spectrum. A White Paper prepared for GSMA, in conjunction with RealWireless, 23 January 2014. Available at http://www.gsma.com/spectrum/wp-content/uploads/2014/02/The-Impacts-of-Licensed-Shared-Use-of-Spectrum.-Deloitte.-Feb-20142.pdf (accessed 12 February 2019).

**15** R. Katz (2014) Assessment of Current and Future Economic Value of Unlicensed Spectrum in the United States. 2014 TPRC Conference Paper. Available at https://ssrn.com/abstract=2418667 (accessed 12 February 2019).

**16** D. Lewin, P. Marks, and S. Nicoletti (2013) Valuing the use of spectrum in the EU. Report prepared by Plum Consulting for GSMA, June 2013. Available at http://www.gsma.com/spectrum/wp-content/uploads/2013/06/Economic-Value-of-Spectrum-Use-in-Europe_Junev4.1.pdf (accessed 12 February 2019).

**17** P. Milgrom, J. Levin, and A. Eilat (2011) The Case for Unlicensed Spectrum. Available at http://ssrn.com/paper=1948257 (accessed 12 February 2019).

**18** R. Thanki (2009) The Economic Value Generated by Current and Future Allocations of Unlicensed Spectrum. A study prepared by Perspective Associates with funding support from Microsoft, 28 September 2009. Available at http://apps.fcc.gov/ecfs/document/view?id=7020039036 (accessed 30 November 2019).

**19** Ofcom (2006) Technology Neutral Spectrum Usage Rights: Final Report. Prepared for OfCom by Aegis Consulting, UK, 10 February 2006. Available at https://www.ofcom.org.uk/__data/assets/pdf_file/0023/36527/final_report.pdf (accessed 12 February 2019).

**20** W. Lehr (2016) Spectrum License Design, Sharing, and Exclusion Rights. *University of Illinois Journal of Law, Technology & Policy*, 2016(1): 1–33.

**21** W. Lehr (2017) Analysis of Proposed Modifications to CBRS PAL License Framework. submitted in response to request for comments on Promoting Investment in the 3.5GHz Band, Before the Federal Communications Commission, GN Docket No. 17-258, 28 December 2017. Available at https://ecfsapi.fcc.gov/file/1228227728544/Lehr%20CBRS%20Comments%2017-258.pdf (accessed 12 February 2019).

**22** PCAST (2012) Report to the President Realizing the Full Potential of Government-Held Spectrum to Spur Economic Growth. *Executive Office of the President, President's Council of Advisors on Science and Technology (PCAST)*, July 2012. Available at https://apps.dtic.mil/dtic/tr/fulltext/u2/a565091.pdf (accessed 12 February 2019).

**23** J. Reed and N. Tripathi (2013) The Value of Spectrum: A response to Professor Jon M. Peha's Paper, Attachment A to Reply Comments of AT&T Inc., WT Docket No. 12-269. Available at http://apps.fcc.gov/ecfs/document/view?id=7022100194 (accessed 12 February 2019).

**24** C. Bazelon and G. McHenry (2013) Spectrum value. *Telecommunications Policy*, 37(9): 737–747.

**25** M. Cave and R. Nicholls (2017) The use of spectrum auctions to attain multiple objectives: Policy implications. *Telecommunications Policy*, 41(5): 367–378.

**26** P. Cramton (2002) Spectrum auctions. Chapter 14 in M. Cave, S. Majumdar, and I. Vogelsang (eds), Handbook of Telecommunications Economics, pp. 605–639. Amsterdam: Elsevier Science.

**27** P. Klemperer (2004) *Auctions: Theory and Practice. Toulouse Lectures in Economics.* Princeton University Press: Princeton.

**28** W. Lehr and J. Musey (2015) Right-Sizing Broadband Spectrum Auction Licenses: The Case for Smaller Geographic License Areas in the TV Broadcast Incentive Auction. *Hastings Communications and Entertainment Law Journal*, 37 : 231–272.

**29** M. Connolly, N. Sa, A. Zaman, C. Roark, and A. Trivedi (2017) The Evolution of US Spectrum Values Over Time. Brandeis Working Paper Series 2018-121, 13 February 2018. Available at http://www.brandeis.edu/economics/RePEc/brd/doc/Brandeis_WP121 .pdf (accessed 12 February 2019).

**30** S. Wallsten (2016) Is there really a spectrum crisis? Disentangling the regulatory, physical, and technological factors affecting spectrum license value. *Information Economics and Policy*, 35: 7–29.

**31** E. Oughton and Z. Frias (2017) The cost, coverage and rollout implications of 5G infrastructure. *Telecommunications Policy*, 42(8): 636–652.

**32** Z. Frias, C. González-Valderrama, and J. Martínez (2017) Assessment of spectrum value: The case of a second digital dividend in Europe. *Telecommunications Policy*, 41(5-6): 518–532.

**33** K Johansson (2007) Cost effective deployment strategies for heterogenous wireless networks. Doctoral Dissertation in Telecommunications, KTH, Stockholm.

**34** C. Bouras, V. Kokkinos, V., A. Kollia, and A. Papazois (2015) Techno-economic analysis of ultra-dense and DAS deployments in mobile 5G. *IEEE International Symposium on Wireless Communications Systems* (ISWCS 2015), August 2015.

**35** M. Gomez and M. Weiss (2013) How Do Limitations in Spectrum Fungibility Impact Spectrum Trading? TPRC 41: The 41st Research Conference on Communication, Information and Internet Policy, September 2013. Available at http://dx.doi.org/10.2139/ssrn .2241731 (accessed 12 February 2019).

**36** R. Marsden, B. Soria, and H. Ihle (2017) Effective Spectrum Pricing: Supporting better quality and more affordable mobile services. Consultancy report prepared by NERA Consulting for GSMA, February 2017. Available at https://www.gsma.com/spectrum/ wp-content/uploads/2017/02/Effective-Spectrum-Pricing-Full-Web.pdf (accessed 18 February 2019).

**37** ITU (2015) Recommendation ITU-R M.2083-0: IMT Vision – Framework and Overall Objectives of the Future Development of IMT for 2020 and Beyond. International Telecommunications Union, September 2015. Available at https://www.itu.int/dms_ pubrec/itu_r/rec/m/R-REC-M.2083-0-201509-I!!PDF-E.pdf (accessed 12 February 2019).

**38** W. Lehr (2019) 5G and the Future of Broadband, in G. Kneips and V. Stocker (eds.), The Future of the Internet – Innovation, Integration, and Sustainability, Baden-Baden: Nomos.

**39** W. Lehr and M. Oliver (2014) Small cells and the mobile broadband ecosystem. 25th European Regional ITS Conference (Euro ITS2014), Brussels, June 2014. Available at http://econpapers.repec.org/paper/zbwitse14/101406.htm (accessed 12 February 2019).

**40** J. Chapin and W. Lehr (2011) Mobile Broadband Growth, Spectrum Scarcity, and Sustainable Competition. TPRC 2011. Available at SSRN: http://ssrn.com/abstract=1992423 (accessed 12 February 2019).

**41** I. Morris (2018) TIP, ORAN Alliance Poised to Join Forces on Open RAN. LightReading, 16 October 2018. Available at https://www.lightreading.com/mobile/fronthaul-c-ran/tip-oran-alliance-poised-to-join-forces-on-open-ran/d/d-id/746815.

**42** J. Mayo and S. Wallsten (2010) Enabling efficient wireless communications: The role of secondary spectrum markets. *Information Economics and Policy*, 22: 61–72.

**43** IDA (2004). A review of approaches to sharing or relinquishing agency-assigned spectrum. The Institute for Defense Analysis (IDA) paper P-5102, January 2014. Available at https://www.ida.org/idamedia/Corporate/Files/Publications/STPIPubs/p5102final.pdf (accessed February 12, 2019).

# 19

# The Future Outlook for Spectrum Sharing

*Richard Womersley*

*LS Telcom, Germany*

## 19.1 Introduction

This book has explored a number of regulatory and technical mechanisms for improving the efficiency of spectrum utilization through sharing between different users and services. The prerequisite for the need for spectrum sharing is the ever-growing demand for wireless connectivity which is mainly driven by the continued appetite for mobile data, whether by consumers or professional organizations such as governments, broadcasting, and various industry verticals. It will come as no surprise, therefore, that much of the examination of the outlook for spectrum sharing focuses on mobile network operators and those who require wide-area broadband communications rather than niche applications which, although desiring radio spectrum, do not place high demands on this limited resource.

Against the backdrop of growing demand for wireless connectivity, however, there has been little evidence of an appetite to implement spectrum sharing mechanisms. For example:

- Television white space has been met with limited success. In the USA, where the Federal Communications Commission (FCC) established multiple TV white-space database providers in 2012, 6 years later there have been just over 1000 transactions, the majority of which (976) have been for fixed, point-to-point connections [1].
- Licensed shared access (LSA) has also been met with some friction. The notion that bands which are primarily for governmental usage (such as 2.3 GHz) could be opened up to commercial services on a basis that potentially requires the second tier of users to switch off or change frequency if the primary user requires access to the band, is less than ideal. Operators need certainty of access and as such very few examples exist of successful LSA deployment.

In addition, the operators of wireless networks have historically been, and continue to be, very protective of their spectrum. They prefer to have individual and exclusive licences which assign frequencies uniquely for their use. This remains the case, even where the specific frequencies assigned to operators are not in full use (i.e. to provide ubiquitous national coverage). Higher frequency mobile bands (i.e., 2.6 or 3.4 GHz) are typically used only to

*Spectrum Sharing: The Next Frontier in Wireless Networks,* First Edition.
Edited by Constantinos B. Papadias, Tharmalingam Ratnarajah, and Dirk T.M. Slock.
© 2020 John Wiley & Sons Ltd. Published 2020 by John Wiley & Sons Ltd.

provide additional capacity in places with dense mobile demand such as urban areas. This is perhaps best evidenced by the fact that a number of new operators, who do not have any pre-existing spectrum portfolio, have secured access to these high-frequency bands and aim to be niche players in these areas (picking on niches where revenues tend to be highest). In rural or remote areas, lower frequencies are used to provide wide-area coverage and the higher frequency bands are often unused. It would therefore make sense for other potential users to be able to access and share unused frequencies in whatever area or region they exist, but the incumbent operators do not encourage such sharing as they are concerned about the consequences of unintentional interference into their network. Nor, it is fair to say, are there often the necessary regulatory mechanisms in place to permit operators to allow secondary access to their spectrum.

Even with moves to millimeter-wave bands for mobile services, in which coverage will be naturally limited to very short ranges due to the propagation characteristics, operators continue to demand individual, exclusive licences despite the fact that such frequencies are only likely to be deployed in a very few locations. Conversely, those responsible for regulating the radio spectrum are becoming increasingly aware of the need to provide a mechanism which permits shared access to harmonized mobile spectrum in order to enable many of the use-cases envisaged for 5G services. For example, the provision of mobile connectivity to high-speed trains is most likely best delivered by the organization responsible for the track-side infrastructure, rather than a traditional operator. The rail infrastructure operator, however, does not need a national, exclusive licence, just spectrum covering a very narrow corridor along which the trains run. Equally, such an entity is unlikely to wish to participate in a spectrum auction as it does not have the financial means to bid.

## 19.2 Share and Share Alike

Spectrum sharing between different services and users is nothing new. Look at any band on any frequency allocation table and there is almost always more than one service allocated to the band, with the different services sharing the same spectrum. Fixed links share with satellite up- and downlinks, maritime services share with land-based services, fixed and mobile services share spectrum in a variety of ways. This is clearly illustrated by the extract from the International Telecommunications Union (ITU) radio regulations [2] shown in Figure 19.1.

Such sharing is the hallmark of spectrum management and comes with a variety of rules concerning which of the services sharing a band takes priority over the others, which in general follows the principle of first-come, first-served. This is enshrined in Article 4.3 of Chapter II of the ITU Radio Regulations:

> "Any new assignment or any change of frequency or other basic characteristic of an existing assignment shall be made in such a way as to avoid causing harmful interference to services rendered by stations using frequencies assigned in accordance with the Table of Frequency Allocations in this Chapter and the other provisions of these Regulations, the characteristics of which assignments are recorded in the Master International Frequency Register."

| Allocation to services | | |
|---|---|---|
| **Region 1** | **Region 2** | **Region 3** |
| 1710–1930 | FIXED<br>MOBILE 5.384A 5.388A 5.388B<br>5.149 5.341 5.385 5.386 5.387 5.388 | |
| 1930–1970<br>FIXED<br>MOBILE 5.388A 5.388B<br><br>5.388 | 1930–1970<br>FIXED<br>MOBILE 5.388A 5.388B<br>Mobile-satellite (Earth-to-space)<br>5.388 | 1930–1970<br>FIXED<br>MOBILE 5.388A 5.388B<br><br>5.388 |
| 1970–1980 | FIXED<br>MOBILE 5.388A 5.388B<br>5.388 | |
| 1980–2010 | FIXED<br>MOBILE<br>MOBILE-SATELLITE (Earth-to-space) 5.351A<br>5.388 5.389A 5.389B 5.389F | |
| 2010–2025<br>FIXED<br>MOBILE 5.388A 5.388B<br><br><br><br>5.388 | 2010–2025<br>FIXED<br>MOBILE<br>MOBILE-SATELLITE<br>(Earth-to-space)<br><br>5.388 5.389C 5.389E | 2010–2025<br>FIXED<br>MOBILE 5.388A 5.388B<br><br><br><br>5.388 |

**Figure 19.1** Extract from the ITU Radio Regulations showing mobile allocations.

This often means that if a user fills a band with transmissions, anyone who follows thereafter will find it impossible to use the spectrum without causing harmful interference to the incumbent user. In essence, whilst sharing is technically and administratively possible, in the end bands often end up being occupied by just a single service.

From the perspective of mobile network operators, sharing is both an opportunity and an inconvenience. The opportunity is that, in principle, any band that has a mobile allocation could be used for mobile service. However, for it to be harmonized for mobile services it also needs to be identified by the ITU for international mobile telecommunications (IMT) services. It is obviously easier for an existing mobile allocation to be identified for IMT than for a band that currently has no mobile allocation.

Attempting to share a band with an IMT service, however, is tantamount to handing that band over to the mobile network operators. It is extremely difficult to share a spectrum band between a mobile network and any other kind of service. The very high density of mobile networks, together with the relatively high power of base stations, leaves little chance of any other service being able to successfully operate unless large geographic distances are left between them. Given the typical density of mobile deployment, even using spatial separation such as might be found between mobile services and fixed point-to-point links does not offer much by way of opportunity for sharing. Whilst the fixed links may use a narrow beam and may be at heights above those of the mobile network subscribers (e.g., those at street level), there are still several interference mechanisms:

- The aggregate signals generated by a multitude of user devices could be sufficient to interfere with the fixed link receivers.

- There could be users in elevated positions (e.g., in tower blocks) who would be directly in line with the fixed links, causing interference to them and receiving interference from them.
- The mobile base stations themselves are often on high towers or on top of tall buildings, and being omni-directional could cause interference to the fixed links even if they are not directly in the path of the links (e.g., through reflections from other structures).

As such, any offer by the mobile industry to share spectrum with other services largely represents a move by the industry to take over the usage of the band. The GSM Association (GSMA) (the organization representing mobile network operators) has endorsed sharing as [3]:

> "a way to help, when clearing a band is not possible, by enabling mobile access to additional bands in areas, and at times, when other services are not using them."

What they seem less keen on is the notion of mobile network operators sharing spectrum amongst themselves, of which it states:

> "Mobile network operators should not be prohibited from voluntarily sharing their spectrum to support faster services, improve coverage and drive innovation."

So whilst the GSMA is keen to open up opportunities for mobile network operators to gain access to bands through sharing, their opinion with regards to operators sharing spectrum between themselves, or indeed with other services, is that it should be voluntary (i.e., not enforced by regulation).

The inconvenience for mobile network operators is that where they are forced to share with other users, the rules around this sharing may present difficulties for the use of the spectrum for mobile services. Whilst the 3.5-GHz band has been licensed in many countries for mobile services, it also supports C-band satellite downlinks. In countries with heavy rainfall, the band is so heavily used for these satellite services that licensing its use for mobile services would cause significant harmful interference to satellite reception.

In countries where the rainfall is not so heavy, however, the band is typically only used at a small number of satellite Earth stations, which often form gateways to satellite capacity between the country concerned and one which is subject to heavy rainfall. There is still a need to protect these Earth stations from interference, and this is often enacted through placing an exclusion or protection zone around the station in question. These can range from 5 km for a micro cell to over 30 km for a macro cell [4]. Whether or not this represents a significant inconvenience for mobile network operators depends on the location of the Earth stations. If these are located in urban centers, the necessary exclusion zone may preclude the use of the band for a large proportion of the population. If they are located in more rural areas, the impact on population coverage may be far less.

Sharing between mobile networks and other services is therefore generally problematic, and in most cases providing exclusive, harmonized spectrum specifically identified for IMT services is preferable to trying to fit services in amongst other uses.

Sharing between non-cellular services is potentially more straightforward. For example, it has been proposed that it may be possible for radiomicrophones to share the spectrum used by aeronautical navigation systems in the frequency range 960–1164 MHz [5]. There are reasons for and against such a move, and at present radiomicrophones tend to use the white-spaces between television transmitters in the 470–790 MHz ultrahigh frequency (UHF) television band. The UHF band is of prime value to the mobile industry and is on the agenda for identification for mobile and IMT services at the 2023 ITU World Radiocommunciations Conference. Whilst the potential for radiomicrophones to share and access new spectrum is of some interest, moving them into a band that is not of interest to the mobile industry, such that it might ease access to the band for IMT services, is of great interest as it would potentially leave the band clear of services (other than television broadcasting), meaning that negotiations over its use would involve just one other party, rather than multiple ones.

This kind of inter-service sharing is likely to become more common as the pressure to maximize the utility of the radio spectrum increases. It may be, though, that it is services other than IMT that find ways to share spectrum to leave other bands empty for mobile networks.

## 19.3   Regulators Recognize the Value of Shared Access

Regulators have historically argued that there has been no impediment to anyone establishing a wireless network for a particular industry or purpose, however the reality of the situation is that if the spectrum that is available is not harmonized for a particular application, the cost of developing equipment in a non-standard band can be excessive.

After the ITU has identified a particular spectrum band for IMT services, there is then a need to determine which parts of that spectrum are used in what way [6]: which parts of the spectrum will be used for uplink and which for downlink. It is these arrangements that are then used by mobile equipment manufacturers to design the infrastructure and handsets. They are known as harmonized bands and are enshrined by the groups who standardize mobile technologies, in particular the 3GPP.

For example, the frequency range 2500–2690 MHz is identified by the ITU for IMT services. However, there are a number of harmonized arrangements within this band, as illustrated in Table 19.1.

The harmonization arrangements also define the channel widths possible (i.e. 1.4, 3, 5, 10 or 20 MHz).

Using frequencies that are not harmonized in this way means that off-the-shelf equipment is not widely available. As an example, the use of frequencies just outside the harmonized 900-MHz band for the Global System for Mobile communications for Railways (GSM-R) services for train control has meant that the rail industry, despite being nearly global in scope, has been restricted to a limited number of vendors, pushing up the cost of both infrastructure and handsets.

This has always been a dilemma for regulators. On the one hand, they wish to encourage, and to ensure, that those who need access to spectrum to establish a wireless network (e.g., the emergency services, defence, or transport industry) should be able to do so. On the other, making available spectrum that is harmonized and thus highly valued by commercial mobile network operators may go counter to the wider governmental expectations concerning maximizing the value of spectrum through raising large sums of money for it at auction.

**Table 19.1** Range of harmonized frequency arrangements in the 2.6 GHz mobile band

| Frequency arrangement | 3GPP reference |
|---|---|
| 2500–2570 MHz (uplink) | Band 7 |
| 2620–2690 MHz (downlink) | |
| 2570–2620 MHz (bidirectional) | Band 38 |
| 2500–2690 MHz (bidirectional) | Band 41 |

Some regulators have made such tough decisions. The Communications Regulatory Authority (CRA) in Qatar has provided the emergency services with spectrum in the 800-MHz band in which to operate a private long-term evolution (LTE) network for public protection and disaster relief (PPDR) users. It is worth noting, however, that Qatar has only two mobile network operators, and with $2 \times 30$ MHz of spectrum being available in the 800-MHz band, each operator, and the PPDR organizations, could each access $2 \times 10$ MHz. In a country with a larger number of mobile network operators, such an arrangement may not be viable.

It is only with the advent of the larger spectrum bands that are being considered for 5G (such as the 3.5-GHz band, which in some countries totals up to 500 MHz of spectrum, and the 24.25–27.5-GHz band, which totals 3.25 GHz) that there may be sufficient spectrum available to consciously set aside spectrum to be shared between smaller users or uses without government treasuries feeling the pinch due to lower income from spectrum auctions. The 3.5-GHz band in particular is being eyed as a potential opportunity for such uses. The propagation characteristics of the band are suited to many of the possible applications allowing reasonable coverage (especially along narrow corridors such as rail and road) whilst having sufficient bandwidth to allow both the mobile network operators and other potential users to have enough spectrum to provide a reasonable quality of service.

Some regulators have begun to identify radio spectrum specifically for 5G use cases. In Germany, for example, the Bundesnetzagentur has set aside 100 MHz of spectrum in the harmonized 3.4–3.8-GHz band for "small and medium-sized enterprises and start-ups" [7] to ensure that such organizations have a mechanism for accessing spectrum as and when they will require it. Such usage is on a fully shared basis, and users will have to coordinate amongst themselves to ensure that any interference between networks is minimized. Given the limited geographic scope of the likely uses, this should not prove excessively onerous.

Of course, some frequency bands operate on a fully shared basis. The so-called licence-exempt bands (such as 433, 915, and 2400 MHz) are shared between a wide variety of users and uses. There are, however, significant restrictions on the use of these bands, in particular the requirement for devices to operate at low power levels, to limit interference potential, and to operate on a "non interference" basis, which is to say that any user should not cause interference to any other and must accept any interference that is caused by those using the band within the regulated parameters. On the whole, the use of such bands is not commercial (though there are paid-for WiFi networks), but there is increasing interest in the opportunities that such shared bands present by commercial operators. Technologies such as LTE in unlicensed spectrum (LTE-U) and MulteFire will bring commercial mobile services into the license-exempt bands, most notably 5.8 GHz. According to the GSA there are nine launches of LTE in unlicensed bands, primarily in the USA and South-East Asia, and over 100 mobile devices, including those from Apple and Samsung, support the necessary technologies [8].

The license-exempt bands are also home to a growing range of Low Power Wide Area Network (LPWAN) Internet of Things (IoT) networks such as Sigfox and LoRa and some of these offer commercial services in these shared bands. There is therefore some evidence that sharing may become more commonplace as a result of innovative technology developments and a willingness for regulators to provide shared access to frequency bands.

The remaining question is therefore, *Is it likely that sharing techniques will extend beyond today's position to become the de-facto norm for the use of the radio spectrum?* This boils down to two key factors:

1. Will demand for wireless connectivity exceed the capability of networks to meet demand without the need for sharing?
2. Will the environment for wireless services change such that operators of systems will be willing to, or have the necessity to, share spectrum?

## 19.4 The True Demand for Spectrum

Mobile infrastructure and handset manufacturers, including Cisco and Ericsson, publish annual statistics and forecasts for the amount of data traffic that will be sent over wireless (primarily cellular) networks. The forecasts present ongoing growth that is exponential, typically rising by around 50% per annum. Growth can never be exponential as, in the limit, it would reach infinity. Current forecasts suggest that the amount of energy needed to transfer 1 bit of data will reduce to around 1 picoJoule by the 2020s (current technologies require 2 or more picoJoules per bit [9]). Based on this figure, if data growth continued to be exponential at a rate of 50% per annum, by 2080 the amount of energy that would be needed to transport the data would equal that which is currently produced by every power-generating plant on the planet. Clearly, therefore, there is a limit to how much data can be consumed. The graph below shows the forecasts produced by Cisco [10], Ericsson [11], the ITU [12], and LS Telcom [13]. Only the LS Telcom forecast suggests any slowdown in growth, suggesting that there may be a flattening out of data traffic by the late 2020s (Figure 19.2).

The ITU's role in convincing national regulators of the relative importance of spectrum requirements that drive international policy on allocation and assignment should not be underestimated, given its highly respected status as a specialist agency of the United Nations. Decisions concerning the necessity for radio spectrum to be used for various wireless technologies and services is often driven by the data published by the ITU, which has continuously over-estimated both traffic and spectrum demand since the first estimates were produced for the World Radiocommunication Conference in 2007. At that time, the ITU [14] forecast that by 2015, 1300 MHz of spectrum would be required for IMT services. Excluding the millimeter-wavelength bands being considered for future 5G services, the total amount of spectrum that has been identified by the ITU for IMT services is generally around 1000 MHz (there are minor variations by ITU region). It is clear, therefore, that the ITU process may be driving unrealistic and excessive expectations of demand for spectrum

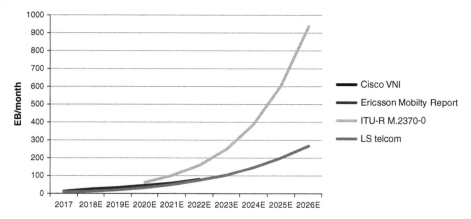

**Figure 19.2** Different forecasts for mobile data traffic growth.

and in the process forcing regulators to consider approaches to spectrum use that are not strictly necessary. Spectrum sharing may fall into that category, with a view being perpetrated that demand will be very high, and thus there is a need for new techniques, whereas in reality, there may be sufficient spectrum to sate demand.

Several organizations have attempted to convert data demand into spectrum demand, and this step is equally fraught with difficulties. The ability of wireless networks to deliver capacity is, of course, dependent on the amount of spectrum that is available to them, but is also driven by the amount of infrastructure (e.g., number of cell sites), the specific technology in use (e.g., 4G versus 5G), and the implementation of advanced spectrum efficiency techniques (e.g., massive multiple input multiple output). There is also the question of traffic offload, from the cellular network to alternatives such as Wi-Fi (IEEE 802.11b, g, n, ac and ax) and WiGig (IEEE 802.11ad and ay) in homes and offices where such networks are available.

Some organizations have used bottom-up methodologies, defining the technical characteristics of mobile networks to develop a model of the operation of the network that leads to an understanding of how it handles traffic. These can range from simplistic models that look only at a limited number of cells (e.g., those handling the peak traffic demand) to those that consider a range of technologies and service types. The ITU's model [15] falls into this latter category, but it has been shown [16] that some of the assumptions used in the calculation of spectrum demand are extremely unrealistic, including very high user density values, high throughput demands per user, and high estimates of the achieved spectrum efficiency of networks.

Other organizations have used top-down methodologies in which the network throughput achieved in the currently available spectrum is extrapolated with data growth and improvements in network efficiency and density to yield a forecast of spectrum demand. Even these methodologies are subject to some error. The FCC in the USA forecast in October 2010 [17] was that 275 MHz of additional spectrum would be required for the country's mobile networks to meet expected demand by 2014. In reality, US wireless operators were able to accommodate all of the traffic growth projected without deploying even the spectrum that was already allocated in 2010 for wireless services, despite the fact that both traffic growth and the number of cell sites deployed have been largely in line with the FCC's projections.

The conversion of data growth forecasts into spectrum demand forecasts is a less than exact science. This is exacerbated by the fact that most demand forecasts focus on countries where traffic is high (e.g., Scandinavia, China, Japan, and South Korea) and where, perhaps not surprisingly, many of the world's mobile equipment manufacturers are based. Whilst it may be the case that the need for spectrum for wireless services in these countries is high, such high demand may not exist in most other countries in the world. Judging the correct reference for spectrum demand is therefore not straightforward.

## 19.5 The Impact of Sharing on Spectrum Demand

A study for the European Commission [18], as part of an examination of the socio-economic data around the roll-out of 5G, also considered the impact of spectrum sharing on spectrum demand. It concluded that:

> "Analysis showed there is a requirement to share spectrum in all the spectrum ranges, particularly in bands below 6 GHz where it is beneficial to share as much spectrum as possible."

In fact, the analysis shows not just that spectrum sharing is beneficial, but that it is *essential*. For one use case (motorways), the report identifies that the demand for spectrum in an environment where no mobile network operators share spectrum is:

- 1.6 GHz of spectrum sub 1 GHz
- 16.2 GHz of spectrum between 1 and 6 GHz
- 38.4 GHz of spectrum above 6 GHz.

Obviously it is not possible to meet the first two of these requirements, even if all of the spectrum in the requisite frequency ranges were made available for mobile networks (i.e., the frequency range 1–6 GHz represents just 5 GHz of available spectrum, so it would not be possible to license 16.2 GHz of it). Figure 19.3 shows the total demand for spectrum based on a number of different sharing scenarios.

The different sharing scenarios envisaged in the European Commission study considered the extent to which operators had exclusive, individual spectrum of their own (0% shared) or where, as shown in the Figure 19.3, 4 operators used the same spectrum as each other (100% shared). In the various use cases, the amount of spectrum needed to support various applications (e.g., healthcare, transport) was calculated and this was then distributed between the operators. In the case where each operator has their own spectrum, it is feasible that all of the subscribers could be on one network, and thus each operator needs enough spectrum to handle all the traffic. Where spectrum is shared between operators, the total of all the traffic can be shared 4 ways.

Disregarding, for the moment, the continued insistence of mobile network operators that they should each have dedicated spectrum, why does spectrum sharing in this scenario make so much difference to the amount of spectrum needed? There is no less traffic being

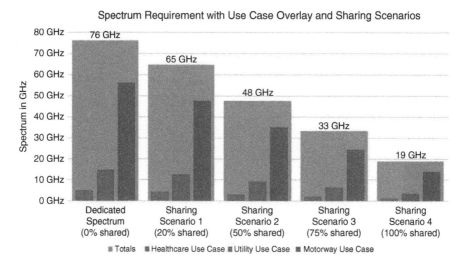

**Figure 19.3** Spectrum requirements for IMT-2020 based on a range of use cases.

transferred across the networks regardless of the sharing scenario. The answer comes from the fact that the lower amount of spectrum (e.g., 19 GHz in Figure 19.3) is sufficient, and perhaps more than sufficient, to carry the expected data traffic. Thus *it is the additional inefficiencies created by overlaying multiple networks on top of each other that drive up spectrum demand in the unshared case.* In the case of there being a number of discrete networks, it has been assumed in the report that each network would be capable of carrying all of the traffic generated in the scenario. In principle this makes some sense, as there is nothing to say that all of the users in the scenario may not be subscribers to a single network, but in reality it is likely that they would be distributed amongst the network operators such that each would see a reduced demand.

The difficulty with having a single network shared between operators in a scenario such as this is the reduction in competitiveness that could result. If all of the mobile network operators shared the same radio access network infrastructure, they could not differentiate themselves on coverage and, depending on the way that the available capacity was divided between them, may also demonstrate identical service levels. Thus, some of the key differentiators between operators are taken away. A few countries have developed national, shared network infrastructure operating in the same spectrum, but the model is one of a wholesale network from which operators can buy capacity, rather than being driven by operators themselves developing a unified network.

The Mexican government licensed *Red Compartida* (Shared Network) in 2016, which offers wholesale mobile coverage and capacity using LTE in the 700 MHz band. Eventually it is expected that the network will cover over 90% of the population. This move has received wide acclaim and the Mexican government was recognized for its leadership at the 2016 Mobile World Congress. However, the desire to build such a network was not driven by spectrum sharing, spectrum efficiency or economic factors, but was instead an attempt to break the near monopoly of Mexico's largest mobile network operator, America Movil. A similar initiative in Rwanda (KTRN) has so far covered 95% of the country's population, but is driven mainly by economics, that is, it is cheaper to roll-out a single network given the commercial reality of providing broadband coverage in countries with low incomes, than to expect multiple operators to provide overlaying coverage. So far, therefore, there does not appear to be a drive by the mobile network operators themselves to roll-out shared networks.

Of course, many mobile network operators already share large proportions of their networks, both passive and active. Site sharing (e.g., the use of a single tower or mast for multiple service providers) is common to reduce costs and results in near-identical coverage for those operators. Similarly, backhaul and electricity supplies may also be shared between operators, meaning that they would have similar levels of availability and reliability due to the common single point of failure. In fact, some countries mandate site sharing (e.g., Austria and some cities in India) or at least have the power to step in and enforce sharing if operators fail to agree commercial terms (e.g., Hong Kong, Jordan, Sweden).

Sharing of the radio access networks (RAN) is less usual, other than national roaming, which is often used to enable an operator who has limited service provision capabilities (e.g., only spectrum for 4G LTE services) to deliver a full range of services (e.g., voice, SMS and data) or to allow a new operator to provide wide-area coverage whilst they grow the coverage of their own network. Even where RAN sharing does happen, mobile network operators will generally deliver services only in their own spectrum assignments, which is not what the authors of the aforementioned Commission report were envisaging. So whilst there appears to be a compelling case for operators to fully share both passive and active elements of their networks, even stretching to spectrum, there are virtually no examples of this taking place at the behest of the operators, rather than when enforced by regulation.

## 19.6 General Authorization needed to Encourage Sharing

One of the reasons why operators may have failed so far to implement the benefits of a fully shared network is the way in which their radio spectrum has been authorized. Almost without exception, the operators of wireless networks (cellular, broadcasting, government, and otherwise) use dedicated spectrum that is individually licensed for their use. Within such a framework there is little opportunity for sharing, and with the exception of licenses which permit spectrum trading, there may be no regulatory or legal mechanism for one operator to provide services in the radio spectrum that has been assigned to another.

There are a wide range of mechanisms for authorizing access to the radio spectrum and Table 19.2 identifies the most common ones.

The individual license approach is the primary mechanism used to license wireless networks. Each licensee has the right to their own piece of radio spectrum, and this spectrum is not shared with anyone else.

The concurrent license approach provides each licensee with access to a dedicated piece of radio spectrum, but there may be more than one licensee with access to the same piece of spectrum. The number of licensees is limited so as to try and enable a reasonable quality of service to be provided. An example of this method of licensing is that used by Ofcom in the UK for the Digital Enhanced Cordless Telecommunications (DECT) guard band which comprises $2 \times 3.3$ MHz of spectrum at the upper end of the 1800-MHz mobile band, directly

**Table 19.2** Different methods for authorizing access to the radio spectrum

| Individual authorization (individual rights of use) | | General authorization (no individual rights of use) | |
| --- | --- | --- | --- |
| Individual license | Concurrent licenses | Light-licensing | License-exempt |
| • Individual frequency planning and co-ordination<br>• Traditional procedure for issuing spectrum licenses | • Individual frequency planning and co-ordination<br>• Simplified licensing procedure<br>• Limitations in the number of users | • No individual frequency planning or co-ordination<br>• Registration and/or notification of use required<br>• No limitation in the number of users | • No individual frequency planning or co-ordination<br>• No registration nor notification of use required |

adjacent to the band used by DECT digital cordless telephones. Ofcom considered that the power output permitted in this band should be reduced to provide interference protection to DECT services and that conversely services in this band may suffer some interference from nearby DECT devices. Ofcom therefore licensed 12 separate organizations to use the spectrum on a low-powered, shared access basis. The specific conditions of usage include a need for the users to co-ordinate their usage:

> "All licensees have the same rights and obligations and they are licensed to use the same frequencies on a shared basis in the whole of the UK. However, to avoid interference each licensee must undertake technical coordination with other licensees." [19]

In the light-licensing approach, licensees are provided with access to specific pieces of spectrum, but the number of possible users of the spectrum is not limited, meaning that any interference between users is not managed. In essence, if users experience interference, it is up to them to change the parameters of their usage (e.g., power, frequency, antenna height) to mitigate the problem. In general, light-licensing either requires users to register their specific usage or may require them to obtain a license. An example of this approach is the *Simple Radio* scheme operated by Ofcom in the UK. This permits licensees access to a pre-defined selection of radio frequencies (VHF and UHF) to be used at relatively low transmitter power (5 W) anywhere in the UK. Users must pay a relatively low fee, and the license is valid for 5 years.

The license-exempt approach permits anyone to use defined pieces of radio spectrum without the need for registration or the payment of any license fee, but is subject to several restrictions (often transmitter power, duty cycle, and emitted bandwidth). No protection from interference is provided. It is also fair to say that many regulators do not patrol these bands or enforce compliance, and there are those who flout the rules with little expectation or chance of being caught or punished. A common example of this is those operating drones who install video links that use transmitter powers far in excess of those permitted.

The license-exempt approach is probably not a recipe for a successful approach to sharing of spectrum by mobile network operators, as there would be little to no guarantee of quality of service. That being said, this would be the primary method used for technologies such as LTE-U and MulteFire, though the light-license approach could be equally valid. What is more likely is that the concurrent license approach could be used to permit multiple operators to share the same spectrum, and in particular support many of the 5G use cases.

In the European Union, for example, the Radio Spectrum Policy Group (RSPG) has recognized the need for a range of approaches to be taken to the licensing of 5G spectrum [20]:

> "Member States will need flexibility in the way they authorise access to spectrum, for example: appropriate geographical areas (e.g. national, regional, city or hyper-local, e.g. for use in a factory), individual licencing or under a general authorisation framework"
>
> "the focus of 5G authorisations in the 26 GHz band should be on an individual licence regime. However, the possibility of a general authorisation regime under sharing conditions that protect the other users of spectrum in this band (e.g. EESS/SRS) is not excluded"
>
> "general authorised frequency use can be an important breeding ground for innovation and contributes towards a dynamic market environment"

Some countries (including Germany, the USA, and the UK) have already begun to consider such an approach to sharing harmonized mobile spectrum for multiple users through a general authorization approach to licensing, and it could be envisaged that this approach will become more common if it is shown to encourage and enable innovative and novel new service provision models.

The USA is perhaps the furthest forward in this respect, with its Citizens Broadband Radio Services (CBRS). This amounts to 150 MHz of spectrum in the 3.6-GHz band (C-band) and is available for licensing to anyone on a number of different bases. On the most basic level, users may access the spectrum at no cost, on an opportunistic and non-interference basis. Above this is priority access, in which users can secure access to a 10-MHz block of spectrum for 3 years and are given some protection from interference. Incumbent users (such as the military and satellite services) are provided with full protection from interference for up to 5 years. The CBRS Alliance, an industry organization established to exploit this spectrum, has over 60 members. Foreseen uses of the CBRS spectrum include in-building (e.g., education, hospitality and healthcare), public spaces (e.g., entertainment and retail), and industrial IoT (e.g., manufacturing, utilities and transportation).

## 19.7 The Long-term Outlook for Spectrum Sharing

The pressure on spectrum bands in and around 1–3 GHz for mobile services is not driven by some odd political or economic principles, but instead is caused by the propagation characteristics of frequencies in that range, which make it ideal spectrum for a range of services, including cellular technologies but also broadcasting, aeronautical, defense, and a variety of others.

Sharing of spectrum between mobile networks and other technologies is problematic, given the dense nature of mobile networks and therefore the large potential for harmful interference, and it seems unlikely that mechanisms to permit this will yield beneficial outcomes insofar as enabling services to operate side by side in a cooperative manner. What seems more likely is that sharing will enable greater spectrum efficiency in two ways:

- non-cellular services will find ways to share spectrum amongst and between themselves such that whole bands can be cleared for IMT services
- mobile network operators will share spectrum amongst themselves.

To a large extent the first of these is already happening. Satellite services often share with fixed links, government services share with civil uses, and a whole host of technologies cooperate in license-exempt bands (including, ironically, mobile networks, though this is not strictly in the first category).

Mobile network operators sharing spectrum between and amongst themselves, however, has yet to become the norm, but the economics of 5G networks and the use of millimeter-wave bands with extremely limited geographic coverage may bring about a change. It may not be economically viable to roll-out multiple 5G networks where adjacent cells may be no more than 100 m apart. In addition, there may not be sufficient infrastructure (e.g., streetlights, traffic lights or other buildings) on which to mount multiple competing networks. The capacity of a 5G cell, using several hundred MegaHertz of spectrum, yet covering no more than a few square meters, is likely to prove sufficient for all the users in that cell in all but the most densely populated areas. It would therefore make sense for the operators to share a single network, and in doing so combine their spectrum assets such that each cell could deliver the widest possible bandwidth. Enforcing such a circumstance through introducing general authorization processes to the licensing of the spectrum to mobile operators may become as commonplace as enforcing passive infrastructure sharing.

In addition, the use of spectrum for delivering 5G services for the various use cases identified amongst a range of industry verticals (including manufacturing, transport, healthcare, and education) will almost certainly be on a shared basis. Though mobile operators claim that they could offer these services through the use of technologies such as network slicing, there are significant doubts over the commercial viability of doing this. Providing, for example, 5G coverage to a factory for use in industrial automation would require a high integrity network, but would have very few revenue generating users. Most use cases are relatively geographically restricted, whether this is along a narrow strip of land that represents a railway, on a university campus or in a hospital. Although there will be cases where such usage may come into close proximity, simple mitigation techniques such as directional antennas ought to be able to permit services to operate without harmful interference.

It may also be possible for such uses to share spectrum with the mobile network operators themselves. There are cases where mobile capacity is under severe pressure, but on a short-term or narrowly focused geographic basis such as at a railway station, in the rush-hour, or at a 2-day rock concert in a remote field where there is usually no need for extensive mobile capacity. Mobile capacity in these cases could easily be supplemented from spectrum shared with the industry verticals, though the additional capacity that may be provided from a relatively small portion of the overall spectrum that is harmonized may not necessarily be sufficient.

The long-term outlook for spectrum sharing looks positive as long as the regulatory measures necessary to encourage and enable sharing are put in place, and the environment in which sharing is considered a beneficial means of improving coverage, services or capacity is embraced by the mobile industry and those with whom it may wish to share spectrum.

## References

**1** Extract from FCC White Space Database, 5 December 2018. https://fcc.gov/general/white-space-database-administration

**2** ITU Radio Regulations, Volume I, 2016.

**3** Spectrum Sharing. GSMA Public Policy Position, November 2018. https://gsma.com/spectrum/wp-content/uploads/2018/11/Spectrum-Sharing-Positions.pdf.

**4** Report ITU-R M. 2019, Sharing studies between IMT-Advanced systems and geostationary satellite networks in the fixed-satellite service in the 3 400-4 200 and 4 500-4 800 MHz frequency bands, 2007. http://itu.int/pub/R-REP-M.2109/ru

**5** Ofcom/CAA joint communication on PMSE sharing in the 960-1164 MHz band. December 2016. https://www.ofcom.org.uk/__data/assets/pdf_file/0026/96254/Ofcom-CAA-joint-communication-on-PMSE-sharing-in-the-960-1164-MHz-band.pdf

**6** Recommendation ITU-R M.1036-4 Frequency arrangements for implementation of the terrestrial component of International Mobile Telecommunications (IMT) in the bands identified for IMT in the Radio Regulations (RR). https://www.itu.int/rec/R-REC-M.1036/en

**7** President's Chamber decision of 14 May 2018 on the order for and choice of proceedings for the award of spectrum in the 2 GHz and 3.6 GHz bands for mobile/fixed communication networks.

**8** LTE in Unlicensed Spectrum –Snapshot November 2018. Global Mobile Suppliers Association. https://gsacom.com/technology/lte-unlicensed/ (subscription required to download the document)

**9** Walker W. and Hidaka Y,. Next Generation Interconnection Research at Fujitsu Laboratories. *Fujitsu Science Tech J*, Vol. 48, No. 2 (April 2012).

**10** Cisco Visual Networking Index: Forecast and Trends, Cisco, 2017–2022.

**11** Ericsson Mobility Report, Ericsson, November 2018.

**12** Report ITU-R M.2370-0 IMT traffic estimates for the years 2020 to 2030. https://www.itu.int/pub/r-rep-m.2370

**13** When will Exponential Mobile Growth Stop? LS Telcom, October 2017.

**14** Report ITU-R M. 2078-0 Estimated spectrum bandwidth requirements for the future development of IMT-2000 and IMT-Advanced. 2006.

**15** Report ITU-R M.2290 Future spectrum requirements estimate for terrestrial IMT. *January* 2014. http://www.itu.int/pub/r-rep-m.2290

**16** Mobile Spectrum Requirement Estimates: Getting The Inputs Right. LS Telcom & TMF Associates, 2014. https://www.satellite-spectrum-initiative.com/files/Mobile Spectrum Forecast final report v106[1].pdf

**17** Federal Communications Commission, Mobile Broadband: The Benefits of Additional Spectrum. October 2010. https://www.satellite-spectrum-initiative.com/files/Mobile Spectrum Forecast final report v106[1].pdf

**18** Identification and quantification of key socio-economic data to support strategic planning for the introduction of 5G in Europe. Tech4i2, realwireless, Trinity College and InterDigital, 2016.

**19** Policy for the DECT Guard Band. Ofcom, September 2016. https://www.ofcom.org.uk/consultations-and-statements/category-1/DECTGB

**20** Strategic Spectrum Roadmap Towards 5G for Europe. Second Opinion on 5G networks. RSPG18-005 Final, January 2018.

# Index

*Spectrum Sharing: The Next Frontier in Wireless Networks,* First Edition.
Edited by Constantinos B. Papadias, Tharmalingam Ratnarajah, and Dirk T.M. Slock.
© 2020 John Wiley & Sons Ltd. Published 2020 by John Wiley & Sons Ltd.